Novel Phytopharmaceutical for Management of Disorders

Novel Phytopharmaceutical for Management of Disorders covers recent advances surrounding phytopharmaceuticals and their potential uses in the management of several disorders as well as in cosmetic benefits. Sections cover the concepts of phytopharmaceuticals and current highlights in disease management. This book also emphasizes phytopharmaceutical drug delivery studies in vivo–in vitro study to various target organs like lungs, liver, and brain. This book provides key information for everyone interested in disease management, drug discovery, and delivery, including medicinal chemists, cosmetic experts, nutritionists, toxicologists, drug formulators, and health care professionals. Students, professors, and researchers working in pharmaceutical sciences and beyond will also find the book useful considering the future prospect of phytodrug delivery for better treatment of various diseases.

We will also focus on patents filed, technology transfer, market potential, and regulatory aspects of herbal nanomedicines, if any. However, research in this area is still at the exploratory stage. But, in the market, there are no such books available that have summarized all the contents on the point of research aspects. This book will be a continuation of the first volume and readers and the target audience will get all the knowledge related to this topic.

Aims and scope:

- To understand the basic information and application of phytoconstituents.
- To understand the various phytoconstituents-based formulations for disease management.
- Herbal drugs used for cosmetic application.
- Phytodrug delivery systems and their applications.
- To provide detailed knowledge about the technical aspects of the formulation part.
- To update readers with the latest research progress in this emerging research area.
- To provide the details about patents filed, technology transfer, market potential, and regulatory aspects for the same.

Novel Phytopharmaceutical for Management of Disorders

Edited by
Durgesh Nandini Chauhan, Madhu Gupta,
Vikas Sharma, and Nagendra Singh Chauhan

CRC Press
Taylor & Francis Group
Boca Raton London New York

CRC Press is an imprint of the
Taylor & Francis Group, an **informa** business

First edition published 2025
by CRC Press
2385 NW Executive Center Drive, Suite 320, Boca Raton FL 33431

and by CRC Press
4 Park Square, Milton Park, Abingdon, Oxon, OX14 4RN

CRC Press is an imprint of Taylor & Francis Group, LLC

ISBN: 978-1-032-27422-5 (hbk)
ISBN: 978-1-032-27427-0 (pbk)
ISBN: 978-1-003-29269-2 (ebk)

DOI: 10.1201/9781003292692

Typeset in Times
by codeMantra

Contents

Preface

Other than nutrients, the chemicals present in the plant-based diet are known as phytopharmaceuticals which might be non-nutritive but play a crucial role in the management of health conditions and are used as therapeutic remedies against various disorders. Initially, plants were used only for food purposes but the medicinal value of plants makes their use in the treatment of different disease conditions. Phytopharmaceuticals may be defined as single compounds or mixture of chemical compounds with medicinal values obtained from the plants, partially purified or solvent fractioned crude plant extracts from one or more medicinal plants. Phytochemicals in the form of phytopharmaceuticals are commercially available as nutraceuticals, food supplements, and health foods.

A thousand types of metabolites are synthesized in plants as intermediate products or products of plant metabolism. Primary and secondary metabolites such as terpenoids, phenolic compounds, flavonoids, anthocyanins, glycosides, saponins, coumarins, and alkaloids present in the different parts of plants are exerting their therapeutic response in the body, also take part in the growth and development of living organisms. Among all these, secondary plant metabolites are gaining more interest from researchers for their medicinal application at the present time. These metabolites play a crucial role in the survival of plants in different disease and pathogenic conditions. They showed remarkable antifungal, anti-microbial, antioxidant, anticancer, and antibiotic activity in plants. Based on these functions, these metabolites are isolated from the plants and used in various chronic disease conditions like diabetes, cancer, hypertension, fungal infections, ulcerative colitis, malaria, typhoid, inflammatory conditions, skin disease, neurodegenerative diseases, aging, immune disorders, etc.

However, although phytopharmaceuticals comprised immense therapeutic efficacy against different disease conditions, they are still having some limitations regarding their solubility in aqueous medium, instability, permeation ability, and toxicity. Different approaches have been carried out for a long time to solve the problems related to the efficacy of phytopharmaceuticals which results in the development of novel carriers used for the phytopharmaceuticals. These carriers are advanced nanosized systems that quickly deliver the increased therapeutic quantity of drug with sustained or controlled release. Due to their advanced properties, these systems are also named novel drug delivery systems (NDDS). These NDDS include nanocarriers like liposomes, ethosome, transferosomes, neosomes, polymeric micellar systems, gel, polymeric nanoparticles, spherical systems, nanocomposites, nanocapsules, etc. Novel phytopharmaceuticals enable site-specific drug delivery that lowers the toxicity of these compounds to healthy body organs.

The book series "Novel Phytopharmaceutical for Management of Disorders" presented herewith is an attempt to describe the research inputs being given by eminent scientists in this direction globally. Novel Phyto-pharmaceutical for the Management of Disorders is a very exciting area however poses so many challenges and pulsating changes happening frequently and this book is an attempt to incorporate the recent trends and potential technologies in the same area.

Chapter 1 reveals the various applications of naringenin that must be a promising therapeutic agent against various diseases. This chapter is a compilation of insight view of naringenin and its alleged therapeutic potential in a number of clinical illnesses, as well as the mechanisms behind it. Neuroprotection, antibacterial anti-diabetic, cardiac disorders, anticancer, antiviral, antioxidant, and anti-inflammatory actions have all been discovered in naringenin. This chapter broadly covers naringenin and its nanocarriers with its potent activities in various disorders.

Chapter 2 focuses on the formulation design of phytopharmaceuticals in the improved therapy of rheumatoid arthritis. This chapter discussed such combinations including hesperidin, indole-3-carbinol, gallic acid, and silibinin that have shown promising anti-arthritic activity. Furthermore, this chapter also discusses the use of NDDS including liposomes, SLNs, micelles, polymeric nanoparticles, nanoemulsions, conjugates for the efficient applicability of natural compounds in rheumatoid arthritis management. The use of such systems for natural compounds exhibiting anti-arthritic, hepatoprotective, and nephroprotective properties in combination with already existing DMARDs may perhaps prove to be an effective strategy in improving rheumatoid arthritis therapy.

Chapter 3 comprehensively compiles the metabolic syndrome and associated diseases, existing treatment approaches along with a special emphasis on evidence-based phytomedicines having beneficial applications in metabolic diseases. Challenges associated with translational herbal product development and future prospects in this arena, discussed in this chapter, will act as a platform for scientists across the globe to focus research for the exploration of phytomedicines in the effective management of metabolic diseases.

Chapter 4 presents the application of curcumin with its nanotechnological interventions in anticancer therapeutics and reviewed the innovative drug nanocarrier systems published for curcumin including micelles, liposomes, polymeric nanoparticles, nanoemulsions, and solid lipid nanoparticles, which provide promising results for curcumin in anticancer application.

Chapter 5 employs the significant potential of lipid-based carriers with phytoconstituents for cancer therapy. Various nanocarrier systems are currently being developed with the aim of improving drug delivery, especially to combat cancer, diabetes, and brain targeting. Among the developed carrier systems, lipid-based nanoparticles are one of the promising drug delivery candidates, based on drug loading, bio-compatibility, and prolonged release.

Chapter 6 describes the role and application of phytodrug delivery in various therapeutic applications for neurodegenerative diseases and offers intriguing alternatives to existing neurodegenerative disease treatment in the form of various newer drug delivery systems that are significantly able to overcome the existing problem of phytopharmaceutical as well.

Chapter 7 discusses the potential benefits of resveratrol especially in the field of ocular delivery. In short, the use of nanotechnology in the field of ophthalmology would allow the benefits of resveratrol that must be optimized. This opens a large window into the spectrum of so-called personalized medicine, which will most likely require new studies with specific designs for populations with individual characteristics.

Chapter 8 discusses the applicability of phenolic compound delivery systems in their anti-aging potential. This chapter brings a recent analysis of the most studied delivery system, incorporating flavonoids, phenolic acids, tannins, stilbenes, and curcuminoids, as well as the respective extracts that contain them, emphasizing their anti-aging potential, especially for the cosmetic market, bringing, also, an approach to the systems already commercialized Furthermore, the use of such system with phenolic compounds becomes a strategy that, in addition to deliver the stable and improved activity inputs, can contribute to the innovative formulation for industries.

Chapter 9 explores the delivery system of cannabidiol to and through the skin: challenges, novel nanotechnology-enabled approaches, and bibliometric analysis. Furthermore, the applicability of topical formulations often seeks the anti-inflammatory activity of cannabidiol. In the scope of nanoformulations, the results suggested that studies are still at the cellular level and that the transdermal route has still been little explored, being a potential for further research.

Chapter 10 exclusively focuses on the wound-healing potential of various phytoconstituents and how the delivery system should be designed to overcome the existing challenges associated with such phytoconstituents. This chapter provides insight knowledge about the evidence-based review of phytopharmaceuticals that are useful in different stages of wound healing, deals with the challenges of herbal formulation design and development, and also mentions the intellectual properties related to phytopharmaceuticals for wound management.

Chapter 11 deals with the present medical treatment, challenges in the present treatment approach, use of phytopharmaceuticals and phytopharmaceuticals-based NDDS for the effective management of psoriasis.

Chapter 12 presents the exploration of the antioxidant and anti-inflammatory activities of the hesperidin-based formulation. Published reports suggested that hesperidin proved the medicinal use of these phytomedicines in various illnesses like cardiovascular, neurological, and psychiatric abnormalities because of having therapeutic efficacy for reducing oxidation, high lipid levels in the body, and inflammation. This book chapter comprises hesperidin and its pharmacokinetics, safety and toxicity, and nanotechnology used for its effective delivery in the body.

Chapter 13 reveals the unexplored facts of the application of phytodrug delivery systems for the treatment and management of ulcerative colitis. A variety of herbal drugs have been explored as supplement therapy while most phytopharmaceuticals reported to reduce oxidative stress and lower the proinflammatory transcriptional factors. This chapter focuses on the applicability of phytochemical-based novel delivery systems, their mechanistic approach, and associated limitations for the treatment of ulcerative colitis.

Chapter 14 deals with effective treatments for skin aging including phytochemicals derived from plants that offer various beneficial qualities for the skin, such as UV protection, antioxidant effects, matrix protection, and hydration. This chapter included various phytoconstituents-based nanoformulations. Lastly, their evaluation techniques, in vitro and in vivo models used to assess phytoconstituents-based products for antiaging effects better are also discussed. Further, to check on their translational concept as a promising strategy for phytochemical and nanotechnology-based cosmetics that could develop an effective antiaging formulation that poses no potential risk to human health with enormous benefits for the skin.

The book series *Novel Phytopharmaceutical for Management of Disorders* is cited about the innovative nanoformulations of phytoconstituents against various disorders. Furthermore, an updated point of view is given to nanoformulations against severe diseases, i.e., cancer, aging, psoriasis, wound-healing neurodegenerative disorders, infectious diseases, ulcerative colitis, etc.

In our opinion, this book provides a consolidated overview on the current status and including the conceptualizations, with their future prospectives. It is a good compilation for giving knowledge not only to non-expert but also to the eminent personalities in the field of Advancement in Novel Phytopharmaceuticals against disorders. Last but not least, I would like to thank all the contributors to the book for the excellent work accomplished. It has been a privilege to work with them.

Durgesh Nandini Chauhan

Madhu Gupta

Vikas Sharma

Nagendra Singh Chauhan

Editor Biography

Durgesh Nandini Chauhan, M.Pharm., has several years of academic (teaching) experience at institutes in India in pharmaceutical sciences working in Columbia College of Pharmacy, Raipur, CG, India. She taught subjects as pharmaceutics, pharmacognosy, traditional concepts of medicinal plants, drug delivery, phytochemistry, cosmetic technology, pharmaceutical engineering, pharmaceutical packaging, quality assurance, dosage form designing, and anatomy and physiology. She is the member of the Association of Pharmaceutical Teachers of India, SILAE: Società Italo-Latinoamericana di Etnomedicina (The Scientific Network on Ethnomedicine, Italy), and others. She has attended several workshops, conferences, and symposiums, including the AICTE-Sponsored Staff Development Program on "Effects of Teaching and Learning Skills in Pharmacy: Tool for Improvement of Young Pharmacy Teachers." She has written more than ten articles published in national and international journals, 15 book chapters, and editors in ten books. She is also active as a reviewer for several international scientific journals and an active participant at national and international conferences, including BhartiyaVigyanSammelan and the International Convention of Society of Pharmacognosy. Presently, she joined Ishita Research Organization, Raipur, India as a freelance writer and guide for students in pharmacy, Ayurvedic, and science in their research projects. She would be interested in data analysis, proposal writing, questionnaire design, designing and developing research questionnaires and policy. She would be actively engaged with accurately monitoring, investigating and analyzing all statistical information, professionally presenting information, reports and investigating and documenting key data and information at the request of senior researchers.

Dr. Madhu Gupta has research experience pertaining to drug delivery to nanoformulations for magical molecule delivery, bioligands for targeting of bioactives and drug moiety, biopolymers, cancer nanomedicine, as well as topical delivery that is carried out in the Department of Pharmaceutical Sciences, Dr. H.S. Gour Central University, Sagar. Dr. Gupta has done her B.Pharm. (Gold Medalist), M.Pharm., Ph.D. in Pharmaceutics with experience of more than 15 years in academics, administrative functions and research in areas of pharmaceutical nanotechnology and targeted drug delivery related to cancer, fungal infection, and psoriasis. Presently, she is Associate Professor at Delhi Pharmaceutical Sciences and Research University. Along with teaching and research, she is the member secretary of World Class skill Center courses, PRO, Website incharge, IQAC incharge with many other managerial work. She is pioneer scientist in the field of nanotechnology and drug delivery field. She has over more than 50 research publications to her credit published in journals of high scientific impact and contributed chapters in various renowned books and to several international and national books. She has supervised 15 M.Pharm. students. Dr. Gupta has availed several prestigious fellowships and awards like SRF AICTE (NDF), JRF (AICTE), Travel grant awards (DST, ICMR, INSA, and DBT, MPCST), Prof. G.P. NAIT Award (2004), and Prof. C.S. Chauhan Award (2004). She has her research work at BioAsia Innovation Award – 2012, Grace India awards, Youth Education Award – 2018, and Young Researcher Award. She has successfully completed one project funded by MPCST Bhopal.

Dr. Vikas Sharma is at Divine International Group of Institutions, Gwalior, MP, India. Dr. Sharma obtained his B.Pharm. from Jiwaji University Gwalior, M.Pharm. in Pharmacognosy and Ph.D. from Dr. H.S. Gour Central University, Sagar, India. Sharma is having teaching and research experience of more than 12 years. He has made significant contributions in the area of medicinal plant research. He has been working on evidence-based validation of medicinal plants and pharmacological screening of herbal drugs. Dr. Sharma works on marker and biomarker analysis of herbs for quality evaluation. He has also made significant contribution on the development of chemical profiling of extracts and formulation through TLC, HPTLC, and LC-MS. He is supervising scientific research of the postgraduation and the doctoral level. He has written several researches and review papers in reputed journals and has contributed several chapters in books on Pharmaceutical and Indigenous drugs. He is a member of various professional and academic bodies like the Association of Pharmaceutical Teachers of India (APTI), and Society of Pharmacognosy.

Dr. Nagendra Singh Chauhan obtained his M.Pharm. and Ph.D. from the Department of Pharmaceutical Sciences, Dr. H.S. Gour University, Sagar, MP, India. He has around 20 years of research experience. He is presently working as a Senior Scientific Officer Grade-I and Government Analyst at Drugs Testing Laboratory Avam Anusandhan Kendra, Raipur, Chhattisgarh, India. He has professional expertise in Natural Product Isolation and Phytopharmacology. He has written more than 75 articles published in national and international journals and 32 book chapters and edited 7 books. He is the recipient of Dr. P. D. Sethi Memorial National Award for the best HPTLC publication from India in the years 2021 and 2022. He is a member of various professional and academic bodies like the Society of Pharmacognosy India, the International Natural Product Sciences Taskforce (INPST) society, SILAE: Società Italo-Latinoamericana di Etnomedicina (The Scientific Network on Ethnomedicine, Italy), Institutional Human ethical committee, Association of Pharmaceutical Teachers of India (APTI), and Indian Pharmacy Graduates Association.

Contributors

Zabeer Ahmed
Pharmacology Division
CSIR – Indian Institute of Integrative Medicine
Jammu and Kashmir, India
Academy of Scientific and Innovative Research (AcSIR)
Ghaziabad, India

Agustina Alaimo
Departamento de Química Biológica, Facultad de Ciencias Exactas y Naturales
Universidad de Buenos Aires (IQUIBICEN-CONICET)
Buenos Aires, Argentina

Daniel A. Allemandi
Departamento de Ciencias Farmacéuticas
Universidad Nacional de Córdoba, UNITEFA- CONICET
Córdoba, Argentina

Fábia Rafaella Silva Alves
Graduate Program in Pharmaceutical Sciences
Federal University of Rio Grande do Norte
Natal, RN, Brazil

Sakshi Bajaj
Department of Pharmacognosy, Delhi Institute of Pharmaceutical Sciences and Research
Delhi Pharmaceutical Sciences and Research University
New Delhi, India

Ashish Baldi
Pharma Innovation Lab, Department of Pharmaceutical Sciences and Technology
MaharajaRanjit Singh Punjab Technical University
Bathinda, Punjab, India

Thaís Nogueira Barradas
Faculdade de Farmácia
Departamento de Farmácia e Medicamentos
Universidade Federal de Juiz de Fora
Minas Gerais, Brazil
Departamento de Ciências Farmacêuticas
UFJF
Juiz de Fora, Minas Gerais, Brazil

Carolina del V. Bessone
Escuela de Ciencias de la Salud
Universidad Nacional de Villa Mercedes (UNViMe)
Villa Mercedes, San Luis, Argentina

Saurabh Bhatia
Natural & Medical Sciences Research Center
University of Nizwa
Nizwa, Oman

Sofía G. Brignone
Departamento de Ciencias Farmacéuticas
Universidad Nacional de Córdoba, UNITEFA- CONICET
Córdoba, Argentina

Stephani Araujo Cardoso
Programa de Pós-Graduação em Ciência e Tecnologia de Polímeros
Instituto de Macromoleculas
Universidade Federal do Rio de Janeiro
Rio de Janeiro, Brazil
Laboratório de Micro e Nanotecnologia (LMN)
Farmanguinhos
Fundação Oswaldo Cruz
Rio de Janeiro, Brazil

Simone Braga Carneiro
Graduate Program in Pharmaceutical Innovation
Federal University of Amazonas
Manaus, AM, Brazil

Durgesh Nandini Chauhan
Columbia College of Pharmacy
Raipur, India

Nagendra Singh Chauhan
Drugs Testing Laboratory Avam Anusandhan Kendra
Raipur, Chhattisgarh, India

Saveena Chauhan
Department of Bachelor of Pharmacy in Ayurveda
Delhi Pharmaceutical Sciences and Research University
New Delhi, India

Gilson Cassiano de Góes Neto
Graduate Program in Pharmaceutical Sciences
Federal University of Rio Grande do Norte
Natal, RN, Brazil

Ádley Antonini Neves de Lima
Pharmacy Department
Federal University of Rio Grande do Norte
Natal, RN, Brazil

Sushma Dev
Department of Pharmaceutical Chemistry, School of Pharmaceutical Sciences
Delhi Pharmaceutical Sciences and Research University
New Delhi, Delhi, India

Fabrício Felippe dos Santos
Programa de Pós-Graduação em Ciências Farmacêuticas
Universidade Federal do Rio de Janeiro
Rio de Janeiro, Brazil
Faculdade de Farmácia, UFRJ
Hospital Universitário Gaffrée e Guinle
Rio de Janeiro, Brazil
Departamento de Farmácia e Medicamentos
Universidade Federal do Estado do Rio de Janeiro
Serviço de Farmácia
Rio de Janeiro, Brazil

Ariel G. Garro
Ministerio de Ciencia y Tecnología de la Provincia de Córdoba
Córdoba, Argentina
Instituto Académico Pedagógico de Ciencias Humanas
Universidad Nacional de Villa María
Villa María, Córdoba, Argentina

Anshul Gupta
Department of Pharmaceutics
Delhi Institute of Pharmaceutical Sciences and Research
Delhi Pharmaceutical Sciences and Research University
New Delhi, India

Madhu Gupta
Department of Pharmaceutics
Delhi Pharmaceutical Sciences and Research University
Pushp Vihar, New Delhi, India

Prem N. Gupta
PK-PD Tox & Formulation Division
CSIR- Indian Institute of Integrative Medicine
Jammu and Kashmir, India
Academy of Scientific and Innovative Research (AcSIR)
Ghaziabad, India

Syed Assim Haq
PK-PD Tox & Formulation Division
CSIR- Indian Institute of Integrative Medicine
Jammu and Kashmir, India
Academy of Scientific and Innovative Research (AcSIR)
Ghaziabad, India

Priti Jain
Department of Pharmaceutical Chemistry, School of Pharmaceutical Sciences
Delhi Pharmaceutical Sciences and Research University
New Delhi, India

Jovita Kanoujia
Amity Institute of Pharmacy
Amity University Madhya Pradesh (AUMP)
Gwalior, Madhya Pradesh, India

M. Karthikeyan
Department of Pharmaceutics
PSG College of Pharmacy
Coimbatore, India

Gurleen Kour
Pharmacology Division
CSIR-Indian Institute of Integrative Medicine
Jammu and Kashmir, India
School of Biotechnology
University of Jammu
Jammu and Kashmir, India

Vijender Kumar
Department of Pharmacognosy & Phytochemistry, School of Pharmaceutical Sciences
Delhi Pharmaceutical Sciences and Research University
New Delhi, India

Arya L. Marisetti
Department of Pharmacognosy & Phytochemistry, School of Pharmaceutical Sciences
Delhi Pharmaceutical Sciences and Research University
New Delhi, India

Sunita Minz
Department of Pharmacy
Indira Gandhi National Tribal University (A Central University)
Amarkamtak, Annupur MP, India

Riya Nagpal
Department of Pharmaceutics
Delhi Pharmaceutical Sciences and Research University
Pushp Vihar, New Delhi, India

Rakesh Pahwa
Institute of Pharmaceutical Sciences
Kurukshetra University
Kurukshetra, Haryana, India

Oscar Perez
Departamento de Química Biológica, Facultad de Ciencias Exactas y Naturales
Universidad de Buenos Aires, (IQUIBICEN-CONICET)
Buenos Aires, Argentina

Asma Praveen
Department of Pharmacognosy & Phytochemistry, School of Pharmaceutical Sciences
Delhi Pharmaceutical Sciences and Research University
New Delhi, India

Bárbara Cristina Silva Holanda Queiroz
Graduate Program in Pharmaceutical Sciences
Federal University of Rio Grande do Norte
Natal, RN, Brazil

Daniela A. Quinteros
Departamento de Ciencias Farmacéuticas
Universidad Nacional de Córdoba, UNITEFA- CONICET
Córdoba, Argentina

Neha Raina
Department of Pharmaceutics
Delhi Pharmaceutical Sciences and Research University
Pushp Vihar, New Delhi, India

Pooja Mongia Raj
Department of Pharmaceutics
Delhi Institute of Pharmaceutical Sciences and Research
Delhi Pharmaceutical Sciences and Research University
New Delhi, India

Rakesh Raj
Meerabai DSEU Maharani Bagh Campus
Delhi Skill and Entrepreneurship University
New Delhi, India

Radha Rani
Department of Pharmaceutics, School of Pharmaceutical Sciences
Delhi Pharmaceutical Sciences and Research University
New Delhi, India

Soledad Ravetti
CIT VM-CONICET, Instituto Académico Pedagógico de Ciencias Humanas
Universidad Nacional de Villa María
Villa María, Córdoba, Argentina

V. Sankar
Department of Pharmaceutics
PSG College of Pharmacy
Coimbatore, India

Ana Beatriz da Silva Araújo Sena
Pharmacy Department
Federal University of Rio Grande do Norte
Natal, RN, Brazil

Ajay Sharma
Department of Phytochemistry
Delhi Pharmaceutical Sciences and Research University
New Delhi, India

Ajay Kumar Sharma
Institute of Nuclear Medicine and Allied Sciences
Defence R&D Organization, Ministry of Defence
New Delhi, India

Pratiksha Sharma
Pharma Innovation Lab, Department of Pharmaceutical Sciences and Technology
Maharaja Ranjit Singh Punjab Technical University
Bathinda, Punjab, India

Priyanka Sharma
Department of Pharmacology
Delhi Pharmaceutical Sciences and Research University
Pushp Vihar, New Delhi, India

Sanyam Sharma
Pharma Innovation Lab, Department of Pharmaceutical Sciences and Technology
Maharaja Ranjit Singh Punjab Technical University
Bathinda, Punjab, India

Vikas Sharma
Department of Pharmacognosy
Divine International Group of Institutions
Gwalior, MP, India

Meghna A. Singh
Department of Pharmaceutics, School of Pharmaceutical Sciences
Delhi Pharmaceutical Sciences and Research University
New Delhi, India

Anil Kumar Teotia
Microbiology Department
Indian Pharmacopoeia Commission, Ministry of Health and Family Welfare
Ghaziabad, India

Cecilia Tettamanti
Departamento de Ciencias Farmacéuticas
Universidad Nacional de Córdoba, UNITEFA- CONICET
Córdoba, Argentina

Anusha Thota
Department of Pharmaceutical Chemistry, School of Pharmaceutical Sciences
Delhi Pharmaceutical Sciences and Research University
New Delhi, India

Prasad Thota
Microbiology Department
Indian Pharmacopoeia Commission, Ministry of Health and Family Welfare
Ghaziabad, India

Raunak Varshney
Institute of Nuclear Medicine & Allied Sciences
Defence Research and Development Organisation
Delhi, India

B. Shanmuganathan Alias Viswesh
Department of Pharmaceutics
PSG College of Pharmacy
Coimbatore, India

Abhishek Wahi
Department of Pharmaceutics
Delhi Pharmaceutical Sciences and Research University
Pushp Vihar, New Delhi, India

Govind Yadav
Pharmacology Division
CSIR- Indian Institute of Integrative Medicine
Jammu and Kashmir, India

Vivek Yadav
Department of Pharmaceutical Chemistry
Delhi Pharmaceutical Sciences and Research University
Pushp Vihar, New Delhi, India

1 Naringenin Promising Therapeutic Agent

Aspects in Various Diseases

Neha Raina, Riya Nagpal, Abhishek Wahi, Priyanka Sharma, Vivek Yadav, Rakesh Pahwa, Vikas Sharma, and Madhu Gupta

1.1 INTRODUCTION

Herbal products provide a wealth of phytoconstituents, such as polyphenols, which gives a powerful antioxidant pool that maintains your health. Based on the structural framework of the cyclic rings of phenols and the functional groups that connect them, polyphenols are categorized into four groups: phenolic acids, stilbenes, lignans, and flavonoids. Furthermore, flavonoids are a part of organic pigments found primarily in plants. They have an intrinsic phenolic configuration (Pandey and Rizvi, 2009; Zobeiri et al., 2018) and are considered traditional medicine around the world because of their medicinal benefits (Patel et al., 2018). Several flavonoids, among over 4,000 discovered so far, have so far been observed to possess a range of therapeutic functions. Amidst this, naringenin (NRG) has recently received a lot of attention because studies have shown that many of its bioactivities may be useful in medicine. Power and Tutin discovered NRG in the year 1907, as chalcone, as a type of flavanone. These hydrophobic molecules are known chemically as 2,3-dihydro-5,7-dihydroxy-2-(4-hydroxyphenyl)-4H-1-benzopyran-4-one. It possesses good solubility in organic solvents including alcohol as well as appears to exist in both the glycol and glycosidic forms. Natural antioxidants can be found in citrus fruits like oranges, lemons, grapefruits as well as in fruits like bergamot and cherries. NRG is an inorganic compound having a melting point of 83 0 c. NRG has 580.5 g/mol of molecular weight, –0.44 value of log p and 1 mg/mL is reported to be its solubility at 40°C in water (Nouri et al., 2019). NRG has immense therapeutic potential and help in inhibiting the growth of cancer cells, and has insulin-like actions in regulating diabetes, including anti-inflammatory properties in disorders like hypertension. Furthermore, it contains anti-fibro genic, anti-atherogenic, antimicrobial, anti-mutagenic, anti-proliferative, immunomodulatory, hepatoprotective, antidiabetic, anti-atherosclerotic, neuroprotective, and oxidative stress reversal abilities (Nagula and Wairkar, 2019). Furthermore, NRG reduces the impact of hydroxyl and superoxide radicals on biomolecules under inflammatory and oxidative conditions. Also, it induces cytotoxicity and apoptosis and inhibits cellular proliferation in a variety of cancer cells (Bhia et al., 2021). Control of diabetes and obesity can be achieved by reducing the total quantity of both the triglyceride as well as cholesterol concentrations in plasma and liver by down-regulating pro-inflammatory mediators such as cyclo-oxygenase-2 (COX-2), intercellular adhesion molecule-1 (ICAM-1), interleukin-6 (IL-6), and tumor necrosis factor-α (TNF-α) (Gattuso et al., 2007). Additionally, it exhibits antibacterial properties against different bacterial strains (Llorach et al., 2008). Despite the hydrophobic nature of NRG, its bioavailability in vivo is very low, which limits its usage practically. Because of its short half-life and fast transformation to crystalline form, only a little amount of it is absorbed by the digestive tract (Llorach et al., 2008; Vallverdu-Queralt et al., 2012). As a result of these limitations in its use, several attempts have been made to improve its bioavailability. Nanotechnology is proving to be one of the most promising alternatives. In addition to protecting the drugs during

DOI: 10.1201/9781003292692-1

storage and after administration, nanomaterials provide quite a lot of advantages for drug delivery. Furthermore, controlled releases can be targeted to only specialized tissues and organs, enhancing their characteristics by escalating their bioavailability as well as lessening unwanted adverse effects. These nanoscale components with engineered surface make them effective at passing via vessels and reaching particular organ (Wang et al., 2020). Therefore, nanocarriers are employed for encapsulating distinct phytochemicals, including the glycosidic form of NRG. The implementation of nanocarriers for NRG delivery has been successful thus far and has improved solubility of NRG in water, and therapeutic efficiency, resulting in dose reductions (Bhia et al., 2021). NRG encapsulated inside these nanocarriers can be released in a controlled manner. Here, the chapter described detailed description of NRG, NRG in treatment of several disorders, some nano formulations for NRG delivery and their therapeutic application to address several different types of NRG bioavailability and delivery challenges.

1.2 NRG: PHARMACOKINETICS AND PHARMACOLOGICAL ACTION

NRG can be found in the range of variable foods. As a consequence of its hydrophobic nature, as evidenced by its poor solubility (ca. 475 mg/L), NRG's bioactivities are hampered, as are extensive gastrointestinal degradation, and restricted membrane transit, which together contribute to a lowered oral bioavailability (Wang et al. 2020). Furthermore, both passive diffusion and active transport are involved in NRG absorption (Joshi et al., 2015; Venkateswara Rao et al., 2017; Hernández-Aquino and Muriel, 2018). NRG is very poorly soluble in water, dissolving at approximately 46 ± 6 g/mL, a factor that limits its oral bioavailability to ~5.81%. It goes through fast hepatic metabolism via its first-pass glucuronide conversion, resulting in a low plasma bioavailability. In human subjects, Kanaze et al. observed that the NRG peaks of absorption, 3 h after oral consumption (135 mg), with. However, despite its rapid absorption, NRG's bioavailability stayed reduced because of the first-pass metabolic activity in the large intestine. After oral administration of naringin or consumption of grapefruit juice, a significant concentration of NRG was observed. Bai et al. (2020), examined the pharmacokinetic profile as well as the metabolic activity of NRG in individuals after consumption of naringin both orally and through intravenous route. It was found that plasma NRG concentration peaks at about 2 h (T_{max}, 2.09 ± 1.15 h) and decreases to 50% of C_{max} at around 3 h ($t_{1/2}$, 2.69 ± 1.77 h). The plasma concentration of NRG increased at a relatively slow rate (3.62 ± 3.19 h) after consuming NRG orally due to intestinal metabolism. In addition, the research looks into naringin and NRG's potencies in food-effect analyses. A high-fat diet did not significantly affect T_{max}, C_{max}, or AUC. In addition, there was a significant difference in their pharmacokinetic parameters. For instance, some of their pharmacokinetic parameters were higher in females than in males (Salehi et al., 2019). Based on several reports, NRG metabolization occurs rapidly in vivo, primarily in the form of glucuronides as well as sulfates, the most frequently circulated derivatives of plasma of NRG. Human liver and kidney microsomes contained approximately a dozen different metabolites of naringin and NRG: neoeriocitrin, hesperetin, naringin, NRG-*o*-glucuronide, eriodyctiol, hesperidin, NRG-*o*-glucoside, NRG and rhoifolin. Although, the flavonoid-like structure of NRG appears to be basic made up of 3 rings of 15 carbon atoms. There is a linkage of two benzene rings linked by a three-carbon chain among them. Chemically, NRG is 2,3-dihydro-5,7-dihydroxy-2-(4-hydroxyphenyl)-4H-1-benzopyran-4-one with a molar mass of 272.3. In order to exhibit its diverse bioactivities, NRG suppresses oxidative stress, which is generated in organisms as a byproduct of aerobic metabolism. NRG's 4-oxo group is connected to a 2,3-double bond, which has been shown to have a crucial role in its antioxidant activity. In recent years, researchers have developed numerous viable NRG formulations by using various techniques and complex inclusion methods. In addition, self-, nanoparticles, nano-emulsifying drug delivery systems (SNEDDS) and nanosuspensions will help to enhance this drug's bioavailability, that will be beneficial to biomedical applications. The LD50 is 5,000 mg/kg for NRG. The main mechanism of NRG's action is the inhibition of some cytochrome P450 isoforms: CYP1A2 and CYP3A4 in humans. Several types of opioid receptors have been successfully antagonistic by NRG (Den et al., 2019; Zeng et al., 2018; Zaidun et al., 2018; Brewer, 2011; Jung et al., 2017) Various studies show that it has, hepatoprotective, anti-carcinogenic, antiatherogenic, antidiabetic, and antioxidant, effects on humans. Flavonoids are considered to be a necessary component of the diet in amounts of a few 100 mg daily. Among its many benefits is its ability to combat inflammation and oxidative injury (Kumar and Tiku, 2020; Md et al., 2020a; Noori et al., 2020; Zhao et al., 2019; Lee et al., 2019; Han et al., 2018; Lim et al., 2017; Bao et al., 2016).

1.3 POTENTIAL APPLICATIONS OF NRG

Various studies show that NRG has, hepatoprotective, anti-carcinogenic, antiatherogenic, antidiabetic, and antioxidant (Figure 1.1) effects. These are discussed in detail in the subsequent sections.

1.3.1 Diabetes

The NRG exhibits anti-hyperglycemic action by averting absorption of glucose from the intestinal region. It has been shown that NRG decreases the levels of glucose in the blood, decreases insulin sensitivity, decreases macrophage migration into adipose tissue, decreases the levels of substrate-2 (IRS-2) (insulin receptor), and increases the concentrations of phosphoinositide 3-kinase (PI3K). Although it is highly water insoluble, its rapid biodegradation and low bioavailability make it unsuitable for oral administration. It may be possible to overcome these limitations by applying nanotechnology to drug delivery. Polymeric nanoparticles can be used here to protect NRG from hydrolytic and enzymatic degradation in the gastrointestinal tract (Prabhakar et al., 2014; Barrett et al., 2010; Priscilla et al., 2014; Ren et al., 2016).

1.3.2 Liver Diseases

Studies have extensively been performed on the potential beneficial impacts of NRG with low toxicity in liver diseases, as it is an organic compound having many medical properties including anti-inflammatory, antitumor, as well as antioxidant. Intoxication by drugs, alcohol, malnutrition, or infection are the most prevalent causes of liver diseases. Often, liver disorders are the result of both inflammatory responses and oxidative stress. Consequently, NRG's exhibited properties may be of major significance in the treatment and prevention of liver disease. In addition, NRG inhibits oxidative stress, the toll-like receptor (TLR) and, non-canonical TGF-β pathways. The NRG also has a vital role in nonalcoholic fatty liver disorders, since it modulates cholesterol and the metabolism of lipids by modulating oxidation as well as synthesis. Studies are conducted on the effects of silymarin as well as of NRG on hepatic injury in rats induced by arsenic. As a result of its ability to bind with and form –SH groups, arsenic can cause hepatic damage because of the production of reactive oxygen species (ROS). Inhibition of free radical-mediated processes is achieved by the phenolic structures. According to the author, chronic arsenic exposure reduced blood GSH levels and increased blood ROS amounts, thereby supporting the hypothesis of oxidative stress induced by arsenic. In addition, administration of both silymarin as well as NRG showed, scavenging ROS, antioxidant properties, stimulating GPX activity, and inhibiting lipid peroxidation (Karim et al., 2018; Hermenean et al., 2014; Chtourou et al., 2015).

1.3.3 Inflammatory Diseases

During inflammatory processes, macrophages make cytokines (e.g., IL-1β, IL-6, TNF-α) as well as some mediators including such like prostaglandins and nitric oxide in response to infection or injury. Numerous chronic inflammatory diseases are associated with excessive production of these substances, including atherosclerosis, asthma, pulmonary fibrosis, osteoarthritis, septic shock, rheumatoid arthritis, and tumors. As a result, inhibiting them is a key target path that leads in the

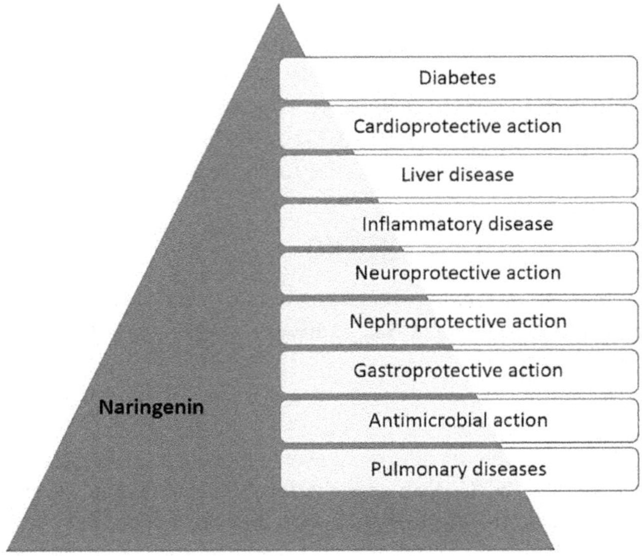

FIGURE 1.1 Potential benefits of naringenin.

treatment of several inflammatory conditions. The antioxidant activity of NRG is due to the presence of hydroxyl group (OH) as substituents on its structural framework, which has a strong reactivity toward reactive nitrogen species (RNS) as well as toward ROS. NRG has a strong potential to scavenge the free radicals and reduces oxidative DNA damage caused by lipid peroxidation, which could be used for treatment of inflammatory disorders (Yoshida et al., 2010; Chen, Siyu et al., 2012; Kanaze et al., 2007; Najmanová et al., 2020; Kunnumakkara et al., 2018).

1.3.4 Neuroprotective Action

Neurons are especially vulnerable to oxidative insults in compared to other bodily tissues for the following reasons. (1) Metal ions and polyunsaturated fatty acids are abundant in brain cells. Metal ions build in the brain with the passage of time and aging, increasing the development of reactive species. (2) Neuronal cells, in contrast to other cells, rely heavily on oxidative phosphorylation as their primary source of energy. (3) Because there are so few antioxidant enzymes in the brain, there aren't many defensive mechanisms in place. These characteristics demonstrate that nerve cells are too sensitive to stress are initiated and instigated by ROS. According to numerous in vitro and in situ investigations, NRG has a high permeability to the blood–brain barrier (BBB). NRG pre-treatment of rats showed considerable neuroprotection against 6-OHDA-induced toxicity. In the rotenone-induced PD paradigm, NRG treatment resulted in significant improvements in motor abilities. The Alzheimer rat model, which was duplicated by A-injection, showed that NRG therapy improved cognitive skills by reducing lipid peroxidation and apoptosis. According to studies, NRG may be useful in restoring cognitive decline associated with Alzheimer's disease by upregulating insulin signaling in the rat brain. The covalent interaction of NRG with lipoic acid produces a new molecule known as "VANL-100." Neuroprotection has been seen in both in vitro and in vivo ischemia/reperfusion models at very low doses of 2 102 M. (100-fold more potent). The improved capacity of nerve cells to chase down free radicals in the presence of "VANL-100" might be the cause. NRG has been shown to be beneficial in the treatment of cerebral ischemia by lowering oxidative damage and NF-B-mediated inflammation. In another investigation, NRG was found to have protective benefits in experimental ischemic stroke models, reducing infarct size and improving neurological insufficiency and brain swelling. Acetylcholinesterase (AChE) activity may be inhibited by an increase in free radical concentration in neuronal cells. AChE levels in rats exposed to iron were reduced, while NRG treatment to iron-exposed rats resulted in a considerable restoration of AChE activity due to its efficient antioxidant capacity. Furthermore, NRG supplementation causes the expression of acetylcholine transferase to be restored, resulting in improved learning and memory in the treated mice. Treatment with NRG has also been shown to improve cholinergic and serotonergic transmission. NRG also has the ability to safeguard cell integrity through modulating antioxidant defense mechanisms, both enzymatic and non-enzymatic. As a consequence, NRG may modulate neuronal cholinergic neurotransmission resting membrane potential in a direct or indirect manner (Feigin et al., 2019; Zhou et al., 2018; Md et al., 2018; Kujawska et al., 2018; Nouri et al., 2019).

1.3.5 Cardioprotective Action

The basic aerobic metabolism of cells produces ROSs and free radicals. Excessive production of ROS, on the other hand, causes serious oxidative damage to the cells' biological components. Numerous investigations have suggested that NRG has potent antioxidant properties against a variety of oxidative insults, and it has been shown to enhance heart function even after damage. The findings of several study groups have backed up the preclinical and clinical data in favor of this. A group of investigators from Cairo University, Egypt, commented on the use of NRG to prevent cardiotoxicity in male mice (Swiss albino) induced by doxorubicin that various enzymes like, catalase (CAT), superoxide dismutase (SOD), and glutathione-S-transferase (GST) had been restored. In this regard, Han et al. (2008) searched into how NRG safeguard cardiomyocytes from cardiac toxicity induced by doxorubicin and found that upregulation of many endogenous antioxidative enzymes via phosphorylation and activation of ERK1/2 and Nrf2, respectively, and also the remobilization to the nucleus probably played an important role in cardiac myocytes protection (Testai et al., 2017; Bala et al., 2020). NRG protects cardiomyocytes against apoptosis by increasing the expression of endogenous antioxidant enzymes (SOD, GST, and CAT) through activation of ERK1/2 and nuclear translocation of Nrf2. According to some other research, NRG suppresses the generation of reactive intermediates, increases the activity of antioxidative enzymes catalases and dismutase (SOD), and suppresses ERK1/2 as well as p38 MAPK activation. Furthermore, it blocks angiotensin II from binding to the AT1 and AT2 angiotensin binding receptors, inhibiting cell proliferation of vascular smooth muscle. NRG's potent cardioprotective effects were established a drop in LDH in coronary effluent to promote regenerative potential, by a decrease in myocardial infarction area, and an enhancement in left ventricular functioning retrieval. Myocardium treated with NRG had a greater level of SOD as well as a reduction in the level of MDA. The ability of NRG to operate on calcium-activated potassium channels may explain its beneficial effects on the heart. NRG prevents calcium from accumulating in the cardiac mitochondrial matrix. There was also a change in functional indices after ischemia, as well as a reduction in the prolonged interval of myocardial damage. The quantity of thallium affects inner mitochondrial membrane depolarization, and after interaction with NRG, isolated rat cardiac mitochondria exhibited transmembrane drifting of thallium (a potassium-mimetic cation). Citrus hybrid fruit like pink grapefruit is created by crossing Clonorchis sinensis with Cucurbita maxima. It's well recognized for having higher amounts of NRG. It prolongs repolarization by increasing temporal cardiac repolarization dispersion. Because of its pro-arrhythmic effects, it may be useful in individuals with structural myocardial abnormalities. In mice with elevated blood pressure, NRG treatment reduced left heart hypertrophy and interstitial fibrosis. It has also been claimed that the left ventricular functions have been altered. Their mechanism of action is exhibited by inhibiting signaling pathways like PI3K/Akt, JNK, and ERK. One of the most important ROSs is hydrogen peroxide (H_2O_2), which is generated in the respiratory cascade of mitochondria and also exhibited in nature as the residue of the basal metabolism. Many organs, including the peroxisome as well as the endoplasmic reticulum, are reported to be the major producers. H_2O_2 has severe negative effects on cardio myoblasts due to its high oxidative activity, which can be reversed by NRG therapies. Stress-induced apoptosis and lipid peroxidation are reduced when antioxidant enzymes (CAT, SOD, and nitric oxide) are synthesized, resulting in a rise in GSH levels. NRG also causes overexpression of the Akt and Nrf2 genes, as well as downregulation of the NF-B genes and Caspase-3. Additionally, NRG treatment improves the concentration of nitric oxide in blood, LDH, and profile of lipid in HCD rats. NRG has been reported to lower the risk of atherosclerosis by reducing the insulin resistance and dyslipidemia, as well as blocking the formation of apolipoprotein B100. According to studies, NRG causes the promotion of oxidation of fatty acids and limits synthesis of cholesterol and esterification, resulting in lower cholesteryl ester as well as hepatic triglycerides; hence, VLDL secretion is restricted (Arafa et al., 2005; Han et al., 2008; Li et al, 2011; Meng et al., 2016; Yu et al., 2019; Yi et al., 2019; Karuppagounder et al., 2015; Arozal et al., 2011).

1.3.6 Nephroprotective Action

Chronic kidney disease (CKD) affects around 35 million individuals in developing and developed nations, according to a WHO estimate. The advancement of kidney impairment is aided by inflammatory cytokines and oxidative stress. Renin–angiotensin system (RAS) activation or Angiotensin II type I receptor (AT1R) and endothelin receptors [58], i.e. ET receptor type A and type B (Arozal et al., 2011; Esteban et al., 2004) have been found to cause other types of kidney injury. Stimulation of such receptors causes injury of the kidney by raising, podocyte autophagy, microvascular pressure, activation of NFB-p65, and oxidative stress, and, all of them causes the generation of inflammatory mediators such as IFN, IL-6, as well as prostaglandins. NRG has been shown to improve nephrotoxicity in a variety of drug-induced models of nephrotoxicity. Following NRG therapy, the technique of protein analysis i.e. Western blotting technique demonstrated activation of PPAR and reduction of signaling pathways like ETAR, AT1R, p-ERK1/2, and p-NFB p65, reducing inflammation as well as oxidative stress in rats having daunorubicin-induced kidney injury. In a NRG-treated group, the histopathological study of the kidney revealed improvement in fibrotic lesions. Immunohistochemical analysis of tissues that was treated with nirangenin revealed considerable downregulation of many biomarkers, like iNOS, caspase-9, and VEGF; nevertheless, survivin expression in the renal tissue is increased. Survivin seems to be an anti-apoptotic protein which prevents cell death by restricting activation-9's activation or by blocking the release of cytochrome-c by suppressing stimulation of ligand Fas/Fas. NRG demonstrated to protect rats from cisplatin-induced renal toxicity, oxidative kidney ill-functioning mediated by cadium, and carbon tetrachloride-induced acute renal toxicity (Esteban et al., 2004; Ozbek et al., 2009; Karuppagounder et al., 2015).

1.3.7 Gastroprotective Action

Because of its high permeability, NRG is absorbed by passive diffusion along the GIT tract, mostly in the small intestine. NRG's anti-inflammatory effects serves as a gastroprotective benefits, that are gained via suppressing activator protein-1,while activating ERK 5 as well as P38 MAPK. IBDs can progress to drastic forms over a period of time, such as ulcerative colitis as well as Crohn's disease (CD). In this type of scenarios, using NRG as a treatment has exhibited anti-ulcerogenic advantages leading to diminished ulceration of tissues, inflammatory processes, and colon necrosis (Schrezenmeier et al., 2017; Orrego-Lagarón et al., 2015; Yang et al., 2011; Al-Rejaie et al., 2013). Furthermore, NRG's positive effects may be achieved by lowering inflammatory cytokines, decreasing pro-inflammatory indicators like iNOS, and reducing the protein levels as well asTLR4 mRNA in the mucosa (Dou et al., 2013). When NRG interacts with antagonists like serotonin (5-HT4) receptor, it causes colonic contractions to rise, allowing stomach emptying to occur (Manabe et al., 2010). Certain endogenous chemicals secreted by the stomach, small intestines, and pancreas, such as Motilin and Ghrelin, increase GI motility by acting on specific receptors (Peeters et al., 2006).

NRG exhibited to act as a Ghrelin receptor agonist, alleviating a variety of GI problems (Jang et al., 2013). NRG-induced apoptosis in the SGC-7901 cell lines while suppressing proliferation, migration, and invasion. The suppression of AKT phosphorylation and decrease in target molecule expression by NRG (Bao et al., 2016) might be the cause. NRG inhibits the growth of cancer cells in the stomach in SGC-7901 cell lines (Zhang et al., 2016a).

1.3.8 NRG in Pulmonary Diseases

Chronic airway inflammatory diseases (CAIDs), like asthma, chronic obstructive pulmonary disease, lung cancer, and pulmonary fibrosis are categorized by an inflammatory reaction of the respiratory system to only certain external factors including such pathogenic organisms, smoke, and dusts, resulting in greater generation of sputum, inflammatory cells as well as the mediators of inflammation, resulting in abnormal pulmonary function (Chin et al., 2020). Various inflammatory indicators, notably interleukins, cytokine production, nuclear factor kappa B, glucocorticoid receptors, and tumor necrosis factors, are considered to mediate the inflammatory response in CAID. These inflammatory mediators cause increased oxidative stress, that activates immunological responses, enhances bronchial hyper-responsiveness, and boosts mucin secretion (Lawrence, 2009). By reducing NF-kB transcription activity, NRG has been shown to suppress the inflammatory pathway implicated in CAID. Furthermore, In the vicinity of pro-inflammatory cytokines such as NF-kB, TNF-a, and IL-1, NRG reduces the inflammatory reaction in the respiratory system, this response is triggered by microbial byproducts. NRG reduces the breakdown of IkB linked to the NF-kB p50/p65 subunits, which decreases the activity of NF-kB (Kim et al., 2020). Furthermore, NRG's therapeutic impact in airway inflammatory illnesses has been demonstrated in multiple in vivo investigations.

1.3.9 Antimicrobial Action

Natural products are currently gaining popularity due to their powerful antibacterial properties and fewer adverse effects than chemically manufactured substances. Natural products are regarded as harmless and have an advantage over antibiotics since they have a lower risk of drug resistance. Due to their diversified action on microbial cells, bacterial strains are less likely to acquire resistance to a chemical of natural origin. Suppression of auto inducer-mediated cell–cell transmission, enzymes inhibition is vital for the synthesis of the bacterial cell membrane, anti-biofilm activities, and involvement with the cell membranes of bacteria and destruction of its integrity are all examples of natural compounds that have antibacterial activity (Górniak et al., 2019). NRG has a broad range of antimicrobial properties, including antifungal and antibacterial properties (Kozłowska et al., 2017; Duda-Madej et al., 2020). NRG has been demonstrated to have significant antibacterial action against resistant strains of many pathogens. NRG inhibits the autoinducer-mediated cell signaling pathway, which has antibacterial properties (Tran Trung et al., 2020; Vikram et al., 2010).

1.4 NANOSCALE DELIVERY SYSTEMS FROM NRG

The implementation of nanocarriers for NRG delivery has been successful thus far and has improved solubility of NRG in water, and therapeutic efficiency, resulting in dose reductions. This section discussed the nanocarriers for NRG delivery (Table 1.1).

1.4.1 Polymeric Nanoparticles

Since the 1980s, drug delivery systems based on nanocarriers like nanogels, nanocapsules, polymeric micelles, various nanovesicles as well as dendrimers have really been employed (Venditti, 2019; Amoabediny et al., 2018). In the context of nanotechnology, nanocarriers constructed of

TABLE 1.1
Different Nanoformulations of Naringenin

S.No	Type of nanoformulation	Method of preparation	Drug release	Particle size	Entrapment efficiency	Zeta potential	Cell lines	Outcomes	References (s)
1	Nanoparticles	High-speed homogenization technique.	>85% at the end of 24 h	176±2.35 to 430±3.32 nm.	61.71±1.11 to 85.23%±1.56%	–13.14±1.22 to –24.32±0.95 mV	Colon-26 (CRC cell line)	Significant tumor suppression, enhancing oral bioavailability and chemotherapeutic efficacy	David et al. (2016)
2	Nanoemulgel	–	Sustained drug release profile	152.03±4.6 nm	–	15 mV	L929 cell line	Biocompatible and nontoxic to the nasal mucosa, enhanced diffusivity, and improved permeation	Nagaraja et al. (2021)
3	Nanoparticles	Nanoprecipitation method	–	242.8±0.72 nm	72.6%±0.42%	10.8±0.012 to –24.24±0.015 mV	A549 cells and J774 macrophage cells	Enhanced anticancer effect and drug uptake by cancer cells, superior cytotoxic effect	Parashar et al. (2018)
4	Nanoparticles	Single emulsion and solvent evaporation technique	–	131.2 nm	32.45%	–25.4 mV	HaCaT cells	Higher SPF value and skin retention	Joshi et al. (2018a)
5	Nanoparticles	–	Sustained release profile	149.69±12 to 302.2±17 nm	91%	–26.3±2.2 to –38.22±1.5 mV	–	Significant hypoglycemic effect, free from toxicity, enhanced bioavailability	Maity et al. (2017)
6	Liposomes	Thin-film dispersion method	98.4% at pH 1.2 after 72 h	70.53 nm	72.8%±0.2%	–37.4±7.3 mV		Improved solubility and bioavailability of naringenin	Wang et al. (2017)
7	Solid lipid nanoparticles	Emulsification and low temperature solidification method	Sustained release profile	98 nm	79.11%	–31.4 mV	A549 cells	Increased bioavailability, nontoxic in nature, good entrapment efficiency	Ji et al. (2016)

(Continued)

TABLE 1.1 (*Continued*)
Different Nanoformulations of Naringenin

S.No	Type of nanoformulation	Method of preparation	Drug release	Particle size	Entrapment efficiency	Zeta potential	Cell lines	Outcomes	References (s)
8	Nanoparticles	Nanoprecipitation technique	72% after 24 h	Around 40–100 nm	–	–	HeLa cancer cells	Better efficacy, minimum side effects, enhanced therapeutic efficiency than free drug	Krishnakumar et al. (2011)
9	Nanocomposites	Modified hummers method	–	16.3±4.4 nm	–	–	–	Better inhibitory effects, improved bactericidal activity	Shanmuganathan et al. (2020)
10	Solid lipid nanoparticles	Emulsification and homogenization technology	79% over 36 h	238–274 nm	86.23%	−10.6±2.1 mV	–	Enhanced oral bioavailability and antifibrotic activity	Yang et al. (2020)
11	Liquid crystalline nanoparticles	–	Sustained release up to 24 h	175.9±2.7 nm	–	−23.8±1.03 mV	BCi-NS1.1, A549 cells	Potent anticancer activity, inhibition of proliferation and migration of cells, induced apoptosis	Hermenean et al. (2014)
12	Nanoparticles	Ionic gelation	Moderate sustained release effect	446.4±112.8 nm	–	+22.5±4.91 mV	–	Better ability to prolong the residence time as compared to suspension, increase bioavailability, reduce administrating frequency	Zhang et al. (2016)
13	Nanostructure lipid carrier	Hot homogenization technique	– Drug release facilitated under acidic conditions	nm	–	−11.4 mV	HT-29 cells	Lower side effect, decline dose chemotherapeutic agent	Raeisi et al. (2019)

(*Continued*)

TABLE 1.1 (*Continued*)
Different Nanoformulations of Naringenin

S.No	Type of nanoformulation	Method of preparation	Drug release	Particle size	Entrapment efficiency	Zeta potential	Cell lines	Outcomes	References (s)
14	Solid lipid nanoparticle	Hot melt emulsification and high-speed homogenization method	70% within 8h	129nm	>80%	23mV	U87MG glioma cells	Better chemoprotective effect over plain drug, higher cytotoxicity, significant improvement in drug release rate and absorption, better cellular uptake and cytotoxic ability	Wang et al. (2021)
15	Nanoemulsion	–	–	113.83±3.35nm	–	12.4±1.05mV	–	Better neuroprotective activity than free drug, small particle size and high zeta potential	Md et al. (2018)
16	Nanocrystal	Wet milling	–	274.3 ± 3.3 nm	–	−12.4±0.4 mV	CaCo-2 cells	Superior dissolution, favorable cellular uptake, improved transcellular ability and significantly enhanced oral bioavailability and anti-inflammatory efficacy	Zhang et al. (2021)
17	Nanoparticles	–	80.02±3.69% drug release in 24h	179.7 ± 2.05 nm	74±3.6%	−9.18±0.78 mV	–	Prolonged circulation time and reduced arthritic inflammation	Mohanty et al. (2021)
18	Nanosuspension	Sonoprecipitation method	91±4.4% of drug release at 60min	117±5nm	–	−14.6±5.6mV	–	Improved dissolution and oral bioavailability, effective permeability	Gera et al. (2017)

(*Continued*)

TABLE 1.1 (*Continued*)
Different Nanoformulations of Naringenin

S.No	Type of nanoformulation	Method of preparation	Drug release	Particle size	Entrapment efficiency	Zeta potential	Cell lines	Outcomes	References (s)
19	Nanosuspension	High-pressure homogenization	89.2±5.4% at 120 min	489±5.5 nm	–	–25.86±1.72 to –40.63±2.23 mV	MCF-7 breast cancer cells	Improved dissolution and antitumor activity of HPMC based nanosuspension as compare to surfactant based	Md et al. (2020b)
20	Nanosuspension	Miniaturized media-milling method	95% within 7 days	216.9 nm	–	−1.41±0.33 mV	–	Enhanced expectorant effect, improved the antitussive effect significantly, improved bioavailability	Dong et al. (2022)
21	Nanoparticles	Solvent evaporation method	–	147–409 nm	22.1%–80.4%	– 13.2 to − 28.9 mV	–	Showed higher efficacy and improved relative bioavailability as compared to zinc/ pectin loaded nanoformulations.	Smruthi et al. (2022)
22	Nanostructured lipid carrier	Emulsion-evaporation, low temperature solidification technique	Improved release rate by 3.5-fold	162.9±11.7 nm	94.5%±5.6%	–6.4±0.4 mV	–	Increased the inhibitory effects on nonalcoholic fatty liver disease, intestinal absorption	Rui et al. (2021)
23	Nanoparticles	Double-emulsion solvent evaporation technique	–	366.8±36.7 nm to 302.2±45.4 nm	80% at 48 h	+35.57±4.25 mV to –4.69±2.66 mV	MCF-7 cells	Improved efficacy and synergistic activity with reduced toxicity	Alshememry et al. (2022)

(*Continued*)

TABLE 1.1 (*Continued*)
Different Nanoformulations of Naringenin

S.No	Type of nanoformulation	Method of preparation	Drug release	Particle size	Entrapment efficiency	Zeta potential	Cell lines	Outcomes	References (s)
24	Nanoparticles	Ionic gelation technique	~15% in 120 min.	407.47 nm	80%	+22.73 mV to +58.43 mV	A549 lung cancer cells, 3T3 cells	Efficient system for delivering NAR with antioxidant and anticancer activities.	Kumar et al. (2015)
25	Nanoemulsion	–	29.118%±2.787% after 8 h	38.70±3.11nm		− 27.4±0.14 mV		Enhancing the uptake of NRG to the brain, avoiding systemic circulation, and hence enhancing the amount of NRG in the brain and improving brain bioavailability	Gaba et al. (2019)
26	Nanoemulsion	Low energy emulsification technique	100% within 12 h	~150 nm		−8.42 and −13.53 mV	NIH-3T3	Accelerate healing of wounds	Akrawi et al. (2020)
27	Nanoparticles	Emulsion diffusion evaporation method	85.67%±6.23%	150.45±12.45 nm	86.59%±3.4%	− 20.5±2.5 mV	ATCC	Improved therapeutic efficiency against pancreatic cancer cell, higher cytotoxic effect	Akhter et al. (2020)
28	Nanosuspension	Precipitation-ultrasonication method	–	118.1±2.7 nm		−12.2±1.2 mV		improve the solubility and bioavailability	Singh et al. (2018)
29	Nanoparticles	Complex coacervation technique	51% at 36 h	~337.2±48.27 nm	94.32%±0.24%	−34.4±7.45 mV	MCF-7 cells	longer duration, better effectivity, significant cytotoxic effect in cancerous cells.	Muralidharan et al. (2021)

(*Continued*)

TABLE 1.1 (*Continued*)
Different Nanoformulations of Naringenin

S.No	Type of nanoformulation	Method of preparation	Drug release	Particle size	Entrapment efficiency	Zeta potential	Cell lines	Outcomes	References (s)
30	Nanoparticles	Nanoprecipitation method	75% in first 48 h and remaining in next 24 h	213.7 nm	–	–	Human mesenchymal stem cell	Cytocompatible against human MSCs, and exhibit a sustained released pattern, neuroprotective against human MSC cells	Ahmad et al. (2018)
31	Nanoemulsion	–	–	98.31 ± 1.17 nm	–	–19.24 mV	–	Improved the neurobehavioral activity, enhanced the bioavailability	Ahmad et al. (2020)
32	Nanoparticles	–	82%–86% in initial hours	180.1 ± 2.6 to 148.4 ± 1.3 nm	20%	–39.1 ± 0.7 to –30.5 ± 0.7 mV	HeLa cells, EA.hy926	Better anticancer activity than free drug, reduced side effects	Fuster et al. (2020)
33	Nanoparticles	Emulsion diffusion evaporation method	–	129 nm in diameter	70%	–		Exhibited higher melting temperature than free naringenin, significant reduction glycated hemoglobin level	Maity et al. (2020)
34	Lipid nanoemulsion	–	–	230 nm	–	–40 mV	C57BL/6	Improved water solubility and bioavailability	Fuior et al. (2020)
35	Nanoemulgel	–	74.62 ± 4.54% in 24 h	145.58 ± 12.5 nm	–	–7.80 ± 3.26 mV	–	Sustained release profile, good spreadability	Yeo et al. (2021)
36	Nanostructured lipid carrier	Solvent diffusion method	82.42% ± 1.88%	84.17–86.71 nm	87.58%–88.21%	–8.15 to –8.21 mV		Better antioxidant potential, good stability	Qizilbash et al. (2022)

biomedical and ecological polymers are known as polymeric nanoparticles. Nanocapsules or nanospheres are the most common form of delivery of these particles, depending on their preparation method (George et al., 2019; Jahangiri et al., 2018). A polymeric nanoparticle delivers a wide variety of molecules, including proteins, plasmids, antisense DNA, and low molecular weight drugs. Polymeric nanoparticles can deliver protein, plasmid, antisense DNA, and low molecular weight drugs. A polymeric nanoparticle's hydrophobicity and hydrophilicity as well as its particle size and surface charge distribution determine whether or not it will reach the desired organ. Comparatively to conventional therapy, the implementation of nanomaterials in drug delivery enhances drug safety, bioavailability, biocompatibility, transparency and durability, and treatment response (Yousefi et al., 2020). Numerous research groups have studied NRG via micelles; delivery. The rate of absorption of NRG enclosed in a mixture of micelle formulation having Tween®80, along with Pluronic F127 was investigated in male SD rats (Song et al., 2015). Poloxamer 407 (Pluronic F127) is a nonionic surfactant with micellar properties that is appropriate for the delivery of drugs (Domínguez-Delgado et al., 2016). Tween®80 is another nonionic surfactant and has been widely recognized in drug delivery due to its lower toxic effects percentage, low price, and good ability to dissolve. A thin-film hydration technique was used to make the mixture of micelles loaded with NRG, with a ratio of 1:10:0.2 (w/w/w) of NRG: Pluronic F127: Tween®80. An additional report highlighting on employing micelles as an NRG delivery system is presented by Li et al. (2020), which investigates NRG-loaded genipin-crosslinked ß-casein micelles (G-CNMs). The biodegradability, biocompatibility, and low toxicity of MPEG-PCL make it attractive as a drug carrier. An evaporation method was used to prepare NRG-loaded nanoparticles in this study. Nanoparticles loaded with NRG were formed by an evaporation method in this study. Nanoparticle-lyophilized powder was made by freeze-drying NRG nanoparticles into stable, re-soluble nanoparticles. The findings exhibited that the particles were tiny and had a higher drug-loading capacity. Using the US Pharmacopeial method, we examined buccal tablets; in vitro release. In comparison with the control group, buccal tablets have NRG release of nanoparticles. NSG faster and more completely over 12 h. It seems that NRG can be used to deliver anti-inflammatory drugs in oral diseases (Li et al., 2020; Wang et al., 2014).

1.4.2 Lipid-Based Nanoparticles

Over the last two decades, lipid-based delivery of drugs seems to be at the forefront of efforts to overcome the shortcomings of pharmaceutical methods, particularly for medications that are insufficiently water soluble (Gordillo-Galeano et al., 2018). Lipids are generally hydrophobic or amphipathic, for example, NRG. Whey protein isolate (WPI) is among the several bioactive peptides that are employed as the biologically active transporters. In one of studies of 2020, Yin and his colleagues, studied the antioxidant properties, stability, and bioaccessibility of hydrophobic α-tocopherols. WPI had no effect on NRG storage and digestion whereas protein-NRG particles had increased antioxidant activity (Ghasemiyeh et al., 2018). Biodegradable and biocompatible, they have very low toxicity profiles, they increase drug solubility, and they can be controlled in their release study. The very first nanomaterials were created by emulsifying solid lipids. Seven years after the discovery of SLNs, the NLCs were generated NLCs contain liquid and solid lipids that can incorporate more drug than SLNs. They improve drug solubility in the lipid matrix and allow for more controlled drug release (Raeisi et al., 2019). They also have a lower burst release due to their compact structure. Using the hot homogenization method, this research encapsulated NRG into NLCs. The nanoparticle size ranged from 50 to 120 nm (mean size = 98). For assessing nanoparticle cytotoxicity, the authors conducted an MTT assay and used the staining method 40,6-diamidino-2-phenylindole (DAPI). Findings exhibited that NRG-NLCs blocked growth of cells very similar to NRG alone after 24 h of incubation. The amount of apoptosis in HT-29 colon cells was increased by using NRG-NLCs and analyzed using flow cytometry. Moreover, the real-time PCR results showed a decrease in anti-apoptotic markers, and an increase in Bid mRNA expression in the pro-apoptotic family. In this study, NLCs were loaded with NRG and it inhibited the development of cancer cells of colon (HT-29). Researchers have found that NRG-NLCs increase anticancer drugs; efficiency, reduce the side effects of chemotherapy, and induce apoptosis (Wang et al., 2017).

For decades, liposomes have been gaining attention for their potential to release drugs in a controlled way, their suitable safety profile, their high content of solubility as well as complete tolerance, which improved drug bioavailability (Zhu et al., 2015; Haghiralsadat et al., 2017). A liposome is a lipid vesicle formed by bilayers of phospholipids. Many times, these lipids are natural phospholipids like soybean phosphatidylcholine (SPC) or chemical phospholipids like dipalmitoyl-phosphatidylcholine (DPPC). A thin-film hydration method was used to prepare NRG-loaded liposomes. The resulting liposomes exhibited high values of zeta potential and good physical stability. On three gastrointestinal media relevant to human digestion, we evaluated in vitrorelease profiles. As predicted by the tissue distribution analysis, higher levels of NRG were found in liver tissue when liposomes were tested. After oral administration of the encapsulated drug in rats, there was a striking increase in drug solubility and oral bioavailability. Based on these findings, liposomes loaded with NRG can improve the bioavailability as well as the solubility of drug, hence providing a desirable method of drug delivery for clinical purposes (Jain et al., 2011).

1.4.3 Nanosuspensions

The nanosuspension method may be used to deliver drugs in the form of particles with a size of 1mm, caused by a pure therapeutic agent and stabilizer. A purified bioactive agent and a stabilizer can be utilized to distribute drugs in the form of 1-mm-sized particles using the method of nanosuspension. Insoluble medicines with low bioavailability and poor pharmacokinetics may be good candidate for this treatment. It can be used for intravascular targeted treatment and solubilization, as well as to inhibit enzymatic degradation in oral delivery (Karadag et al., 2014). Its formulation is simple, in addition to improving medication pharmacokinetics. NRG, like other flavonoids like quercetin (Mauludin et al., 2013) rutin (Jacob et al., 2020), and others, may be made into a nanosuspension to increase therapeutic efficacy. Tween® 80, pluronics (F68 and F127), polyvinyl alcohols (PVA), vitamin E-TPGS or D-tocopherol polyethylene glycol 1000 succinate, polyethylene glycols (PEG), surfactants, poloxamer-188, polymers and other stabilizers are among the most prevalently used stabilizers, and polymers in nanosuspensions (Singh et al., 2018).

Kumar Singh et al. created nanosuspensions of NRG that are surfactant-stabilized in order to improve their pharmacokinetic and physiochemical properties. They used a precipitation-ultrasonication process to make distinct nanosuspensions with variable amounts of different surfactants and polymers . They also tried a variety of stabilizers, including sodium lauryl sulfate (SLS), sodium cholate (SC), Tween®80, vitamin E-TPGS, polyethylene glycol 4000 (PEG), and poloxamer-188, before settling on the final one as the best for NRG. Due to its increased emulsification efficiency in contrast to other surfactants, TPNS improved NRG solubility, which was monitored by the micellar property of TPGS with the change from crystalline to amorphous form of NRG. When compared to the pure NRG, the studies revealed substantially increased bioavailability due to significant reduction in size and increase in solubility as well as dissolution. Throughout a 6-month period, it also displayed a high level of stabilization. The oral bioavailability of NRG coated with TPGS by homogenization under high levels of pressure leads to a significant enhancement in the bioactivity and, also minimized hepatotoxicity, according to (Sumathi et al., 2017). In the gastrointestinal tract, this formulation exhibited remarkable NRG permeability. Polymers, in addition to surfactants, could be used to stabilize nanosuspensions. Using a high-pressure homogenization approach, ten distinct nanosuspensions of NRG were produced with diversified modifiers and TPGS was used as a co-stabilizer in a recent study (Khan et al., 2015). Overall oral bioavailability, solubility, potency, and lowering of erythrocyte hemolysis of NRG-nanosuspension were mostly enhanced.

1.4.4 Nanoemulsions

Nanoemulsions give a plethora of possibilities for enhancing the bioavailability of poorly solubility drugs. Surfactants, co-solvents, oils, and hydrophilic solvents that have the specialized ability to manufacture colloidal dispersions of type oil-in-water (o/w) are used to generate nanoemulsions. Several publications suggested employing nanoemulsions for delivery of drugs, in case of NRG so as to promote oral bioavailability and solubility. Tween®80 (surfactant), transcutol HP (co-surfactant) and Triacetin (oily phase), were used to create an NRG self-emulsifying nanoemulsion that improved NRG pharmacokinetics. This technology was modified to obtain a high dissolution rate, appropriate polydispersity, and rapid and comprehensive drug release (Fuior et al., 2020; Fuior et al., 2019; Kanaze et al., 2010).

1.4.5 Co-Delivery Systems

Because of the obvious complex nature of various disorders, a singular bioactive agent is inadequate; therefore, a treatment that combines numerous processes might improve some procedures. Kanaz et al. used a solvent evaporation platform to make nanodispersions of NRG-hesperetin (another glycan flavonoid) utilizing PVP in 2010 (Kerdudo et al., 2014).

Recrystallization of a drug in a nanocarrier system can compromise the physiochemical properties of targeted drug delivery. As a consequence, interactions among polymers and drugs can strengthen the stability of nanoformulations, particularly during storage. PVP reduces medication crystallization by forming hydrogen bonds with the drugs. As an outcome, employing nanocarriers of PVP can help to enhance the stability of formulation while being in storage. Storage period is observed at 40°C for 3 months, the nanodispersion system of PVP exhibits good physiochemical durability, without any alterations in sustained release or nordrug crystallization. To strengthen the antioxidant activities of some flavonoids, including NRG, Kerdudo et al. enclosed NRG and rutin in onion-type lipid-based multilamellar vesicles (MLVs) (Sandhu et al., 2017). NRG shows primarily adsorption on the outer surface of nanocarriers, as according to their observations. After 30 days, it showed outstanding stability with no leakage. Tamoxifen-NRG nanoemulsion was developed in another research to enhance tamoxifen, antitumor activities against breast cancer (Sandhu et al., 2017). Polyunsaturated fatty acids (PUFAs) were applied as the lipid phase in this nanoemulsion. Prior research has found that PUFAs increase distribution and absorption of drugs. In this experiment, the researchers developed nanoemulsions by simply combining the formulation constituents. In both in vitro and in vivo trials, this nanoemulsion was extremely successful at speeding up medication absorption in inhibiting tumor growth.

1.5 CONCLUSIONS AND FUTURE PERSPECTIVE

We have provided extensive pharmacological characteristics of NRG based on the most recent research in this review. According to evidence gathered from the literature, NRG possesses various noteworthy pharmacological effects, including antioxidant, anti-inflammatory, antibacterial, antidiabetic, and anticancer activities. The pleiotropic effects of NRG are mediated by a variety of molecular mechanisms that entail the fusion of cellular signaling pathways at various stages of illness. Despite its enormous promise for treating a variety of ailments, NRG, like other polyphenols, is hampered by a bioavailability issue. To improve the efficacy of NRG as a treatment, methods to increase its bioavailability must be researched. To explore the intriguing pharmacological properties of NRG in the development of improved human wellness, further active study is required while the parallel research is being carried out.

REFERENCES

Ahmad, Anas, Eram Fauzia, Manish Kumar, Rakesh Kumar Mishra, Ajay Kumar, Mohsin Ali Khan, Syed Shadab Raza, and Rehan Khan. "Gelatin-coated polycaprolactone nanoparticle-mediated naringenin delivery rescue human mesenchymal stem cells from oxygen glucose deprivation-induced inflammatory stress." *ACS Biomaterials Science & Engineering* 5, no. 2 (2018): 683–695, doi:10.1021/acsbiomaterials.8b01081.

Ahmad, Niyaz, Rizwan Ahmad, Farhan Jalees Ahmad, Wasim Ahmad, Md Aftab Alam, Mohd Amir, and Abuzer Ali. "Poloxamer-chitosan-based Naringenin nanoformulation used in brain targeting for the treatment of cerebral ischemia." *Saudi Journal of Biological Sciences* 27, no. 1 (2020): 500–517.

Akhter, Md Habban, Sandeep Kumar, and Shibli Nomani. "Sonication tailored enhance cytotoxicity of naringenin nanoparticle in pancreatic cancer: Design, optimization, and in vitro studies." *Drug Development and Industrial Pharmacy* 46, no. 4 (2020): 659–672, doi:10.1080/03639045.2020.1747485.

Akrawi, Sabah H., Bapi Gorain, Anroop B. Nair, Hira Choudhury, Manisha Pandey, Jigar N. Shah, and Katharigatta N. Venugopala. "Development and optimization of naringenin-loaded chitosan-coated nanoemulsion for topical therapy in wound healing." *Pharmaceutics* 12, no. 9 (2020): 893, doi:10.3390/pharmaceutics 12090893.

Al-Rejaie, Salim S., Hatem M. Abuohashish, Maher M. Al-Enazi, Abdullah H. Al-Assaf, Mihir Y. Parmar, and Mohammed M. Ahmed. "Protective effect of naringenin on acetic acid-induced ulcerative colitis in rats." *World Journal of Gastroenterology: WJG* 19, no. 34 (2013): 5633.

Alshememry, Abdullah, Mohd Abul Kalam, Abdulhadi Almoghrabi, Abdulwahab Alzahrani, Mudassar Shahid, Azmat Ali Khan, Anzarul Haque et al. "Chitosan-coated poly (lactic-co-glycolide) nanoparticles for dual delivery of doxorubicin and naringin against MCF-7 cells." *Journal of Drug Delivery Science and Technology* 68 (2022): 103036, doi:10.1016/j.jddst.2021.103036.

Amoabediny, Ghasem, Fateme Haghiralsadat, Samira Naderinezhad, Marco N. Helder, Elham Akhoundi Kharanaghi, Javad Mohammadnejad Arough, and Behrouz Zandieh-Doulabi. "Overview of preparation methods of polymeric and lipid-based (niosome, solid lipid, liposome) nanoparticles: A comprehensive review." *International Journal of Polymeric Materials and Polymeric Biomaterials* 67, no. 6 (2018): 383–400.

Arafa, Hossam M., Mohamed F. Abd-Ellah, and Hafez F. Hafez. "Abatement by naringenin of doxorubicin-induced cardiac toxicity in rats." *Journal of the Egyptian National Cancer Institute* 17, no. 4 (2005): 291–300.

Arozal, Wawaimuli, Kenichi Watanabe, Punniyakoti T. Veeraveedu, Meilei Ma, Rajarajan A. Thandavarayan, Vijayakumar Sukumaran, Kenji Suzuki, Makoto Kodama, and Yoshifusa Aizawa. "Telmisartan prevents the progression of renal injury in daunorubicin rats with the alteration of angiotensin II and endothelin-1 receptor expression associated with its PPAR-γ agonist actions." *Toxicology* 279, no. 1-3 (2011): 91–99.

Bai, Yang, Wei Peng, Cuiping Yang, Wei Zou, Menghua Liu, Hao Wu, Loudi Fan, Peibo Li, Xuan Zeng, and Weiwei Su. "Pharmacokinetics and metabolism of naringin and active metabolite naringenin in rats, dogs, humans, and the differences between species." *Frontiers in Pharmacology* 11 (2020): 364.

Bala, Kiran, Islam Husain, and Anjana Sharma. "Arginine deaminase from Pseudomonas aeruginosa PS2: Purification, biochemical characterization and in-vitro evaluation of anticancer activity." *3 Biotech* 10, no. 5 (2020): 1–17.

Bao, Lei, Feng Liu, Huai-bin Guo, Yong Li, Bi-bo Tan, Wan-xing Zhang, and Yan-hui Peng. "Naringenin inhibits proliferation, migration, and invasion as well as induces apoptosis of gastric cancer SGC7901 cell line by downregulation of AKT pathway." *Tumor Biology* 37, no. 8 (2016): 11365–11374.

Barrett, Jacqueline S., Richard B. Gearry, Jane G. Muir, Peter M. Irving, Rosemary Rose, Ourania Rosella, M. L. Haines, Susan J. Shepherd, and Peter R. Gibson. "Dietary poorly absorbed, short-chain carbohydrates increase delivery of water and fermentable substrates to the proximal colon." *Alimentary Pharmacology & Therapeutics* 31, no. 8 (2010): 874–882.

Bhia, Mohammed, Mahzad Motallebi, Banafshe Abadi, Atefeh Zarepour, Miguel Pereira-Silva, Farinaz Saremnejad, Ana Cláudia Santos et al. "Naringenin nano-delivery systems and their therapeutic applications." *Pharmaceutics* 13, no. 2 (2021): 291.

Brewer, M. S. "Natural antioxidants: Sources, compounds, mechanisms of action, and potential applications." *Comprehensive Reviews in Food Science and Food Safety* 10, no. 4 (2011): 221–247.

Chen, Siyu, Yan Ding, Weiwei Tao, Wenxiang Zhang, Tingming Liang, and Chang Liu. "Naringenin inhibits TNF-α induced VSMC proliferation and migration via induction of HO-1." *Food and Chemical Toxicology* 50, no. 9 (2012): 3025–3031.

Chin, Li Hian, Chian Ming Hon, Dinesh Kumar Chellappan, Jestin Chellian, Thiagarajan Madheswaran, Farrukh Zeeshan, Rajendra Awasthi et al. "Molecular mechanisms of action of naringenin in chronic airway diseases." *European Journal of Pharmacology* 879 (2020): 173139.

Chtourou, Yassine, Hamadi Fetoui, Raoua Jemai, Ahlem Ben Slima, Mohamed Makni, and Radhouane Gdoura. "Naringenin reduces cholesterol-induced hepatic inflammation in rats by modulating matrix metalloproteinases-2, 9 via inhibition of nuclear factor κB pathway." *European Journal of Pharmacology* 746 (2015): 96–105.

David, Christopher AW, Andrew Owen, and Neill J. Liptrott. "Determining the relationship between nanoparticle characteristics and immunotoxicity: Key challenges and approaches." *Nanomedicine* 11, no. 11 (2016): 1447–1464.

Den Hartogh, Danja J., and Evangelia Tsiani. "Antidiabetic properties of naringenin: A citrus fruit polyphenol." *Biomolecules* 9, no. 3 (2019): 99.

Domínguez-Delgado, Clara Luisa, Enrique Fuentes-Prado, José Juan Escobar-Chávez, Gustavo Vidal-Romero, I. Rodríguez Cruz, and Roberto Díaz-Torres. "Chitosan and pluronic(r) F-127: Pharmaceutical applications." edited by Munmaya Mishra *Encyclopedia of Biomedical Polymers and Polymeric Biomaterials*; Taylor and Francis: New York, NY, USA (2016): 1513–1535.

Dong, Zhengqi, Rui Wang, Mingyue Wang, Zheng Meng, Xiaotong Wang, Meihua Han, Yifei Guo, and Xiangtao Wang. "Preparation of naringenin nanosuspension and its antitussive and expectorant effects." *Molecules* 27, no. 3 (2022): 741. doi:10.3390/molecules27030741.

Dou, Wei, Jingjing Zhang, Aning Sun, Eryun Zhang, Lili Ding, Subhajit Mukherjee, Xiaohui Wei, Guixin Chou, Zheng-Tao Wang, and Sridhar Mani. "Protective effect of naringenin against experimental colitis via suppression of Toll-like receptor 4/NF-κB signalling." *British Journal of Nutrition* 110, no. 4 (2013): 599–608.

Duda-Madej, Anna, Joanna Kozłowska, Paweł Krzyżek, Mirosław Anioł, Alicja Seniuk, Katarzyna Jermakow, and Ewa Dworniczek. "Antimicrobial o-alkyl derivatives of naringenin and their oximes against multidrug-resistant bacteria." *Molecules* 25, no. 16 (2020): 3642.

Esteban, Vanesa, Oscar Lorenzo, Mónica Rupérez, Yusuke Suzuki, Sergio Mezzano, Julia Blanco, Mathias Kretzler, Takeshi Sugaya, Jesús Egido, and Marta Ruiz-Ortega. "Angiotensin II, via AT1 and AT2 receptors and NF-κB pathway, regulates the inflammatory response in unilateral ureteral obstruction." *Journal of the American Society of Nephrology* 15, no. 6 (2004): 1514–1529.

Feigin, Valery L., Emma Nichols, Tahiya Alam, Marlena S. Bannick, Ettore Beghi, Natacha Blake, William J. Culpepper et al. "Global, regional, and national burden of neurological disorders, 1990-2016: A systematic analysis for the Global Burden of Disease Study 2016." *The Lancet Neurology* 18, no. 5 (2019): 459–480.

Fuior, Elena Valeria, Cristina Ana Mocanu, Mariana Deleanu, Geanina Voicu, Maria Anghelache, Daniela Rebleanu, Maya Simionescu, and Manuela Calin. "Evaluation of VCAM-1 targeted naringenin/indocyanine green-loaded lipid nanoemulsions as theranostic nanoplatforms in inflammation." *Pharmaceutics* 12, no. 11 (2020): 1066.

Fuior, Elena Valeria, Mariana Deleanu, Cristina Ana Constantinescu, Daniela Rebleanu, Geanina Voicu, Maya Simionescu, and Manuela Calin. "Functional role of VCAM-1 targeted flavonoid-loaded lipid nanoemulsions in reducing endothelium inflammation." *Pharmaceutics* 11, no. 8 (2019): 391.

Fuster, Marta G., Guzmán Carissimi, Mercedes G. Montalbán, and Gloria Víllora. "Improving anticancer therapy with naringenin-loaded silk fibroin nanoparticles." *Nanomaterials* 10, no. 4 (2020): 718.

Gaba, Bharti, Tahira Khan, Md Faheem Haider, Tausif Alam, Sanjula Baboota, Suhel Parvez, and Javed Ali. "Vitamin E loaded naringenin nanoemulsion via intranasal delivery for the management of oxidative stress in a 6-OHDA Parkinson's disease model." *BioMed Research International* 2019 (2019), doi:10.1155/2019/2382563.

Gattuso, Giuseppe, Davide Barreca, Claudia Gargiulli, Ugo Leuzzi, and Corrado Caristi. "Flavonoid composition of citrus juices." *Molecules* 12, no. 8 (2007): 1641–1673.

George, Archana, Priyanka A. Shah, and Pranav S. Shrivastav. "Natural biodegradable polymers based nano-formulations for drug delivery: A review." *International Journal of Pharmaceutics* 561 (2019): 244–264.

Gera, Sonia, Sreekanth Talluri, Nagarjun Rangaraj, and Sunitha Sampathi. "Formulation and evaluation of naringenin nanosuspensions for bioavailability enhancement." *Aaps Pharmscitech* 18, no. 8 (2017): 3151–3162.

Ghasemiyeh, Parisa, and Soliman Mohammadi-Samani. "Solid lipid nanoparticles and nanostructured lipid carriers as novel drug delivery systems: Applications, advantages and disadvantages." *Research in Pharmaceutical Sciences* 13, no. 4 (2018): 288.

Gordillo-Galeano, Aldemar, and Claudia Elizabeth Mora-Huertas. "Solid lipid nanoparticles and nanostructured lipid carriers: A review emphasizing on particle structure and drug release." *European Journal of Pharmaceutics and Biopharmaceutics* 133 (2018): 285–308.

Górniak, Ireneusz, Rafał Bartoszewski, and Jarosław Króliczewski. "Comprehensive review of antimicrobial activities of plant flavonoids." *Phytochemistry Reviews* 18, no. 1 (2019): 241–272.

Haghiralsadat, Fateme, Ghasem Amoabediny, Mohammad Hasan Sheikhha, Behrouz Zandieh-doulabi, Samira Naderinezhad, Marco N. Helder, and Tymour Forouzanfar. "New liposomal doxorubicin nanoformulation for osteosarcoma: Drug release kinetic study based on thermo and pH sensitivity." *Chemical Biology & Drug Design* 90, no. 3 (2017): 368–379.

Han, Kuei-Yang, Pei-Ni Chen, Ming-Chian Hong, You-Cheng Hseu, Ke-Ming Chen, Li-Sung Hsu, and Wei-Jen Chen. "Naringenin attenuated prostate cancer invasion via reversal of epithelial-to-mesenchymal transition and inhibited uPA activity." *Anticancer Research* 38, no. 12 (2018): 6753–6758.

Han, Xiuzhen, Jihong Pan, Dongmei Ren, Yanna Cheng, Peihong Fan, and Hongxiang Lou. "Naringenin-7-O-glucoside protects against doxorubicin-induced toxicity in H9c2 cardiomyocytes by induction of endogenous antioxidant enzymes." *Food and Chemical Toxicology* 46, no. 9 (2008): 3140–3146.

Hermenean, Anca, Aurel Ardelean, Miruna Stan, Nicoleta Hadaruga, Ciprian-Valentin Mihali, Marieta Costache, and Anca Dinischiotu. "Antioxidant and hepatoprotective effects of naringenin and its β-cyclodextrin formulation in mice intoxicated with carbon tetrachloride: A comparative study." *Journal of Medicinal Food* 17, no. 6 (2014): 670–677.

Hernández-Aquino, Erika, and Pablo Muriel. "Beneficial effects of naringenin in liver diseases: Molecular mechanisms." *World Journal of Gastroenterology* 24, no. 16 (2018): 1679.

Hu, Rui, Shu Liu, Gulinigaer Anwaier, Qinyu Wang, Wanli Shen, Qiang Shen, and Rong Qi. "Formulation and intestinal absorption of naringenin loaded nanostructured lipid carrier and its inhibitory effects on nonalcoholic fatty liver disease." *Nanomedicine: Nanotechnology, Biology and Medicine* 32 (2021): 102310, doi:10.1016/j.nano.2020.102310.

Jacob, Shery, Anroop B. Nair, and Jigar Shah. "Emerging role of nanosuspensions in drug delivery systems." *Biomaterials Research* 24, no. 1 (2020): 1–16.

Jahangiri, Azin, and Leila Barghi. "Polymeric nanoparticles: Review of synthesis methods and applications in drug delivery." *Journal of Advanced Chemical and Pharmaceutical Materials (JACPM)* 1, no. 2 (2018): 38–47.

Jain, Amit K., Nitin K. Swarnakar, Chandraiah Godugu, Raman P. Singh, and Sanyog Jain. "The effect of the oral administration of polymeric nanoparticles on the efficacy and toxicity of tamoxifen." *Biomaterials* 32, no. 2 (2011): 503–515.

Jang, Yongwoo, Su-Won Kim, Jungeun Oh, Gyu-Sang Hong, Eun-Kyoung Seo, Uhtaek Oh, and Won-Sik Shim. "Ghrelin receptor is activated by naringin and naringenin, constituents of a prokinetic agent Poncirus fructus." *Journal of Ethnopharmacology* 148, no. 2 (2013): 459–465.

Ji, Peng, Tong Yu, Ying Liu, Jie Jiang, Jie Xu, Ying Zhao, Yanna Hao, Yang Qiu, Wenming Zhao, and Chao Wu. "Naringenin-loaded solid lipid nanoparticles: Preparation, controlled delivery, cellular uptake, and pulmonary pharmacokinetics." *Drug Design, Development and Therapy* 10 (2016): 911.

Joshi, Haritima, Aswathi R. Hegde, Pallavi K. Shetty, Hemanth Gollavilli, Renuka S. Managuli, Guruprasad Kalthur, and Srinivas Mutalik. "Sunscreen creams containing naringenin nanoparticles: Formulation development and in vitro and in vivo evaluations." *Photodermatology, Photoimmunology & Photomedicine* 34, no. 1 (2018a): 69–81.

Jung, Hyun Ah, Pradeep Paudel, Su Hui Seong, Byung-Sun Min, and Jae Sue Choi. "Structure-related protein tyrosine phosphatase 1B inhibition by naringenin derivatives." *Bioorganic & Medicinal Chemistry Letters* 27, no. 11 (2017): 2274–2280.

Kanaze, F. I., M. I. Bounartzi, M. Georgarakis, and I. Niopas. "Pharmacokinetics of the citrus flavanone aglycones hesperetin and naringenin after single oral administration in human subjects." *European Journal of Clinical Nutrition* 61, no. 4 (2007): 472–477.

Kanaze, F. I., E. Kokkalou, Ioannis Niopas, P. Barmpalexis, E. Georgarakis, and Dimitrios Bikiaris. "Dissolution rate and stability study of flavanone aglycones, naringenin and hesperetin, by drug delivery systems based on polyvinylpyrrolidone (PVP) nanodispersions." *Drug Development and Industrial Pharmacy* 36, no. 3 (2010): 292–301.

Karadag, Ayse, Beraat Ozcelik, and Qingrong Huang. "Quercetin nanosuspensions produced by high-pressure homogenization." *Journal of Agricultural and Food Chemistry* 62, no. 8 (2014): 1852–1859.

Karim, Naymul, Zhenquan Jia, Xiaodong Zheng, Sunliang Cui, and Wei Chen. "A recent review of citrus flavanone naringenin on metabolic diseases and its potential sources for high yield-production." *Trends in Food Science & Technology* 79 (2018): 35–54.

Karuppagounder, Vengadeshprabhu, Somasundaram Arumugam, Rajarajan Amirthalingam Thandavarayan, Vigneshwaran Pitchaimani, Remya Sreedhar, Rejina Afrin, Meilei Harima et al. "Naringenin ameliorates daunorubicin induced nephrotoxicity by mitigating AT1R, ERK1/2-NFκB p65 mediated inflammation." *International Immunopharmacology* 28, no. 1 (2015): 154–159.

Kerdudo, Audrey, Alexandre Dingas, Xavier Fernandez, and Chrystel Faure. "Encapsulation of rutin and naringenin in multilamellar vesicles for optimum antioxidant activity." *Food Chemistry* 159 (2014): 12–19.

Khan, Abdul Wadood, Sabna Kotta, Shahid Husain Ansari, Rakesh Kumar Sharma, and Javed Ali. "Self-nanoemulsifying drug delivery system (SNEDDS) of the poorly water-soluble grapefruit flavonoid Naringenin: Design, characterization, in vitro and in vivo evaluation." *Drug Delivery* 22, no. 4 (2015): 552–561.

Kim, Tae-muk, Keshav Raj Paudel, and Dong-Wook Kim. "Eriobotrya japonica leaf extract attenuates airway inflammation in ovalbumin-induced mice model of asthma." *Journal of Ethnopharmacology* 253 (2020): 112082.

Kozłowska, Joanna, Bartłomiej Potaniec, Barbara Żarowska, and Mirosław Anioł. "Synthesis and biological activity of novel O-alkyl derivatives of naringenin and their oximes." *Molecules* 22, no. 9 (2017): 1485.

Krishnakumar, Narendran, Nechikkad Sulfikkarali, Nagarajan RajendraPrasad, and Subburayan Karthikeyan. "Enhanced anticancer activity of naringenin-loaded nanoparticles in human cervical (HeLa) cancer cells." *Biomedicine & Preventive Nutrition* 1, no. 4 (2011): 223–231.

Kujawska, Małgorzata, and Jadwiga Jodynis-Liebert. "Polyphenols in Parkinson's disease: A systematic review of in vivo studies." *Nutrients* 10, no. 5 (2018): 642.

Kumar, Raj, and Ashu Bhan Tiku. "Naringenin suppresses chemically induced skin cancer in two-stage skin carcinogenesis mouse model." *Nutrition and Cancer* 72, no. 6 (2020): 976–983.

Kumar, Sekar Praveen, Kannan Birundha, Kannan Kaveri, and KT Ramya Devi. "Antioxidant studies of chitosan nanoparticles containing naringenin and their cytotoxicity effects in lung cancer cells." *International Journal of Biological Macromolecules* 78 (2015): 87–95. doi:10.1016/j.ijbiomac.2015.03.045.

Kunnumakkara, Ajaikumar B., Bethsebie L. Sailo, Kishore Banik, Choudhary Harsha, Sahdeo Prasad, Subash Chandra Gupta, Alok Chandra Bharti, and Bharat B. Aggarwal. "Chronic diseases, inflammation, and spices: How are they linked?" *Journal of Translational Medicine* 16, no. 1 (2018): 1–25.

Lawrence, T. "The nuclear factor NF-kappaB pathway in inflammation." *Cold Spring Harbor Perspectives in Biology* 1, no. (6) (2009): a001651.

Lee, Jungwhoi, Da-Hye Kim, and Jae Hoon Kim. "Combined administration of naringenin and hesperetin with optimal ratio maximizes the anti-cancer effect in human pancreatic cancer via down regulation of FAK and p38 signaling pathway." *Phytomedicine* 58 (2019): 152762.

Li, Li, Peng Gao, Huina Zhang, Houzao Chen, Wei Zheng, Xiang Lv, Tingting Xu, Yusheng Wei, Depei Liu, and Chihchuan Liang. "SIRT1 inhibits angiotensin II-induced vascular smooth muscle cell hypertrophy." *Acta Biochimica et Biophysica Sinica* 43, no. 2 (2011): 103–109.

Li, Mo, Kunli Wang, Yuxiao Wang, Qianyun Han, Yuanying Ni, and Xin Wen. "Effects of genipin concentration on cross-linked β-casein micelles as nanocarrier of naringenin: Colloidal properties, structural characterization and controlled release." *Food Hydrocolloids* 108 (2020): 105989.

Lim, Whasun, Sunwoo Park, Fuller W. Bazer, and Gwonhwa Song. "Naringenin-induced apoptotic cell death in prostate cancer cells is mediated via the PI3K/AKT and MAPK signaling pathways." *Journal of Cellular Biochemistry* 118, no. 5 (2017): 1118–1131.

Llorach, Rafael, Ascensión Martínez-Sánchez, Francisco A. Tomás-Barberán, María I. Gil, and Federico Ferreres. "Characterisation of polyphenols and antioxidant properties of five lettuce varieties and escarole." *Food Chemistry* 108, no. 3 (2008): 1028–1038.

Maity, Subhajit, and Abhay Sankar Chakraborti. "Formulation, physico-chemical characterization and antidiabetic potential of naringenin-loaded poly D, L lactide-co-glycolide (N-PLGA) nanoparticles." *European Polymer Journal* 134 (2020): 109818.

Maity, Subhajit, Piyasi Mukhopadhyay, Patit Paban Kundu, and Abhay Sankar Chakraborti. "Alginate coated chitosan core-shell nanoparticles for efficient oral delivery of naringenin in diabetic animals-an in vitro and in vivo approach." *Carbohydrate Polymers* 170 (2017): 124–132.

Manabe, Noriaki, Banny S. Wong, and Michael Camilleri. "New-generation 5-HT4 receptor agonists: Potential for treatment of gastrointestinal motility disorders." *Expert Opinion on Investigational Drugs* 19, no. 6 (2010): 765–775.

Mauludin, Rachmat, and Rainer H. Müller. "Preparation and storage stability of rutin nanosuspensions." *Journal of Pharmaceutical Investigation* 43, no. 5 (2013): 395–404.

Md, Shadab, Nabil A. Alhakamy, Hibah M. Aldawsari, Mohammad Husain, Sabna Kotta, Samaa T. Abdullah, Usama A Fahmy, Mohamed A. Alfaleh, and Hani Z. Asfour. "Formulation design, statistical optimization, and in vitro evaluation of a naringenin nanoemulsion to enhance apoptotic activity in a549 lung cancer cells." *Pharmaceuticals* 13, no. 7 (2020a): 152, doi:10.3390/ph13070152.

Md, Shadab, Nabil A. Alhakamy, Sohail Akhter, Zuhier AY Awan, Hibah M. Aldawsari, Waleed S. Alharbi, Anzarul Haque, Hira Choudhury, and Ponnurengam Malliappan Sivakumar. "Development of polymer and surfactant based naringenin nanosuspension for improvement of stability, antioxidant, and antitumour activity." *Journal of Chemistry* 2020 (2020b).

Md, Shadab, Sook Yee Gan, Yong Hong Haw, Chee Liang Ho, Sinrun Wong, and Hira Choudhury. "In vitro neuroprotective effects of naringenin nanoemulsion against β-amyloid toxicity through the regulation of amyloidogenesis and tau phosphorylation." *International Journal of Biological Macromolecules* 118 (2018): 1211–1219.

Meng, Li-Min, Hui-Jie Ma, Hui Guo, Qian-Qian Kong, and Yi Zhang. "The cardioprotective effect of naringenin against ischemia-reperfusion injury through activation of ATP-sensitive potassium channel in rat." *Canadian Journal of Physiology and Pharmacology* 94, no. 9 (2016): 973–978.

Mohanty, Sangeeta, V. Badireenath Konkimalla, Abhisek Pal, Tripti Sharma, and Sudam Chandra Si. "Naringin as sustained delivery nanoparticles ameliorates the anti-inflammatory activity in a Freund's complete adjuvant-induced arthritis model." *ACS Omega* 6, no. 43 (2021): 28630–28641.

Muralidharan, Shruthi, and Kumaran Shanmugam. "Synthesis and characterization of naringenin-loaded chitosan-dextran sulfate nanocarrier." *Journal of Pharmaceutical Innovation* 16, no. 2 (2021): 269–278, doi:10.1007/s12247-020-09444-2.

Nagaraja, Sreeharsha, Girish Meravanige Basavarajappa, Ranjith Kumar Karnati, Esam Mohamed Bakir, and Swati Pund. "Ion-triggered in situ gelling nanoemulgel as a platform for nose-to-brain delivery of small lipophilic molecules." *Pharmaceutics* 13, no. 8 (2021): 1216.

Nagula, Ruchika L., and Sarika Wairkar. "Recent advances in topical delivery of flavonoids: A review." *Journal of Controlled Release* 296 (2019): 190–201.

Najmanová, Iveta, Marie Vopršalová, Luciano Saso, and Přemysl Mladěnka. "The pharmacokinetics of flavanones." *Critical reviews in Food Science and Nutrition* 60, no. 18 (2020): 3155–3171.

Noori, Shokoofe, Mostafa Rezaei Tavirani, Niloofar Deravi, Mohammad Ismaiil Mahboobi Rabbani, and Afshin Zarghi. "Naringenin enhances the anti-cancer effect of cyclophosphamide against MDA-MB-231 breast cancer cells via targeting the STAT3 signaling pathway." *Iranian Journal of Pharmaceutical Research: IJPR* 19, no. 3 (2020): 122.

Nouri, Zeinab, Sajad Fakhri, Fardous F. El-Senduny, Nima Sanadgol, Ghada E. Abd-ElGhani, Mohammad Hosein Farzaei, and Jen-Tsung Chen. "On the neuroprotective effects of naringenin: Pharmacological targets, signaling pathways, molecular mechanisms, and clinical perspective." *Biomolecules* 9, no. 11 (2019): 690.

Orrego-Lagarón, Naiara, Miriam Martínez-Huélamo, Anna Vallverdú-Queralt, Rosa M. Lamuela-Raventos, and Elvira Escribano-Ferrer. "High gastrointestinal permeability and local metabolism of naringenin: Influence of antibiotic treatment on absorption and metabolism." *British Journal of Nutrition* 114, no. 2 (2015): 169–180.

Ozbek, Emin, Yusuf Ozlem Ilbey, Mustafa Ozbek, Abdulmuttalip Simsek, Mustafa Cekmen, and Adnan Somay. "Melatonin attenuates unilateral ureteral obstruction-induced renal injury by reducing oxidative stress, iNOS, MAPK, and NF-kB expression." *Journal of Endourology* 23, no. 7 (2009): 1165–1173.

Pandey, Kanti Bhooshan, and Syed Ibrahim Rizvi. "Plant polyphenols as dietary antioxidants in human health and disease." *Oxidative Medicine and Cellular Longevity* 2, no. 5 (2009): 270–278.

Parashar, Poonam, Meena Rathor, Monika Dwivedi, and Shubhini A. Saraf. "Hyaluronic acid decorated naringenin nanoparticles: Appraisal of chemopreventive and curative potential for lung cancer." *Pharmaceutics* 10, no. 1 (2018): 33.

Patel, Kanika, Gireesh Kumar Singh, and Dinesh Kumar Patel. "A review on pharmacological and analytical aspects of naringenin." *Chinese Journal of Integrative Medicine* 24, no. 7 (2018): 551–560.

Peeters, Theo Louis. "Potential of ghrelin as a therapeutic approach for gastrointestinal motility disorders." *Current Opinion in Pharmacology* 6, no. 6 (2006): 553–558.

Prabhakar, P. K., Anil Kumar, and Mukesh Doble. "Combination therapy: A new strategy to manage diabetes and its complications." *Phytomedicine* 21, no. 2 (2014): 123–130.

Priscilla, David Hansi, Debasish Roy, Aishwarya Suresh, Vinod Kumar, and Kavitha Thirumurugan. "Naringenin inhibits α-glucosidase activity: A promising strategy for the regulation of postprandial hyperglycemia in high fat diet fed streptozotocin induced diabetic rats." *Chemico-Biological Interactions* 210 (2014): 77–85.

Qizilbash, Farheen Fatima, Muhammad Usama Ashhar, Ameeduzzafar Zafar, Zufika Qamar, Javed Ali, Sanjula Baboota, Mohammed M. Ghoneim, Sultan Alshehri, and Asgar Ali. "Thymoquinone-enriched naringenin-loaded nanostructured lipid carrier for brain delivery via nasal route: In vitro prospect and in vivo therapeutic efficacy for the treatment of depression." *Pharmaceutics* 14, no. 3 (2022): 656, doi:10.3390/pharmaceutics14030656.

Raeisi, Sina, Hadi Chavoshi, Maryam Mohammadi, Marjan Ghorbani, Mehdi Sabzichi, and Fatemeh Ramezani. "Naringenin-loaded nano-structured lipid carrier fortifies oxaliplatin-dependent apoptosis in HT-29 cell line." *Process Biochemistry* 83 (2019): 168–175.

Ren, Bei, Weiwei Qin, Feihua Wu, Shanshan Wang, Cheng Pan, Liying Wang, Biao Zeng, Shiping Ma, and Jingyu Liang. "Apigenin and naringenin regulate glucose and lipid metabolism, and ameliorate vascular dysfunction in type 2 diabetic rats." *European Journal of Pharmacology* 773 (2016): 13–23.

Salehi, Bahare, Patrick Valere Tsouh Fokou, Mehdi Sharifi-Rad, Paolo Zucca, Raffaele Pezzani, Natália Martins, and Javad Sharifi-Rad. "The therapeutic potential of naringenin: A review of clinical trials." *Pharmaceuticals* 12, no. 1 (2019): 11.

Sandhu, Premjeet Singh, Rajendra Kumar, Sarwar Beg, Sanyog Jain, Varun Kushwah, O. P. Katare, and Bhupinder Singh. "Natural lipids enriched self-nano-emulsifying systems for effective co-delivery of tamoxifen and naringenin: Systematic approach for improved breast cancer therapeutics." *Nanomedicine: Nanotechnology, Biology and Medicine* 13, no. 5 (2017): 1703–1713.

Schrezenmeier, E. V., J. Barasch, K. Budde, T. Westhoff, and K. M. Schmidt-Ott. "Biomarkers in acute kidney injury-pathophysiological basis and clinical performance." *Acta Physiologica* 219, no. 3 (2017): 556–574.

Shanmuganathan, Rajasree, Gnanasekar Sathishkumar, Kathirvel Brindhadevi, and Arivalagan Pugazhendhi. "Fabrication of naringenin functionalized-Ag/RGO nanocomposites for potential bactericidal effects." *Journal of Materials Research and Technology* 9, no. 4 (2020): 7013–7019.

Singh, Mayank Kumar, Deep Pooja, Halley Gora Ravuri, Anusha Gunukula, Hitesh Kulhari, and Ramakrishna Sistla. "Fabrication of surfactant-stabilized nanosuspension of naringenin to surpass its poor physiochemical properties and low oral bioavailability." *Phytomedicine* 40 (2018): 48–54.

Smruthi, M. R., Ilaiyaraja Nallamuthu, and T. Anand. "A comparative study of optimized naringenin nanoformulations using nano-carriers (PLA/PVA and zein/pectin) for improvement of bioavailability." *Food Chemistry* 369 (2022): 130950, doi:10.1016/j. foodchem.2021.130950.

Song, Im-Sook, Jin-Sun Cha, and Min-Koo Choi. "Enhanced oral bioavailability of naringenin administered in a mixed micelle formulation with Pluronic F127 and Tween 80 in rats." *Journal of Pharmaceutical Investigation* 45, no. 7 (2015): 633–640.

Sumathi, R., S. Tamizharasi, and T. Sivakumar. "Formulation and evaluation of polymeric nanosuspension of naringenin." *International Journal of Applied Pharmaceutics* 9, no. 6 (2017): 60–70.

Testai, Lara, Eleonora Da Pozzo, Ilaria Piano, Luisa Pistelli, Claudia Gargini, Maria Cristina Breschi, Alessandra Braca, Claudia Martini, Alma Martelli, and Vincenzo Calderone. "The citrus flavanone naringenin produces cardioprotective effects in hearts from 1 year old rat, through activation of mitoBK channels." *Frontiers in Pharmacology* 8 (2017): 71.

Tran Trung, Hieu, Hoa Truong Thi Huynh, Linh Nguyen Thi Thuy, Hoang Nguyen Van Minh, My-Nuong Thi Nguyen, and My Ngan Luong Thi. "Growth-Inhibiting, Bactericidal, Antibiofilm, and Urease Inhibitory Activities of Hibiscus rosa sinensis L. Flower Constituents toward Antibiotic Sensitive-and Resistant-Strains of Helicobacter pylori." *ACS Omega* 5, no. 32 (2020): 20080–20089.

Vallverdu-Queralt, Anna, Isabel Odriozola-Serrano, Gemma Oms-Oliu, Rosa Maria Lamuela-Raventós, Pedro Elez-Martinez, and Olga Martin-Belloso. "Changes in the polyphenol profile of tomato juices processed by pulsed electric fields." *Journal of Agricultural and Food Chemistry* 60, no. 38 (2012): 9667–9672.

Venditti, Iole. "Morphologies and functionalities of polymeric nanocarriers as chemical tools for drug delivery: A review." *Journal of King Saud University-Science* 31, no. 3 (2019): 398–411.

Venkateswara Rao, P., S. D. V. S. Kiran, P. Rohini, and P. Bhagyasree. "Flavonoid: A review on Naringenin." *Journal of Pharmacognosy and Phytochemistry* 6 (2017): 2778–2783.

Vikram, Amit, Guddadarangavvanahally K. Jayaprakasha, P. R. Jesudhasan, S. D. Pillai, and B. S. Patil. "Suppression of bacterial cell-cell signalling, biofilm formation and type III secretion system by citrus flavonoids." *Journal of Applied Microbiology* 109, no. 2 (2010): 515–527.

Wang, Hui, Yuanlong He, Yuzhen Hou, Yiwan Geng, and Xianggen Wu. "Novel self-nanomicellizing formulation based on Rebaudioside A: A potential nanoplatform for oral delivery of naringenin." *Materials Science and Engineering: C* 112 (2020): 110926.

Wang, Ke, Tingting Liu, Rong Lin, Bing Liu, Guangde Yang, Xin Bu, Weirong Wang, Peipei Zhang, Li Zhou, and Jiye Zhang. "Preparation and in vitro release of buccal tablets of naringenin-loaded MPEG-PCL nanoparticles." *RSC Advances* 4, no. 64 (2014): 33672–33679.

Wang, Liying, Xiangbo Wang, Lina Shen, Majed Alrobaian, Sunil K. Panda, Hussain A. Almasmoum, Mazen M. Ghaith et al. "Paclitaxel and naringenin-loaded solid lipid nanoparticles surface modified with cyclic peptides with improved tumor targeting ability in glioblastoma multiforme." *Biomedicine & Pharmacotherapy* 138 (2021): 111461, doi:10.1016/j.biopha.2021.111461.

Wang, Yuanwen, Shicheng Wang, Caleb Kesse Firempong, Huiyun Zhang, Miaomiao Wang, Ya Zhang, Yuan Zhu, Jiangnan Yu, and Ximing Xu. "Enhanced solubility and bioavailability of naringenin via liposomal nanoformulation: Preparation and in vitro and in vivo evaluations." *AAPS Pharmscitech* 18, no. 3 (2017): 586–594.

Yang, Fan, Shenghe Hu, Xun Sheng, and Yu Liu. "Naringenin loaded multifunctional nanoparticles to enhance the chemotherapeutic efficacy in hepatic fibrosis." *Biomedical Microdevices* 22, no. 4 (2020): 1–9.

Yang, J., Q. Li, X. D. Zhou, V. P. Kolosov, and J. M. Perelman. "Naringenin attenuates mucous hypersecretion by modulating reactive oxygen species production and inhibiting NF-κB activity via EGFR-PI3K-Akt/ERK MAPKinase signaling in human airway epithelial cells." *Molecular and Cellular Biochemistry* 351, no. 1 (2011): 29–40.

Yeo, Eileen, Clement Jia Yew Chieng, Hira Choudhury, Manisha Pandey, and Bapi Gorain. "Tocotrienols-rich naringenin nanoemulgel for the management of diabetic wound: Fabrication, characterization and comparative in vitro evaluations." *Current Research in Pharmacology and Drug Discovery* 2 (2021): 100019, doi:10.1016/j.crphar.2021.100019.

Yi, Hao, Wu Bin, and Li ChunGuang. "Cardioprotective efficacy of naringenin against isoproterenol induced chronic heart failure in a rat model." *International Journal of Pharmacology* 15, no. 6 (2019): 759–765.

Yoshida, Hiroki, Norito Takamura, Tsuyoshi Shuto, Kenji Ogata, Jin Tokunaga, Keiichi Kawai, and Hirofumi Kai. "The citrus flavonoids hesperetin and naringenin block the lipolytic actions of TNF-α in mouse adipocytes." *Biochemical and Biophysical Research Communications* 394, no. 3 (2010): 728–732.

Yousefi, Mohammad, Asghar Narmani, and Seid Mahdi Jafari. "Dendrimers as efficient nanocarriers for the protection and delivery of bioactive phytochemicals." *Advances in Colloid and Interface Science* 278 (2020): 102125.

Yu, Li-Ming, Xue Dong, Jian Zhang, Zhi Li, Xiao-Dong Xue, Hong-Jiang Wu, Zhong-Lu Yang, Yang Yang, and Hui-Shan Wang. "Naringenin attenuates myocardial ischemia-reperfusion injury via cGMP-PKGIα signaling and in vivo and in vitro studies." *Oxidative Medicine and Cellular Longevity* 2019 (2019). doi:10.1155/2019/7670854.

Zaidun, Nurul Hannim, Zar Chi Thent, and Azian Abd Latiff. "Combating oxidative stress disorders with citrus flavonoid: Naringenin." *Life Sciences* 208 (2018): 111–122.

Zeng, Wenfeng, Lingtao Jin, Fayun Zhang, Chunling Zhang, and Wei Liang. "Naringenin as a potential immunomodulator in therapeutics." *Pharmacological Research* 135 (2018): 122–126.

Zhang, Guangshuai, Guangyuan Sun, Haishan Guan, Mo Li, Yanhua Liu, Baocheng Tian, Zhonggui He, and Qiang Fu. "Naringenin nanocrystals for improving anti-rheumatoid arthritis activity." *Asian Journal of Pharmaceutical Sciences* 16, no. 6 (2021): 816–825.

Zhang, Haiyang, Xia Zhong, Xiao Zhang, Deya Shang, Y. I. Zhou, and Chunqing Zhang. "Enhanced anticancer effect of ABT-737 in combination with naringenin on gastric cancer cells." *Experimental and Therapeutic Medicine* 11, no. 2 (2016a): 669–673.

Zhang, Ping, Xin Liu, Wenjing Hu, Yan Bai, and Liangke Zhang. "Preparation and evaluation of naringenin-loaded sulfobutylether-β-cyclodextrin/chitosan nanoparticles for ocular drug delivery." *Carbohydrate Polymers* 149 (2016b): 224–230, doi:10.1016/j.carbpol.2016.04.115.

Zhao, Zhenjiang, Guoguo Jin, Yinghui Ge, and Zhiping Guo. "Naringenin inhibits migration of breast cancer cells via inflammatory and apoptosis cell signaling pathways." *Inflammopharmacology* 27, no. 5 (2019): 1021–1036.

Zhou, Yiqun, Zhili Peng, Elif S. Seven, and Roger M. Leblanc. "Crossing the blood-brain barrier with nanoparticles." *Journal of Controlled Release* 270 (2018): 290–303.

Zhu, Yuan, Miaomiao Wang, Jiajia Zhang, Wei Peng, Caleb Kesse Firempong, Wenwen Deng, Qilong Wang et al. "Improved oral bioavailability of capsaicin via liposomal nanoformulation: Preparation, in vitro drug release and pharmacokinetics in rats." *Archives of Pharmacal Research* 38, no. 4 (2015): 512–521.

Zobeiri, Mehdi, Tarun Belwal, Fatemeh Parvizi, Rozita Naseri, Mohammad H. Farzaei, Seyed F. Nabavi, Antoni Sureda, and Seyed M. Nabavi. "Naringenin and its nano-formulations for fatty liver: Cellular modes of action and clinical perspective." *Current Pharmaceutical Biotechnology* 19, no. 3 (2018): 196–205.

2 Novel Delivery Systems for Natural Products in Improved Therapy of Rheumatoid Arthritis

Gurleen Kour, Syed Assim Haq, Govind Yadav, Prem N. Gupta, and Zabeer Ahmed

2.1 INTRODUCTION

RA is a chronic inflammatory autoimmune disease characterized by pain in the symmetrical joints eventually causing cartilage and bone erosion. In most cases, the pathogenesis of the disease occurs prior to articular inflammation develops and the stage is called preclinical RA. It is characterized by the occurrence of antibodies and chemokines and pro-inflammatory cytokines years before established RA, indicating immune activation. Autoantibodies to citrullinated proteins and immunoglobulin G called anticitrullinated protein antibodies (ACPAs) and rheumatoid factor (RF), respectively, are the characteristics of the disease often used as diagnostic markers in RA. Thus, early diagnosis is crucial as well as challenging, and the use of disease-modifying antirheumatic drugs (DMARDs) forms the basis of RA treatment in attaining disease remission. DMARDs are pharmacologic therapies including synthetic, biological, and targeted DMARDs that control inflammation consequently preventing bone and cartilage damage (Smolen et al. 2018; Scott, Wolfe, and Huizinga 2010).

The occurrence of RA has a lot of ethnic, regional, and geographic variation. The prevalence rate reported in white individuals ranges from 0.5% to 1%, that is, 1.5 million people approximately. However, among the indigenous North American populations, a significantly high prevalence rate has been reported (Myasoedova et al. 2010; McDougall, Hurd, and Barnabe 2017). On the contrary, the prevalence percentage of RA in the general black population in Kinshasa, Democratic Republic of the Congo, is 0.6% (Malemba et al. 2012). Studies have revealed that RA is more widespread in females and older populations than among men. Furthermore, female hormonal factors and decline in ovarian function during menopause also contribute to increased risk of RA in women. Additionally, according to a study, a healthier diet reduced the development of autoimmunity associated with RA in women, specifically seropositive RA (Alpizar-Rodriguez et al. 2017; Hu et al. 2017). In this autoimmune, chronic disorder, the synovium holds significant importance. The outer layer of the synovium called the subintima is composed of blood vessels, fibroblasts, and immune cells, and the inner layer called the intima is composed of macrophage-like synoviocytes (MLS) and fibroblast-like synoviocytes (FLS). The primary function of the synovium is the production of synovial fluid that lubricates and nourishes the cartilage. In RA, synoviocytes are activated which results in the release of cytokines, for example, TNF-α, IL-1, IL-6, and proteases by MLS, and matrix metalloproteinases (MMPs), prostaglandins, and leukotrienes by FLS. This synovial invasion and immunological events ultimately lead to cartilage and bone damage (McInnes and Schett 2011; Bartok and Firestein 2010). Furthermore, the transmigration of the activated FLS contributes to propagating the disease (Lefèvre et al. 2009). The synovium of RA patients contains T and B cells clustered in aggregates and allied with follicular dendritic cell networks that closely resemble

DOI: 10.1201/9781003292692- 2

germinal centers thus called ectopic lymphoid structures. The infiltration of adaptive immune cells further contributes to ACPAs and RF production within the joint itself, thus endorsing autoimmunity (Silverman and Carson 2003; Humby et al. 2009).

2.2 THERAPY IN RA

DMARDs are the major therapeutic agents for treating RA and have gained significant importance in the last two decades because of their efficiency in attenuating disease activity, reducing synovitis, limit progressive joint and cartilage damage thus improving function. These can be categorized into orally administered conventional synthetic DMARDs and parenterally administered biologic DMARDs. The first-line agents for treating RA, that is, traditional DMARDs include methotrexate (MTX), leflunomide (LEF), hydroxychloroquine (HCQ), and sulfasalazine (SSZ). However, MTX remains the top-drawer first-line agent because of its significant efficacy and good safety profile (Donahue et al. 2008). Besides these, biologic DMARDs which are molecules that specifically target immunological pathways have revolutionized RA therapy. The US FDA has approved 10 biologic DMARDs (bDMARDs) for RA treatment, however are only recommended in case of insufficient response to MTX monotherapy/combination DMARDs therapy (Smolen et al. 2017; Singh et al. 2016). Among these five are TNF-α inhibitors viz infliximab, etanercept, adalimumab, certolizumab, and golimumab. TNF-α is the principal cytokine in disseminating the inflammatory response in RA. TNF-α inhibitors prevent the principal cytokine, TNF-α disseminating the inflammatory response thus acting as potential therapeutic agents in reversing bone erosion and cartilage degradation. However, these are also accompanied by a higher risk of infections and adverse events (Criscione and St Clair 2002). Further biological DMARDs include Anakinra, antagonist to IL-1 receptor; Tocilizumab and Sarilumab-mAbs to IL-6 receptor; Rituximab-mAb to CD-20 on B cells; Abatacept-inhibits T cells. Other DMARDs include targeted small molecules, and inhibitors of the JAK/STAT signaling pathway including Tofacitinib, Baricitinib, and Upadacitinib (Schwartz et al. 2017). However, all bDMARDs are high-priced and are linked with risks of serious infections (Upchurch and Kay 2012).

The American College of Rheumatology (ACR) and The European League Against Rheumatism have devised a classification criterion to diagnose patients in the early stages of RA with the aim of disease prevention before the onset of irreversible complications. The crucial factors that indicate RA according to this criterion are symmetrical arthritis of small and large joints; positive RF and ACPA; abnormal levels of acute-phase reactants; and duration of symptoms lasting longer than 6 weeks. According to this criterion, a total score of 6 or greater in the above-mentioned domains along with the incidence of synovitis in at least one joint is classified as "definite RA." Furthermore, ACR 20/50/70 criteria define the improvement levels as 20%, 50%, and 70%, respectively, based on parameters including joint count both swollen and tender, patient's and physician global assessment, and inflammatory biomarkers like erythrocyte sedimentation rate (ESR) and C-reactive proteins (CRP) (Aletaha et al. 2010; Smolen et al. 2014; Ramiro et al. 2017).

NSAIDs and glucocorticoids are generally used to relieve pain and stiffness since they have a rapid onset of action. Low-dose glucocorticoids are often used as "bridging therapy" before employing DMARDs during episodes of acute flares of the disease. Studies have revealed that prednisone at low doses results in a substantial decrease in morning stiffness without causing any adverse effects (Bakker et al. 2012; Hoes et al. 2010). A clinical trial (SEMIRA) was conducted to study the effects of continuing prednisone therapy versus tapering prednisone therapy. The study was conducted in patients with low disease activity receiving tocilizumab and synthetic DMARDs for 24 weeks. The results endorsed that continuing prednisone 5 mg daily demonstrated better control of disease than narrowing prednisone from 5 to 0 mg daily over 24 weeks (Buttgereit 2020). Furthermore, better clinical outcomes have been seen with prednisone bridging therapy along with MTX monotherapy in the early stages of RA (van der Goes, Jacobs, and Bijlsma 2014; Brilman 2018).

Thus, the first-line treatment for RA takes the use of short-term glucocorticoids and MTX monotherapy to attain the target within 6 months. When the patient responds inadequately to MTX monotherapy, an alternative strategy using a combination with other traditional DMARDs is employed. "Triple therapy" is considered in patients involving the combination of MTX, SSZ, and HCQ. In a randomized controlled trial, it was discovered that triple therapy was significantly more durable than the MTX–etanercept combination. Thus, it was concluded that the combination of traditional DMARDs should be predominantly used immediately after MTX's inadequate response (Peper et al. 2017). If high disease activity persists, second-line therapy of RA involving the combination of traditional and biological DMARDs along with Jak inhibitors is used thereafter. The eventual target of the RA therapy using traditional and biological DMARDs including small molecule JAK/STAT inhibitors is to achieve sustained remission.

2.3 METHOTREXATE: A CORNERSTONE IN RA THERAPY

MTX is a first-line DMARD predominantly used in RA patients because of its good efficacy alone and with other DMARDs. It is administered at a low dose of 5 mg/week with slow escalation to a highest of 25 mg/week depending on the patient's response to the therapy and intolerance. Subcutaneous injection of MTX has gained substantial importance for the reason of its fewer side effects over oral administration. Furthermore, superior clinical efficacy and tolerability have been observed in subcutaneous injection (Visser and van der Heijde 2009). MTX has been proposed to target the enzyme dihydrofolate reductase involved in the de novo synthesis of purines and pyrimidines consequently inhibiting the proliferation of cells responsible for inflammation in RA. Low-dose MTX has shown a cytostatic effect on mitogen-stimulated human T-lymphocytes *in vitro*. Apart from this, studies have shown that treatment with MTX augments the concentration of adenosine, a potent anti-inflammatory mediator (Brown, Pratt, and Isaacs 2016; Riksen et al. 2006). Furthermore, it has been observed that the anti-inflammatory and immunosuppressive effect of MTX primarily relies on the production of reactive oxygen species (ROS) that reduce monocyte adhesion to the endothelial cells (Phillips, Woollard, and Griffiths 2003). It has been reported that MTX decreases the production of cytokines including TNF-α, IL-1β, and IL-6 primarily by the suppression of the pro-inflammatory factor NF-κB in T cells and FLS (Spurlock III et al. 2015). Additionally, MTX has an inhibitory effect on the production of MMPs, primarily involved in joint destruction via downregulation of IL-1β. Additionally, MTX downregulates the production of MMPs and eicosanoids (Cronstein and Aune 2020).

Although MTX is the gold standard for treating RA with excellent efficacy and cost-effectiveness, the toxicity linked with long-term treatment is still a concern and the adverse effects related to it sometimes lead to discontinuation of the therapy. Serious adverse events that are associated with MTX treatment are gastrointestinal (GI) toxicity and hepatotoxicity which sometimes lead to discontinuation of the therapy or treatment withdrawal (Romao et al. 2014). However, adverse effects caused by GI and liver toxicities can be managed with folic acid supplementation (Liu et al. 2019). Another adverse effect associated with MTX therapy is the development of accelerated nodulosis; the accumulation of multinucleated cells generally observed in 1%–10% of patients, primarily mediated by adenosine (Motegi and Ishikawa 2014). Another adverse effect is pulmonary infiltrates triggered by MTX pulmonary toxicity even at low doses. This MTX-induced pneumonitis represents a hypersensitive reaction contributing to inflammatory lung disease (Lateef, Shakoor, and Balk 2005). Furthermore, MTX treatment also involves mild neurotoxicity primarily because of the accumulation of adenosine in the CNS or the toxic effect on the blood–brain barrier. However, MTX-induced leukoencephalopathy can develop even at low doses, especially with intravenous or intrathecal administration, and rarely in patients receiving long-term low-dose oral administration. Although, MTX-induced encephalopathy is infrequent but is imperative because therein lies the possibility of neurological sequelae or death (González-Suárez et al. 2014). Despite the observed adverse drug reactions of MTX, it is still the drug of choice and is comprehensively used in RA therapy.

2.4 RISKS COUPLED TO SYNTHETIC AND BIOLOGICAL DMARDS

According to the ACR and EULAR guidelines, MTX is the most appropriate first-line agent along with being safe, well-tolerated in most RA patients, and cost-effective, thus being the gold standard of RA therapy. Side effects of MTX include nausea, stomatitis, fatigue, elevations in liver enzymes, alopecia, and pneumonitis. Regardless of the risks, it remains the initial choice for RA therapy. Other traditional DMARDs widely used are LEF, SSZ, and HCQ. LEF is another first-line DMARD considered in RA patients who have an unsuccessful first-line MTX therapy. A recent study was conducted on RA patients to assess the difference in efficacy of LEF monotherapy versus LEF in combination with other synthetic DMARDs. The results suggested that both the groups showed similar efficacy in terms of DAS28 score, ESR, and CRP levels after treatment for 3 months clearly demonstrating that LEF monotherapy was equally effective to combination therapy. Furthermore, LEF at 20 mg/daily shows similar efficacy to MTX in all clinical, imaging, and laboratory parameters (Deng et al. 2020; Li, Tam, and Tomlinson 2004). Besides, the common side effects of nausea, hypertension, diarrhea, and other adverse effects linked with LEF are liver toxicity, teratogenicity, and hematological toxicity. As a consequence of increased ALT/AST levels with LEF administration, it is endorsed that the patients receiving LEF along with other hepatotoxic drugs should be thoroughly monitored for increased liver transaminases (Maddison et al. 2005; Li, Tam, and Tomlinson 2004).

The possibility of toxicity with SSZ administration is relatively lower. Side effects of nausea, diarrhea, and elevations in LFTs are common and are usually dose-related. Furthermore, infertility, nephrotoxicity, hepatotoxicity, allergic reactions, and hematologic adverse effects have been observed (Linares, Alonso, and Domingo 2011). Additionally, HCQ has mild antirheumatic effects and is used in combination with other DMARDs to cure moderate to severe disease activity. It is generally well-tolerated; however, over long-term use may cause retinopathy. Furthermore, GI and renal toxicities along with adverse effects on the central nervous system have been observed (Kim et al. 2017; Melles and Marmor 2014).

The emergence of bDMARDs has transformed the RA treatment by targeting specific signaling pathways/proteins involved in RA. The first among this class are the TNF-α inhibitors. However, these are related to the high risk of infections during the initial 6 months of the therapy (Galloway et al. 2011). Injection site reactions and drug-induced lupus are common. Furthermore, observations have demonstrated risks of infections including *Mycobacterium tuberculosis, pneumocystis* pneumonia, and herpes zoster with TNF-α inhibitors (Ruderman 2012). Another increasing concern with biological DMARD therapy is Hepatitis B, particularly in patients receiving rituximab. In a recent investigation, the occurrence of hepatitis B reactivation was evaluated in $HBsAg^-$/$HBcAb^+$ patients (hepatitis B virus surface antigen, hepatitis B virus core antibody) who had received rituximab therapy for RA. The results suggested that hepatitis B reactivation occurred in 8% of patients with acute hepatitis in 4% of patients. The results suggested that RA patients administered with rituximab were at higher risk of hepatitis B reactivation commonly in patients with the surface antibody at baseline $(HBsAb)^-$ thus contemplating the use of antiviral therapy (Kuo et al. 2020).

Tofacitinib and baricitinib are the small molecules targeted DMARDs (JAK inhibitors) currently used for RA. They are usually well-tolerated with side effects of kidney damage, liver enzyme elevation, anemia, thrombosis, and serious infections. Other adverse events accompanying JAK inhibitors include respiratory and urinary tract infections and cardiovascular adverse events (Huang and Luo 2018; Cohen et al. 2017). Therefore, it is suggested that the patient must be examined for any heart, pulmonary, and interstitial lung diseases prior to the selection of DMARDs (Jani et al. 2014). This is because RA patients are at higher risk of cardiovascular diseases, hypertension, atherosclerotic damage, a stroke which may sometimes lead to heart failure (Anyfanti et al. 2020; Bartoloni, Alunno, and Gerli 2018). Additionally, studies have revealed that certain risk factors like age, diabetes, and the use of corticosteroid therapy in RA patients accelerated subclinical atherosclerosis. Thus, clinicians recommend ultrasonography for the detection of atherosclerotic plaques in the carotid arteries for risk prevention (Dalbeni et al. 2020; Semb et al. 2020). The 2015 ACR guidelines provide recommendations on the usage of traditional and biological DMARDs in RA patients with serious infections and cardiovascular diseases. Furthermore, it focuses on the use of vaccines and screening for tuberculosis infection prior to starting treatment with DMARDs. Furthermore, it addresses the 'treat to target approach' involving tapering and suspended administration of medications to active low disease activity or remission (Singh et al. 2016).

2.5 NATURAL COMPOUNDS FOR THE TREATMENT OF RA

The intricacy involved with the administration of both synthetic and bDMARDs urges the use of plant-based compounds/phytochemicals with potential antiarthritic activity and minimal side effects. There is enormous scientific evidence of natural compounds including flavonoids, isoflavones, alkaloids, terpenoids, and phytoestrogens with antioxidant, immunomodulatory, anti-inflammatory, and antiarthritic potential (Dudics et al. 2018; Rengasamy et al. 2019). These natural compounds exhibit antiarthritic effect through various mechanisms viz. inhibition NF-κB, MAPK, JAK/STAT pathways; inhibition of TNF-α, IL-1β, IFN-γ, TGF-β, IL-6; induction of anti-inflammatory cytokines; induction of apoptosis in synovium; modulation of Th17/Treg balance and enzymes activities like COX, LOX, iNOS, and radical scavenging properties (Li et al. 2019). Examples of natural agents with potential antiarthritic activity include quercetin, resveratrol, epigallocatechin, silymarin, curcumin, berberine, and triptolide.

2.6 NATURAL COMPOUNDS AND DMARDS CO-THERAPY FOR IMPROVED RA ACTIVITY

Considering the potential of natural compounds in RA, the combination of natural agents with DMARDs can prove to be beneficial while reducing the toxic consequences of DMARDs. Such co-therapies have gained substantial importance in recent years and hold the potential to decrease the risks of long-term RA therapy. Such combinations that have been investigated in *in vitro* and animal models have been discussed in this section. Table 2.1 illustrates the structure of natural compounds with DMARDs with antiarthritic potential and their mechanisms of action.

2.6.1 Hesperidin and MTX/DMARD Therapy

Hesperidin is a bioflavonoid found abundantly in citrus fruits with medicinal properties. It possesses a number of pharmacological properties including radical scavenging, anti-inflammatory, anticancer, antiviral, and antiallergic. The antioxidant activity of hesperidin is not only restricted to the inhibition of ROS and free radicals. However, it significantly augments the ERK/Nrf2 signaling pathway which further upregulates the heme oxygenase 1 (HO-1) expression thus decreasing the intracellular pro-oxidants and enhancing the levels of antioxidant enzymes like superoxide dismutases (SOD), glutathione S-transferases (GST), catalase (CAT), and bilirubin production. Furthermore, hesperidin exhibits anti-inflammatory activities by attenuating the expression of iNOS and modulating the JAK/STAT signaling pathway. It attenuates the production of COX-2, prostaglandin E2, NO, and cytokine-like TNF-α and IL-1β (Parhiz et al. 2015). The therapeutic efficacy of hesperidin has been investigated on a mouse model of RA. The results demonstrated a significant suppression of paw edema, synovial infiltration, and inflammatory cells. Furthermore, treatment with hesperidin resulted in downregulation of mRNA expressions of TNF-α in lesions (Kawaguchi et al. 2006).

Transglycosylation of hesperidin results in the formation of α-glucosyl hesperidin (HspG) with enhanced aqueous solubility and permeability. Thus, the antipotential of HspG was investigated in mice and humans. In the mice model of collagen-induced arthritis (CIA), the results proposed that oral administration of HspG resulted in significant suppression of interchondral joint damage

TABLE 2.1
Natural Product Combination with MTX for Improved RA Activity

Combination Study	Structure of Natural Compound and Rationale	Results	References
Hesperidin + Standard DMARD therapy Detralex (Diosmin and hesperidin) + MTX	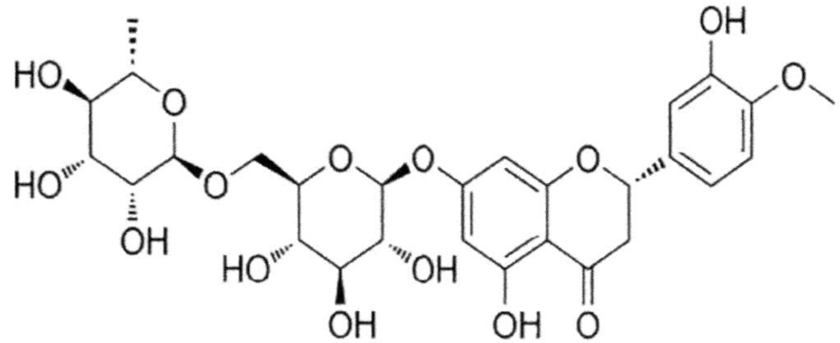 • Anti-inflammatory potential • Antioxidant properties • Regulates ERK/Nrf2 and JAK/STAT pathways • Attenuates iNOS expression	Clinicalstudy-HspG+standardRAtherapy(3months)- • ↑ tolerability • ↓ adverse effects • ↓CRP values In AIA rat model Detralex + MTX- • ↓arthritic scores • ↑ bone mineral density	Kometani et al. (2008) (Rovenský et al. 2009)
Indole-3-Carbinol + MTX	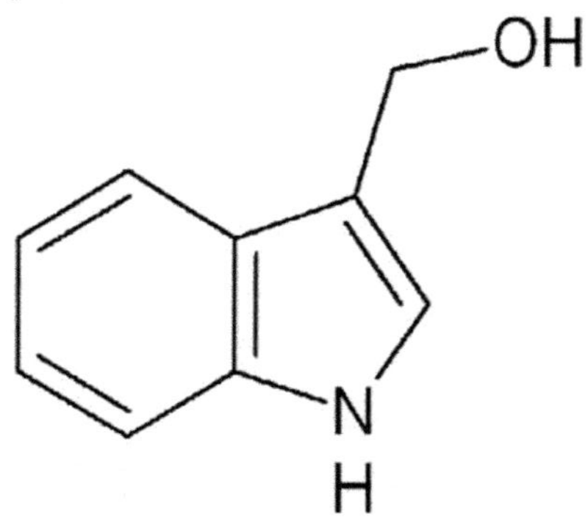 • Anti-inflammatoryand hepatoprotective properties • Antiviral potential • Radical scavenging activity • Inhibits NF-κB signaling • Activates Nrf2 signaling pathway	AIA rats I3C + MTX • ↓erythema • ↓serum TNF-α levels and ESR • AST and ALT levels decreased by 1.5 folds compared to MTX monotherapy • ↓ liver injury	Hasan (et al. 2018)

(Continued)

TABLE 2.1 (*Continued*)
Natural Product Combination with MTX for Improved RA Activity

Combination Study	Structure of Natural Compound and Rationale	Results	References
Gallic acid + MTX	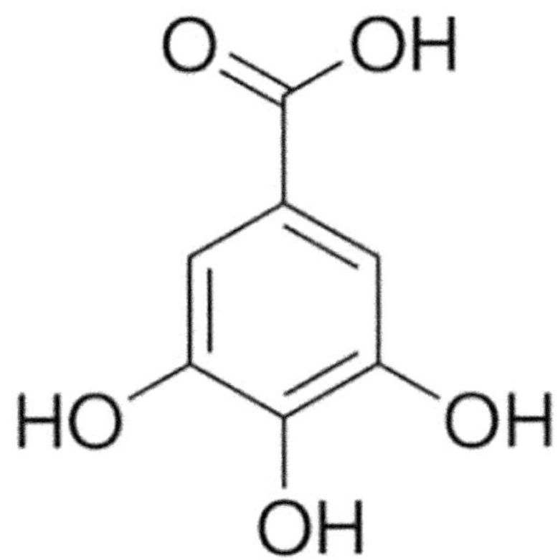 • Anti-inflammatory, antimicrobial and antiulcerative, radical scavenging properties	Combination therapy of Gallic acid+MTX in rats • ↓ MTX-induced BUN, creatinine and uric acid • Reduced kidney hemorrhages and tubular cell necrosis • ↓ expression of TNF-α and PGE-2	Asci et al. (2017)
Silibinin+MTX	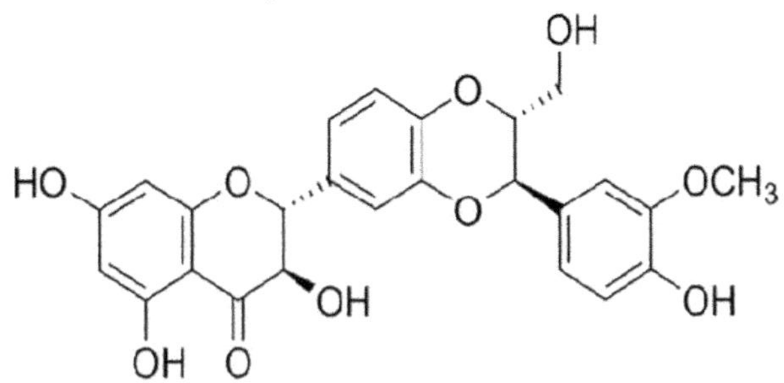 Anti-inflammatory property Hepatoprotective potential Regulates estrogen receptor (ERβ)	Clinical study in RA patients- MTX (12 mg/week)+Silibinin (120 mg twice daily) for 4 months. ↓SJC and TJC ↓ IL-6, IL-8, TNF-α ↓ESR and CRP	Hussain, Mortada, et al. (2016)

(*Continued*)

TABLE 2.1 (*Continued*)
Natural Product Combination with MTX for Improved RA Activity

Combination Study	Structure of Natural Compound and Rationale	Results	References
Resveratrol+MTX	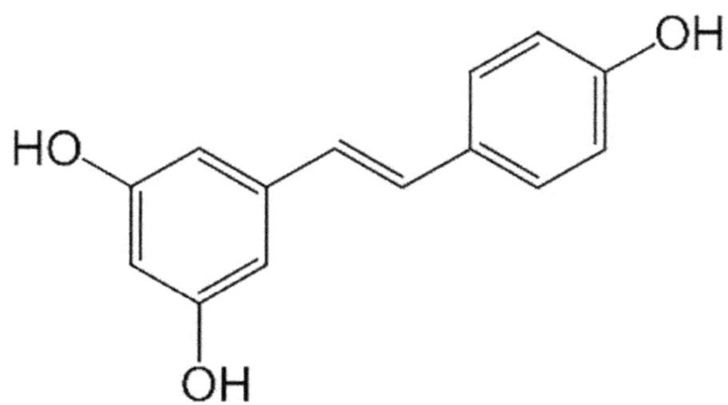 • Anti-inflammatory and antioxidant potential • Neuroprotective and nephroprotective properties • ↓ iNOS and IL-6 • Modulates NF-κB and JAK/STAT signaling pathways.	RSV+MTX combination in Wistar rats- • ↑ Intestinal absorption of MTX through P-gp/MRP2 inhibition • RSV↓ MTX renal elimination and MTX-induced nephrotoxicity via inhibition of OAT1 and OAT3 Clinical study (100 RA patients) 1. Group 1 (50 patients)-RSV (1 g/day)+standard treatment 2. Group 2 (50 patients)- Regular treatment Co-therapy resulted in- • ↓ DAS28, TJC-28, SJC-28 • ↓ CRP, ESR, RF, MMP-3 • ↓ TNF-α, IL-6	Jia et al. (2016) Hedrich et al. (2008)

(*Continued*)

TABLE 2.1 (*Continued*)
Natural Product Combination with MTX for Improved RA Activity

Combination Study	Structure of Natural Compound and Rationale	Results	References
MgIG+MTX	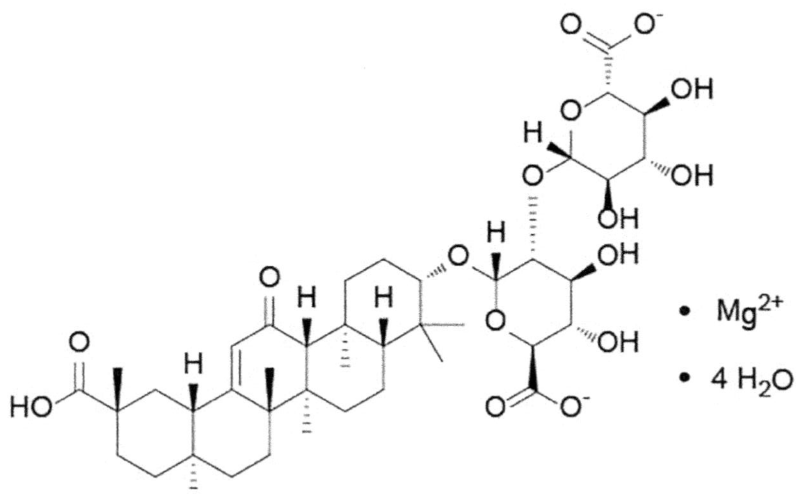 • Antioxidant and hepatoprotective properties • Anti-inflammatory potential • Modulates IL-6 and Nrf2 pathways	Pretreatment with MgIG- • ↓ AST/ALT levels • ↓pyknosis • ↑ antioxidant activity • ↓inflammatory infiltrates • ↓COX-2 expression	(Cao et al. 2019)

and pannus formation. Furthermore, a significant amelioration in the production of TNF-α was witnessed. In the clinical study, 3 g of HspG was administered to RA patients along with concomitant standard RA therapy for a period of 3 months. Results of the ACR20 evaluation suggested an improvement in RA in the HspG group. Furthermore, the plasma CRP value decreased in the HspG-administered group (Kometani et al. 2008).

In another study, the effect of a flavonoid fraction was investigated alone and in combination with MTX in AIA rats. The purified fraction called detralex contains two flavonoids: diosmin (90%) and hesperidine (10%). Diosmin (flavonoid) present in citrus fruits exhibits antioxidant and anti-inflammatory properties. In a mice model, diosmin has demonstrated a protective effect against MTX-induced hepatic, renal, and cardiac injuries via attenuating the AST and levels, cytokines-IL-1β, TNF-α, and IL-6, and tissue levels of malondialdehyde (MDA) and NO (Abdel-Daim et al. 2017). The results suggested that the decline in the hind paw swelling and arthritic scores was prominent in rats treated with detralex+MTX in comparison to MTX monotherapy. The combination therapy significantly inhibited the decrease in bone mineral density. Furthermore, a significant drop in the serum nitrite/nitrate levels was found in the group treated with the combination. The findings clearly suggest that detralex potentiated the beneficial effects of MTX in AIA models (Rovenský et al. 2009).

2.6.2 Indole-3-Carbinol and MTX

Indole-3-carbinol (I3C) is a phytochemical found in the *Brassica* genus and exhibits immunomodulatory, anticancer, radical scavenging, and hepatoprotective properties. The protective effect of I3C has been witnessed in ethanol-induced liver injury mice models. I3C attenuated ethanol-induced injury by decreasing hepatic lipid peroxidation (MDA) and restoring the hepatic antioxidant capacity. Furthermore, it also suppressed pro-inflammatory cytokines like IL-1β and neutrophil infiltration in the liver (Choi, Abdelmegeed, and Song 2018). I3C has displayed potent anti-inflammatory properties by suppressing the production of IL-6, IL-1β, and nitric oxide (NO) levels in LPS-induced RAW264.7 and THP-1 cells.

The anti-inflammatory, antiarthritic, and hepatoprotective potential of I3C has been investigated in combination with MTX. The results revealed a significant reduction in paw inflammation, erythema, and serum TNF-α in rats treated with MTX and I3C. The arthritic rats treated with MTX and I3C showed a more than twofold decrease in the ESR. The pathological assessment of the tibiotarsal joint revealed a significant reduction in synovium hyperplasia, cellular infiltration, and pannus formation in the group administered with the combination. The levels of ALT and AST were evaluated on days 23 and 44 as a measure of liver dysfunction. The group treated with MTX monotherapy showed elevated levels of ALT and AST on both the 23rd and 44th days. However, with the combination treatment, a significant reduction (1.5-folds) was found on day 44 in comparison to the MTX-treated group. Additionally, the combination treatment inhibited lipid peroxidation and restored the glutathione (GSH) levels and CAT and SOD activities thus upregulating the antioxidant defense. Histopathological studies revealed the liver injury in the group treated with MTX evident by a severe portal, lobular inflammation, and necrosis. Combination treatment effectively reduced the MTX-induced liver injury demonstrated by mild portal inflammation, and minimum cellular infiltration with no signs of necrosis (Hasan et al. 2018). In conclusion, I3C effectively abrogated the oxidative stress and liver injury caused by MTX treatment in arthritic rats. Thus, it can act as a novel therapeutic agent in combination with MTX with reduced hepatotoxicity in RA treatment.

2.6.3 Gallic Acid and MTX

Gallic acid (GA) is a polyphenol found in tea leaves, grapes, and sumac (*Rhuscoriaria* L.). It has shown innumerable biological potential such as anticancerous, anti-inflammatory, antimicrobial, and antiulcerative. GA is a strong chelating agent and exhibits excellent free-radical scavenging potential. It maintains the antioxidant defense system and is a strong inhibitor of myeloperoxidase, a crucial enzyme involved in infection and inflammation (Badhani, Sharma, and Kakkar 2015). The nephroprotective activity of GA has been seen in glyoxal-induced renal fibrosis in rats. Treatment with GA attenuated the mRNA expression of matrix metalloproteinases (MMP-2, MMP-9) and NOX elevated by glyoxal. Furthermore, the levels of kidney damage markers viz. blood urea nitrogen, creatinine, and alkaline phosphatase were significantly reduced. Also, co-treatment with GA restored the SOD activity thereby decreasing ROS generation (Yousuf and Vellaichamy 2015).

The protective effect of GA has also been seen in MTX-induced nephrotoxicity in experimental rats. The combination treatment group was administered MTX at 20 mg/kg/week and GA was administered at 100 mg/kg orally for a week. On day 8, blood was collected for biochemical analysis and the kidneys were excised for further histopathological and immunohistochemical examinations. The biochemical analysis result revealed that MTX-induced nephrotoxicity as evidenced by elevated levels of BUN, creatinine, and uric acid levels. Also, the levels of oxidative stress markers in the kidney homogenate viz. total oxidant status and oxidative stress index were substantially increased by MTX. Co-administration of GA ameliorated all these markers of kidney damage. Histopathological analysis of the kidney revealed normal histology in the group treated with GA, with significant reversal of tubular cell necrosis, hemorrhages, and glomerulosclerosis, and proteinous material in the tubules induced by MTX. The immunohistochemical examination further confirmed the reduction of elevated inflammation markers TNF-α and prostaglandin E2 in the tubular epithelial cells of kidneys with GA. Furthermore, no immunoreaction with serum amyloid A (SAA) and CRP was observed which are the major acute-phase indicators of inflammation. Thus, the findings suggested the nephroprotective potential of GA on MTX-induced kidney damage attributed to the antioxidant capacity and anti-inflammatory potential of GA (Asci et al. 2017).

2.6.4 Silibinin and MTX

Silibinin is a flavonolignan, a major active constituent isolated from *Silybummarianum* (milk thistle). It exhibits extensive biological and pharmacological potential through antioxidant, immunomodulatory, and hepatoprotective potential. The hepatoprotective potential of silibinin has been confirmed in a number of liver ailments including hepatitis B and C, liver cirrhosis, and malignant hepatoma (Federico, Dallio, and Loguercio 2017). Furthermore, the ability of silibinin to modify the expression of ERβ (estrogen receptor) has been evaluated in human T-lymphocytes. It has been proposed that the downregulation of this receptor is associated with high disease activity in inflammatory disorders. Naturally occurring phytoestrogens like silibinin act as ERβ agonist ligands thus having the potential to dampen the inflammatory response. The study revealed that silibinin upregulated ERβ expression while downregulating IL-17 and TNF-α in T cells. Furthermore, silibinin induced apoptosis and downregulated the expression of microRNA-155, which exhibits a primary role in T-cell immunity and the pathogenesis of RA (Dupuis et al. 2018).

The combination of silibinin and MTX was also investigated in clinical studies in 60 RA patients maintained on an MTX dose of 12 mg/week for 3 months. These patients were administered 120 mg of silibinin twice a day for 16 weeks. The results suggested that there was an improvement in the clinical scores as demonstrated by a decrease in tender and swollen joint count and visual analog scale with silibinin co-administration in contrast to MTX monotherapy. Furthermore, serum levels of pro-inflammatory cytokines, CRP, ACPA, and ESR were significantly reduced. Additionally, augmented hemoglobin levels and anti-inflammatory cytokines IL-10, and IL-2 were observed with the combination therapy (Hussain, Mortada, et al. 2016). Overall, the results of the clinical scores and the biochemical parameters insinuated that silibinin improved the antirheumatic activity of MTX thus having the potential to be used as an adjuvant with MTX monotherapy.

2.6.5 Resveratrol and MTX

Belonging to the family of stilbenes, resveratrol (RSV) is a polyphenol found in different species of plants such as pine, grapevine, grapes, and Chinese knotweed. *In vitro* data have demonstrated the anticatabolic, anti-inflammatory, antiapoptotic, and antioxidant activity of RSV in various cell types including synoviocytes, and it also exhibits an immunomodulatory effect on T and B cells (Nguyen et al. 2017). Findings in murine macrophage cells have shown that the anti-inflammatory potential of RSV is due to its ability to reduce the release of high-mobility group box 1 (HMGB1) from the nucleus and by modulating the Janus kinase/STAT and NF-κB pathways (Ma et al. 2015). In a CIA DBA/1 mouse model, RSV exhibited an antiarthritic effect by suppressing key cellular and humoral responses including the production of Th17 cells, IL-17, and autoantibodies (Xuzhu et al. 2012). In a study conducted by Yongming et al., the potential of RSV in altering the pharmacokinetic profile of MTX was investigated. Oral administration of the combination of RSV and MTX in Wistar rats significantly increased the area under the curve and maximum serum concentration of MTX by about 1.1- and 1.47-folds, respectively (Figure 2.1). To further verify the results, everted gut sac model was used which also showed an increase in MTX concentration (serosal). Overall, it was found that MTX absorption was enhanced by RSV in the intestine in addition to decreased renal elimination of MTX by inhibition of MRP2, P-gp, OAT1, and OAT3 both *in vivo* and *in vitro*. Also, without increasing intestinal toxicity, renal damage due to MTX was shown to be improved by RSV in rats (Jia et al. 2016). In another study, synovial fluid mononuclear cells isolated from RA patients were treated with RSV, alone or in combination (RSV+MTX) and (RSV+adalimumab). A substantial additive effect of RSV in combination with MTX was found in low disease activity patients as well as in lymphocyte-dominated synovial fluid (Lomholt et al. 2018). In a clinical study involving 100 RA patients (32 males, 68 females), the therapeutic effects of RSV were evaluated. Out of 100, 50 patients received RSV capsules of 1 g for 3 months along with standard treatment, whereas other patients received regular treatment. The results revealed a significant reduction in DAS28, clinical markers (tender and swollen joint count-28), and biochemical markers (CRP, ESR, RF, MMP-3, TNF-α, IL-6, and undercarboxylated osteocalcin) in patients treated with a combination therapy (Hedrich et al. 2008). In general, the studies insinuated that RSV can be a suitable candidate to be used in combination with MTX for improved RA therapy.

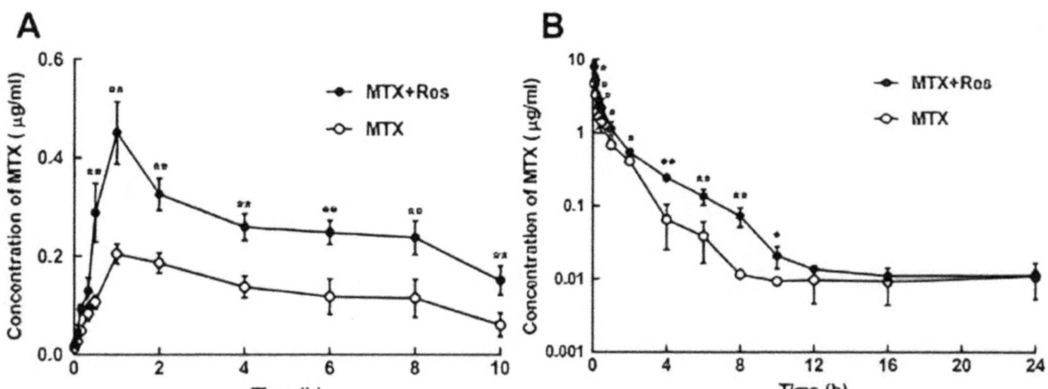

FIGURE 2.1 Mean plasma concentration-time curves of MTX by oral and intravenous administration in rats. (a) Plasma concentration of MTX after oral co-administration with MTX (5 mg/kg) and RSV (100 mg/kg). (b) Plasma concentration of MTX after intravenous administration with MTX (5 mg/kg) and Res (10 mg/kg). Statistical differences between each set of points were compared with those for the control group (MTX only group) by a two-tailed unpaired *t*-test, with $p<0.05$ as the limit of significance. (*$p<0.05$, **$p<0.01$; mean±S.D., n=3–5). Reprinted from Jia *et.al* (2016) with permission from Elsevier.

2.6.6 Magnesium Glycyrrhizinate and MTX

Magnesium glycyrrhizinate (MgIG) is a major bioactive triterpene glycoside found in the roots of *Glycyrrhiza glabra* and exhibits potential immunomodulatory and hepatoprotective properties. *In vitro* studies in mouse macrophages have shown that MgIG-suppressed LPS (lipopolysaccharide) stimulated the activation of phospholipase A2 in the arachidonic acid pathway. Also, it inhibited LPS-stimulated PGI2 synthase, TBX synthase, and arachidonic acid metabolizing enzymes (COX-1, 5-LOX, and COX-2) (Xie et al. 2015). Protective effects of MgIG have also been seen in oxaliplatin-induced and triptolide-induced liver toxicity via modulation of IL-6 and Nrf2 signaling pathways, respectively (Tan et al. 2018; Zou et al. 2018).

In another study, the effect of MgIG on MTX-induced liver and intestinal toxicity was studied. Results showed that MgIG pretreatment reduced serum AST/ALT levels, hepatic fibrosis, and hepatocytes apoptosis as well as increased SOD, Gpx activities, and GSH levels. Also, MgIG was able to reduce MTX-induced COX-2 expression and intestinal permeability (Figure 2.2) (Cao et al. 2019).

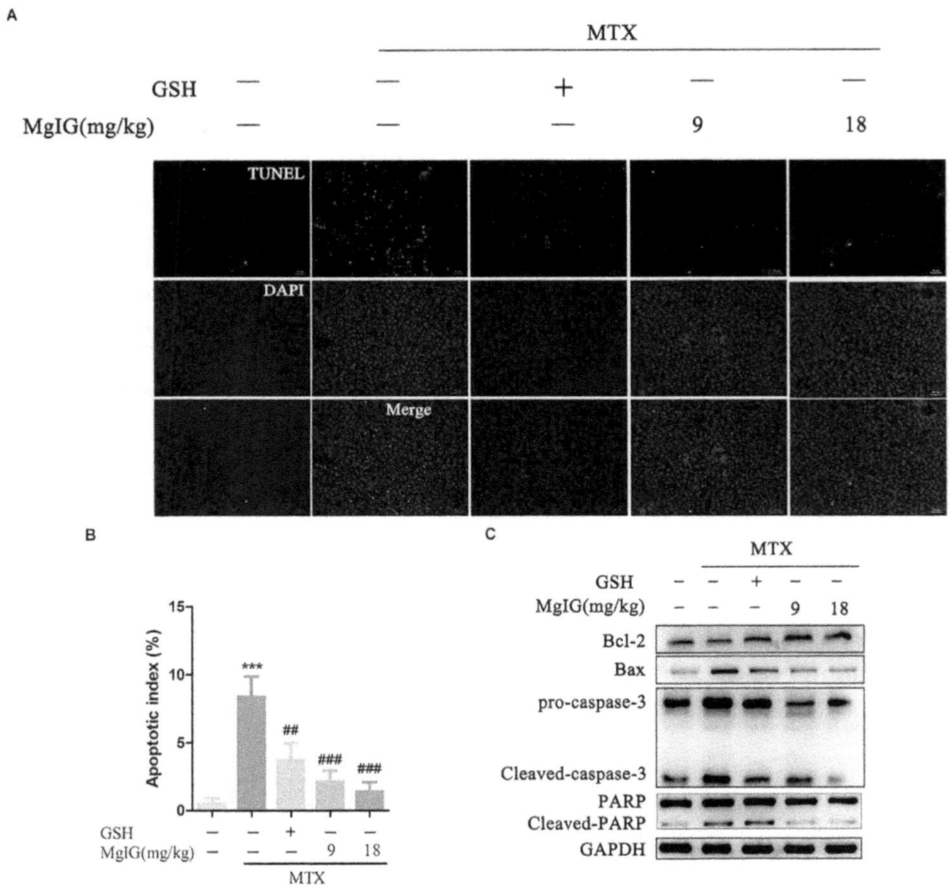

FIGURE 2.2 Effects of MgIG on MTX-induced apoptosis in hepatic tissue: (a) Liver tissue sections from different treatment groups were stained with TUNEL (green) and counterstained with DAPI to localize the nuclei (blue), and colocalized green and blue cells in the same view indicated TUNEL-positive cells, scale bar=50 μm; (b) semiquantitative analysis of TUNEL-positive cells, $n=6$; (c) western blotting of liver caspase-3, cleaved-caspase-3, Bax, Bcl-2, PARP and cleaved-PARP protein expression. The results are shown as mean±*SD*, $n=3$. Compared with control group: $^{*}P<0.05$, $^{**}P<0.01$, $^{***}P<0.001$; compared with MTX group: $^{\#}P<0.05$, $^{\#\#}P<0.01$, $^{\#\#\#}P<0.001$. Reprinted from Cao *et.al.* (2019) under the terms of the Creative Commons Attribution License (CC BY).

In conclusion, the results suggested that MgIG can be an effective adjuvant therapy in RA while improving the intestinal and hepatic side effects of MTX.

2.7 DRUG DELIVERY STRATEGIES FOR THE EFFECTIVE APPLICATION OF NATURAL PRODUCTS IN RA

Natural products have a broad range of therapeutic potential; however, various physicochemical limitations restrict the clinical application of natural products in treating diseases. Chemical complexity, poor solubility, efflux by transporters including P-gp, rapid metabolism, and poor bioavailability are some of the challenges that limit the successful implication of natural compounds in clinical settings (Medina, Zhu, and Kairemo 2004). Natural products need to be chemically stable, exhibit good solubility, and permeability for adequate absorption across the intestinal epithelium. After the absorption process, the molecule needs to endure the first-pass metabolism before entering the systemic circulation, where it is said to be bioavailable (Song et al. 2013). The limited solubility and permeability of natural agents in addition to increased first-pass metabolism, account for their lower bioavailability which requires them to be consumed in a more enormous amount to achieve therapeutic efficacy (Efferth and Kaina 2011; Byard 2010; Kim et al. 2013). Consequently, to improve the therapeutic efficacy, and to decrease the toxicity of the natural products, novel drug delivery systems (DDSs) are employed (Watkins et al. 2015; Shi et al. 2010). DDSs transform drug release, pharmacokinetics, and pharmacodynamic profile of the drug. It should involve the delivery of the drug at the desired rate in the body over a treatment period and specifically target the site of action. Developing various delivery systems like liposomes, nanoemulsion, solid lipid nanoparticles (SLNs), polymeric nanoparticles (PNs), micelles, conjugates, nanocapsules, dendrimers, and phytosomes offers several advantages for natural products, including lowering of toxicity, sustained delivery, and enhancement of stability, solubility, bioavailability, and therapeutic efficacy (Figure 2.3) (Medina, Zhu, and Kairemo 2004; Fang et al. 2019). The following section discusses the DDSs that have been used for the effective delivery of natural agents in RA (Table 2.2).

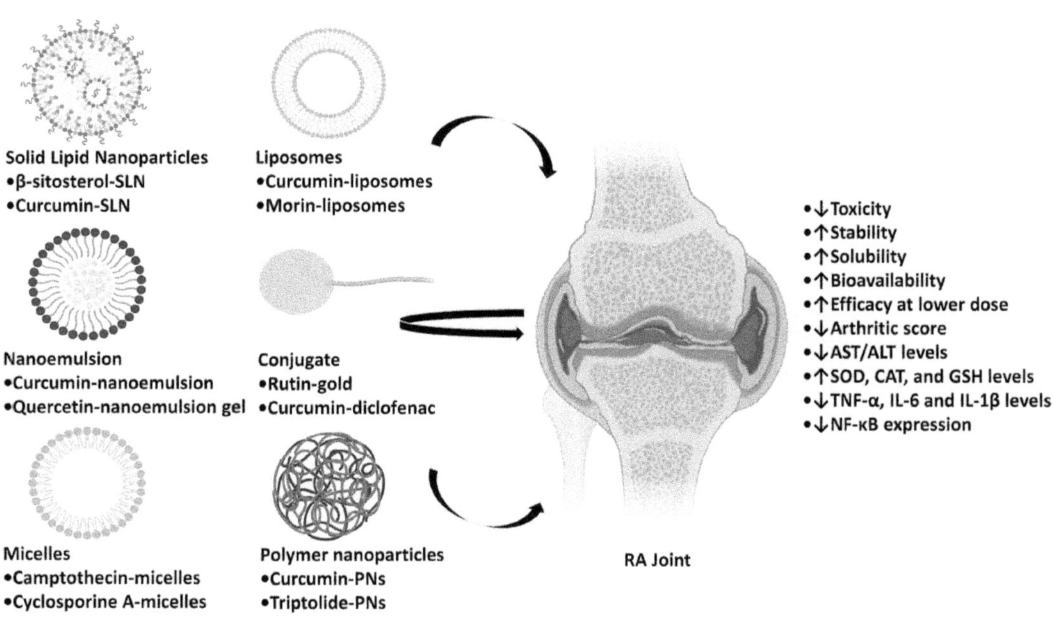

FIGURE 2.3 DDSs that have been used for the effective delivery of natural compounds in RA.

TABLE 2.2
Drug Delivery Strategies for the Effective Application of Natural Products in RA

Drug Delivery System	Natural Compound	Study Model	Findings	References
Liposomes	Morin Formulation-Mannosylated liposomal morin 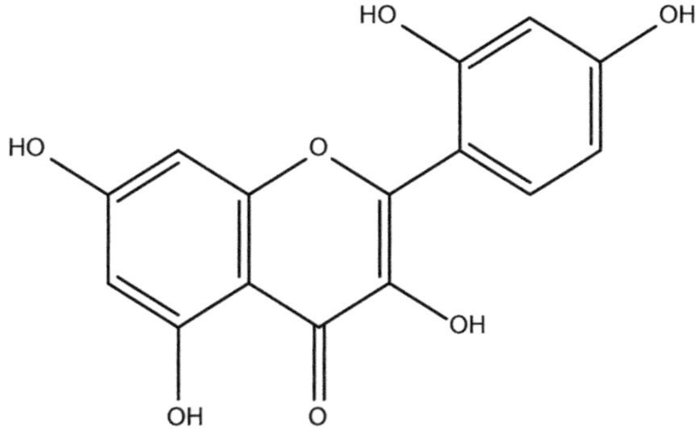 	AIA	ML-Morin: • ↑ Uptake and preferential internalization in synovial and spleen macrophages • Effective inhibition of the inflammatory immune response and osteoclastogenesis	Sultana, Neog, and Rasool (2017)

(*Continued*)

TABLE 2.2 (*Continued*)
Drug Delivery Strategies for the Effective Application of Natural Products in RA

Drug Delivery System	Natural Compound	Study Model	Findings	References
SLNs	β-Sitosterol Formulation-β-sitosterol loaded SLN 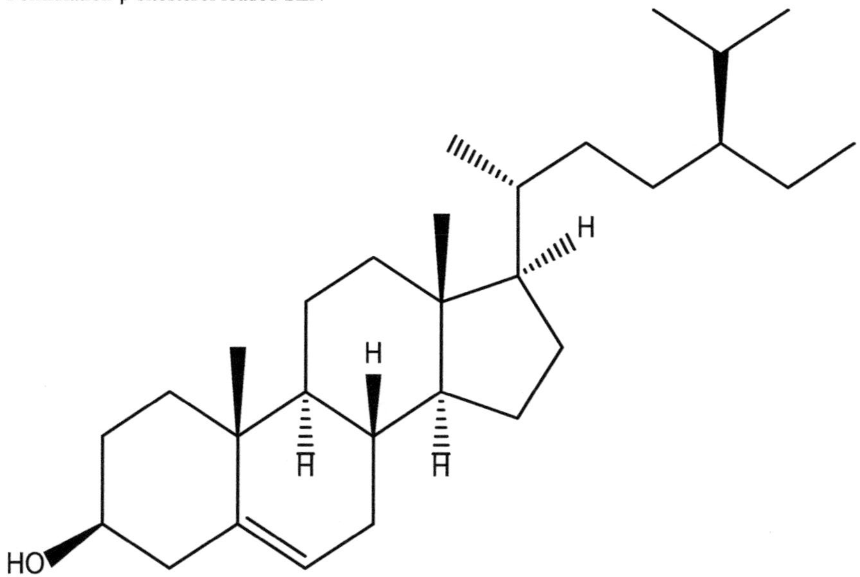	AIA	β-Sitosterol-SLNs: • Showed better antiarthritic activity than β-sitosterol • Significantly ↓ cytokines and inflammatory mediators • Acted via NF-κB suppression and HO-1/Nrf-2 pathway activation	Zhang et al. (2020)

(*Continued*)

TABLE 2.2 (*Continued*)
Drug Delivery Strategies for the Effective Application of Natural Products in RA

Drug Delivery System	Natural Compound	Study Model	Findings	References
Micelles	Camptothecin Formulation-camptotheCinloaded steriCally stabilized miCelles with vasoaCtive intestinal peptides on the surfaCe 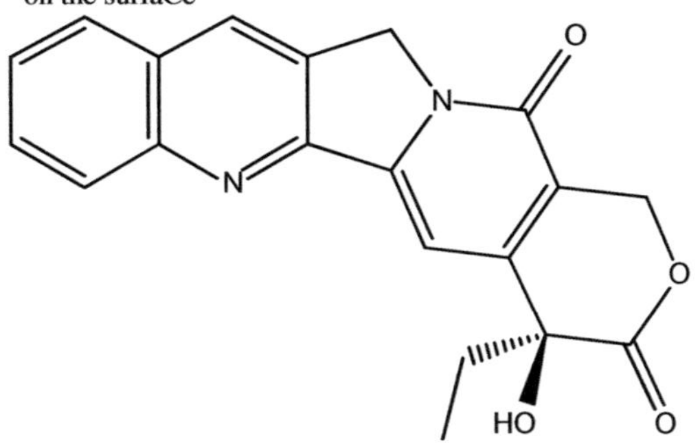	CIA	CPT-SSM-VIP (camptothecin, 0.1 mg/kg): • Significantly ↓ arthritic joint macrophages, T-lymphocytes, and synoviocytes Ten-folds increase in CPT dose was required to attain the same effect	Koo, Rubinstein, and Önyüksel (2011)
PNs	Triptolide Formulation-TP-loaded poly-γ-glutamiC aCid imbedded L-phenylalanine ethyl ester Copolymer 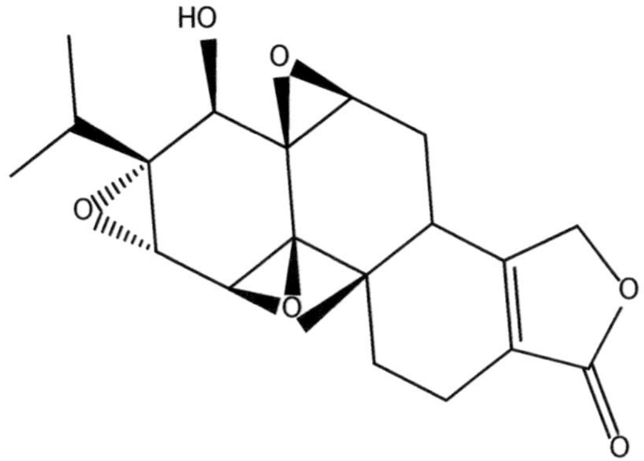	Toxicity studies in RAW 264.7 cells and C57/B6 mice	PPT ↓ TP-induced cytotoxicity, apoptosis and ROS generation in RAW 264.7 cells PTT: • ↑ Survival rate of C57/B6 mice • Significantly ↓ TP-induced toxicity on kidney, liver, and spleen	Zhang et al. (2016)

(*Continued*)

TABLE 2.2 (*Continued*)
Drug Delivery Strategies for the Effective Application of Natural Products in RA

Drug Delivery System	Natural Compound	Study Model	Findings	References
Nanoemulsions	CurCumin Formulation- CurCumin-based nanoemulsion 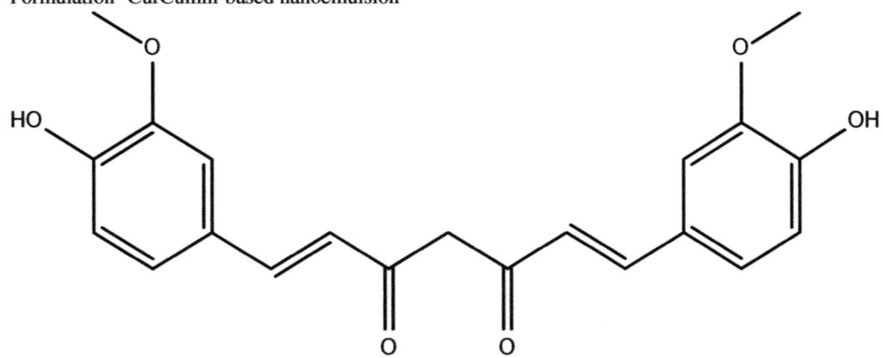	AIA	CM-Ns was found to be more effective than curcumin suspension. CM-Ns- • ↑ Stability and Bioavailability • ↓ Serum and synovial fluid levels of TNF-α and IL-1β	Zheng et al. (2015)
Conjugates	Rutin Formulation-Rutin-gold nanopartiCle Conjugates 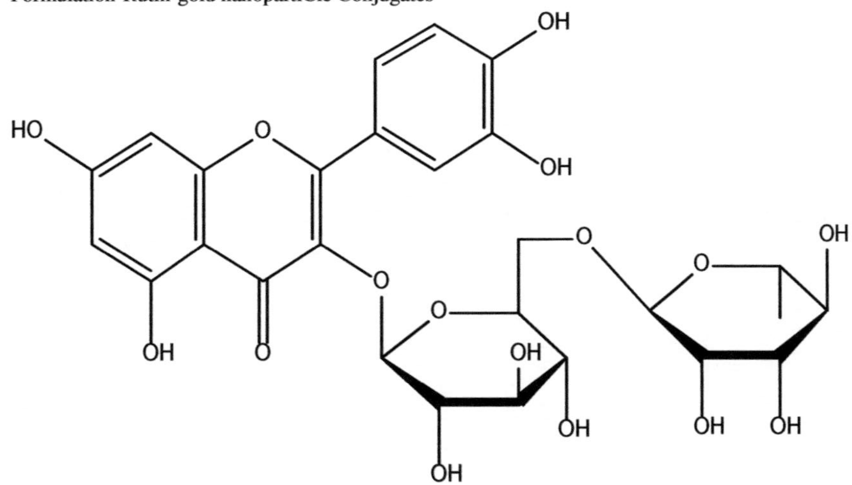	CIA	R-AuNPs (12.5 mg/kg) showed comparable results to that of rutin (50 mg/kg) R-AuNPs- • ↓ Arthritic score, serum NO and PO levels • ↓ NF-κB and iNOS expression	Gul et al. (2018)

2.7.1 Liposomes

Liposomes are nanocarriers, mainly composed of phospholipids with an arrangement like that of a lipid bilayer. They are used for the successful delivery of pharmacologically active agents, which can transport both hydrophobic and hydrophilic drugs (Allen and Cullis 2013; Fenske and Cullis 2008). Liposomes improve stability, reduce toxicity, and enable high loading of the drug (Hattori et al. 2006).

In a study for targeted delivery of morin (bioflavonoid), mannosylated liposomal formulation (ML-Morin) with decorated mannose was prepared using a thin-film hydration method. The efficacy was evaluated for RA treatment in the AIA rat model. Results showed an increase in uptake and preferential internalization in macrophages of the synovium and spleen in addition to the downregulation of pro-inflammatory cytokines, transcription factor (NF-κB-p65), angiogenic factor (VEGF), and iNOS. Furthermore, a significant decrease in protein expression of TNF-α, IL-17, IL-6, IL-1β, STAT-3, p-STAT-3, and receptor activator of nuclear factor kappa beta ligand (RANKL) along with increased osteoprotegerin expression in macrophages of the spleen and the synovium. Additionally, the formulation effectively inhibited the inflammatory immune response while inhibiting osteoclastogenesis (Sultana, Neog, and Rasool 2017). In another study, a mannosylated liposomal formulation of p-coumaric acid was used to study its targeted delivery to the synovial activated macrophages of AIA rats to lower the rate of osteoclastogenesis and bone resorption. It demonstrated superior efficiency in targeting and internalization into the activated macrophages and exhibited significant inhibition of differentiation and proliferation of osteoclast MMP-9 and pro-inflammatory cytokines (Neog, Rasool, and Biopharmaceutics 2018). Rinkunaite et al. demonstrated the effect of liposomal curcumin on AIA in Wistar rats. Liposomal curcumin reduced the levels of IL-6, TNF-α, and IL-1β, leukocytes and neutrophils effectively as compared to other formulations (Rinkunaite et al. 2021). Overall, the studies suggested the effective delivery of natural agents using liposome-based formulations.

2.7.2 Solid Lipid Nanoparticles

SLNs are nanosized lipid-based carriers with colloidal delivery systems having an average size of about 50–1,000 nm. They have been widely exploited for targeted delivery of a large number of therapeutic agents. SLNs have high loading capacity, large surface area, prolonged stability, low toxicity, and outstanding biocompatibility (Mehnert and Mäder 2001; Puglia et al. 2012). These are made by dispersing liquefied lipid molecules and active therapeutic drugs in an emulsifier containing an aqueous medium. An emulsifier is used to solidify the lipid phase (saturated fatty acids or lipid-like triglycerides) (Ekambaram, Sathali, and Priyanka, 2012). In comparison to other lipid-based delivery systems, SLNs display advantages viz. high stability, loading capacity, encapsulation efficiency, and bioavailability, prolonged-release profile, and target-specific release properties. Also, SLNs are produced in an inexpensive, simple, and scalable manner with industrial applicability (Katouzian et al. 2017).

In the AIA rat model, *Tripterygium wilfordii* Hook f. (TWHF)-loaded SLNs exhibited enhanced anti-inflammatory activity along with a protective effect against TWHF-induced hepatotoxicity. In this study, TWHF-SLN significantly decreased arthritic paw volume and serum PG (prostaglandins) levels, the serum levels of ALP, ALT, AST, and gamma-glutamyl transferase. However, liver MDA, glutathione peroxidase, and xanthine oxidase were not elevated significantly. From histopathology findings, it was further confirmed that TWHF caused more liver damage than TWHF-SLN, showing the superiority of SLNs over plain TWHF (Xue et al. 2012). Another study evaluated the protective antiarthritic effect of curcumin and curcumin-loaded SLNs (C-SLNs) in the AIA rat model in Wistar rats. Compared to free curcumin, C-SLNs dose-dependently and significantly improved symptoms of arthritic rats (thermal hyperalgesia, mechanical allodynia, mechanical hyperalgesia, joint hyperalgesia, mobility score, joint stiffness, and paw volume). Furthermore, it showed significant improvement in WBC count and serum biochemical markers (MDA, nitrite levels, reduced GSH, SOD activity, catalase activity, TNF-α levels, and C-RP levels). C-SLNs also showed enhanced amelioration of radiological alterations in arthritic joints. It was found that in most of the biomarkers tested in arthritic rats, the activity of C-SLNs (30 mg/kg) was comparable with that of the standard drug used (naproxen, 25 mg/kg) (Arora et al. 2015). In a study by Zhang et al., β-sitosterol-loaded SLN (β-sitosterol-SLNs) prepared by double emulsion solvent displacement method, evaluated against the AIA model in rats showed antiarthritic effect via NF-κB suppression and HO-1/Nrf-2 pathway activation. β-sitosterol-SLNs showed better antiarthritic activity as compared to β-sitosterol. β-sitosterol-SLNs significantly decreased the arthritic index, paw edema and increased the body weight of arthritic rats. It significantly decreased the thiobarbituric acid reactive substances (TBARS) and increased CAT, SOD, and GSH levels in ankle tissue homogenate. β-sitosterol-SLNs significantly reduced serum level of cytokines (TNF-α, IL-2, IL-1β, IL-6, IL-16, IL-17), PGE2, COX-2, VEGF, NF-κBand increased TGF-β and IL-10 levels. Additionally, gene expression results showed a substantial increase in HO-1 and Nrf2 expression, whereas there was a decrease in the expression of RANKL, STAT3, and NF-κB in β-sitosterol-SLNs treated arthritic rats (Zhang et al. 2020). Overall, the study demonstrates the potential of SLNs in the field of drug delivery.

2.7.3 Micelles

Micelles are structures that are made up of amphiphilic copolymers with a size of 10–100 nm. They naturally form aggregates or structures having core–shell spherical form when polymeric chains are dissolved in an aqueous solution. A critical micelle concentration (CMC) of polymer is required above which micelle production takes place. The molecular weight and chemical composition of polymer are the basis of the CMC of polymers. Amphiphilic copolymers that makeup micelles associate at a certain threshold temperature called the critical micelle temperature, CMT (Lukyanov and Torchilin 2004; Zhang, Huang, and Li 2014). Below CMC and CMT, micellar polymers remain as monomers. The central region of micelles forms a hydrophobic core whereas the outer shell is hydrophilic. Because of their nanosized structure, enhanced solubilization, colloidal stability, sustained release, high loading, and increased plasma half-life, micelles have attained remarkable recognition in DDSs (Kedar et al. 2010; Cabral and Kataoka 2014; Yokoyama 2010).

In a study by Koo et al., a single, low-dose injection (subcutaneous) of CPT (camptothecin, 0.1 mg/kg) in the form of CPT-SSM-VIP (camptothecin-loaded sterically stabilized micelles with vasoactive intestinal peptide on the surface) directed after induction of CIA in mice reduced joint-inflammation for a lengthy period subsequently without any systemic toxicity. CPT-SSM-VIP significantly reduced macrophages, T-lymphocytes, and synoviocytes in the arthritic joints of CIA mice. About ten-folds higher dose of CPT was required to attain the same effect, although some vacuole formation was observed in liver histology (Koo, Rubinstein, and Önyüksel 2011). Wilson et al. investigated polysilicon acid-polycaprolactone (PSA–PCL)-loaded cyclosporine A (CyA) micelles to target inflamed synovial cells. *In vitro* studies confirmed the internalization of CyA-loaded micelles by rheumatoid arthritis synovial fibroblasts. Also, CyA upon uptake was released from the micelle and later separated into the phospholipid membrane, signifying a mode of action that commences with membrane disruption (Wilson et al. 2014). In the AIA rat model, the anti-inflammatory activity of pure and micellar curcumin was evaluated. Micellar curcumin exhibited a marked reduction in paw edema, serum CRP, myeloperoxidase (MPO), TNF-α and IL-6, and TBARS levels compared to native curcumin (Khayyal et al. 2018).

2.7.4 Polymer Nanoparticles

PNs are submicronic or nanosized colloidal particles (1–1,000 nm) having active therapeutic agents encapsulated or absorbed into macromolecular polymers. Modification of surface properties of PNs can be done using various polymer end groups or by linking specific polymers to the designed PN. PNs are one of the extensively used delivery systems for the efficient delivery of drugs, vaccines, and biological macromolecules (Shao et al. 2016; Hussain, Sahudin, et al. 2016). Small molecules, peptides, or antibodies targeting ligands can also be fixed to the surface of PNs in order to accelerate specific cellular interactions with receptors or tissue components. PNs are biocompatible, and biodegradable with minimal immunogenicity, having the ability to boost solubility, stability, and therapeutic efficacy, protect active therapeutic agents from physio-chemical degradation, limit toxicity, promote the sustained release and increased interaction with biological tissues. PNs as nanomedicines with specific features like safety, viability, targetability, biodegradability, compatibility, ideal size, surface charge, encapsulation efficiency, and morphology can be designed by modifying methods and formulation parameters (Shao et al. 2016; Hussain, Sahudin, et al. 2016; Banik, Fattahi, and Brown 2016; Hussain, Thu, Ng, et al. 2017; Hussain, Thu, Amjad, et al. 2017).

In a study by Dewangan et al., curcuminladen carboxymethyl cellulose acetate butyrate (Cu-CMCAB) nanoparticles were prepared by rapid precipitation technique with 73.41%±2.4% loading efficiency and better release pattern than pure curcumin. The effect of Cu-CMCAB nanoparticles on different disease parameters involving thermal hyperalgesia, mechanical allodynia, mechanical hyperalgesia, paw volume, joint stiffness, mobility score, locomotor activity, blood leukocyte count, and radiography was evaluated using the AIA rat model. Increased efficacy of Cu-CMCAB nanoparticles was observed over pure curcumin in most of the disease parameters demonstrating the increased efficacy of nanoparticle-based delivery systems (Dewangan et al. 2017). In a study, to lessen the side effects linked to antirheumatic agent triptolide (TP), TP-loaded PN [poly-γ-glutamic acid (PGA)-grafted L-phenylalanine ethyl ester copolymer] [PPT] were developed having controlled release pattern. *In vitro* flow cytometry, cytotoxicity assay (MTT), and ROS detection in RAW 264.7 cells showed that the developed PNs effectively reduced the apoptosis, toxicity, and cellular ROS, respectively, induced by pure TP. Furthermore, *in vivo* toxicity studies demonstrated that PPT improved the survival rate of C57/B6 mice as compared to pure TP while having no side effects on the spleen and thymic index. From the biochemical analysis of blood serum, it was found that as compared to TP, PPT caused less toxicity to the liver, kidney, and blood cells (WBCs, RBCs, and lymphocytes). Also, PTT showed a significant reduction in toxicity of TP in the kidney, liver, and spleen specified by HE staining analysis and TUNEL assay. PTT-treated group showed a decrease in the toxic effects of TP on the spleen whereas less alteration in cellular structures of the liver and kidney was found in the PTT-treated group as compared to TP-treated group. Furthermore, TP-induced apoptosis of the kidney, liver, and spleen was slightly observed in the PTT-treated group demonstrating the effectiveness of formulation over plain drug (Zhang et al. 2016). In another study, TP-loaded PGA-imbedded di-*tert*-butyl L-aspartate hydrochloride (PAT) nanoparticles were prepared for evaluation of TP efficacy and reduced toxicity. Flow cytometry and cytotoxicity assay showed that PAT was able to reduce the apoptosis, and toxicity induced by pure TP in RAW 264.7 cells. It reduced LPS/IFNγ-stimulated cytokines (TNFα, IL-1β, IL-6, iNOS, and β-actin) expression of macrophage. *In vivo* toxicity study using TNFα-transgenic (TNFα-Tg) mice revealed that as compared to TP, PAT treated group showed better survival rate and lower toxicity of kidney, liver, and spleen. PAT was also able to decrease inflammatory synovium and bone and cartilage degradation in mice model (Zhang et al. 2018).

2.7.5 Nanoemulsions

Nanoemulsion is another colloidal delivery system that has upright stability, controlled release, fast digestibility, high loading capacity, degradation resistance, and the potential to ameliorate the bioavailability of active therapeutic agents. Nanoemulsions generally consist of dispersed oil phase systems in an aqueous system or vice versa, forming nanosized (20–200 nm) droplets. These are thermodynamically unstable and can be made stable for longer durations, by using suitable surfactants (Singh et al. 2017; Hörmann and Zimmer 2016; Yukuyama et al. 2017). By increasing the surface-to-volume ratio of nanoemulsions, they show a prolonged release of drugs and improve the transport of drugs across cell membranes, thereby increasing penetrability (Chen et al. 2012). Furthermore, since nanoemulsions exhibit similarities to low-density lipoproteins, they have been broadly used for better penetrability and targeted delivery-specific sites in proliferating tissues including activated macrophages in inflamed synovium (Padoveze et al. 2009).

In a study by Zheng et al., the activity of curcumin-based nanoemulsion (CM-N) and free curcumin were evaluated in the AIA rat model. CM-Ns (150 nm) prepared using a high-pressure homogenizing method showed a marked decrease in the degradation of curcumin in GI fluids thereby enhancing its stability. Also, an oral bioavailability study showed a more than three-fold increase in the maximum plasma concentration of curcumin for CM-N as compared to suspension formulation. Results obtained after oral dosing of CM-N and curcumin suspension in AIA rats showed that there was a decrease in levels of TNF-α and IL-1β both in synovial fluid and blood serum, however, CM-Ns was found to be more effective than curcumin suspension (Zheng et al. 2015). In another study, a nanoemulsion-based gel of quercetin (QCT-NE) was formulated using a spontaneous emulsification technique for effective RA therapy. Cytotoxicity study showed no toxic effects of QCT-NE in synoviocytes HIG-82 cells, whereas it showed a strong inhibitory effect on LPS-stimulated TNF-α production in RAW264.7 cells. As compared to QCT gel, it showed better drug permeation. QCT-NE gel significantly decreased paw inflammation, arthritic index, and stiffness, improved various hematological parameters, and also was found to be safe (non-irritating) for tropical use on rat skin (Gokhale et al. 2019).

2.7.6 Drug Conjugates

Conjugation involves covalent linking interaction of drugs/prodrugs to various carriers like polymers and antibodies for pharmacological applications. It is an important tool for targeted drug delivery and control release of various active therapeutic agents. Conjugation of an active pharmacological agent to a carrier improves stability, toxicity, and pharmacodynamic and pharmacokinetic profiles. In polymer–drug conjugates, the drug is generally attached to the biocompatible, water-soluble polymeric molecule. These drug delivery carriers offer several advantages over conventional drug use including controlled release, targeted delivery, and improved stability, solubility, plasma circulation time, and therapeutic efficacy of a drug (Kerlin and Li 2013).

A study using Rutin-gold nanoparticles (R-AuNPs, 12.5 mg/kg) versus rutin (50 mg/kg) was conducted in the CIA rat model. Marked reduction in arthritic score, serum NO and peroxide levels, and NF-κB and iNOS expression levels were found in treated groups. R-AuNPs given at 12.5 mg/kg dose showed comparable results to that of rutin given at a higher dose (50 mg/kg), demonstrating the effectiveness of conjugate (Gul et al. 2018). In another study, as compared to curcumin, curcumin-diclofenac conjugate exhibited improved stability, higher bioavailability (more than five-folds), and significantly improved the arthritis symptoms (loss of mobility and inflammation) of Balb/c mice (Jain et al. 2014).

2.7.7 Miscellaneous Drug Delivery Strategies

For RA management and effective delivery of quercetin, a controlled release DDS (quercetin-loaded polycaprolactone microspheres) was investigated in *in vitro* and *in vivo* systems. The microspheres were found to be non-toxic to synovial cells (HIG-82 cells) and also showed biocompatibility with the synovium of Wistar rats on intra-articular injection (Natarajan et al. 2011). In another study, selenium-deposited tripterinephytosomes (Se@Tri-PTs) were prepared, where phytosomes were found to be chiefly facilitating the transepithelial transport of tripterine in Caco-2 cells. Furthermore, Se coating on phytosomes improved the drug release pattern and antiarthritic activity of tripterine. In the AIA rat model, Se@Tri-PTs significantly improved arthritic symptoms (arthritic score, mobility, synovial thickening, and bone damage) and reduced inflammatory factors (NO, TNF-α, and

IL-6) (Zhu et al. 2020). In a study by Coradini et al., resveratrol and curcumin, co-encapsulated into lipid-core nanocapsules were investigated in the AIA rat model. Co-encapsulated nanocapsules (1.7 mg/kg, twice a day, i.p.) showed a marked decrease in paw edema. However, at the same dose, resveratrol and curcumin in solution showed no decrease in paw edema. Co-encapsulated nanocapsules also minimized histological changes including synovium fibrosis, bone and cartilage loss in arthritic rats (Coradini et al. 2015). A study involving thioglycolic acid (TGA) covered cadmium telluride (CdTe) quantum dots (QDs) of quercetin, investigated in the AIA rat model showed a substantial decrease in inflammation along with enhancement in the regeneration of cartilage. QDs of quercetin significantly increased antioxidant enzyme activity (CAT, GPx, GSH, SOD) and significantly restored levels of ESR, ACRP, RF, and WBC count to normal in arthritic rats (Jeyadevi et al. 2013). For targeted and improved efficacy of hesperidin (HP) against RA, hesperidin-loaded green silver nanoparticles (AgNPs) stabilized by gum acacia (GA) were investigated in the AIA rat model. HP-GA-AgNPs weighing 1 mg/kg demonstrated mild-to-moderate swelling, slight arthritic score, and decreased degenerative changes. Furthermore, histopathological studies of ankle joint tissue showed reduced granulomatous inflammation and influx of inflammatory cells in arthritic rats. There was also a significant reduction in Toll-like receptors (TLRs) expression in HP-GA-AgNPs-treated arthritic group. As compared to pure compounds, the developed formulation showed a marked increase in the effectiveness of hesperidin (Rao et al. 2018).

2.8 CONCLUSION

Natural compounds have long been used to treat various autoimmune diseases owing to their extensive biological and pharmacological activities. These include flavonoids, phytoestrogens, alkaloids, and other plant-derived natural agents with enormous therapeutic potential often underrated. The treatment in RA involves the practice of conventional synthetic and biological DMARDs including the recent small molecule, JAK/STAT inhibitors. However, these are all associated with several adverse effects including hepatic, renal, and GI toxicities and dangers of grave infections including pneumococcus influenza, herpes zoster, and hepatitis, and reactivation of latent tuberculosis. The adverse effects can lead to the cessation of the therapy. Recent *in vitro* and *in vivo* experimental studies reveal a relatively new strategy involving the use of DMARDs in combination with natural compounds as immunomodulating, hepatoprotective, and nephroprotective agents. This chapter focuses on such combinations including hesperidin, indole-3-carbinol, GA, and silibinin with MTX that have shown promising antiarthritic activity. Furthermore, this chapter also discusses the use of DDSs including liposomes, SLNs, micelles, PNs, nanoemulsions, conjugates, etc., for the efficient applicability of natural compounds in RA management. Overall, the article explains the potential of natural compounds in combination with DMARDs and different drug delivery strategies for their efficient application in the treatment of RA. The use of such delivery systems for natural compounds exhibiting antiarthritic, hepatoprotective, and nephroprotective properties in combination with already existing DMARDs may perhaps prove to be an effective strategy in improving the RA therapy.

ABBREVIATIONS

ACPA	Anticitrullinated protein antibodies
ACR	American College of Rheumatology
AIA	Adjuvant induced arthritis
ALP	Alkaline phosphatase
ALT	Alanine aminotransferase
AST	Aspartate aminotransferase
COX	Cyclooxygenase
CRP	C-reactive protein

DMARDs	Disease-modifying antirheumatic drugs
ESR	Erythrocyte sedimentation rate
EULAR	European League Against Rheumatism
FLS	Fibroblast-like synoviocytes
HCQ	Hydroxychloroquine
IFN-γ	Interferon-gamma
IL-1β	Interleukin-1β
IL-6	Interleukin 6
iNOS	Inducible nitric oxide synthase
JNK-c	Jun N-terminal kinase
LEF	Leflunomide
LPS	Lipopolysaccharide
MAPK	Mitogenactivated protein kinase
MMPs	Matrix metalloproteinases
MTX	Methotrexate
NF-κB	Nuclear factor kappa-light-chain-enhancer of activated B cells
NO	Nitric oxide
NSAIDs	Nonsteroidal anti-inflammatory drugs
PGs	Prostaglandins
PNs	Polymeric nanoparticles
RA	Rheumatoid arthritis
RANKL	Receptor activator of nuclear factor kappa beta ligand
ROS	Reactive oxygen species
STAT	Signal transducer and activator of transcription
SLNs	Solid lipid nanoparticles
SSZ	Sulfasalazine
TP	Triptolide
TNFα	Tumor necrosis factor α
WBC	White blood cells

REFERENCES

Abdel-Daim, Mohamed M, Hesham A Khalifa, Abdelrahman Ibrahim Abushouk, Mohamed A Dkhil, Saleh AAl-Quraishy. 2017. Diosmin attenuates methotrexate-induced hepatic, renal, and cardiac injury: a biochemical and histopathological study in mice. *Oxidative Medicine and Cellular Longevity.* 2017:3281670.

Aletaha, Daniel, Tuhina Neogi, Alan J Silman, et al. 2010. 2010 Rheumatoid arthritis classification criteria: an American College of Rheumatology/European League against Rheumatism collaborative initiative. *Arthritis & Rheumatism* 62 (9):2569–2581.

Allen, Theresa M, and Pieter R Cullis. 2013. Liposomal drug delivery systems: from concept to clinical applications. *Advanced Drug Delivery Reviews* 65 (1):36–48.

Alpizar-Rodriguez, Deshiré, Rüdiger B Mueller, Burkhard Möller, et al. 2017. Female hormonal factors and the development of anti-citrullinated protein antibodies in women at risk of rheumatoid arthritis. *Rheumatology* 56 (9):1579–1585.

Anyfanti, Panagiota, Eugenia Gkaliagkousi, Areti Triantafyllou, et al. 2020. Hypertension in rheumatic diseases: prevalence, awareness, treatment, and control rates according to current hypertension guidelines. *Journal of Human Hypertension* 35:419–427.

Arora, Rajesh, Anurag Kuhad, Indu Pal Kaur, et al. 2015. Curcumin loaded solid lipid nanoparticles ameliorate adjuvant-induced arthritis in rats. *European Journal of Pain* 19 (7):940–952.

Asci, Halil, Ozlem Ozmen, Hamit Yasar Ellidag, Bunyamin Aydin, Ercan Bas, and Necat Yilmaz. 2017. The impact of gallic acid on the methotrexate-induced kidney damage in rats. *Journal of Food and Drug Analysis* 25 (4):890–897.

Badhani, Bharti, Neha Sharma, and Rita Kakkar. 2015. Gallic acid: a versatile antioxidant with promising therapeutic and industrial applications. *RSC Advances* 5 (35):27540–27557.

Bakker, Marije F, Johannes WG Jacobs, Paco MJ Welsing, et al. 2012. Low-dose prednisone inclusion in a methotrexate-based, tight control strategy for early rheumatoid arthritis: a randomized trial. *Annals of Internal Medicine* 156 (5):329–339.

Banik, Brittany L, Pouria Fattahi, and Justin L Brown. 2016. Polymeric nanoparticles: the future of nanomedicine. *Wiley Interdisciplinary Reviews: Nanomedicine, and Nanobiotechnology* 8 (2):271–299.

Bartok, Beatrix, and Gary S Firestein. 2010. Fibroblast-like synoviocytes: key effector cells in rheumatoid arthritis. *Immunological Reviews* 233 (1):233–255.

Bartoloni, Elena, Alessia Alunno, and Roberto Gerli. 2018. Hypertension as a cardiovascular risk factor in autoimmune rheumatic diseases. *Nature Reviews Cardiology* 15 (1):33.

Brilman, Elisabeth G. 2018. Corticosteroid bridging strategies with methotrexate monotherapy in early rheumatoid and undifferentiated arthritis; A comparison of efficacy and toxicity in 2 clinical trials. *In 2018 ACR/ARHP Annual Meeting*. Chicago, Illinois, United States of America.

Brown, Philip M, Arthur G Pratt, and John D Isaacs. 2016. Mechanism of action of methotrexate in rheumatoid arthritis, and the search for biomarkers. *Nature Reviews Rheumatology* 12 (12):731–742.

Buttgereit, Frank. 2020. Views on glucocorticoid therapy in rheumatology: the age of convergence. *Nature Reviews Rheumatology* 16:239–246.

Byard, Roger W. 2010. A review of the potential forensic significance of traditional herbal medicines. *Journal of Forensic Sciences* 55 (1):89–92.

Cabral, Horacio, and Kazunori Kataoka. 2014. Progress of drug-loaded polymeric micelles into clinical studies. *Journal of Controlled Release* 190:465–476.

Cao, Yuzhu, Hang Shi, Zhiguang Sun, et al. 2019. Protective effects of magnesium glycyrrhizinate on methotrexate-induced hepatotoxicity and intestinal toxicity may be by reducing COX-2. *Frontiers in Pharmacology* 10:119.

Chen, Yu, Yu Gao, Hangrong Chen, et al. 2012. Engineering inorganic nanoemulsions/nanoliposomes by fluoride-silica chemistry for efficient delivery/co-delivery of hydrophobic agents. *Advanced Functional Materials* 22 (8):1586–1597.

Choi, Youngshim, Mohamed A Abdelmegeed, and Byoung-Joon Song. 2018. Preventive effects of indole-3-carbinol against alcohol-induced liver injury in mice via antioxidant, anti-inflammatory, and anti-apoptotic mechanisms: role of gut-liver-adipose tissue axis. *The Journal of Nutritional Biochemistry* 55:12–25.

Cohen, Stanley B., Yoshiya Tanaka, Xavier Mariette, et al. 2017. Long-term safety of tofacitinib for the treatment of rheumatoid arthritis up to 8.5 years: integrated analysis of data from the global clinical trials. *Annals of the Rheumatic Diseases* 76 (7):1253–1262.

Coradini, Karine, Rossana B Friedrich, Francisco N Fonseca, et al. 2015. A novel approach to arthritis treatment based on resveratrol and curcumin co-encapsulated in lipid-core nanocapsules: in vivo studies. *European Journal of Pharmaceutical Sciences* 78:163–170.

Criscione, Lisa G, and E William St Clair. 2002. Tumor necrosis factor-α antagonists for the treatment of rheumatic diseases. *Current Opinion in Rheumatology* 14 (3):204–211.

Cronstein, Bruce N, and Thomas M Aune. 2020. Methotrexate and its mechanisms of action in inflammatory arthritis. *Nature Reviews Rheumatology* 16:145–154.

Dalbeni, Andrea, Alessandro Giollo, Michele Bevilacqua, et al. 2020. Traditional cardiovascular risk factors and residual disease activity are associated with atherosclerosis progression in rheumatoid arthritis patients. *Hypertension Research* 43:922–928.

Deng, Daihua, Jun Zhou, Min Li, et al. 2020. Leflunomide monotherapy versus combination therapy with conventional synthetic disease-modifying antirheumatic drugs for rheumatoid arthritis: a retrospective study. *Scientific Reports* 10 (1):1–8.

Dewangan, Ashish K, Yamini Perumal, Naresh Pavurala, Kanwaljit Chopra, and Sonal Mazumder. 2017. Preparation, characterization and anti-inflammatory effects of curcumin loaded carboxymethyl cellulose acetate butyrate nanoparticles on adjuvant induced arthritis in rats. *Journal of Drug Delivery Science and Technology* 41:269–279.

Donahue, Katrina E, Gerald Gartlehner, Daniel E Jonas, et al. 2008. Systematic review: comparative effectiveness and harms of disease-modifying medications for rheumatoid arthritis. *Annals of Internal Medicine* 148 (2):124–134.

Dudics, Steven, David Langan, Rakeshchandra R. Meka, et al. 2018. Natural products for the treatment of autoimmune arthritis: their mechanisms of action, targeted delivery, and interplay with the host microbiome. *International Journal of Molecular Sciences* 19 (9):2508.

Dupuis, Maria Luisa, Fabrizio Conti, Angela Maselli, et al. 2018. The natural agonist of estrogen receptor β silibinin plays an immunosuppressive role representing a potential therapeutic tool in rheumatoid arthritis. *Frontiers in* Immunology 9:1903.

Efferth, Thomas, and Bernd Kaina. 2011. Toxicities by herbal medicines with emphasis to traditional Chinese medicine. *Current Drug Metabolism* 12 (10):989–996.

Ekambaram, Prabhakaran, Abdul Hasan Sathali, and Karunanidhi Priyanka. 2012. Solid lipid nanoparticles: a review. *Scientific Reviews and Chemical Communications* 2 (1):80–102.

Fang, Gang, Qinghuai Zhang, Yuzhou Pang, Hnin Ei Thu, and Zahid Hussain. 2019. Nanomedicines for improved targetability to inflamed synovium for treatment of rheumatoid arthritis: multi-functionalization as an emerging strategy to optimize therapeutic efficacy. *Journal of Controlled Release* 303:181–208.

Federico, Alessandro, Marcello Dallio, and Carmelina Loguercio. 2017. Silymarin/silybin and chronic liver disease: a marriage of many years. *Molecules* 22 (2):191.

Fenske, David B, and Pieter Rutter Cullis. 2008. Liposomal nanomedicines. *Expert Opinion on Drug Delivery* 5 (1):25–44.

Galloway, James B, Kimme L Hyrich, Louise K Mercer, et al. 2011. Anti-TNF therapy is associated with an increased risk of serious infections in patients with rheumatoid arthritis especially in the first 6 months of treatment: updated results from the British Society for Rheumatology Biologics Register with special emphasis on risks in the elderly. *Rheumatology* 50 (1):124–131.

Gokhale, Jayanti P, Hitendra S Mahajan, and Sanjay J Surana. 2019. Quercetin loaded nanoemulsion-based gel for rheumatoid arthritis: in vivo and in vitro studies. *Biomedicine and Pharmacotherapy* 112:108622.

González-Suárez, Inés, M José Aguilar-Amat, María Trigueros, et al. 2014. Leukoencephalopathy due to oral methotrexate. *The Cerebellum* 13 (1):178–183.

Gul, Anum, Bimal Kunwar, Maryam Mazhar, et al. 2018. Rutin and rutin-conjugated gold nanoparticles ameliorate collagen-induced arthritis in rats through inhibition of NF-κB and iNOS activation. *International Immunopharmacology* 59:310–317.

Hasan, Hiba, Hanan Ismail, Youmna El-Orfali, and Ghada Khawaja. 2018. Therapeutic benefits of Indole-3-Carbinol in adjuvant-induced arthritis and its protective effect against methotrexate induced-hepatic toxicity. *BMC Complementary and Alternative Medicine* 18 (1):337.

Hattori, Yoshiyuki, Shigeru Kawakami, Kazumi Nakamura, Fumiyoshi Yamashita, and Mitsuru Hashida. 2006. Efficient gene transfer into macrophages and dendritic cells by in vivo gene delivery with mannosylated lipoplex via the intraperitoneal route. *Journal of Pharmacology and Experimental Therapeutics* 318 (2):828–834.

Hedrich, Christian Michael, Barbara Fiebig, Fabian H Hauck, et al. 2008. Chilblain lupus erythematosus – a review of literature. *Clinical Rheumatology* 27 (8):949–954.

Hoes, Jos N, Johannes WG Jacobs, Frank Buttgereit, and Johannes WJ Bijlsma. 2010. Current view of glucocorticoid co-therapy with DMARDs in rheumatoid arthritis. *Nature Reviews Rheumatology* 6 (12):693.

Hörmann, Karl, and Andreas Zimmer. 2016. Drug delivery and drug targeting with parenteral lipid nanoemulsions-a review. *Journal of Controlled Release* 223:85–98.

Hu, Yang, Jeffrey A Sparks, Susan Malspeis, et al. 2017. Long-term dietary quality and risk of developing rheumatoid arthritis in women. *Annals of the Rheumatic Diseases* 76 (8):1357–1364.

Huang, Feng, and Zu-Chun Luo. 2018. Risk of adverse drug events observed with baricitinib 2 mg versus baricitinib 4 mg once daily for the treatment of rheumatoid arthritis: a systematic review and meta-analysis of randomized controlled trials. *BioDrugs* 32 (5):415–423.

Humby, Frances, Michele Bombardieri, Antonio Manzo, et al. 2009. Ectopic lymphoid structures support ongoing production of class-switched autoantibodies in rheumatoid synovium. *PLoS Medicine* 6 (1):e1.

Hussain, Saad Abdulrahman, Ahmed Hashem Mortada, Nazar Abdulateef Jasim, and Faiq Isho Gorial. 2016. Silibinin improves the effects of methotrexate in patients with active rheumatoid arthritis: pilot clinical study. *Oman Medical Journal* 31 (4):263.

Hussain, Zahid, Shariza Sahudin, Hnin Ei Thu, Ahmad Nazrun Shuid, Syed Nasir Abbas Bukhari, and Endang Kumolosasi. 2016. Recent advances in pharmacotherapeutic paradigm of mild to recalcitrant atopic dermatitis. *Critical Reviews(tm) in Therapeutic Drug Carrier Systems* 33 (3): 213–263.

Hussain, Zahid, Hnin Ei Thu, Muhammad Wahab Amjad, et al. 2017. Exploring recent developments to improve antioxidant, anti-inflammatory and antimicrobial efficacy of curcumin: a review of new trends and future perspectives. *Materials Science and Engineering: C* 77:1316–1326.

Hussain, Zahid, Hnin Ei Thu, Shiow-Fern Ng, Shahzeb Khan, and Haliza Katas. 2017. Nanoencapsulation, an efficient and promising approach to maximize wound healing efficacy of curcumin: a review of new trends and state-of-the-art. *Colloids and Surfaces B: Biointerfaces* 150:223–241.

Jain, SK, MS Gill, HS Pawar, and Sarasija Suresh. 2014. Novel curcumin diclofenac conjugate enhanced curcumin bioavailability and efficacy in streptococcal cell wall-induced arthritis. *Indian Journal of Pharmaceutical Sciences* 76 (5):415.

Jani, Meghna, Nik Hirani, Eric L Matteson, and William G Dixon. 2014. The safety of biologic therapies in RA-associated interstitial lung disease. *Nature Reviews Rheumatology* 10 (5):284.

Jeyadevi, R, T Sivasudha, A Rameshkumar, et al. 2013. Enhancement of anti arthritic effect of quercetin using thioglycolic acid-capped cadmium telluride quantum dots as nanocarrier in adjuvant induced arthritic Wistar rats. *Colloidsand Surfaces B: Biointerfaces* 112:255–263.

Jia, Yongming, Zhihao Liu, Changyuan Wang, et al. 2016. P-gp, MRP2 and OAT1/OAT3 mediate the drug-drug interaction between resveratrol and methotrexate. *Toxicology and Applied Pharmacology* 306:27–35.

Katouzian, Iman, Afshin Faridi Esfanjani, Seid Mahdi Jafari, and Sahar Akhavan. 2017. Formulation and application of a new generation of lipid nano-carriers for the food bioactive ingredients. *Trends in Food Science and Technology* 68:14–25.

Kawaguchi, Kiichiro, Hiroko Maruyama, Takashi Kometani, and Yoshio Kumazawa. 2006. Suppression of collagen-induced arthritis by oral administration of the citrus flavonoid hesperidin. *Planta Medica* 72 (05):477–479.

Kedar, Uttam, Prasanna Phutane, Supriya Shidhaye and Vilasrao Kadam, Biology. 2010. Advances in polymeric micelles for drug delivery and tumor targeting. *Nanomedicine: Nanotechnologyand Medicine* 6 (6):714–729.

Kerlin, Roy L, and Xiantang Li. 2013. Pathology in non-clinical drug safety assessment. In edited by Wanda M. Haschek, Colin G. Rousseaux, and Matthew A. Wallig *Haschek and Rousseaux's Handbook of Toxicologic* Pathology, II, 725–750. Academic Press.

Khayyal, Mohamed T, Rania M El-Hazek, Walaa A El-Sabbagh, Jan Frank, Dariush Behnam, and Mona Abdel-Tawab. 2018. Micellar solubilisation enhances the antiinflammatory activities of curcumin and boswellic acids in rats with adjuvant-induced arthritis. *Nutrition* 54:189–196.

Kim, Ellie JY, Yuling Chen, Johnson Q Huang, et al. 2013. Evidence-based toxicity evaluation and scheduling of Chinese herbal medicines. *Journal of Ethnopharmacology* 146 (1):40–61.

Kim, Ji-Won, Yoon Young Kim, Hwajeong Lee, Sung-Hoon Park, Seong-Kyu Kim, and Jung-Yoon Choe. 2017. Risk of retinal toxicity in longterm users of hydroxychloroquine. *The Journal of Rheumatology* 44 (11):1674–1679.

Kometani, Takashi, Takaaki Fukuda, Tatsuyuki Kakuma, et al. 2008. Effects of α-glucosylhesperidin, a bioactive food material, on collagen-induced arthritis in mice and rheumatoid arthritis in humans. *Immunopharmacology and Immunotoxicology* 30 (1):117–134.

Koo, Otilia May Yue, Israel Rubinstein, and Hayat Önyüksel. 2011. Actively targeted low-dose camptothecin as a safe, long-acting, disease-modifying nanomedicine for rheumatoid arthritis. *Pharmaceutical Research* 28 (4):776–787.

Kuo, Meng Hsuan, Chih-Wei Tseng, Chi-Hui Lee, Chien-Hsueh Tung, Kuo-Chih Tseng, and Ning-Sheng Lai. 2020. Moderate risk of hepatitis B virus reactivation in HBsAg–/HBcAb+ carriers receiving rituximab for rheumatoid arthritis. *Scientific Reports* 10 (1):1–9.

Lateef, Omar, Najia Shakoor, and Robert A Balk. 2005. Methotrexate pulmonary toxicity. *Expert Opinion on Drug Safety* 4 (4):723–730.

Lefèvre, Stephanie, Anette Knedla, Christoph Tennie, et al. 2009. Synovial fibroblasts spread rheumatoid arthritis to unaffected joints. *Nature Medicine* 15 (12):1414–1420.

Li, Edmund K, Lai-Shan Tam, and Brian Tomlinson. 2004. Leflunomide in the treatment of rheumatoid arthritis. *Clinical therapeutics* 26 (4):447–459.

Li, Tai-ping, Ai-hua Zhang, Jian-hua Miao, et al. 2019. Applications and potential mechanisms of herbal medicines for rheumatoid arthritis treatment: a systematic review. *RSC Advances* 9 (45):26381–26392.

Linares, Victoria, Virginia Alonso, and José L Domingo. 2011. Oxidative stress as a mechanism underlying sulfasalazine-induced toxicity. *Expert Opinion on Drug Safety* 10 (2):253–263.

Liu, Lijun, Shengyun Liu, Cong Wang, et al. 2019. Folate supplementation for methotrexate therapy in patients with rheumatoid arthritis: a systematic review. *JCR: Journal of Clinical Rheumatology* 25 (5):197–202.

Lomholt, Søren, Anders Mellemkjaer, Marie Beck Iversen, et al. 2018. Resveratrol displays anti-inflammatory properties in an ex vivo model of immune mediated inflammatory arthritis. *BMC Rheumatology* 2 (1):1–10.

Lukyanov, Anatoly N, and Vladimir P Torchilin. 2004. Micelles from lipid derivatives of water-soluble polymers as delivery systems for poorly soluble drugs. *Advanced Drug Delivery Reviews* 56 (9):1273–1289.

Ma, Chunfang, Yin Wang, Lei Dong, Minjing Li, and Wanru Cai. 2015. Anti-inflammatory effect of resveratrol through the suppression of NF-κB and JAK/STAT signaling pathways. *Acta biochimica et biophysica Sinica* 47 (3):207–213.

Maddison, Peter, Patrick Kiely, Bruce Kirkham, et al. 2005. Leflunomide in rheumatoid arthritis: recommendations through a process of consensus. *Rheumatology* 44 (3):280–286.

Malemba, Jean J, Jean M Mbuyi-Muamba, Jean Mukaya, Xavier Bossuyt, Patrick Verschueren, and Rene Westhovens. 2012. The epidemiology of rheumatoid arthritis in Kinshasa, Democratic Republic of Congo-a population-based study. *Rheumatology* 51 (9):1644–1647.

McDougall, Cairistin, Kelle Hurd, and Cheryl Barnabe. 2017. Systematic review of rheumatic disease epidemiology in the indigenous populations of Canada, the United States, Australia, and New Zealand. *Seminars in Arthritis and Rheumatism* 46 (5):675–686.

McInnes, Iain B, and Georg Schett. 2011. The pathogenesis of rheumatoid arthritis. *New England Journal of Medicine* 365 (23):2205–2219.

Medina, Oula P, Ying Zhu, and Kalevi Kairemo. 2004. Targeted liposomal drug delivery in cancer. *Current Pharmaceutical Design* 10 (24):2981–2989.

Mehnert, Wolfgang, and Karsten Mäder. 2001. Solid lipid nanoparticles: production, characterization and applications. *Advanced Drug Delivery Reviews* 47 (2–3):165–196.

Melles, Ronald B, and Michael F Marmor. 2014. The risk of toxic retinopathy in patients on long-term hydroxychloroquine therapy. *JAMA Ophthalmology* 132 (12):1453–1460.

Motegi, Sei-ichiro, and Osamu Ishikawa. 2014. Methotrexate-induced accelerated nodulosis in a patient with rheumatoid arthritis and scleroderma. *Acta Dermato-Venereologica* 94 (3):357–358.

Myasoedova, Elena, Cynthia S Crowson, Hilal Maradit Kremers, Terry M Therneau, and Sherine E Gabriel. 2010. Is the incidence of rheumatoid arthritis rising? Results from Olmsted County, Minnesota, 1955–2007. *Arthritis & Rheumatism* 62 (6):1576–1582.

Natarajan, Venkatachalam, Natarajan Krithica, Balaraman Madhan, and Praveen Kumar Sehgal. 2011. Formulation and evaluation of quercetin polycaprolactone microspheres for the treatment of rheumatoid arthritis. *Journal of Pharmaceutical Sciences* 100 (1):195–205.

Neog, Manoj Kumar, Mahaboobkhan Rasool, and Biopharmaceutics. 2018. Targeted delivery of p-coumaric acid encapsulated mannosylated liposomes to the synovial macrophages inhibits osteoclast formation and bone resorption in the rheumatoid arthritis animal model. *European Journal of Pharmaceutics* 133:162–175.

Nguyen, Christelle, Jean-François Savouret, Magdalena Widerak, Marie-Thérèse Corvol, and François Rannou. 2017. Resveratrol, potential therapeutic interest in joint disorders: a critical narrative review. *Nutrients* 9 (1):45.

Padoveze, Amanda Felippe, Fernanda Maniero, Tatiane Vanessa de Oliveira, et al. 2009. Effect of a cholesterol-rich diet on the metabolism of the free and esterified cholesterol components of a nanoemulsion that resembles LDL in rabbits. *Brazilian Journal of Medical and Biological Research* 42 (2):172–178.

Parhiz, Hamideh, Ali Roohbakhsh, Fatemeh Soltani, Ramin Rezaee, and Mehrdad Iranshahi. 2015. Antioxidant and anti-inflammatory properties of the citrus flavonoids hesperidin and hesperetin: an updated review of their molecular mechanisms and experimental models. *Phytotherapy Research* 29 (3):323–331.

Peper, Shana M, Robert Lew, Ted Mikuls, et al. 2017. Rheumatoid arthritis treatment after methotrexate: the durability of triple therapy versus etanercept. *Arthritis Care & Research* 69 (10):1467–1472.

Phillips, Darren C, Kevin J Woollard, and Helen R Griffiths. 2003. The anti-inflammatory actions of methotrexate are critically dependent upon the production of reactive oxygen species. *British Journal of Pharmacology* 138 (3):501–511.

Puglia, Carmelo, Francesco Bonina, Luisa Rizza, et al. 2012. Lipid nanoparticles as carrier for octyl-methoxycinnamate: in vitro percutaneous absorption and photostability studies. *Journal of Pharmaceutical Sciences* 101 (1):301–311.

Ramiro, Sofia, Alexandre Sepriano, Katerina Chatzidionysiou, et al. 2017. Safety of synthetic and biological DMARDs: a systematic literature review informing the 2016 update of the EULAR recommendations for management of rheumatoid arthritis. *Annals of the Rheumatic Diseases 76* (6):1101–1136.

Rao, Komal, Sabahat Aziz, Talat Roome, et al. 2018. Gum acacia stabilized silver nanoparticles based nano-cargo for enhanced anti-arthritic potentials of hesperidin in adjuvant induced arthritic rats. *Artificial Cells, Nanomedicine, and Biotechnology* 46 (sup1):597–607.

Rengasamy, Kannan RR, Haroon Khan, Shanmugaraj Gowrishankar, et al. 2019. The role of flavonoids in autoimmune diseases: therapeutic updates. *Pharmacology & Therapeutics* 194:107–131.

Riksen, Niels Peter, Pilar Barrera, Petra HH van den Broek, PLCM Van Riel, Paul Smits, and GA Rongen. 2006. Methotrexate modulates the kinetics of adenosine in humans in vivo. *Annals of the Rheumatic Diseases* 65 (4):465–470.

Rinkunaite, Ieva, Egidijus Simoliunas, Milda Alksne, Dominyka Dapkute and Virginija Bukelskiene. 2021. Anti-inflammatory effect of different curcumin preparations on adjuvant-induced arthritis in rats. *BMC Complementary Medicine and Therapies* 21 (1):1–12.

Romao, Vasco C, Aurea Lima, Miguel Bernardes, Helena Canhao, and Joao Eurico Fonseca. 2014. Three decades of low-dose methotrexate in rheumatoid arthritis: can we predict toxicity? *Immunologic Research* 60 (2–3):289–310.

Rovenský, Jozef, Mária Stančíková, Emília Rovenská, Svetoslav Štvrtina, Viera Štvrtinová, and Karol Švík. 2009. Treatment of rat adjuvant arthritis with flavonoid (Detralex(r)), methotrexate, and their combination. *Annals of the New York Academy of Sciences* 1173 (1):798–804.

Ruderman, Eric M. 2012. Overview of safety of non-biologic and biologic DMARDs. *Rheumatology* 51 (suppl_6):vi37–vi43.

Schwartz, Daniella M, Yuka Kanno, Alejandro Villarino, Michael Ward, Massimo Gadina, and John J O'Shea. 2017. JAK inhibition as a therapeutic strategy for immune and inflammatory diseases. *Nature Reviews Drug Discovery* 16 (12):843–862.

Scott, David L, F Wolfe, and TW Huizinga. 2010. Rheumatoid arthritis *Lancet* 376:1094–108.

Semb, Anne Grete, Eirik Ikdahl, Grunde Wibetoe, Cynthia Crowson, and Silvia Rollefstad. 2020. Atherosclerotic cardiovascular disease prevention in rheumatoid arthritis. *Nature Reviews Rheumatology* 16:361–379.

Shao, Mei, Zahid Hussain, Hnin Ei Thu, et al. 2016. Drug nanocarrier, the future of atopic diseases: advanced drug delivery systems and smart management of disease. *Colloids and Surfaces B: Biointerfaces* 147:475–491.

Shi, Jinjun, Alexander R Votruba, Omid C Farokhzad, and Robert Langer. 2010. Nanotechnology in drug delivery and tissue engineering: from discovery to applications. *Nano Letters* 10 (9):3223–3230.

Silverman, Gregg J, and Dennis A Carson. 2003. Roles of B cells in rheumatoid arthritis. *Arthritis Research & Therapy* 5 (4):1–6.

Singh, Jasvinder A, Kenneth G Saag, S Louis Bridges Jr, et al. 2016. 2015 American College of Rheumatology guideline for the treatment of rheumatoid arthritis. *Arthritis & Rheumatology* 68 (1):1–26.

Singh, Yuvraj, Jaya Gopal Meher, Kavit Raval, et al. 2017. Nanoemulsion: concepts, development and applications in drug delivery. *The Journal of Controlled Release* 252:28–49.

Smolen, Josef S, Daniel Aletaha, Anne Barton, et al. 2018. Rheumatoid arthritis. *Nature Reviews Disease Primers* 4:1–23.

Smolen, Josef S, Robert Landewé, Ferdinand C Breedveld, et al. 2014. EULAR recommendations for the management of rheumatoid arthritis with synthetic and biological disease-modifying antirheumatic drugs: 2013 update. *Annals of the Rheumatic Diseases* 73 (3):492–509.

Smolen, Josef S, Robert Landewé, Johannes Bijlsma, et al. 2017. EULAR recommendations for the management of rheumatoid arthritis with synthetic and biological disease-modifying antirheumatic drugs: 2016 update. *Annals of the Rheumatic Diseases* 76 (6):960–977.

Song, Gao, Basu Sumit, Yang Guangyi, Deb Arijita, and Hu Ming. 2013. Oral bioavailability challenges of natural products used in cancer chemoprevention. *Progress in Chemistry* 25 (09):1553.

Spurlock III, Charles F, Henry M Gass IV, Carson J Bryant, Benjamin C Wells, Nancy J Olsen, and Thomas M Aune. 2015. Methotrexate-mediated inhibition of nuclear factor κB activation by distinct pathways in T cells and fibroblast-like synoviocytes. *Rheumatology* 54 (1):178–187.

Sultana, Farhath, Manoj Kumar Neog, and Mahaboob Khan Rasool. 2017. Targeted delivery of morin, a dietary bioflavanol encapsulated mannosylated liposomes to the macrophages of adjuvant-induced arthritis rats inhibits inflammatory immune response and osteoclastogenesis. *European Journal of Pharmaceuticsand Biopharmaceutics* 115:229–242.

Tan, Qin-You, Qian Hu, Sheng-Nan Zhu, et al. 2018. Licorice root extract and magnesium isoglycyrrhizinate protect against triptolide-induced hepatotoxicity via up-regulation of the Nrf2 pathway. *Drug Delivery* 25 (1):1213–1223.

Upchurch, Katherine S, and Jonathan Kay. 2012. Evolution of treatment for rheumatoid arthritis. *Rheumatology* 51 (suppl_6):vi28–vi36.

van der Goes, Marlies C, Johannes W Jacobs, and Johannes W Bijlsma. 2014. The value of glucocorticoid co-therapy in different rheumatic diseases-positive and adverse effects. *Arthritis Research & Therapy* 16 (2):1–13.

Visser, Karen, and Desiree van der Heijde. 2009. Optimal dosage and route of administration of methotrexate in rheumatoid arthritis: a systematic review of the literature. *Annals of the Rheumatic Diseases* 68 (7):1094–1099.

Watkins, Rebekah, Ling Wu, Chenming Zhang, Richey M Davis, and Bin Xu. 2015. Natural product-based nanomedicine: recent advances and issues. *International Journal of Nanomedicine* 10:6055.

Wilson, David R, Nan Zhang, Angela L Silvers, Martin B Forstner, and Rebecca A Bader. 2014. Synthesis and evaluation of cyclosporine A-loaded polysialic acid-polycaprolactone micelles for rheumatoid arthritis. *European Journal of Pharmaceutical Sciences* 51:146–156.

Xie, Chunfeng, Xiaoting Li, Jieshu Wu, et al. 2015. Anti-inflammatory activity of magnesium isoglycyrrhizinate through inhibition of phospholipase A2/arachidonic acid pathway. *Inflammation* 38 (4):1639–1648.

Xue, Mei, Zhen-zhou Jiang, Tao Wu, et al. 2012. Anti-inflammatory effects and hepatotoxicity of Tripterygium-loaded solid lipid nanoparticles on adjuvant-induced arthritis in rats. *Phytomedicine* 19 (11):998–1006.

Xuzhu, Gao, Mousa Komai-Koma, Bernard P Leung, et al. 2012. Resveratrol modulates murine collagen-induced arthritis by inhibiting Th17 and B-cell function. *Annals of the Rheumatic Diseases* 71 (1):129–135.

Yokoyama, Masayuki. 2010. Polymeric micelles as a new drug carrier system and their required considerations for clinical trials. *Expert Opinion on Drug Delivery* 7 (2):145–158.

Yousuf, Mohammed Jainuddin, and Elangovan Vellaichamy. 2015. Protective activity of gallic acid against glyoxal-induced renal fibrosis in experimental rats. *Toxicology Reports* 2:1246–1254.

Yukuyama, Nishitani Megumi, Edna Tomiko Myiake Kato, Raimar Lobenberg, and Nadia Araci Bou-Chacra. 2017. Challenges and future prospects of nanoemulsion as a drug delivery system. *Current Pharmaceutical Design* 23 (3):495–508.

Zhang, Feng, Zhiyu Liu, Xijing He, Zhanqi Li, Bin Shi, and Fengmei Cai. 2020. β-Sitosterol-loaded solid lipid nanoparticles ameliorate complete Freund's adjuvant-induced arthritis in rats: involvement of NF-κB and HO-1/Nrf-2 pathway. *Drug Delivery* 27 (1):1329–1341.

Zhang, Li, Junli Chang, Yongjian Zhao, et al. 2018. Fabrication of a triptolide-loaded and poly-γ-glutamic acid-based amphiphilic nanoparticle for the treatment of rheumatoid arthritis. *International Journal of Nanomedicine* 13:2051.

Zhang, Li, Tengteng Wang, Qiang Li, et al. 2016. Fabrication of novel vesicles of triptolide for antirheumatoid activity with reduced toxicity in vitro and in vivo. *International Journal of Nanomedicine* 11:2663.

Zhang, Yifei, Yixian Huang, and Song Li. 2014. Polymeric micelles: nanocarriers for cancer-targeted drug delivery. *Aaps Pharmscitech* 15 (4):862–871.

Zheng, Zhaoling, YanHua Sun, Ziliang Liu, et al. 2015. The effect of curcumin and its nanoformulation on adjuvant-induced arthritis in rats. *Drug Design, Development and Therapy* 9:4931.

Zhu, Shiping, Chunhua Luo, Weifeng Feng, et al. 2020. Selenium-deposited tripterine phytosomes ameliorate the antiarthritic efficacy of the phytomedicine via a synergistic sensitization. *International Journal of Pharmaceutics* 578:119104.

Zou, Xueqing, Yongmei Wang, Cheng Peng, et al. 2018. Magnesium isoglycyrrhizinate has hepatoprotective effects in an oxaliplatin-induced model of liver injury. *International Journal of Molecular Medicine* 42 (4):2020–2030.

3 Phytoarmamentarium to Manage Metabolic Syndrome
Perspectives and Prospects

Pratiksha Sharma, Sanyam Sharma, and Ashish Baldi

3.1 INTRODUCTION

Metabolic syndromes (MetS) are complicated pathological conditions that affect public health on a global scale. MetS significantly raise the risk factor of cardiovascular disease, cognitive impairment, non-alcoholic fatty liver disease (NAFLD), stroke, type-2 diabetes, and obesity. Physical inactivity and an atherogenic diet are the main risk factors for this syndrome's pandemic. The central component of MetS is adiposity or abdominal obesity (Dandona et al., 2005). Risks associated with MetS are depicted in Figure 3.1. Patients typically also have high triglycerides, poor high-density lipoprotein (HDL) cholesterol, high low-density lipoprotein (LDL) cholesterol, abnormal levels, of other lipids, increased fasting blood sugar, obesity indices, and hypertension. With a decrease in fibrinolysis and an increase, in clot formation, there is additional evidence of a vulnerability to thrombotic events in such patients (Morita et al., 2005).

3.2 PHARMACOLOGICAL APPROACHES TO TREAT METABOLIC SYNDROME

MetS is a condition of chronic or long-term low-grade inflammation with serious systemic implications. Diagnosis, management, and care of patients with this syndrome are significant to initiate attempts to provide therapeutics to decrease the risk of associated diseases (Wong, 2005). Actual preventative efforts include lifestyle modifications, with bodyweight reduction, nutrition, and exercise being the three main ones. Pharmacological medicines should be used properly for therapy to lower the specific risk factors. For patients, with minimal or no decrease in risk factors with lifestyle adjustments and preventative measures, pharmacological therapy should be taken into consideration (Deen, 2004). Therefore, most doctors treat each MetS component separately, focusing special attention on the parts that are most receptive to pharmacological therapy. In fact, prescribing a medication to lower blood glucose, blood pressure, and triglycerides is simpler than starting a long-term plan to transform people's lifestyles (better eating habits and initiate more exercise) in the aspirations that they will eventually lose body weight and blood glucose lower blood pressure, triglycerides, fatty liver (Kaur, 2014). Drugs widely administered to treat various MetS are shown in Figure 3.2.

3.3 PHYTOMEDICINE: AN INCREASING TREND

Despite the ongoing global recession over the past few years, the market for herbal medicines has been booming worldwide. The sector is increasingly moving toward a robust value chain structure, with medicinal and aromatic plants and plant derivatives in growth and size at the base of the pyramid. The markets for both product categories of the herbal medicine business are thought to be quite inelastic because it is commonly believed that value addition at these two production phases has been relatively minimal. Even while herbal remedies are frequently used,

 DOI: 10.1201/9781003292692-3

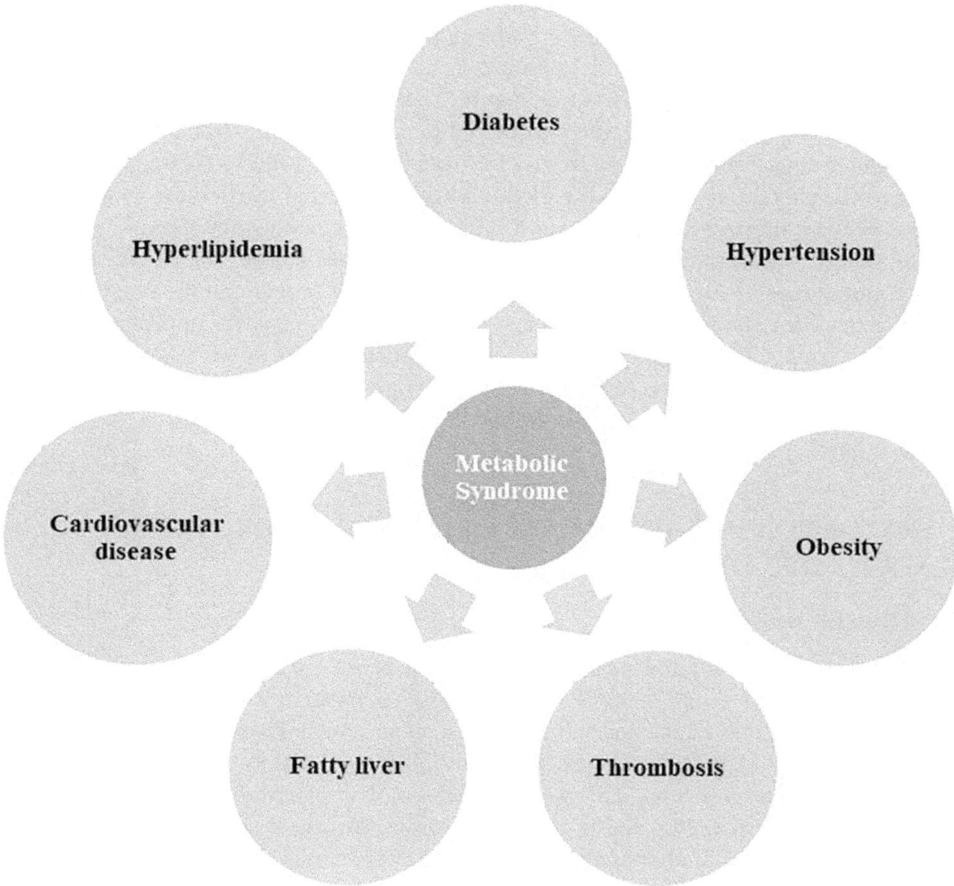

FIGURE 3.1 Risks associated with metabolic syndrome.

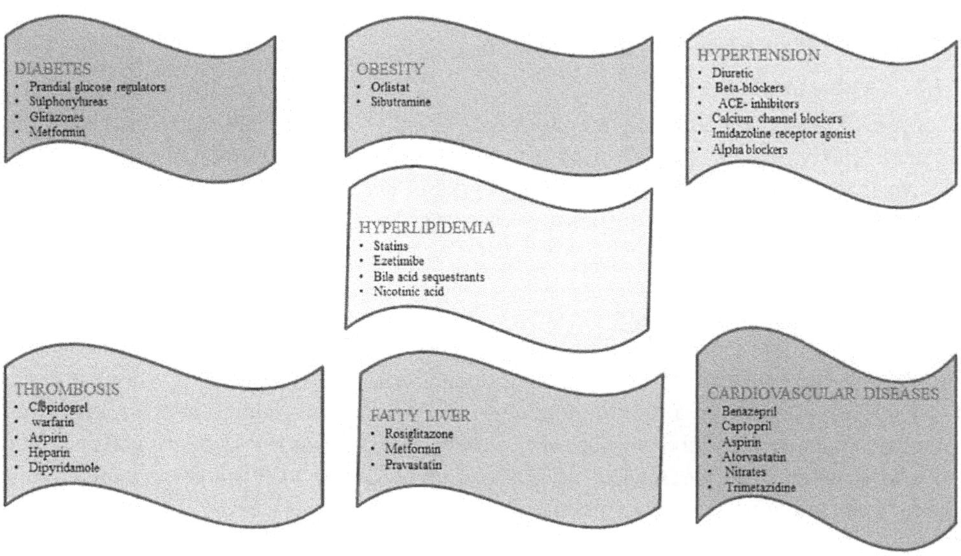

FIGURE 3.2 Major drugs used to treat metabolic syndrome.

they are also sold there as herbal supplements. The market of industrialized countries trades a lot in herbal supplements. In the markets of developed nations, herbal supplements, sometimes known as Western herbalism, are widely traded. Leading nations like the USA allow the selling of herbal supplements as final medical treatments. The magnitude of the sector may be reflected in the global trade of these often known as herbal plants and plant extract. The global trade in medicinal and aromatic plants ranged from 3.1 billion to 3.1 billion US$ between 2014 and 2020, according to the current analysis. In the global market for herbal medicines, monopolistic suppliers have become more prevalent. Similar to this, there is a monopsony trend in a buyer's market due to a high concentration of a small number of international buyers. These trends include upcoming global trade patterns as well as the consistency of herbal commodity prices through time. While the USA, Japan, Hong Kong, and Germany are the top net importers of medicinal and aromatic plants, China, India, and Egypt have reported net positive trade balances in their exports of these plants. The trade research revealed net positive demand from developed nations, which may be the key factor driving the global market for herbal medications and dietary supplements (Behera et al., 2010).

3.4 PHYTOMEDICINES TO MANAGE METABOLIC SYNDROME-ASSOCIATED DISEASES

3.4.1 Diabetes mellitus

A metabolic condition called diabetes mellitus is characterized by hyperglycemia, often known as high blood sugar (Gupta et al., 2008). Polyuria (excessive urine) is a sign of diabetes, along with polydipsia (increased fluid intake and thirst; Paolisso et al., 1993). Diabetes is a condition that can be treated, but when the danger of contracting certain illnesses is not under control diseases, particularly cardiovascular problems, will rise (Goyal & Baldi, 2011). Additionally, obesity is a major risk factor for cardiac disorders and is linked to insulin resistance (Bagher et al., 2005). It is a long-lasting condition that can have major multisystem implications such as ischemic heart disease, stroke, retinopathy, neuropathy, nephropathy, etc. Diabetes-associated complications are shown in Figure 3.3. Comparing individuals with diabetes to healthy people diabetic patients report worse physical functioning and wellbeing, even if they do better than those with other long-term diseases (Raoufi et al., 2015).

3.4.1.1 Herbal Plants Used to Treat Diabetes

3.4.1.1.1 Acacia arabica

Babul or prickly acacia are common names for *A. arabica*. *A. arabica* belongs to the *Fabaceae* family. As per a review, Acacia is widely used to cure cancer, high cholesterol, diabetes, and other conditions. Normal rabbits received a dosage of 2–4 g/kg of body weight of *A. arabica* and *Caralluma edulis* roots, had a hypoglycemic response by starting the release of insulin from the cells of the pancreas (Baldi et al., 2013).

3.4.1.1.2 Allium cepa

Different ether-soluble extracts of dried onion powder as well as its insoluble fractions exhibited antihyperglycemic effects in diabetes induced in rabbits (Mathew & Augusti, 1975). *A. cepa* has hypolipidemic and antioxidant properties. When administered to the alloxan-induced diabetic rats, S-methyl cysteine sulphoxide (SMCS) (200 mg/kg for 45 days), sulfur-containing amino acid from *A. cepa*, decreased blood sugar levels along with lipids in tissues and serum that leads to normal glucose 6-phosphatase, liver hexokinase and HMG Co-A reductase activity (Bordoloi & Dutta, 2014).

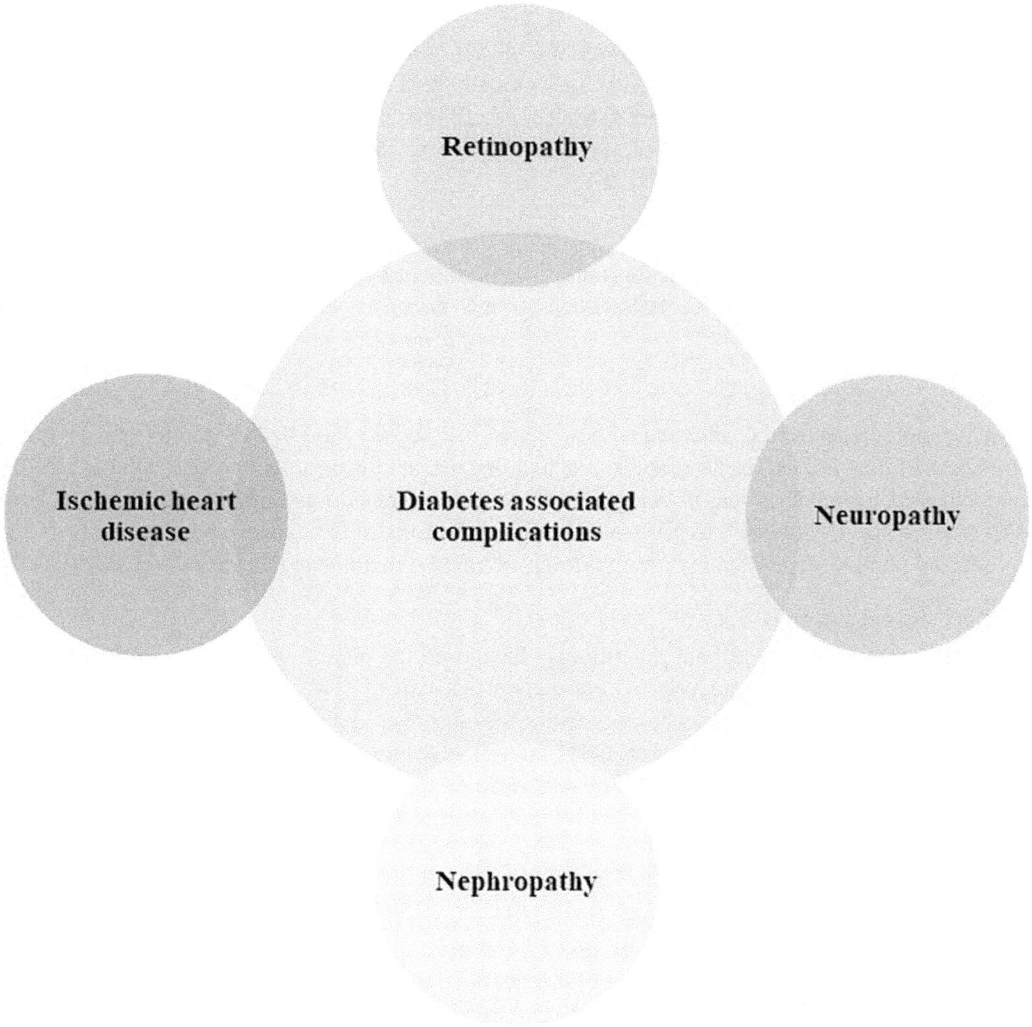

FIGURE 3.3 Major complications associated with diabetes.

3.4.1.1.3 Allium sativum

The *Alliaceae* family includes the herb garlic, *A. sativum*. The antidiabetic activities of ethanolic extract from *A. sativum* were evaluated in rats (both healthy and streptozotocin-induced diabetic). The ethanolic extract of the plant was taken orally for 14 days. The glucose level in the blood, triglycerides, cholesterol, creatinine, urea, uric acid, aspartate aminotransferase (AST), and alanine transaminase (ALT) were decreased (Eidi et al., 2006). Daily oral administration of 100 mg/kg of garlic extract significantly decreased plasma glucose levels by increasing plasma insulin levels (Kazi, 2014).

3.4.1.1.4 Aloe barbadensis

A. barbadensis, the most popular species of aloe, is known by the common name *Aloe vera* (Tanaka et al., 2006). *A. vera* is used to treat diabetes, mouth ulcers, burns, and other conditions (Yagi et al., 1985). Lectins, anthraquinones, and polysaccharides with antidiabetic potential have also been found in *A. vera* (Eshun & He, 2004).

3.4.1.1.5 *Azadirachta indica*

Leprosy, piles, and urinary diseases are treated with *A. indica* (neem), a large tree found across India. Oil extracted from leaves and nuts works locally as a stimulant, antiseptic, and insecticide (Satyavati et al., n.d.). Neem leaf water extract was administered in alloxan-induced diabetic rabbits for 25 days. With respect to control, the blood sugar level decreased significantly in a dose and time-dependent pattern (Al-Awar et al., 2018).

3.4.1.1.6 *Barleria cristata*

B. cristata popularly referred to as Kala Bansa, is a member of the *Acanthaceae* family. *B. cristata* is usually used as a treatment for diabetes, gingivitis, nocturnal ejaculation, wounds, and burns (Quattrocchi, 2012).

3.4.1.1.7 *Cucurbita maxima*

The active ingredients from *C. maxima* with hypoglycemic activity have been isolated from seeds and fruit pulp, and they are useful for diabetic rats and diabetes mellitus humans (Conforti et al., 2005). The polyherbal formulation that is being used by traditional practitioners to treat diabetes mellitus (Baldi et al., 2010). *C. ficifolia* was given at a dosage of 4 mL/kg of fruit juice. Consumption of 1–2 g/kg powdered pumpkin considerably lowered levels of glucose in diabetic rats (Asgary et al., 2011).

3.4.1.1.8 *Eugenia jambolana*

Over many generations, traditional practitioners have used the plant *E. jambolana*, which belongs to the *Myrtaceae* family, extensively to cure diabetes (Satyavati et al., n.d.). Jamun helped significantly in the treatment of diabetes before insulin was discovered and is a key component of some alternative medical systems. A very well-reported hypoglycemic or blood sugar-lowering, effects of *E. jambolana* exist (Achrekar et al., 1991).

3.4.1.1.9 *Helicteres isora*

It grows rapidly in forests across India. The extract of *H. isora* fruits in hot water exhibited significant antioxidant and modest antidiabetic effects of *H. Isora* (Suthar et al., 2009). It exhibited glucose uptake activity at a dosage of 200 mg/mL and was shown to be as active as metformin and insulin (Gupta et al., 2009). The ethanolic extract may also be used to treat type 2 diabetes since it exhibits insulin-sensitizing and hypolipidemic properties (Chakrabarti et al., 2002).

3.4.1.1.10 *Lepidium sativum*

Locals refer to the natural shrub *L. sativum* as "habarrachad." It is a member of the *Brassicaceae* family. *L. sativum* seeds are utilized to cure diabetes mellitus (Eddouks et al., 2002). A single injection of aqueous *L. sativum* extract (20 mg/kg) in healthy rats resulted in a considerable drop in blood sugar levels within the first hour of treatment (Eddouks et al., 2002, 2003).

3.4.1.1.11 *Mangifera indica*

Mango belongs to the *Anacardiaceae* family. When compared to an oral dosage of chlorpropamide, the aqueous extract of *M. indica* reduced blood sugar levels in mice with normal glycemia and hyperglycemia brought on by glucose but had no impact on streptozotocin-induced diabetic mice. The outcome demonstrated that the aqueous extract of *M. indica* exhibited significant hypoglycemic effects (Aderibigbe et al., 2001).

3.4.1.1.12 *Myrcia bella*

M. bella, from the *Myrtaceae* family, is used to cure diabetes mellitus, hypertension, and cholesterolemia. Diabetic mice treated with *M. bella* fraction (300 or 600 mg/kg body weight) as evident from reduced fasting blood glucose (Vareda et al., 2014).

3.4.2 Hypertension

Hypertension is defined by the WHO as having a blood pressure level of more than 140/90 mm mercury (Lifton, 1996). Due to its asymptomatic nature and role as a prominent risk factor for cardiovascular-related deaths, it has acquired the name "silent killer" (Gavras, 2009). Hypertension affected 26.4% population of the world in 2000, and by 2025, that percentage is expected to have increased by 60% (Kearney et al., 2005). Hypertension is mainly of two types:

i. Primary or essential hypertension (90%–95%) which has an unknown source for raised blood pressure directly, and
ii. Secondary hypertension (5%–10%) which is triggered by underlying conditions affecting the kidneys, heart, arteries, and endocrine system, such as renal damage, pheochromocytoma, muscle problems, etc. (Mayo, 2008). In India, cardiovascular disease has become a significant public health issue (Schrader, 2009). Numerous studies have shown that using properly selected herbal medicines and nutritional supplements can help to decrease blood pressure and enhance the general health of the heart, arteries, and complete cardiovascular system (Chan, 2003).

3.4.2.1 Herbal Plants Used in the Treatment of Hypertension

3.4.2.1.1 A. sativum

Garlic powder is used in the treatment of hypertension. Garlic extracts have antihypertensive effects and significantly lower diastolic and systolic blood pressure (Bongiorno et al., 2008; Silagy & Neil, 1994).

3.4.2.1.2 Anacardium occidentale

Africa uses the herb *A. occidentale,* belonging to the *Anacardiaceae* family, extensively to treat many disorders. It is frequently utilized to cure hypertension (Tchikaya et al., 2011). Utilizing the isolated organ method, *A. occidentale* leaf extract demonstrated *in vitro* antihypertensive properties (Nugroho et al., 2013).

3.4.2.1.3 Acorus calamus

A. calamus (Vacha) a plant from the *Acoraceae* family, is used to cure obesity, hypertension, and other conditions (Quattrocchi, 2012). Rats given a dose of 250 mg/kg b.w. extract from rhizomes had lower diastolic and systolic blood pressure, mean blood urea nitrogen, plasma renin activity, and creatinine levels (Patel et al., 2012).

3.4.2.1.4 Aloe barbadensis

*A. barbadensis*is a member of the *Liliaceae* family. The plant aloe is used to treat hypertension, diabetes, obesity, and cardiovascular disorders. In rats with tail clips, aloe extract at 50 mg/kg body reduced systolic blood pressure (Shakib et al., 2019).

3.4.2.1.5 Blumea balsamifera

Locally known as Kukur Sunga, *B. balsamifera* is an *Asteraceae* plant, used to cure hypertension, obesity, and other conditions (Bhuiyan et al., 2009).

3.4.2.1.6 Camellia sinensis

C. sinensis (green tea) is a high source of polyphenols particularly flavonoids, effective in treatments for several types of cancer, arthritis, and cardiovascular diseases (Higdon & Frei, 2003). Crude extract of *C. sinesis* produced a dose-dependent decrease in blood pressure and minor bradycardia in anesthetized rats (Gilani & Aftab, 1994).

3.4.2.1.7 *Coriandrum sativum*

A member of the *Umbelliferae* family, coriander (*C. sativum*) has both culinary and therapeutic uses. The fruits are used as astringents, aromatics, anthelmintics, emollients, anti-inflammatory, carminative, anti-bilious, digestive, appetizer, constipating, diuretic, antipyretic, stimulants, aphrodisiacs, refrigerants, tonics, expectorants, anodyne, and in the cure of dyspepsia and hyperlipidemia in various traditional systems of medicine (Hosseini et al., 2007).

3.4.2.1.8 *Hibiscus sabdariffa*

Obesity and hypertension are both reported to be treated by *H. sabdariffa.* Calyx extract (500 and 1,000 mg/kg) of this plant improved urine production and serum uric acid while lowering blood pressure (Onyenekwe et al., 1999).

3.4.2.1.9 *Rauwolfia serpentina*

R. serpentinia is an evergreen herb from *Apocynaceae* family (Endress & Bruyns, 2000). It is widely used in the treatment of hypertension (Freis, 1956).

3.4.3 Obesity

The increasing intake of high-calorie meals is the primary cause of obesity, which is regarded as a global epidemic and an energetic imbalance. Other variables include the lack of physical activity, socioeconomic and environmental changes, especially the rise in purchasing power and education levels, as well as the influence of others on food consumption (Cohen-Cole & Fletcher, 2008). Obesity is linked to dyslipidemia, diabetes, osteoarthritis, and some kinds of cancer, including endometrial, colon and breast cancer (Klein et al., 2007; Weir, 2009). In addition to having clinically apparent high blood pressure, glucose intolerance, insulin resistance, and a high body mass index, obese people also have these risk factors for heart disease (Kopelman, 2000).

3.4.3.1 Herbal Plant Used in the Treatment of Obesity

3.4.3.1.1 *A. sativum*

A. sativum, belonging to the family, *Alliaceae* is used to treat diabetes, hypertension, obesity, and other conditions. It is also useful for weight loss (Sharifi et al., 2010). Administration in mice fed with a high-fat diet (HFD), cause obesity and with the use of *A. sativum* stem extract at a dose of 250 mg/kg insulin resistance obesity and oxidative damage were decreased (Kim et al., 2013).

3.4.3.1.2 *Bixa orellana*

Small trees in the *Bixaceae* family, including the *B. Orellana*, is used in the treatment of diabetes, snakebite, and weight loss products (Chandel et al., 2014).

3.4.3.1.3 *Bombax ceiba*

The stem bark of the *B. ceiba*, a member of the *Bombacaceae* family, possesses substantial anti-obesity activity against experimental obesity, presumably as a result of modulating PTP-1B and FAS signaling in Wistar rats because of the existence of lupeol, and active flavonoids respectively (Gupta et al., 2013).

3.4.3.1.4 *Ligustrum robustum*

The leaves of *L. robustum* were frequently used to treat hyperlipidemia and obesity. To examine the anti-obesity effect phenylpropanoid glycosides from the leaves of plant were given to C57BL/6J mice, fed with fatty diet (Yang et al., 2015).

3.4.4 Hyperlipidemia

A group of metabolic variables that raise the risk of atherosclerotic cardiovascular disease were identified by the National Cholesterol Education Program Adult Treatment Panel III recommendations in 2001 (Schuppan et al., 2010). In sensitive people who gain weight, tiny dense LDL, (the most atherogenic subfraction of LDL), has been denoted as a critical factor in relation to raised triglyceride and low HDL cholesterol (HU, 2003).In developed nations, hypercholesterolemia and coronary heart disease (CHD) are common and CHD is currently the leading cause of mortality worldwide (Hasani-Ranjbar et al., 2010).

3.4.4.1 Herbal Plants Used in the Treatment of Hyperlipidemia

3.4.4.1.1 A. sativum

A. sativum is used to treat conditions of the digestive system, high blood pressure, hypertension, diabetes, asthma, and cardiovascular problems (Shirzad et al., 2011). A typical daily intake of 900 mg of garlic powder, or around 12 mg of alliin, contains 1.3% alliin. Alliin, (10 mg) which corresponding to 4–5 mg of allicin, is the recommended dose that has been found through clinical tests. The benefits of fresh garlic can be obtained by consuming 1–7 g of alliin-free garlic everyday (Koscielny et al., 1999).

3.4.4.1.2 Apium graveolens

The aqueous extract of *A. graveolens* (celery) considerably reduced the LDL and total cholesterol (TC) measures in rats that had been given an 8-week HFD to cause hyperlipidemia. The systolic blood pressure was dramatically reduced by an aqueous celery extract (Leung, 1980).

3.4.4.1.3 Asparagus racemosus

A. racemosus is a member of the *Asparagaceae* family. An aqueous extract of its roots has hypocholesterolemic action by raising catalase, superoxide dismutases, and ascorbic acid levels in hypercholesterolemic rats (Visavadiya & Narasimhacharya, 2011).

3.4.4.1.4 Bauhinia variegata

B. variegata, often known as kachnar, belonging to *Fabaceae* family. At 200 and 400 mg/kg b.w. doses, *B. variegata* root-water extract and ethanolic fraction were found effective in lipid lowering (Rajani & Ashok, 2009).

3.4.4.1.5 Cassia auriculata

C. auriculata, is a member of *Pinaceae* family (Pari & Latha, 2002).This herb is used to treat hyperlipidemia, diabetes, and high blood pressure. At 450 mg/kg/b.w. its ethanolic extract lowered the lipid profile in liver tissue and found comparable to standard drug (lovastatin). There was no significant difference in both liver lipid profiles, and serum in Group I and Group II rats. In both groups profiles of serum and lipids in liver remained almost similar (Chakrabarti et al., 2002).

3.4.4.1.6 Commiphora mukul

C. mukul, is a type of sticky gum. In Indian traditional medicine, this substance is blended with other plants and tested to treat a number of diseases including oral infections, rheumatic aches, obesity, diabetes, digestive issues, and menstrual irregularities. The recommended dosage is typically 100 mg/day, depending on the blood lipid level and guggul concentration (Satyavati, 1988).

3.4.4.1.7 Curcuma longa

Curcumin, a substance derived from the spice turmeric, which belongs *Zingiberaceae* family, has long been utilized as an immunomodulator in conventional and Ayurvedic medicine (Mollazadeh et al., 2019). Curcumin has cholesterol-lowering properties at all levels of the lipid-lowering process, including lowering total and LDL cholesterol (by 11%), raising HDL cholesterol (by 29%), and reducing lipid peroxidation, which prevent LDL and cholesterol from being over oxidized in rats (Akram et al., 2010).

3.4.4.1.8 Emblica officinalis

E. officinalis, also known as Amla and Indian gooseberry, is used to treat a number of ailments including diabetes, atherosclerosis, and liver damage. It also has a very effective hypocholesterolaemic effect. In both healthy and cholesterol-induced hypercholesterolemic rabbits, *E. officinalis* was found to significantly lower serum aortic cholesterol, TC, and hepatic cholesterol. It had no effect on serum triglyceride levels (Augusti et al., 2001).

3.4.5 Thrombosis

In a variety of cardiovascular disorders, thrombosis is one of the major pathogenic causes of morbidity and mortality (American Heart Association, 2000). The adhesion of circulating platelets to the damaged blood vessel walls causes the formation of thrombus (Ofosu et al., 1995). When exposed to subendothelial surfaces at vascular damage sites, fluid within the vasculature can rapidly clot. Both thrombosis and hemorrhage are often prevented by a careful balance between coagulation and fibrinolysis. Vaso-occlusive events are a major cause of death and involve serious vascular illnesses like myocardial infraction and ischemic stroke, unstable angina. Thrombosis is the pathological formation of a "hemostatic" plug in the blood vessels in absentia of bleeding is known as thrombosis. Blood must remain of coagulation in order for thrombosis to occur. Blood vessels such as arteries or veins may develop thrombi, which are made up of platelet aggregates, fibrin, and entrapped red blood cells (Rosendaal, 1999).

3.4.5.1 Herbal Plants Used in the Treatment of Thrombosis

3.4.5.1.1 A. sativum

Garlic, has been linked to lower blood pressure, platelet aggregation, and anti-atheromatous effects (Awe et al., 1998). Using a butterfly needle, 0.1 mL of a platelet-aggregating solution containing adenosine di-phosphate and adrenaline at doses of 1.8 mg/mouse and 260 mg were injected into the mice's tail veins at a rate of 20 mL/s. *A. sativum* (100 and 200 mg/kg) was tested in mice for evaluating selected antithrombotic drug's capacity to shield the animals from a lethal or paralyzing effect and found effective (Majewski, 2014).

3.4.5.1.2 Angelica sinensis

A popular traditional Chinese medication, *Radix A. sinensis* (RAS); also called as *Danggui* in Chinese, has long been a part of several traditional Sino-Japanese herbal formulas for the treatment of ischemia conditions, and also clinically utilized to cure female menstruation problems by replenishing the blood and energizing its circulation. At dosages of 100 or 200 mg/kg, sodium ferulate, an aqueous soluble component of RAS, was shown to significantly reduce platelet aggregation. The antithrombotic effect of lymphocyte immune globulin was assessed in a rat model of arteriovenous thrombosis. For 3 days, rats were administered lymphocyte immune globulin, aspirin (40 mg/kg), or vehicle by oral administration (Wu & Hsieh, 2011).

3.4.5.1.3 C. sinensis

Green tea is extracted from *C. sinensis*, an evergreen shrub or small tree belonging to the family *Theaceae*. Epigallocatechin gallate (EGCG), is found in green tea, and catechin derivative from the flavonoid subgroup, is found in green tea. One of the studies assessed the effects of EGCG on thrombus formation at an anastomotic site as well as its impact on vessel tips applied in microvascular procedures in adult Wistar rats (Tabassum & Ahmad, 2011).

3.4.5.1.4 Cyamopsis tetragonoloba

The seeds of *C. tetragonoloba*, member of the *Fabaceae* family, are utilized for various pharmacological action having Saponins, polyphenols, and gum remnants as chemical components. Anticoagulant and antibacterial activity are examples of exhibited pharmacological activity by this plant (Mestechkina et al., 2008). All fractions examined the capacity for platelet aggregation. *In vitro* platelet aggregation was demonstrated by all extracts. *C. tetragonoloba* should not be used in individuals receiving anticoagulant treatment since it may increase the anticoagulant effects of other drugs (Gholkar et al., 2020).

3.4.5.1.5 Cinnamomum cassia

Chinese cassia, commonly referred to as *C. cassia*, is a member of the *Lauraceae* family. Typically, this plant's bark is utilized for therapeutic and condiment purposes. The plant's extract demonstrated potent antiplatelet and anticoagulant properties (Fakhrudin et al., 2020).

3.4.5.1.6 Codium fragile

The common names for *C. fragile* are green sea finger, dead man finger. These are all varieties of seaweed from the *Codiaceae* family. Using *in vivo* arterial thrombus formation caused by ferric chloride, the antithrombotic impact was examined in mice. The ability of *C. fragile* was assessed using the transection tail bleeding time. The therapeutic effect was observed due to inhibition of primary hemostasis (Kim et al., 2021).

3.4.5.1.7 Curcuma longa

Turmeric, also called as *C. longa*, which belongs to the family of *Zingiberaceae*. Turmeric's primary component is curcumin (diferuloylmethane; Ngo et al., 2019). In experimental animal models, curcuma oil has been found possess have neuroprotective, anticancer, antioxidant, and anti-hyperlipidemic properties. Curcuma was given orally to Wistar rats that had been (300 mg/kg) given for 12 weeks to investigate its potential for protection. Blood sugar, insulin, platelet aggregation, thrombosis, and inflammatory markers were measured and effective control of selected markers was observed upon treatment with curcuma (Prakash et al., 2011).

3.4.5.1.8 Filipendula ulmaria

F. ulmaria is a member of the *Rosaceae* family. The components that are utilized include the leaves, roots, and flowers. *F. ulmaria* was tested to treat peptic ulcers, arthritis, pneumonia, inflammation and diphtheria and has anti-inflammatory and antithrombotic pharmacological actions. Flavonoids, tannins vitamins, phenolic acids, polyphenols, ellagitannins, rugosins, and methyl gallate 3-O-beta-glucoside are some chemical components present in this herb. Heparin-like substances found in *F. ulmaria* bind to the protein in complex form. After being given to animals, it exhibited anticoagulant and anti-fibrinolytic properties (Kudriashov et al., 1990).

3.4.5.1.9 Gingko biloba

The dioecious tree *G. biloba*, belonging to *Ginkgoaceae* family, provides the ginkgo leaves. *G. biloba* as a herbal supplement, has been shown to have antiplatelet antioxidant, anti-apoptotic, anti-inflammatory, fibrinolytic, and vasoactive characteristics (Sasaki et al., 2002).

3.4.6 Cardiovascular disease

Multiple significant metabolic risk factors for CHD and stroke are frequently present in people with these illnesses. A group of metabolic illnesses is responsible for the MetS (Kannel et al., 1976). The main reason for mortality in the US is cardiovascular disease (CVD). In fact, 38% of adults with CVD have risk factors like diabetes, high blood pressure, and elevated serum lipid levels. These risk factors come together to form a single mechanism including inflammation and oxidation in the wall of the artery which over a period result in the development of the recognizable fatty-fibrous. The main contributor to cardiovascular disease is atherosclerosis, which is also the source of the heart failure pandemic that is currently on the rise (Naman et al., 2022; Skålén et al., 2002).

3.4.6.1 Herbal Plants Used in the Treatment of Cardiovascular Disease

3.4.6.1.1 Adonis aestivalis

A. aestivalis is used as a cough suppressant, diuretic, and spasmolytic medication. Adonis species are utilized to make drugs that stimulate cardiac function (Al-Snafi, 2016). One of the cardenolides, obtained from *A. aestivalis*, is strophenthidin aglycone. Similar to other plants that contain cardiac glycosides, these substances have a direct impact by blockage of the sodium-potassium adenosine triphosphatase enzyme system pump. This raises vagal tone, which slows the pace of depolarization of the sinoatrial node. Intoxication causes ventricular arrhythmias, bradycardia, different degrees of atrioventricular block, and ventricular fibrillation to appear on electrocardiograms (Al-Snafi, 2016).

3.4.6.1.2 A. sativum

A. sativum, or garlic has long been valued as a spice and a common treatment for a variety of illnesses and physiological conditions. Garlic is primarily used to prevent and treat cardiovascular diseases by lowering cholesterol and blood pressure, as an antimicrobial, and cancer preventative (Singh & Singh, 2008).

3.4.6.1.3 Hydrastis canadensis

In addition to being a common alkaloid to be found and being used in the management of CHF, berberine is also thought to have antiarrhythmic and antihypertensive properties. In a recent placebo-controlled study, individuals with substantial ventricular ectopy and heart failure were given berberine 1.2–2.0 mg/day for up to 1 year. When compared with the placebo control study it was determined that there was a gradual decrease in mortality and ventricular ectopy (Zeng et al., 2003).

3.4.6.1.4 Rauwolfia serpentina

Also known as snakeroot *R. serpentina* being a source of reserpine alkaloids, is used for the cure and management of hypertension and psychoses (Sharma et al., 2006).

3.4.7 Herbal Plants Used in the Treatment of Fatty Liver

3.4.7.1 Fatty Liver

Human health and disease etiologic are directly impacted by dietary choices and lifestyle. Around the world, chronic diseases, such as fatty liver disease, are a widespread issue. As the incidence of type 2 diabetes mellitus and obesity increases, NAFLD is quickly evolving into a severe worldwide health issue (Schuppan et al., 2010). Specifically, among people who do not drink alcohol excessively amounts of alcohol, the term "NAFLD" denotes a plethora of liver illnesses, from hepatic steatosis to non-alcoholic cirrhosis and steatohepatitis (Bhatia et al., 2012). There isn't a pharmaceutical agent being used to treat NAFLD that has received regulatory approval yet. For starters, lifestyle adjustment, including calorie restriction and increased physical exercise, is the suggested intervention for NAFLD (Angulo, 2002).

3.4.7.2 *Adenosma glutinosum*

Adenom species, one of the indigenous genera, of the *Scrophulariaceae* family, have been and continue to be employed in traditional Vietnamese medicine for Jaundice in viral hepatitis, oliguria (low urine output), fever, ophthalmoplegia and dyspepsia in pregnant women (Dũng et al., 1996).

3.4.7.3 *Curcuma longa*

A perennial herb with rhizomes, *C. longa* is a tropical plant, and the dried, powdered rhizomes are used to make the spice, turmeric. *C. longa* exhibited antioxidant, anticancer, wound healing, antibacterial, anti-inflammatory, lipid-reducing, chemopreventive, gastroprotective, and immunomodulatory properties and the potential to treat fatty liver (Kocaadam & Şanlier, 2017).

3.4.7.4 *Picrorhiza kurroa*

This herbal medicine, a member of the *Scrophulariaceae* family, is administered to treat liver and bronchial issues (Stuppner & Wagner, 1989). *P. kurroa* is used as a hepatoprotective, anti-inflammatory, antidiabetic, immunomodulator, antimicrobial and antioxidant agent (Kant et al., 2013).

3.5 FUTURE PROSPECTS OF PHYTOPHARMACEUTICALS IN THE TREATMENT OF METABOLIC DISORDERS

According to data from the WHO, 25% of currently sold drugs have an herb origin. Different synthetic medications are chemical analogs of elements found in plants. Phytomedicines' stability, drug release, effectiveness, and efficiency have all been shown to be enhanced due to recent advances in nanotechnology research. Currently, popular plant-based nanoparticles related to nanocarriers include nanostructured lipid carriers, metallic nanoparticles, liposomes, polymeric nanoparticles, and niosomes. Site-specific and target action are significant benefits of herbal medicine nanoformulations (Rajasekaran et al., 2006). Berberine has been improved with nanoformulations made utilizing liquid polymers and chitosan-coated liposomes to treat hyperglycemia. When compared to the administration of natural berberine, oral nanoformulations minimized insulin resistance and body weight gain in type 2 diabetic mice (Rajasekaran et al., 2006). Numerous herbal and synthetic medications are used to treat diabetes, but their effectiveness and ability to reach the target site are limited by a lack of required pharmacokinetic qualities. Target-based medication delivery strategies have evolved and are used for metabolic diseases in order to provide effective therapy. Currently, nanosuspensions, nanostructured lipid carriers, solid–liquid nanoparticles, metal nanoparticles, chitosan nanoparticles, polymeric nanoparticles, and phytochemical and plant extract-based formulations are used (Paul et al., 2021). In addition, these nanoherbal formulations result in a dearth of clinical trials comparing the *in vivo* and *in vitro* methods. The affinity for the targeted receptor, the ambient pH, the penetration behavior, the efficient method to decrease the level of blood sugar levels, and the removal of nanocarriers from the body should all be taken into consideration when developing or creating nano-herbal formulations. As per the FDA's recommendations for clinical trials, toxicity-based investigations are now essential for nano-herbal medicines. By demonstrating the positive results of the clinical trials, these in-depth investigations may in the future examine the application of natural-based nanoformulations for the efficient treatment of diabetes (Paul et al., 2021). A vasodilatory effect was seen following the administration of *Psiadia punctulate* self-nanoemulsifying drug delivery system (SNEDDS) formulations in comparison to a hypertensive rat model produced by angiotensin. More phytochemical components have been identified as a result of ongoing improvements in analytical techniques, which may be used to locate previously unidentified phytochemical components that may be beneficial in the management of various illnesses and disorders. With increased pharmacological qualities, the present phytopharmaceutical research trends provide a promising future for treating metabolic disorders (Khan & Ahmad, 2019). Hence standardization of herbal extracts in terms of phytomarkers and established evidence-based pharmacology along with clarity of mono and poly-therapy possibilities with mechanistic understandings are need of the hour for further acceptance of phytopharmaceuticals as effective armamentarium to cure treat and manage a diverse range of diseases (Singh et al., 2021) (Figure 3.4).

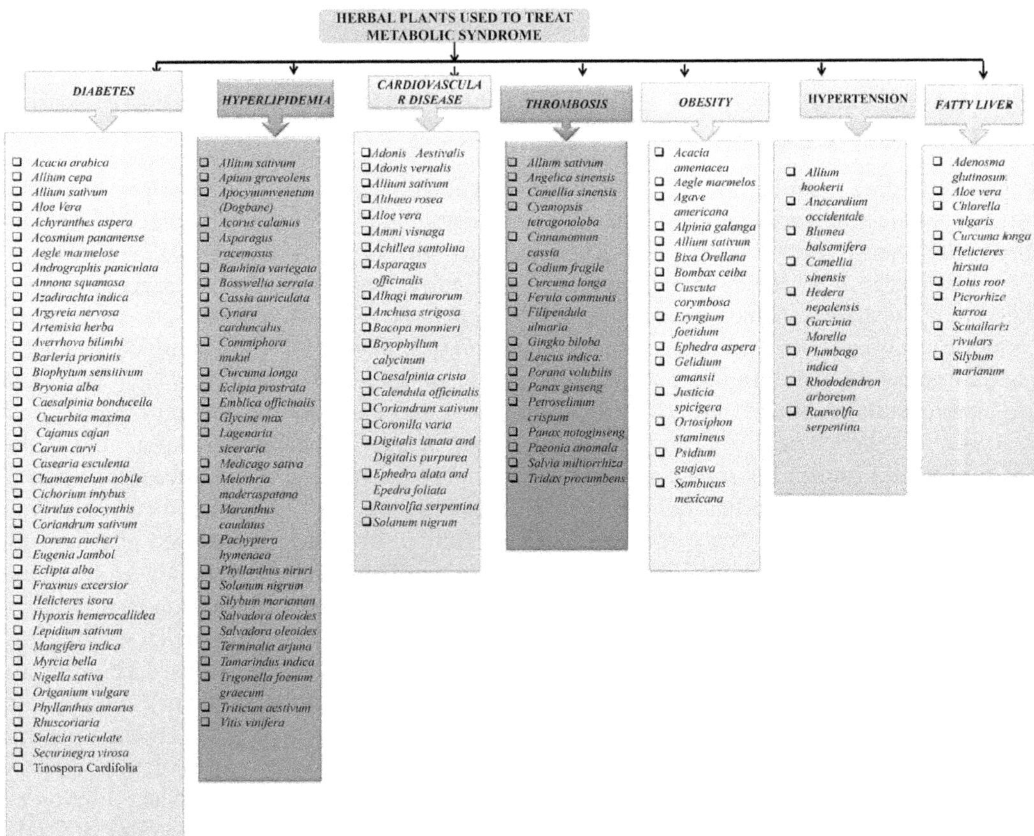

FIGURE 3.4 Herbal plants used to treat metabolic syndrome.

Herbal Plants Used in the Treatment of Diabetes Mellitus

Acacia arabica

Allium cepa

Allium sativum

Aloe barbadensis

Acosmium panamense

Achyranthes aspera

Aegle marmelose

Andrographis paniculata

Annona squamosa

Azadirachta indica

Argyreia nervosa

Artemisia herba

(*Continued*)

(*Continued*)

Herbal Plants Used in the Treatment of Diabetes Mellitus

Averrhoa bilimbi

Barleria prionitis

Biophytum sensitivu

Bryonia alba

Cucurbita maxima

Caesalpinia bonducella

Cajanus cajan

Carum carvi

Casearia esculenta

Chamaemelum nobile

Cichorium intybus

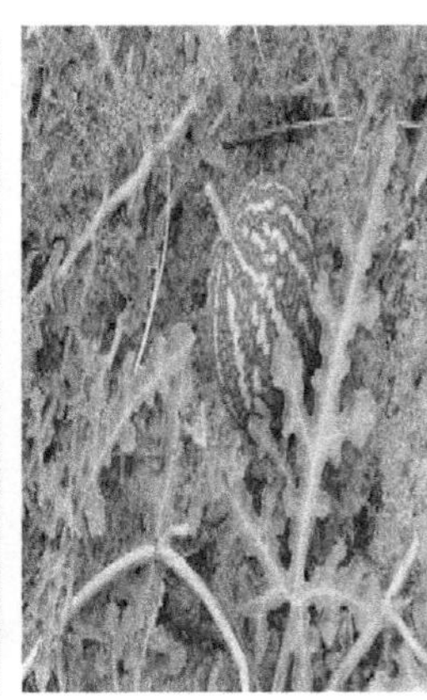

Citrulus colocynthis

(*Continued*)

(*Continued*)

Herbal Plants Used in the Treatment of Diabetes Mellitus

Coriandrum sativum

Dorema aucheri

Eugenia jambolana

Eclipta alba

Fraxinus excersior

Ginseng

Gymnema sylvestre

Helicteres isora

Hypoxishemero callidea

Lagerstroemia speciosa

Lepidium sativum

Mangifera indica

(*Continued*)

(*Continued*)

Herbal Plants Used in the Treatment of Diabetes Mellitus

Myrcia bella

Momordica charantia

Nigella sativa

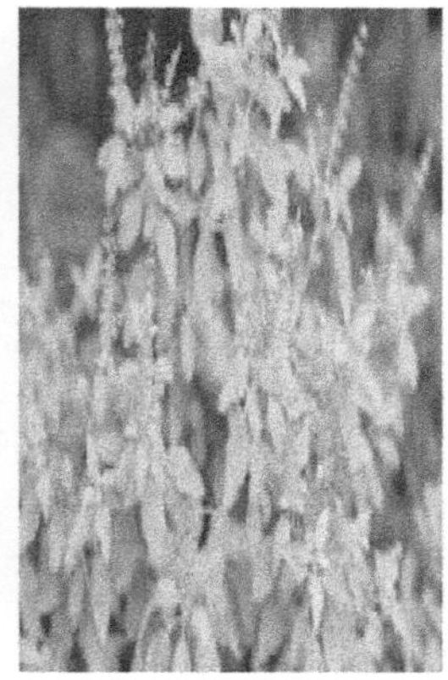

Ocium sanctum

Origanium vulgare

Phyllanthus amarus

Genus Pinus

Rhus coriaria

Silybum marianum

Salacia reticulate

Securinegra virosa

Tinospora cardifolia

Herbal Plants Used in the Treatment of Thrombosis

Allium sativum

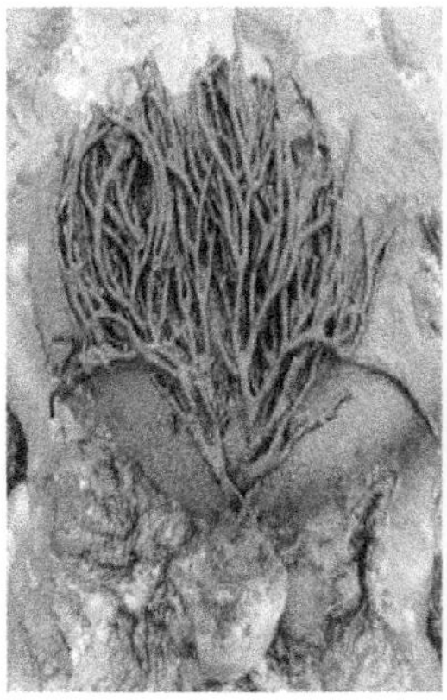

Codium fragile

Curcuma longa

Cinnamomum cassia

Camellia sinensis

Ferula communis

Fili pendulaulmaria

Gingko biloba

Leucus indica

Paeonia anomala

Panax notoginseng

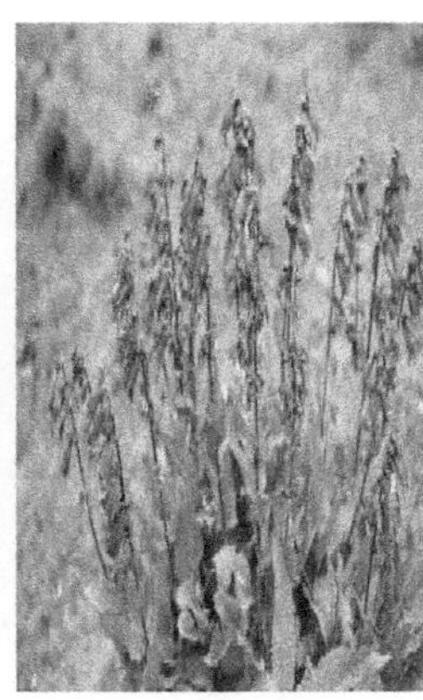

Salvia miltiorrhiza

(*Continued*)

(Continued)

Herbal Plants Used in the Treatment of Thrombosis

Porana volubilis

Panax ginseng

Petroselinum

Tridax procumbens

Tulbaghia violaceae

Herbal Plants Used in the Treatment of Fatty Liver

Adenosma glutinosum

Aloe barbadensis

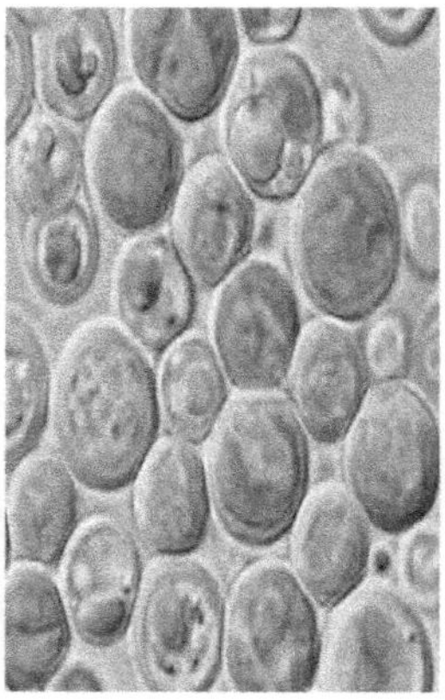

Chlorella vulgaris

Curcuma longa

(*Continued*)

(Continued)

Herbal Plants Used in the Treatment of Fatty Liver

Helicteres hirsuta

Lotus root

Picrorhiza kurroa

Scutallaria rivularis

Silybum marianum

Herbal Plants Used in the Treatment of Hypertension

Allium hookerii

Anacardium occidentale

Acorus calamus

Aloe barbadensis

(*Continued*)

(*Continued*)

Herbal Plants Used in the Treatment of Hypertension

Blumea balsamifera

Coriandrum sativm

Camellia sinensis

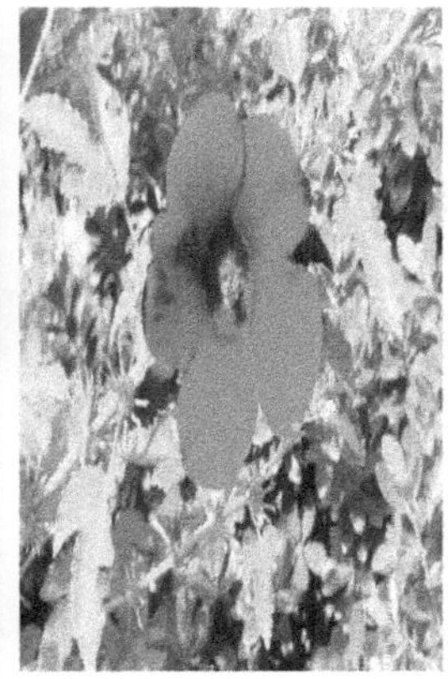

Hibiscus sabdariffa

Hedera nepalensis

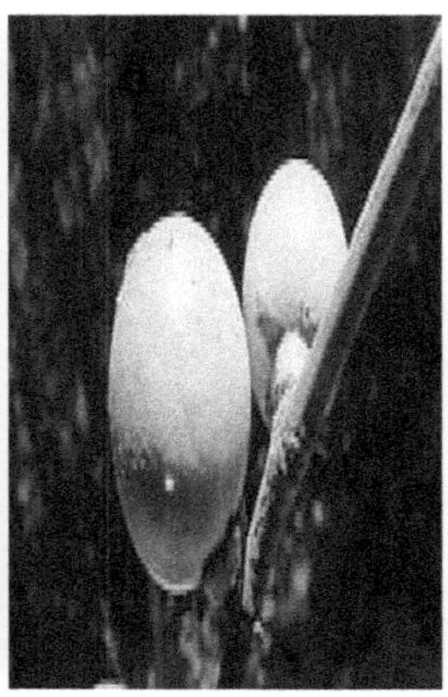

Garcinia morella

Plumbago indica

Rhododendron arboreum

Rauwolfia serpentina

Herbal Plants Used in the Treatment of Hyperlipidemia

Allium sativum

Apium graveolens

Apocynum venetum

Acorus calamus

Asparagus racemosus

Bauhinia variegata

Bosswellia serrata

Cassia auriculata

Cynara cardunculus

Commiphora mukul

Curcuma longa

Eclipta prostrata

(*Continued*)

(*Continued*)

Herbal Plants Used in the Treatment of Hyperlipidemia

Emblica officinalis

Glycine max

Lagenaria siceraria

Medicago sativa

Melothria maderaspatana

Maranthus caudatus

Pachyptera hymenaea

Phyllanthus niruri

Solanum nigrum

Silybum marianum

Salvadora oleoides

Terminalia arjuna

(*Continued*)

(Continued)

Herbal Plants Used in the Treatment of Hyperlipidemia

Tamarindus indica

Trigonellfoenum graecum

Triticum aestivum

Vitis vinifera

Herbal Plants Used in the Treatment of Obesity

Acacia amentacea

Aegle marmelos

Agave americana

Alpinia galanga

Allium sativum

Bixa orellana

Bombax ceiba

Cuscuta corymbosa

(*Continued*)

(*Continued*)

Herbal Plants Used in the Treatment of Obesity

Eryngium foetidum

Ephedra aspera

Gelidium amansii

Justicia spicigera

Ortosiphon stamineus

Psidium guajava

Sambus mexicana

Herbal Plants Used in the Treatment of Cardiovascular Disease

Adonis Aestivalis

Adonis vernalis

Allium sativum

Althaea rosea

(*Continued*)

(*Continued*)

Herbal Plants Used in the Treatment of Cardiovascular Disease

Aloebarbadensis

Ammi visnaga

Achillea santolina

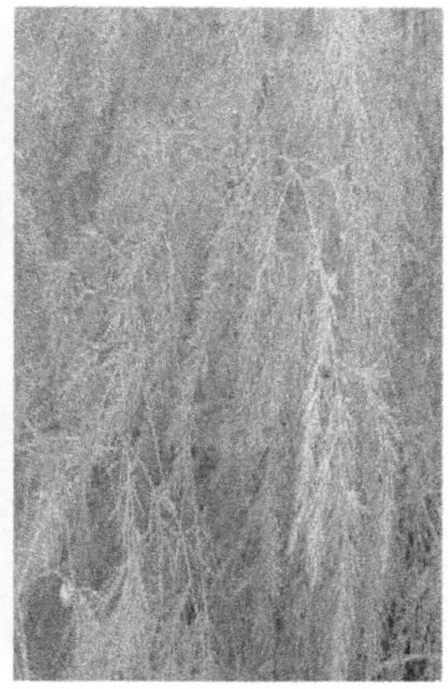

Asparagus officinalis

Alhagi maurorum

Anchusa strigosa

Bacopa monnieri

Bryophyllum calycinum

Hydrastis canadensis

Calendula officinalis

Coriandrum sativum

Coronilla varia

(*Continued*)

(Continued)

Herbal Plants Used in the Treatment of Cardiovascular Disease

Digitalis lanata

Ephedra alata

Ocimum sanctum

Rauvolfia serpentina

Solanum nigrum

3.6 CONCLUSION

MetS is a complicated pathophysiological condition linked to cancer, CVS diseases, diabetes mellitus, and obesity. Scientific signs suggests that nutrition, along with inactive behavior and other risk factors, contributes to the onset of MetS. Medicinal herbs also provide a wide range of incredibly potent medications that are used to cure a broad range of diseases. Some of theplant-derived bioactive substances have demonstrated preventative or therapeutic benefits against MetS associated conditions. Present chapter will survey provide an understanding to the readers about various herbal drugs explored for MetS and related pathological conditions to explore further in order to find an effective therapeutic armamentarium for the same. In-depth investigation into this topic will reveal many possible mysteries of current medications and uncover unexplored uses for well-known medicinal plants in treating metabolic diseases.

REFERENCES

Achrekar S, Kaklij GS, Pote MS, Kelkar SM. Hypoglycemic activity of *Eugenia jambolana* and *Ficus bengalensis*: Mechanism of action. *In Vivo*. 1991;5(2):143–177.

Aderibigbe AO, Emudianughe TS, Lawal BAS. Evaluation of the antidiabetic action *of Mangifera indica* in mice. *Phyther Res*. 2001;15(5):456–458.

Akram M, Shahab-Uddin AA, Usmanghani K, Hannan A, Mohiuddin E, Asif M. *Curcuma longa* and curcumin: A review article. *Rom J Biol Plant Biol.* 2010;55(2):65–70.

Al-Awar MSA, Muaqeb AAA, Salih EMA. Antihyperglycemic and hypolipidemic effect of *Azadirachta indica* leaves aqueous extract in alloxan-induced diabetic male rabbits. *Int J Pharm Biol Arch.* 2018;9(1):47–51.

Al-Snafi AE. *Adonis aestivalis*: Pharmacological and toxicological activities-A review. *Asian J Pharm Sci Technol.* 2016;6(2):96–102.

Angulo P. Nonalcoholic fatty liver disease. *N Engl J Med.* 2002;346(16):1221–1231.

Asgary S, Moshtaghian SJ, Setorki M, Kazemi S, Rafieian-Kopaei M, Adelnia A, Shamsi Fet. Hypoglycaemic and hypolipidemic effects of pumpkin (*Cucurbita pepo L.*) on alloxan-induced diabetic rats. *African J Pharm Pharmacol.* 2011;5(23):2620–2626.

Augusti KT, Arathy SL, Asha R, Ramakrishanan J, Zaira J, Lekha V, Smitha S, Georges, Vijayasree VM. A comparative study on the beneficial effects of garlic (*Allium sativum Linn*), amla (*Emblica OfficinalisGaertn*) and onion (*Allium cepa Linn*) on the hyperlipidemia induced by butter fat and beef fat in rats. 2001;39(08):760–766

Awe SO, Olajide OA, Makinde JM. Effects of *Allium sativum* and Vernonia amygdalina on thrombosis in mice. *Phyther Res.* 1998;12(1):57–58.

Bagher L, Farid A, Ozra T. Prevalence of diabetes mellitus in Iran in 2000. *Iran J Public Health.* 2005;4(3):75–83.

Baldi A, Choudhary N, Kumar S. Nutraceuticals as therapeutic agents for holistic treatment of diabetes. *Int J Green Pharm.* 2013;7(4):278–287.

Baldi A, Chaudhary N, Maru J, Joshi R. Effect of pumpkin concentrate on alloxan induced diabetic rats. *J Glob Pharma Technol.* 2010;2(10):24–27.

Behera AL, Sahoo SK, Patil S V. Enhancement of solubility: A pharmaceutical overview. *Der Pharm Lett.* 2010;2(2):310–318.

Bhatia LS, Curzen NP, Calder PC, Byrne CD. Non-alcoholic fatty liver disease: A new and important cardiovascular risk factor? *Eur Heart J.* 2012;33(10):1190–2000.

Bhuiyan MNI, Chowdhury JU, Begum J. Chemical components in volatile oil from *Blumea balsamifera (L.)* DC. *Bangladesh J Bot.* 2009;38(1):107–109.

Bongiorno PB, Fratellone PM, LoGiudice P. Potential health benefits of garlic (*Allium sativum*): A narrative review. *J Complement. Integr Med.* 2008;5(1):1–24.

Bordoloi R, Dutta KN. A review on herbs used in the treatment of diabetes mellitus. *J Pharm Chem Biol Sci.* 2014;2(2):86–92.

Chakrabarti R, Vikramadithyan RK, Mullangi R, Sharma VM, Jagadheshan H, Rao YN, Siaram P, Rajagopalpan R. Antidiabetic and hypolipidemic activity of *Helicteres isora* in animal models. *J Ethnopharmacol.* 2002;81(3):343–349.

Chan K. Some aspects of toxic contaminants in herbal medicines. *Chemosphere.* 2003;52(9):1361–1371.

Chandel U, Begum T, Syedy M. Pharmacological studies of annatto (*Bixa orellana L.*). *Int J Phar Biomedi Res.* 2014;1(1):1720.

Cohen-Cole E, Fletcher JM. Is obesity contagious? Social networks vs. environmental factors in the obesity epidemic. *J Health Econ.* 2008;27(5):1382–1387.

Conforti F, Statti G, Loizzo MR, Sacchetti G, Poli F, Menichini F. *In vitro* antioxidant effect and inhibition of α-amylase of two varieties of *Amaranthus caudatus* seeds. *Biol Pharm Bull.* 2005;28(6):1098–1102.

Dandona P, Aljada A, Chaudhuri A, Mohanty P, Garg R. Metabolic syndrome: A comprehensive perspective based on interactions between obesity, diabetes, and inflammation. *Circulation.* 2005;111(11):1448–1454.

Deen D. Metabolic syndrome: Time for action. *Am Fam Physician.* 2004;69(12):2875–2882.

Dũng NX, Van Hac L, Leclercq PA. A new chemotype of *Adenosma glutinosum (L.)* Druce var. caeruleum (R. Br.) Tsoong from Vietnam. *J Essent Oil Res.* 1996;8(4):359–362.

Eddouks M, Jouad H, Maghrani M, Lemhadri A, Burcelin R. Inhibition of endogenous glucose production accounts for hypoglycemic effect of *Spergularia purpurea* in streptozotocin mice. *Phytomedicine.* 2003;10(6–7):594–599.

Eddouks M, Maghrani M, Lemhadri A, Ouahidi M-L, Jouad H. Ethnopharmacological survey of medicinal plants used for the treatment of diabetes mellitus, hypertension and cardiac diseases in the south-east region of Morocco (Tafilalet). *J Ethnopharmacol.* 2002;82(2–3):97–103.

Eidi A, Eidi M, Esmaeili E. Antidiabetic effect of garlic (*Allium sativum L.)* in normal and streptozotocin-induced diabetic rats. *Phytomedicine.* 2006;13(9–10):624–629.

Endress ME, Bruyns PV. A revised classification of the *Apocynaceae*. *Bot Rev.* 2000;66(1):1–56.

Eshun K, He Q. *Aloe vera*: A valuable ingredient for the food, pharmaceutical and cosmetic industries-A review. *Crit Rev Food Sci Nutr.* 2004;44(2):91–96.

Fakhrudin N, Wiyono T, Putra AR, Nurrochmad A, Widyarini S. The evaluation on anti-platelet and antithrombosis activities of *Cinnamomum sintoc* bark extract. *Thai J Pharm Sci.* 2020;43(4):219–226.

Fries ED. A method for administering the antihypertensive agents. *AMA Arch Intern Med.* 1956;98(4):444–448.
Gavras H. Pathogenesis of hypertension: A review. *J Med Sci.* 2009;2(1):25–28.
Gholkar AA, Nikam YP, Zambare KK, Reddy KV, Ghorpade AD. Potential anticoagulant herbal plants: A review. *Asian J Res Pharm Sci.* 2020;10(1):51–55.
Gilani AH, Aftab K. Hypotensive and spasmolytic activities of ethanolic extract of *Capparis cartilaginea. Phyther Res.* 1994;8(3):145–148.
Goyal S, Baldi A. Hypoglycemic effect of polyherbal formulation in alloxan induced diabetic rats. *Pharmacologyonline.* 2011;3:764–773.
Gupta P, Goyal R, Chauhan Y, Sharma PL. Possible modulation of FAS and PTP-1B signaling in ameliorative potential of *Bombax ceiba* against high fat diet induced obesity. *BMC Complement Altern Med.* 2013;13(1):1–9.
Gupta R, Bajpai KG, Johri S, Saxena AM. An overview of Indian novel traditional medicinal plants with anti-diabetic potentials. *African J Tradit Complement Altern Med.* 2008;5(1):1–7.
Gupta RN, Pareek A, Suthar M, Rathore GS, Basniwal PK, Jain D. Study of glucose uptake activity of *Helicteres isora Linn.* fruits in L-6 cell lines. *Int J Diabetes Dev Ctries.* 2009;29(4):170–173.
Hasani-Ranjbar S, Nayebi N, Moradi L, Mehri A, Larijani B, Abdollahi M. The efficacy and safety of herbal medicines used in the treatment of hyperlipidemia: A systematic review. *Curr Pharm Des.* 2010;16(26):2935–2947.
Heart and Stroke Statistical Update. Dallas, TX: American Heart Association, 2000. Available at: www.americanheart.org/statistics. Accessibility verified July 26, 2001.
Higdon JV, Frei B. Tea catechins and polyphenols: Health effects, metabolism, and antioxidant functions. 2003;43(1):89–143.
Hosseini M, Shafiee SM, Baluchnejadmojarad T. Garlic extract reduces serum angiotensin converting enzyme (ACE) activity in nondiabetic and streptozotocin-diabetic rats. *Pathophysiology.* 2007;14(2):109–112.
Hu D. The status quo of Chinese prevention and treatment of dyslipidemia. *Chinese J Pract Intern Med.* 2003;12:564786.
Kannel WB, McGee D, Gordon T. A general cardiovascular risk profile: The Framingham Study. *Am J Cardiol.* 1976;38(1):46–51.
Kant K, Walia M, Agnihotri VK, Pathania V, Singh B. Evaluation of antioxidant activity of *Picrorhiza kurroa* (leaves) extracts. *Indian J Pharm Sci.* 2013;75(3):324–329.
Kaur J. A comprehensive review on metabolic syndrome. *Cardiol Res Pract.* 2014;1–21.
Kazi S. Use of traditional plants in diabetes mellitus. *Int J Pharm.* 2014;4(4):283–289.
Kearney PM, Whelton M, Reynolds K, Muntner P, Whelton PK, He J. Global burden of hypertension: Analysis of worldwide data. *Lancet.* 2005;365(9455):217–223.
Khan MSA, Ahmad I. Herbal medicine: Current trends and future prospects. In: edited by Mohd Sajjad Ahmad Khan, Iqbal Ahmad, and Debprasad Chattopadhyay *New Look to Phytomedicine.* Elsevier, 2019. pp. 3–13.
Kim I, Kim H, Kim J, Om A. Beneficial effects of *Allium sativum L.* stem extract on lipid metabolism and antioxidant status in obese mice fed a high-fat diet. *J Sci Food Agric.* 2013;93(11):2749–2757.
Kim TI, Kim Y-J, Kim K. Extract of seaweed *Codium fragile* inhibits integrin αIIbβ3-induced outside-in signaling and arterial thrombosis. *Front Pharmacol.* 2021;12:685948
Klein G, Kim J, Himmeldirk K, Cao Y, Chen X. Antidiabetes and anti-obesity activity of *Lagerstroemia speciosa. Evid Based Complement Altern Med.* 2007;4(4):401–407.
Kocaadam B, Şanlier N. Curcumin, an active component of turmeric (*Curcuma longa*), and its effects on health. *Crit Rev Food Sci Nutr.* 2017;57(13):2889–2895.
Kopelman PG. Obesity as a medical problem. *Nature.* 2000;404(6778):635–643.
Koscielny J, Klüssendorf D, Latza R, Schmitt R, Radtke H, Siegel G, Kiesewetter H. The antiatherosclerotic effect of *Allium sativum. Atherosclerosis.* 1999;144(1):237–249.
Kudriashov BA, Liapina LA, Azieva LD. The content of a heparin-like anticoagulant in the flowers of the meadowsweet (*Filipendula ulmaria*). *Farmakol Toksikol.* 1990;53(4):39–41.
Leung's *Encyclopedia of Common Natural Ingredients Used in Food Drugs and Cosmetics.* John Wiley & Sons, New Jersey, USA, 1980.
Lifton RP. Molecular genetics of human blood pressure variation. *Science.* 1996;272(5262):676–680.
Majewski M. *Allium sativum*: Facts and myths regarding human health. *Rocz Państwowego Zakładu Hig.* 2014;65(1):1–8.
Mathew PT, Augusti KT. Hypoglycaemic effects of onion, *Allium cepa Linn.* on diabetes mellitus – A preliminary report. *Indian J Physiol Pharmacol.* 1975;19(4):213–217.
Mayo M. *Foundation for Medical Education and Research on Secondary Hypertension.* London, UK: McGraw-Hill, 2008.

Mestechkina NM, Shcherbukhin VD, Bannikova GE, Varlamov VP, Drozd NN, Tolstenkov AS. Anticoagulant activity of low-molecular-weight sulfated derivatives of galactomannan from *Cyamopsis tetragonoloba (L.)* seeds. *Appl Biochem Microbiol.* 2008;44(1):98–103.

Mollazadeh H, Mahdian D, Hosseinzadeh H. Medicinal plants in treatment of hypertriglyceridemia: A review based on their mechanisms and effectiveness. *Phytomedicine.* 2019;53:43–52.

Morita T, Tabata S, Mineshita M, Mizoue T, Moore MA, Kono S. The metabolic syndrome is associated with increased risk of colorectal adenoma development: The self-defense forces health study. *Asian Pacific J Cancer Prev.* 2005;6(4):485–489.

Naman S, Naryal S, Palliwal R, Paliwal SR, Baldi A. Combating atherosclerosis with nanodrug delivery approaches: From bench side to commercialization. In: edited by Harish Dureja, Narasimha Murthy, Peter Wich, and Kamal Dua *Drug Delivery Systems for Metabolic Disorders.* Elsevier, 2022. p. 97–136.

Ngo T, Kim K, Bian Y, An GJ, Bae ON, Lim KM, Chung JH. Cyclocurcumin from *Curcuma longa* selectively inhibits shear stress-induced platelet aggregation. *J Funct Foods.* 2019;61:103462

Nugroho AE, Malik A, Pramono S. Total phenolic and flavonoid contents, and *in vitro* antihypertension activity of purified extract of Indonesian cashew leaves (*Anacardium occidentale L.*). *Int Food Res J.* 2013;20(1):299–305.

Ofosu FA, Buchanan MR, Brister SJ. Thrombin-catalyzed amplification and inhibitory reaction of blood coagulation. In: MR Buchanan, SJ Brister, and FA Ofosu. *Thrombin Its Key Role Thrombogenesis-Implications Its Inhibition Clinically.* CRC Press, Boca Raton, FL, 1995, pp. 1–127.

Onyenekwe PC, Ajani EO, Ameh DA, Gamaniel KS. Antihypertensive effect of roselle (*Hibiscus sabdariffa*) calyx infusion in spontaneously hypertensive rats and a comparison of its toxicity with that in Wistar rats. *Cell Biochem Funct Cell.* 1999;17(3):199–206.

Paolisso G, D'Amore A, Di Maro G, Galzerano D, Tesauro P, Varricchio MD, Onofrio F. Evidence for a relationship between free radicals and insulin action in the elderly. *Metabolism.* 1993;42(5):659–663.

Pari L, Latha M. Effect of *Cassia auriculata* flowers on blood sugar levels, serum and tissue lipids in streptozotocin-diabetic rats. *Singapore Med J.* 2002;43(12):617–621.

Patel P, Vaghasiya J, Thakor A, Jariwala J. Antihypertensive effect of rhizome part of *Acorus calamus* on renal artery occlusion induced hypertension in rats. *Asian Pacific J Trop Dis.* 2012;2:S6–10.

Paul RK, Kesharwani P, Raza K. Recent update on nano-phytopharmaceuticals in the management of diabetes. *J Biomater Sci Polym Ed.* 2021;32(15):2046–2068.

Prakash P, Misra A, Surin WR, Jain M, Bhatta RS, Pal R, Raj K, Barthwai MK, Dikshit M. Anti-platelet effects of *Curcuma* oil in experimental models of myocardial ischemia-reperfusion and thrombosis. *Thromb Res.* 2011;127(2):111–118.

Quattrocchi U. CRC World Dictionary of Medicinal and Poisonous Plants: Common Names, Scientific Names, Eponyms, Synonyms, and Etymology (5 Volume Set). CRC Press, Boca Raton, FL, 2012.

Rajani GP, Ashok P. *In vitro* antioxidant and antihyperlipidemic activities of *Bauhinia variegata* Linn. *Indian J Pharmacol.* 2009;41(5):227–232.

Rajasekaran S, Ravi K, Sivagnanam K, Subramanian S. Beneficial effects of *Aloe vera* leaf gel extract on lipid profile status in rats with streptozotocin diabetes. *Clin Exp Pharmacol Physiol.* 2006;33(3):232–237.

Raoufi S, Baluchnejadmojarad T, Roghani M, Ghazanfari T, Khojasteh F, Mansouri M. Antidiabetic potential of salvianolic acid B in multiple low-dose streptozotocin-induced diabetes. *Pharm Biol.* 2015;53(12):1803–1809.

Rosendaal FR. Venous thrombosis: A multicausal disease. *Lancet.* 1999;353(9159):1167–1173.

Sasaki Y, Noguchi T, Yamamoto E, Giddings JC, Ikeda K, Yamori Y, Yamamoto J. Effects of *Ginkgo biloba* extract (EGb 761) on cerebral thrombosis and blood pressure in stroke-prone spontaneously hypertensive rats. *Clinical Exp Pharmacol Physiol.* 2002;29(11):963–967.

Satyavati GV. Gum guggul (*Commiphora mukul*) The success story of an ancient insight leading to a modern discovery. *Indian J Med Res.* 1988;87:327–335.

Satyavati GV, Raina MK, Sharma M. Medicinal Plants of India, 1976, 1, Indian Council of Medical Research New Delhi, 1976, Volume 1, pp. 201–206.

Schrader J. Stroke and hypertension. *Internist (Berl).* 2009;50(4):423–432.

Schuppan D, Gorrell MD, Klein T, Mark M, Afdhal NH. The challenge of developing novel pharmacological therapies for non-alcoholic steatohepatitis. *Liver Int.* 2010;30(6):795–808.

Shakib Z, Shahraki N, Razavi BM, Hosseinzadeh H. *Aloe vera* as an herbal medicine in the treatment of metabolic syndrome: A review. *Phyther Res.* 2019;33(10):2649–2660.

Sharifi F, Sheikhi A, Behdad M, Mousavinasab N. Effect of garlic on serum adiponectin and interleukin levels in women with metabolic syndrome. *Int J Endocrinol Metab.* 2010;8(2):68–73.

Sharma SB, Nasir A, Prabhu KM, Murthy PS. Antihyperglycemic effect of the fruit-pulp of *Eugenia jambolana* in experimental diabetes mellitus. *J Ethnopharmacol.* 2006;104(3):367–373.

Shirzad H, Taji F, Rafieian-Kopaei M. Correlation between antioxidant activity of garlic extracts and WEHI-164 fibrosarcoma tumor growth in BALB/c mice. *J Med Food.* 2011;14(9):969–974.

Silagy CA, Neil HA. A meta-analysis of the effect of garlic on blood pressure. *J Hypertens.* 1994;12(4):463–468.

Singh DK, Singh VK. Pharmacological effects of *Allium sativum L.* (Garlic). *Ann Rev Biomed Sci.* 2008;10:6–26.

Singh PA, Bajwa N, Baldi A. A comparative review on the standard quality parameters of turmeric. *Indian J Nat Prod.* 2021;35(1):2–8.

Skålén K, Gustafsson M, Rydberg EK, Hultén LM, Wiklund O, Innerarity TL, Boren J. Subendothelial retention of atherogenic lipoproteins in early atherosclerosis. *Nature.* 2002;417(6890):750–754.

Stuppner H, Wagner H. New cucurbitacin glycosides from *Picrorhiza kurrooa. Planta Med.* 1989;55(06):559–563.

Suthar M, Rathore GS, Pareek A. Antioxidant and antidiabetic activity of *Helicteres isora (L.)* fruits. *Indian J Pharm Sci.* 2009;71(6):695–699.

Tabassum N, Ahmad F. Role of natural herbs in the treatment of hypertension. *Pharmacogn Rev.* 2011;5(9):30–40.

Tanaka M, Misawa E, Ito Y, Habara N, Nomaguchi K, Yamada M, Toida T, Hayasawa H, Takase M, Inagaki M, Higuchi R. Identification of five phytosterols from *Aloe vera*gel as anti-diabetic compounds. *Biol Pharm Bull.* 2006;29(7):1418–1422.

Tchikaya FO, Bantsielé GB, Kouakou-Siransy G, Datté JY, Datté PA, Zirihi NG, Offoumou MA. Anacardium occidentale Linn. *(Anacardiaceae)* stem bark extract induces hypotensive and cardio-inhibitory effects in experimental animal models. *African J Tradit Complement Altern Med.* 2011;8(4):452–461.

Vareda PMP, Saldanha LL, Camaforte NA de P, Violato NM, Dokkedal AL, Bosqueiro JR. *Myrcia bella* leaf extract presents hypoglycemic activity via PI3k/Akt insulin signaling pathway. *Evid Based Complement Altern Med.* 2014;16(9):1–11.

Visavadiya NP, Narasimhacharya A. Ameliorative effects of herbal combinations in hyperlipidemia. *Oxid Med Cell Longev.* 2011;2011:1–8.

Weir MR. The obesity paradox: Impact of obesity on the prevalence and prognosis of cardiovascular diseases. *Postgrad Med.* 2009;121(1):164–165.

Wong ND. Intensified screening and treatment of the metabolic syndrome for cardiovascular risk reduction. *Prev Cardiol.* 2005;8(1):47–54.

Wu Y-C, Hsieh C-L. Pharmacological effects of *Radix Angelica Sinensis* (Danggui) on cerebral infarction. *Chin Med.* 2011;6(1):1–5.

Yagi A, Machii K, Nishimura H, Shida T, Nishioka I. Effect of aloe lectin on deoxyribonucleic acid synthesis in baby hamster kidney cells. *Experientia.* 1985;41(5):669–671.

Yang RM, Liu F, He ZD, Ji M, Chu XX, Kang ZY, Cai DY, Gao NN. Anti-obesity effect of total phenylpropanoid glycosides from *Ligustrum robustum Blume* in fatty diet-fed mice via up-regulating leptin. *J Ethnopharmacol.* 2015;169:459–465.

Zeng X-H, Zeng X-J, Li Y-Y. Efficacy and safety of berberine for congestive heart failure secondary to ischemic or idiopathic dilated cardiomyopathy. *Am J Cardiol.* 2003;92(2):173–176.

4 Role of Curcumin and Its Nanotechnological Innovations in Cancer Therapeutics

Soledad Ravetti, Ariel G. Garro, Sofía G. Brignone, Carolina del V. Bessone, Daniel A. Allemandi, and Daniela A. Quinteros

4.1 INTRODUCTION

Cancer is a generic term that refers to a set of multifaceted diseases where normal cells begin to be hyper-proliferative, the result of mutations or abnormal changes in the genes that regulate the growth of healthy cells, initiating the expression of genes involved in cell proliferation, invasion, survival, metastasis, and angiogenesis (Hanahan and Weinberg 2000, 2011). More than 200 different types are known and can affect any part of the body. Each of these has a particular origin and characteristics, with a different evolution, prognosis, and treatment.

Despite the significant progress made in recent years to understand and contain this disease, according to statistical data, cancer continues to be one of the most devastating causes of death worldwide. In 2020, there were an estimated 19.2 million new cancer cases and almost 10 million cancer-related deaths in the world. Furthermore, this trend is projected to continue to increase at the international level, reaching 13.1 million by 2030 (American Cancer Society 2015; Stewart and Wild 2014). This is mainly due to the complexity of this disease, which makes the systematic and effective application of conventional treatments difficult, including new and promising therapies. And, what is even more complex: cancer evolves, which calls into question the search for a single cure and makes it clear that multiple solutions are needed to attack cancer.

Of all the available treatments, cytostatic or traditional chemotherapy continues to play a fundamental role in cancer therapy, interfering in the different phases of the cell cycle (mitosis), in order to achieve death or slow the growth of cancer cells. However, these cytostatic drugs are linked with adverse side effects, which significantly limit their efficacy, while also markedly decreasing the quality of life of patients. In addition, these drugs are often highly expensive and are beyond the reach of the vast majority of the world's population. Therefore, the search and development of anticancer agents that do not present these drawbacks continues to be urgent and necessary.

One focuses on the search for biologically active molecules directed at tumor molecular targets, while the other approaches improve the efficacy of existing drugs by developing innovative transport systems. The molecules under investigation must aim to interfere with one or more of the distinctive capabilities that tumoral cells acquire during their transformation, such as the ability to generate their own growth signals; insensitivity or evasion to inhibitory growth signals; evading of programmed cell death (apoptosis); unlimited replicative potential, the ability to invade tissues generating metastasis, the ability to modify or reprogram their energy metabolism and the ability to evade the immune system, in order to help neoplastic proliferation (Hanahan and Weinberg 2000, 2011).

This approach is being carried out mainly, through the screening of natural products (Sarker et al. 2020; Huang, Lu, and Ding 2021), synthesis of low molecular weight compounds via traditional chemistry (Zhong et al. 2021; Hoelder, Clarke, and Workman 2012; Cui et al. 2020), or from

DOI: 10.1201/9781003292692-4

the identification of natural macromolecules, including antibodies, proteins and oligonucleotides with inherent biological activity (Imai and Takaoka 2006; Marqus, Pirogova, and Piva 2017). The use of a multitargeted active molecule in cancer chemotherapy may be instrumental in curtailing therapy-related side effects while preserving therapeutic efficacy.

The second approach is aimed at improving the efficacy of existing anticancer agents, through the development of innovative transport systems (Kemp and Kwon 2021; Etheridge et al. 2013; Wicki et al. 2015), which allows modifying the pharmacokinetic and biodistribution characteristics of the active principles. These systems aim to increase the specific accumulation of the drug in the tumor area and increase the drug-malignant cell exposure time while reducing the exposure of normal tissues to it. Even, the development of innovative transport systems can be very useful to improve obstacles presented by potential therapeutic agents, similar to conventional drugs, such as poor water solubility, pharmacokinetic limitations, adverse side effects, poor therapy index, drug resistance, and other related drawbacks.

To date, there are numerous systems designed from a variety of materials including polymers (nanoparticles, micelles, conjugates, or dendrimers), lipids (microemulsions, nanoparticles, liposomes, micelles), viruses (viral nanoparticles), cyclodextrins or organometallic compounds (nanoparticles of silica, gold, nanotubes) (Sarker et al. 2020; Huang, Lu, and Ding 2021). From simple systems such as liposome-encapsulated drugs to more complex ones, known as multifunctional nanoplatforms, which not only contain the active agent but also other elements that allow orientation toward specific therapeutic objectives, drug visualization, and monitoring of its activity.

4.1.1 Physical-Chemical Properties of Curcumin and Structural Features

In the search for biologically active molecules, it is well known that plants have been, and will continue to be, important sources of therapeutic products (nutraceuticals and pharmaceuticals). They offer enormous structural diversity and, in some cases, great biological potency, making it difficult for synthetic chemistry to replace cellular biochemistry as a source of new compounds. It is estimated that 25% of the total drugs used in the clinic correspond to active compounds isolated from higher plants and to semi-synthetic drugs obtained from these natural precursors.

Among the herbal bioactive with enormous therapeutic potential, curcumin (CUR) is a promising candidate for the treatment of numerous chronic diseases (Kotha and Luthria 2019; Kunnumakkara et al. 2016; Shehzad et al. 2017; Shimizu et al. 2019; Kunnumakkara et al. 2017).

CUR is the major biologically active polyphenolic compound derived from the rhizomes of Curcuma longa (turmeric), a yellow spice of the ginger family, *Zingiberaceae*, widely used in Ayurvedic and folk medicine for centuries. The turmeric plant is an herbaceous perennial that originated from India and is extensively cultivated in the tropical areas of Asia and to a lesser extent is currently grown in several other parts of the world, including Southeast Asia, China, and Latin America. In India, it is popularly known as haldi, called turmeric in English, and ukon in Japanese.

CUR makes up between 3% and 5% of turmeric compounds and is considered the main cause of the yellow/golden color of turmeric, and it has also been identified as responsible for most of the properties of turmeric.

Traditionally, this spice and preparations derived from it have been used for the treatment of such ailments as gynecological and gastric problems, hepatic disorders, infectious diseases, and blood disorders. Turmeric paste is also available in many Indian homes for the treatment of inflammation and wounds. The crude and often dried plant material is widely consumed as a food additive, as part of curry spices, which typically contain numerous other ingredients. Various CUR-based products are now marketed in several countries including the United States, India, Japan, Korea, Thailand, China, Turkey, South Africa, Nepal, and Pakistan in the form of capsules, tablets, ointments, energy drinks, soaps, and cosmetics (Gupta, Kismali, and Aggarwal 2013).

CUR was discovered for the first time about two centuries ago by Vogel and Pelletier, from Harvard College Laboratory from the rhizomes of C. longa (turmeric) (Vogel 1815). In 1842, it was purely acquired by Vogel (1842). Subsequently, the chemical structure of CUR was reported by Milobedzka, Kostanecki, and Lampe (1910). Subsequently, Lampe and Milobedzka (1913) reported the synthesis of CUR, however, Srinivasan (1953) was the first to determine the existence of other compounds called curcuminoids in addition to CUR by chromatographic analysis of turmeric.

In addition, this molecule is considered "generally recognized as safe" (GRAS) by the US Food Drug Administration (US-FDA), given that good tolerability and safety profiles have been shown by clinical trials, even at doses up to 12,000 mg/day of 95% concentration of three curcuminoids: CUR, bisdemethoxycurcumin, and demethoxycurcumin.

CUR is a bis-α, β-unsaturated β-diketone. Contains a conjugated phenolic group substituted by methoxy, hydroxyl, and carbonyl groups. Its chemical name is dipheruloylmethane or 1,7-bis-(4-hydroxy-3-methoxyphenyl)-hepta-1,6-diene-3,5-dione; while its chemical formula is $C_{21}H_{20}O_6$ ("PubChem Compound Summary for CID 969516, Curcumin," n.d.; Nelson et al. 2017). As a result of this structure, CUR can exist in at least two tautomeric forms, such as a keto-enol and diketone form, as shown in Figure 4.1. The molecular weight of CUR is 368.38 g/mol, with a melting point of 179°C–183°C, and a pKa value of 8.54.

Commercial CUR, mostly used in preclinical studies and clinical trials is usually a mixture of three curcuminoids: CUR (75%–80%), demethoxycurcumin (15%–20%), and bisdemethoxycurcumin (3%–5%).

According to its physical-chemical properties, there are several factors that significantly affect the efficient use of CUR as a therapeutic molecule; in order to treat disease conditions (Sabet et al. 2021), these include:

i. *low water solubility and intrinsic dissolution rate(s).* CUR is practically insoluble in water at acidic and neutral conditions. This has a crucial role in the bio-accessibility of the compound within the intestinal fluids of the GIT, limiting its oral bioavailability;
ii. *low physicochemical stability.* One of the factors affecting the stability of CUR is pH. The stability of CUR is pH-dependent, being reasonably stable at pH 1–6 and unstable at pH >7. Color changes in the CUR solution at different pH values demonstrate its instability. Under physiological pH conditions, such as 0.1 M phosphate buffer (pH 7.2) at 37°C, over 90%

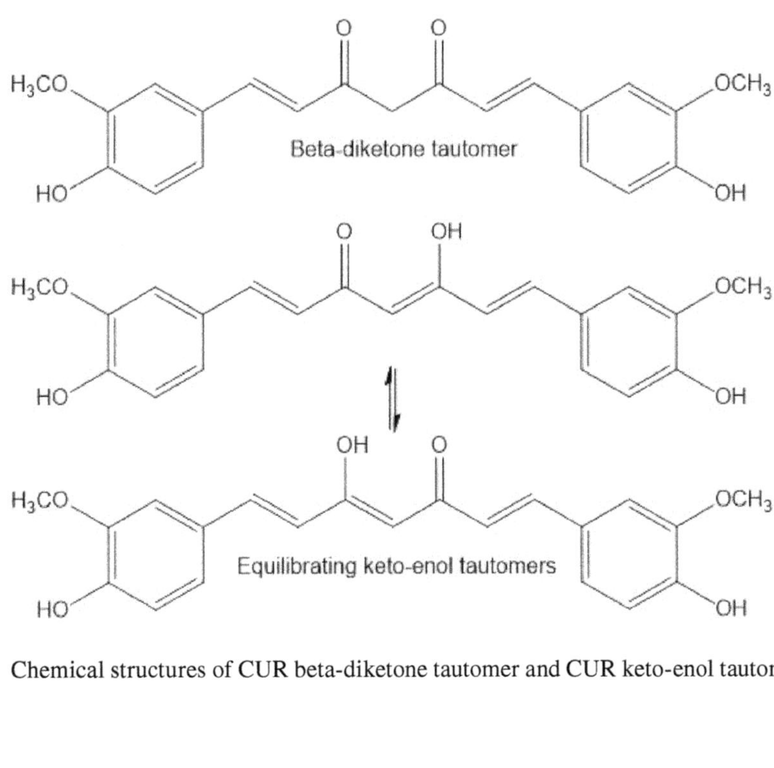

FIGURE 4.1 Chemical structures of CUR beta-diketone tautomer and CUR keto-enol tautomers.

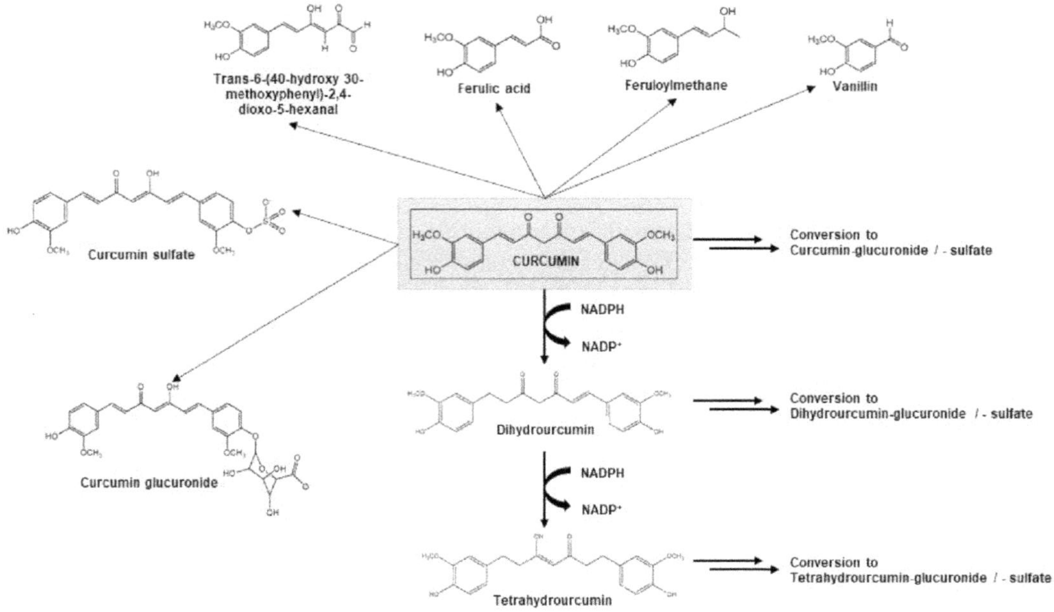

FIGURE 4.2 Chemical structures of curcumin, its degradation products and metabolites.

of curcumin is degraded within 30 min to trans-6-(40-hydroxy 30-methoxyphenyl)-2,4-dioxo-5-hexanal, feruloylmethane, ferulic acid and vanillin (Kurita and Makino 2013) (see Figure 4.2).

Under acidic conditions, the degradation is much slower, with less than 20% of total CUR decomposed at 1 h. The presence of fetal calf serum, human blood, or the addition of antioxidants successfully slows down the degradation of CUR in culture media or phosphate buffer above pH 7(Lestari and Indrayanto 2014; Heger et al. 2014). Though many of these degradation products are biologically active, they are substantially less active compared to CUR, supporting that chemical degradation has a limited contribution to the biological activities of CUR (Zhu et al. 2017).

It is also known that CUR is sensitive to light and therefore it is recommended that biological samples containing CUR are to be protected (Prasad et al. 2014). In dried powder, it is more stable against light exposure than in solution. The neutral and alkaline hydrolysis as well as the presence of oxygen and light are factors that promote CUR autoxidation (Kurita and Makino 2013). Autoxidation creates a bicyclopentadione that is the major degradation product in aqueous conditions.

In relation to the temperature, some studies show that the chemical stability of CUR is temperature-dependent; being thermostable up to 70°C for 10 min. Above this temperature, it starts to decompose, and at 100°C, its degradation increases (Lestari and Indrayanto 2014);

iii. *rapid metabolization.* Preliminarily animal studies demonstrate that CUR is rapidly metabolized and conjugated in the liver (and to some extent in the intestine mucosa and kidney), and then excreted in feces with limited systemic bioavailability. Two major pathways have been identified in CUR metabolism, like O-conjugation and reduction. After oral administration, the predominant metabolites found in the plasma were the glucuronides/sulfates of CUR (see Figure 4.2), reaching their peak concentration only 1 h. Comparing these metabolic products with the degradation products, it appears that simple hydrolytic degradation is prevented in biofluids, probably because specific enzymatic reactions are faster and do not allow slow hydrolytic degradation, so the latter process cannot compete with the first reaction. It has also been reported that these conjugated metabolites, like chemical degradation products, are less active or inactive, compared with CUR (Priyadarsini 2014). A fast metabolism is considered the main reason for CUR's poor bioavailability when it is orally ingested, suggesting that the therapeutic potential of oral CUR is limited;

iv. *low absorption from the gut and rapid systemic elimination.* Consequently, high concentrations of CUR cannot be achieved and maintained in plasma and tissues after oral ingestion. Interestingly, there is also low bioavailability with intravenous administration. However, despite its demonstrated effects, the potential health benefits of CUR are limited by its poor solubility, low absorption in the gut, rapid metabolism, and rapid systemic elimination. Since most of the ingested CUR is excreted through the feces unmetabolized, the percentage that is absorbed is largely converted to its water-soluble metabolites, glucuronides, and sulfates. The microbial metabolism of CUR produces dihydrocurcumin and tetrahydrocurcumin through NADPH-dependent reduction. CUR and its reduced metabolites are conjugated to the monoglucuronide via beta-glucuronidase, to a monosulfate via sulfatase enzymes, and to a mixed sulfate/glucuronide, resulting in CUR glucuronide, dihydrocurcumin glucuronide, tetrahydrocurcumin glucuronide or the corresponding monosulfate and mixed sulfate/glucuronide (Wahlström and Blennow 1978; Hassaninasab et al. 2011) (see Figure 4.2).

4.2 ROLE OF CUR IN HEALTH BENEFITS

Scientific interest in CUR has grown and its health benefits have increasingly been discovered. To date, CUR is one of the most extensively studied phytoconstituents during the last two or three decades for its possible therapeutic effects and health benefits. Different analyses of the documented scientific literature account for this. Searching for peer-reviewed publications using the term "curcumin" in PubMed, ScienceDirect, and Scopus published during 2010–2020 results in 34.4, 20.4, and 26.6 thousand publications, respectively, of which 74.6%, 62.4%, and 33.4% are cancer-related papers (Araya-Sibaja et al. 2021). There are several highly cited reviews that summarize the above-mentioned important research findings about the health-beneficial effects of CUR (Araya-Sibaja et al. 2021; Gupta et al. 2013; Prasad et al. 2014; Kocaadam and Şanlier 2017; Kotha and Luthria 2019; Kunnumakkara et al. 2016). In addition, various clinical trials that are underway or have been completed support the potential role of CUR in the treatment of various chronic disorders (http://clinicaltrials.gov/ct2/results?term=curcumin). The extensive research from *in vitro* and animal models, including some clinical trials, over the last few years, provides strong scientific evidence for the biological activities of CUR and provides a rationale for the therapeutic potential of this molecule against numerous diseases.

It has been shown that CUR possesses a wide range of multi-mechanistic medicinally beneficial properties, including antiinflammation, anticarcinogenesis, antioxidant, wound healing, and antimicrobial effects, which allows it to help in the management of several pathological conditions such as diabetes, neurodegenerative diseases, obesity, cardiovascular diseases, allergy and bronchial asthma, rheumatoid arthritis and osteoarthritis, psoriasis, renal ischemia, pancreatitis, inflammatory bowel, metabolic syndrome, inflammatory and degenerative eye conditions, certain types of cancers (Hewlings and Kalman 2017; Shehzad et al. 2017). While there appear to be countless therapeutic benefits to CUR, most of these benefits are related to its antioxidant and anti-inflammatory effects (Hewlings and Kalman 2017).

The chemistry of its structure is the basis of its several biological activities. CUR, through its a-, b-unsaturated b-diketone moiety, carbonyl and enolic groups of the b-diketone moiety, methoxy, and phenolic hydroxyl groups, and phenyl rings, has been shown to directly interact with numerous signaling molecules.

Important chemical reactions associated with the biological activity of CUR are the hydrogen donation reactions leading to oxidation of CUR, reversible and irreversible nucleophilic addition reactions (Michael reaction), hydrolysis, degradation, and enzymatic reactions. All these have significant roles in different biological activities of CUR (Priyadarsini 2014).

At the molecular level, CUR has been shown to modulate numerous signaling molecules either directly by binding to them or indirectly by modulating the expression of other proteins. The indirect targets of CUR include transcription factors, enzymes, growth factors, receptors, inflammatory mediators, protein kinases, drug resistance proteins, adhesion molecules, cell survival proteins, chemokines and chemokine receptors, invasive and angiogenic proteins, and cell cycle regulatory proteins (Figure 4.3).

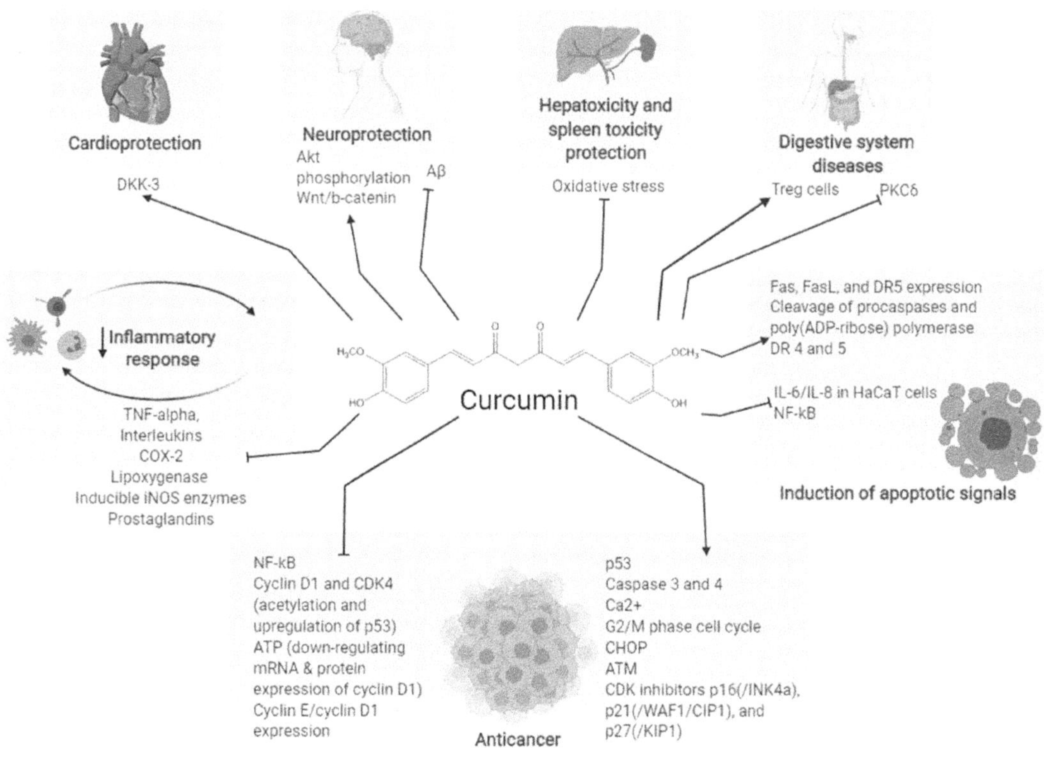

FIGURE 4.3 Role of CUR in different diseases. Abbreviations: **DKK-3**, Dickkopf Wnt Signaling Pathway Inhibitor 3 or Dickkopf-related protein 3; **AKT**, Protein kinase B or PKB; **Wnt/β-catenin**, Signal transduction pathways: Wnt family of secreted glycolipoproteins via the transcription co-activator β-catenin; **Aβ**, Amyloid-β Peptide; **Treg cells**, Regulatory T cells; **PKC δ**, Protein kinase C, δ isoform; **Fas**, Apoptosis antigen 1 (APO-1 or APT), cluster of differentiation 95 (CD95) or tumor necrosis factor receptor superfamily member 6 (TNFRSF6); **FasL**, Fas Ligand; **DR4 or 5**, Death receptor 4 or 5; **IL-6**, Interleucina-6; **IL-8**, Interleucina-8; **HaCat cells**, Human epidermal keratinocyte line cells; **NF-κB**, Nuclear Factor kappa-light-chain-enhancer of activated B cells; **p53**, Tumor protein P53, cellular tumor antigen p53, phosphoprotein p53, tumor suppressor p53, antigen NY-CO-13, or transformation-related protein 53 (TRP53); **Ca2+**, calcium ion; **CHOP**, CCAAT-enhancer-binding protein homologous protein; **ATM**: Serine/threonine protein kinase; **p16INK4a**, inhibitor of cyclin-dependent kinases 4 and 6 (CDK4 and 6); **p21(WAF1,CIP1)**: Cyclin-dependent kinase inhibitor 1 or CDK-interacting protein 1; **p27(kip1)**: inhibitor of cyclin-dependent kinases (CDKs); **CDK4 or 6**, Cyclin-dependent kinase 4 or 6; **ATP**: Adenosine triphosphate; **TNF alpha**, tumor necrosis factor alpha; **COX-2**: Ciclooxigenase2; **iNOS**, Inducible nitric oxide synthase (Gupta et al. 2013; Prasad et al. 2014; Kocaadam and Şanlier 2017; Kotha and Luthria 2019; Kunnumakkara et al. 2016; Shehzad et al. 2017).

4.3 ROLE OF CUR IN CANCER THERAPEUTICS

The anticancer activity of CUR was demonstrated for the first time in the 1980s by Kuttan and colleagues in both *in vitro* and *in vivo* models (Kuttan et al. 1985; Soudamini and Kuttan 1988). Since then, the pharmacology and putative anticancer properties of CUR have been the subject of several review articles published (Tomeh, Hadianamrei, and Zhao 2019; Mansouri et al. 2020; Giordano and Tommonaro 2019; Kunnumakkara et al. 2017; Shanmugam et al. 2015).

As a result of recent literature research, it was found that 37% of published articles on CUR have cancer as the main targeted disease (Giordano and Tommonaro 2019). In this scenario, CUR has been found to be effective in many stages of cancer development, to suppress tumor transformation, initiation, development and invasion, angiogenesis, and metastasis.

Numerous lines of evidence demonstrate that CUR exerts its anticancer effects through a variety of mechanisms, due to its ability to interact with different molecules, modulating multiple molecular pathways and their target.

These anticancer effects are predominantly mediated through its negative regulation of various transcription factors, growth factors, inflammatory cytokines, protein kinases, and other oncogenic molecules (Figure 4.4).

In summary, the extensive data accumulated, mainly research studies, demonstrate that CUR can modulate various stages of cancer such as uncontrolled cell proliferation, cancer cell death, cancer-associated inflammation, signaling pathways, cancer angiogenesis, and metastasis. Therapeutically, CUR can interact with most of the targets modulated by FDA-approved cancer drugs.

Available clinical trials support the potential role of CUR for the management of a broad spectrum of cancers as pancreatic, hepatocellular, gastric, breast, prostate, skin, lung, and colon cancer, as well as multiple myeloma (Shehzad et al. 2017; Fadus et al. 2017; Salehi et al. 2019; Gupta,

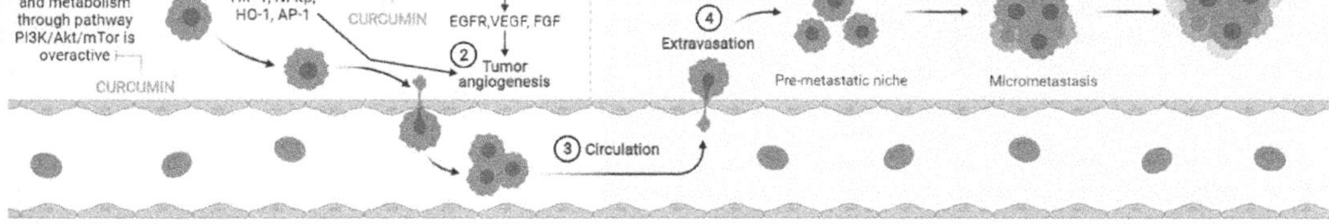

FIGURE 4.4 Molecular mechanisms of CUR in cancer diseases. Abbreviations: **MMPs**, Matrix Metalloproteinases; **u-PA**, Urokinase-type plasminogen activator; **RNA**, Ribonucleic Acid; **JAK**, Janus kinase; **STAT**, Signal Transducer and Activator of Transcription proteins; **MAPK**, Mitogen-activated protein kinase; **p53**, Tumor protein P53, cellular tumor antigen p53 (UniProt name), phosphoprotein p53, tumor suppressor p53, antigen NY-CO-13, or transformation-related protein 53 (TRP53); **ICAM**, Intercellular Adhesion Molecule; **VCAM**, Vascular Cell Adhesion Molecule; **TGF-β**, Transforming growth factor beta; **EGFR**, Epidermal growth factor receptor; **VEGF**, Vascular Endothelial Growth Factor; **FGF**, Fibroblast Growth Factor; **HIF-1**, Hypoxia-inducible factor-1; **HO-1**, Heme oxygenase-1; **AP-1**, Activator protein-1; **NF-κB**, Nuclear Factor kappa-light-chain-enhancer of activated B cells; **AKT**, Protein kinase B or PKB; **PI3K**, Phosphoinositide 3-kinase; **mTor**, mammalian Target of Rapamycin (Tomeh, Hadianamrei, and Zhao 2019; Mansouri et al. 2020; Giordano and Tommonaro 2019; Kunnumakkara et al. 2017; Shanmugam et al. 2015; Akbari et al. 2021; Zhou et al. 2017; Wang et al. 2019).

Patchva, and Aggarwal 2013), demonstrating tolerability safety, non-toxicity and efficacy. These studies provide a solid foundation for more well-controlled studies in larger cohorts, as well as open avenues for future drug development.

In addition, CUR has been shown to have radiodensities, radioprotective, and chemosensitizing effects. At this point, promising results describe the ability of CUR to increase the effectiveness of radiotherapy, being able to act as a radiosensitizer for various types of cancer, aside from radioprotective for some tissues, especially tissues epithelial cells, potentially owing to direct contact with at-risk surfaces (Sak 2020; Verma 2016; Goel and Aggarwal 2010). In the same way, several results suggest that CUR can be used as an effective combination with standard chemotherapeutic agents to inhibit drug resistance and sensitize cancer cells to these agents (Saha et al. n.d.; Bordoloi et al. 2016; Shaikh et al. 2021; Tan and Norhaizan 2019). CUR may have a potential role as a dose-modifying agent that can be used either to sensitize resistant cells to therapy or to reduce the dose of these therapeutic agents.

Recently, an exhaustive systematic review of articles published in international databases further confirmed that CUR reduces the side effects of chemotherapy or radiotherapy, resulting in improving patients' quality of life. They also highlight that several studies analyzed, CUR has increased patient survival time and decreased tumor markers' level (Mansouri et al. 2020).

However, evidence suggests that the therapeutic potential of oral CUR is low even in cancers from the gastrointestinal tract. As previously described, its clinical application is severely injured like its poor aqueous buffer solubility and low bioavailability, instability in body fluids, and rapid metabolism that greatly compromise its pharmaceutical use. *In vitro* studies have clearly established that cancer cells do not undergo apoptosis in the presence of CUR unless this dietary agent is at concentrations of approximately 5–50 μM for several hours. These concentrations of CUR are not achieved outside the gastrointestinal tract through the oral route. In the gastrointestinal tract, these concentrations cannot be kept for several hours, suggesting that the therapeutic effects of oral CUR are limited. Thus, it is imperative to improve the bioavailability and intracellular accumulation of CUR *in vivo*. Several approaches could be applied to overcome these limitations. The utilization of adjuvants like piperine, structural analogs, parenteral administration route, and the development of delivery vehicles (e.g., micelles, liposomes, and nanoparticles (López-Lázaro 2008) are promising strategies to improve CUR problems described previously.

4.4 CURCUMIN NANO-FORMULATIONS FOR CANCER TREATMENTS

Nanoparticles such as liposomes, polymeric nanoparticles, solid lipid nanoparticles, nanoemulsions, and micelles are emerging as one of the useful alternatives that have been shown to deliver therapeutic concentrations of CUR. The use of novel drug delivery systems has improved the main problems of CUR such as low solubility, instability, poor bioavailability, and rapid metabolism in cancers.

Our focus in this work is on demonstrating and comparing the results obtained from the encapsulation of CUR in nanoparticles (Ns), such as size (nm); polydispersity index (PDI); zeta potential (mV); drug loading (%DL); encapsulation efficiency (%EE); CUR release profile; and *in vitro/in vivo* studies against tumor cells (Figure 4.5).

4.4.1 Micelles

Micelles (Ms) are conventionally defined as self-assemblies of amphiphilic diblock polymers in morphologies, where a more hydrophobic block forms the core and a more hydrophilic block forms the crown of the Ms with a typical diameter < 50 nm (Kulkarni and Shaw 2016). The hydrophobic segments of the polymers/copolymers form the Ms core that acts as a reservoir for the hydrophobic drugs, while the hydrophilic head groups form the micellar shell that can enhance colloidal stability and prevent protein adsorption, thus prolonging circulation time (Gou et al. 2011; Zhao, Sun, and Dong 2020).

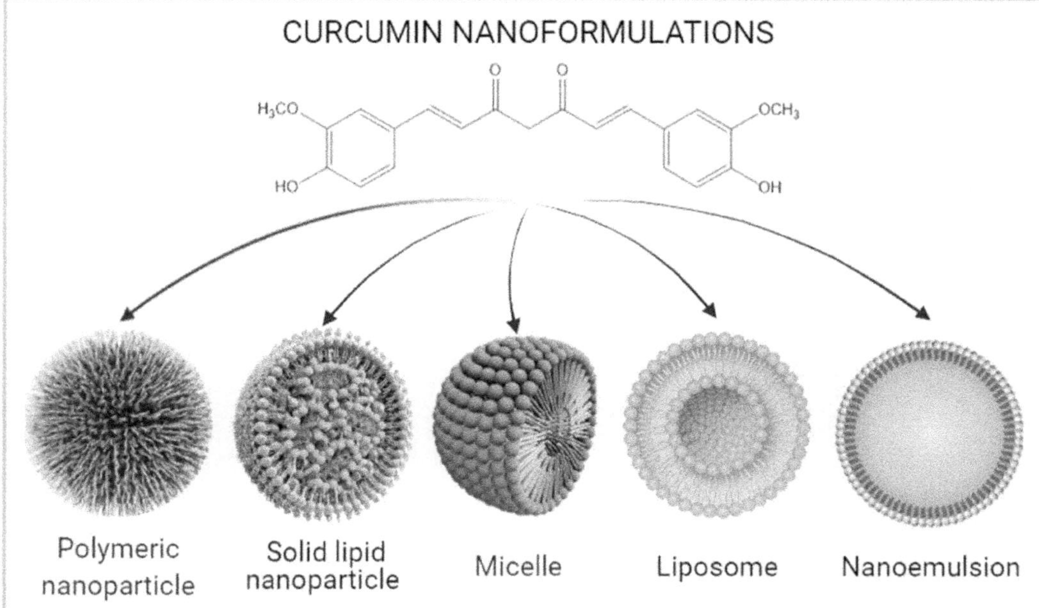

FIGURE 4.5 Curcumin nano-formulations for cancer treatments.

Polymeric micelles (PMs) are defined as organized self-assemblies formed in a liquid, which are composed of amphiphilic macromolecules, typically di- or triblock amphiphilic copolymers composed of solvo- and solvo-phobic blocks, can exceed 100 nm in size and still be considered Ms (Dewangan et al. 2012; Cabral et al. 2011).

PMs are structures that form spontaneously by self-assembly of amphiphilic polymeric molecules (usually block copolymers), once the critical micellar concentration (CMC) is exceeded, and can reach sizes of up to several hundreds of nanometers. The structures of these self-assembled Ms can protect insoluble hydrophobic drugs, which mimic aspects of biological transport systems in terms of structure and function (Feng et al. 2006).

Ms and MPs have been widely used as drug delivery carriers for a number of different molecules, including low molecular mass hydrophobic drugs, proteins, and genes. In fact, they are widely used in the delivery of poorly water-soluble drugs, including CUR.

MPs have attracted great interest as drug delivery systems due to their low toxicity, good biocompatibility, modified release pattern, longer blood circulation time, and their ability to solubilize hydrophobic drugs such as CUR in the micelle core (Akbar et al. 2018). This is why we are interested in discussing some studies of the last 8 years related to CUR within the micelles system applicable to different types of cancer.

Pluronic® copolymers is a material used as a biocompatible complexing agent to improve the water solubility, stability, and bioavailability of poorly water-soluble drugs, which contain hydrophobic poly(propylene oxide) (PPO) and hydrophilic poly(ethylene oxide) (PEO) blocks arranged in a triblock structure. These attractive mixed PMs can be used as pharmaceutical carriers for CUR and other hydrophobic drugs (Akbar et al. 2018).

Kurnik et al. (2021) described an alternative environmentally friendly method to form PMs composed of the Pluronic® L35 (PEO11-PPO16-PEO11) polymer by using aqueous two-phase systems micellar systems (ATPMS) with cholinium-based ionic liquids (ILs). This method showed a novel approach for the formation of PM nanostructures under safer and more eco-friendly conditions being CUR as a model compound to demonstrate the encapsulation and drug delivery efficiencies of these new systems (see Table 4.1). Besides, its potential as an antitumor compound CUR was

TABLE 4.1
Summary of Curcumin-Loaded Polimeric Micelles Approach for Cancer Therapy Currently Reported

Type of System	Size (nm)/PDI	ZP (mV)	%EE	%DL	Release Results	Cell Lines Evaluated	*In Vivo* Model	References
CUR-composed by PL35	42.5–45.6/0.110–0.150	−18.4 −14.6	94.1–96.0	14.8–16.2	*In vitro* drug release studies demonstrated that the interactions of the CUR with ILs have a great impact on their drug release behavior	Human prostatic carcinoma (PC3)	–	Kurnik et al. (2021)
CUR-Bisphosphonate (ALN) conjugated PF127Ms	~27 N/A	+2.87	N/A	Cur-NPs: 5,7 Cur-ALN-NPs: 4	–	–	–	Kamble et al. (2020)
CUR in folate (FA) modified (MPEG–PCL) Ms	30.47/0.17	−3.55	98	10	Free CUR: 91.60 h Nano-CUR: 44.53 h FA/Nano-CUR: 46.32 100 h	–	Mice model of heterotopic CRC by inoculating CT26 cells	Hu et al. (2020)
DOX and CUR co-conjugated onto a pCB	164.2/0.43	+32.1	–	11.7 µg/mL (CUR) 64.2 µg/mL (DOX) – µg/mL (free DOX)	89.7% of CUR after 30 h and 92.4% of DOX after 30 h (both from CUR-pCB-DOX)	Human mammary adenocarcinoma (MCF-7 and MCF7/Adr)	–	Zhao et al. (2020)
CUR-Amphiphilic Pluronic F-127 Ms	25.4/0.08	–	91.76	9.09	–	Human gastric cancer cell lines SGC-7901 and BGC-823	BCG-823 tumor model	Lin et al (2020) and Li et al (2014)

(*Continued*)

TABLE 4.1 (*Continued*)
Summary of Curcumin-Loaded Polimeric Micelles Approach for Cancer Therapy Currently Reported

Type of System	Size (nm)/PDI	ZP (mV)	%EE	%DL	Release Results	Cell Lines Evaluated	*In Vivo* Model	References
CUR-Bioconjugate of sodium alginate - crosslinked with calcium	178–200/0.200–0.290	–53.0 –25.9	N/A	N/A	95%~4 h	Mammary carcinoma (4T1), melanoma (B16F10), and colon carcinoma (CT26-CEA and MC38-CEA)	–	Lachowicz et al. (2019)
CUR-Hydrophilic hyaluronic acid-SS-functionalized with Tween80	74.2/0.160	–30.2	94.1	8.9	43.2% after 24 h in GSH (pH 7.4) 22.7% after 24 h without GSH (pH 5.0) 43.2% after 24 h in GSH (pH 5.0)	Glioblastoma (G422)	–	Tian et al. (2018)
CUR-Pluronics P123 and PF127 (Best formulations: F2 and F10)	16.93 (F2) 16.39 (F10)	–8.51 (F2) –18.7 (F10)	~96 (F2) ~99 (F10)	~10 (F2) ~10 (F10)	–	Human cervical cancer (HeLa) cells	–	Akbar et al.(2018)
CUR-Monomethoxy poly(ethylene glycol)- poly(lactide) copolymers	~33/0.120	–3.3	98.5	9.85	25% after 24 h >40% after 120 h	Glioma (C6 and U251)	–	Zheng et al. (2016)
CUR and Monomethoxy poly(ethylene glycol)-poly(ε-caprolactone)	28.2/0.136	–0.41	14.84	98.91	In free CUR group, 77% or 96% of CUR were released in the first 24 h or 7 days. In contrast, only 21% or 57% of CUR were released from CUR-M in the first 24 h or 7 days	–	Subcutaneous 4T1 breast tumor mouse model	Liu et al. (2013)

Abbreviations: ZP, zeta potential; PL35, Pluronic® L35; ILs, cholinium-based ionic liquids; PMs, polymeric micelles; ALN, alendronato; PF127, Pluronic F127; MPEG–PCL, methoxypoly (ethylene glycol)-poly (ε-caprolactone); pCB, poly(carboxybetaine); DOX, doxorubicin; AA, sodium alginate, N/A, not available; GSH, overexpressed glutathione.

chosen because of its current encapsulation challenges, which include poor solubility in aqueous solutions and low bioavailability. In this study, each treatment group showed a significant decrease in the percentage of cells compared to the untreated controls, and the highest decrease was observed in the CUR-Ms group. CUR-PMs showed an increased dose-dependent inhibition of cell proliferation, with half-maximal inhibitory concentration (IC_{50}) values of 14.93 μg/mL in BGC-823 cells and 14.95 μg/mL in SGC-7901 cells. Moreover, CUR-PMs decreased the expression of the anti-apoptotic protein Bcl-2 and downregulated the activity of cleaved caspase-9, followed by cleaved caspase-3 activation in gastric cancer cells. Previous results complement these studies, indicating that CUR-PMs inhibit gastric cancer cell proliferation and induce apoptosis, which enhances its antitumor effect compared to common CUR (Li et al. 2014).

Another recent study developed a drug delivery system using bisphosphonate (alendronate, ALN) conjugated Pluronic F127 Ms. Kamble et al. (2020) described that Ms could efficiently target, and specifically deliver CUR to the osteolytic tumor microenvironment in the bone (CUR-ALN-NPs). The Ms have a nano-range size with a positive surface charge, 4% of DL, and the other characteristics are presented in Table 4.1. The alendronate-conjugated Ms showed significant bone-targeting potential. Rapid binding of the Ms to hydroxyapatite surface suggested these Ns as promising carriers for effective and targeted delivery of CUR n to osteolytic tumors in the bone. The affinity of CUR-ALN-NPs to the bones was demonstrated using an *in vitro* hydroxyapatite (HA) binding affinity study.

Previously, Akbar et al. (2018) prepared, optimized, and characterized the CUR-loaded biodegradable and biocompatible, self-assembled mixed PMs of Pluronic F123 and Pluronic F127. The DL and entrapment efficiency were significantly increased when CUR was encapsulated with pluronic-based mixed PMd as compared to that of pluronic-based Ms alone. *In vitro* studies showed that pluronic-based mixed PMs significantly increased anticancer when compared to pure CUR and/or Pluronic-based Ms alone.

Lin et al. (2020) investigated the antitumor efficacy of amphiphilic Pluronic F-127 Ms and their active mechanism. Gastric cancer cells were treated with different concentrations of CUR-Ms. CellTiter 96® AQueous One Solution Cell Proliferation Assay, flow cytometry, real-time cellular analysis, and nude mice xenografts were used to evaluate the effects of CUR-Ms on gastric cancer cell growth *in vitro* and *in vivo*. CUR-Ms significantly inhibited proliferation and colony formation and induced apoptosis in gastric tumors. Furthermore, CUR-Ms demonstrated that these NPs acted on mitochondrial proteins, causing changes in mitochondrial function and affecting mitochondrial bioenergetics. Furthermore, CUR-Ms decreased mitochondrial membrane potential, increased reactive oxygen species (ROS) generation, and disrupted redox equilibrium. The nude mouse model verified that CUR-M treatment significantly attenuated tumor growth *in vivo*. These results demonstrated that CUR-Ms suppressed gastric tumor cell growth *in vitro* and *in vivo*.

Zhao, Sun, and Dong (2020) used a central composite design to optimize a formulation of mixed micelles composed of Pluronics P123 and F68. The average size of the mixed micelles was 68 nm, and the EE and loading capacity for curcumin were 87% and 7%, respectively. Other NP characteristics are described in Table 4.1. It was shown that 50% of the loaded curcumin was released from the micelles in 72 h demonstrating that this formulation had sustained release properties.

Hu et al. (2020) prepared a new injectable dosage form by encapsulating CUR in folate (FA) modified methoxypoly (ethylene glycol)-poly (ε-caprolactone) (MPEG–PCL) Ms. Preclinical's study in colorectal cancer in mice model has shown that FA/Nano-CUR-Ms exhibited a significantly better therapeutic effect than free CUR and Nano-CUR. FA/Nano-CUR treatment induced tumor cell apoptosis while significantly reducing tumor proliferation and angiogenesis, which might be attributed to improved bioavailability and enhanced tumor targeting effect. Taken together, FA/Nano-CUR is a promising therapeutic agent for colorectal cancer treatment with low toxicity and better efficacy as well as good biodegradability.

FA/Nano-CUR was suggested to have the capability to target the FA receptors on tumor cells, thus enhancing tumor-targeted drug delivery. The physical and chemical characterization and the *in vitro* drug release study are described in Table 4.1. *In vivo,* drug release study results showed elongated t1/2 and increased Cmax as well as enlarged area under curve (AUC) by Nano-CUR and FA/Nano-CUR-Ms. Other studies showed that the nanoformulation of CUR significantly extended the drug duration of the agent.

Zheng et al. (2016) studied biodegradable Ms using the copolymer monomethoxy poly(ethylene glycol)-poly(lactide) (MPEG-PLA). These CUR-MPEG-PLA-PMs were applied against C6 and U251 glioma cells *in vitro* and *in vivo* to treat glioma cancer. Compared with free CUR, the cellular uptake and cytotoxicity were increased *in vitro* after CUR was encapsulated into PMs. The characterization of these systems is found in Table 4.1. The IC_{50} values of the nanoformulation on U251 cells were 29.02 and 6.82 μg/mL after 24 and 48 h, respectively, while against C6 cells were 4.38 and 1.57 μg/mL after 24 and 48 h, respectively. There was an increase of 40% in the amount of CUR released from the system after 120 h, while the free CUR exhibited a high release level (40%) at 2 h. CUR-MPEG-PLA-Ms showed significant effectiveness in suppressing glioma growth by inducing more cell apoptosis and inhibiting more cell proliferation and angiogenesis compared to free CUR. The authors considered that these formulations had the potential for clinical applications in glioma therapy.

To overcome the multidrug resistance in cancer cells, a novel redox-sensitive micellar system was described by (Zhao, Sun, and Dong 2020). Doxorubicin (DOX) and CUR were co-conjugated onto a zwitterionic polymer, poly(carboxybetaine) (pCB), to form CUR-pCB-DOX that self-assembled into stable Ms. These systems nanometric described in Table 4.1 exerted significantly stronger antitumor activity against human mammary adenocarcinoma MCF-7 and MCF-7/Adr *in vitro* than individual delivery formulation (i.e., pCB-DOX+pCB-CUR) at the same concentrations of both drugs. Moreover, there was a significant reduction in the IC_{50} value of pCB-DOX (14.1 μg/mL) compared to the free DOX value (437.2 μg/mL), and co-delivery of pCB-DOX with pCB-CUR further reduced the IC50 value to 8.9 μg/mL. Also, CUR-pCB-DOX exhibited the strongest cytotoxicity effect against MCF-7/Adr cells (5.87 μg/mL). The conjugate cellular uptake was enhanced due to the increased permeability and retention effect of tumor cells on the micellar conjugate.

Lachowicz et al. (2019) obtained a bioconjugate of sodium alginate (AA) and CUR (AA-CUR) and used this to prepare stable calcium crosslinked spherical Ms serving as a delivery vehicle for CUR for the treatment of gastric cancer. The AA-CUR-PMs showed a particle size of 200 nm, with high colloidal stability described in Table 4.1. Prolonged release of CUR was achieved for PMs during 5 h under physiological conditions (pH 7.4, at 37°C), with ~62% and 95% of CUR being released after 1 and ~4 h, respectively. A dose-dependent sensitivity of all tested cancer cells to AA-CUR was observed. The IC_{50} values were 0.35, 0.175, 0.175, and 0.35 mg/mL for 4T1, B16F10, CT26-CEA, and MC38-CEA cell lines, respectively.

Another redox-sensitive Ms based on hydrophilic hyaluronic acid (HA)-ss-hydrophobic CUR conjugate (HSC) was designed by Tian et al. (2018) as a novel delivery system for glioma targeting. Ms obtained were further functionalized with Tween 80 (CUR-THSC) for better brain penetration. Spherical Ns with a mean particle size of 74.2 nm, negative zeta potential (–30.25 mV), high EE (94.12%) and DL (8.9%) were obtained (see Table 4.1). The highest cytotoxicity in G422 cells was observed compared to the free CUR group and non-sensitive Ms group. CUR-THSC-Ms could emerge as a promising platform for the delivery of CUR in the treatment of gliomas.

Li et al. (2014) revealed through the *in vitro* cytotoxicity assay that CUR-Ms were reasonably more efficient compared to native CUR against multiple cancer cell lines because of the improved cellular uptake of CUR, which further resulted in the apoptosis of cancer cell lines. In addition, increased apoptosis of S-65 cancer cells via CUR-Ms was observed because of the downregulation of p-Akt, Blc-2, and p-Rb and activation of caspase-9. It was also exhibited that intraperitoneal administration of CUR nanoformulation (25 mg/kg) may markedly suppress the growth of tumors in comparison with the treatment with native CUR, along with the reduced expression of vascular endothelial growth factor in tumor tissue and markedly elevated apoptosis of tumor cells.

Liu et al. (2013), in a breast tumor model, utilized a one-step solid dispersion method to prepare CUR encapsulated PMs (CUR-PMs) to overcome poor water solubility of Cur and to meet the requirement of intravenous administration (see Table 4.1).

Other data of interest were that encapsulation of CUR in PMs led to a release of Cur-M for at least 1 week, while cytotoxicity of CUR-MS for at least 1 week, while cytotoxicity remained comparable to that of free CUR.

CUR-Ms were more effective in inhibiting tumor growth and spontaneous pulmonary metastasis and prolonged survival in the 4T1 breast tumor model. Immunofluorescent and immunohistochemical studies also demonstrated the improved antitumor and anti-metastasis activity of CUR-MS.

The findings of this study also suggested that CUR-Ms might be effective in treating pulmonary carcinoma.

4.4.2 Liposomes

Liposomes (LPs) are colloidal nano-to-micro size vesicle carrier systems with a spherical shape that comprises at least a lipid bilayer that encloses an aqueous core. The nature of the phospholipids purified from natural or synthetic helps to encapsulate hydrophilic, hydrophobic, and photo-unstable drugs. The size and number of bilayers can be used to control the LP properties conveniently and get several different functions arranged simultaneously, especially to assist CUR delivery in cancer areas and overcoming drawbacks in conventional pharmacotherapy. The liposomal system has been demonstrated to enhance bioavailability, increase the concentration at the site of action, and the half-life, and reduce the photo-instability that hindered CUR therapeutic effectiveness. Probably, the LPs are the most used structures for CUR encapsulation, after Ms. In this context, we discussed here some works of the last 5 years related to CUR within the liposomal system applicable to different types of cancer. Ruttala and Ko (2015) formulated CUR (an effective NF-κB inhibitor) with paclitaxel (PTX) in a polymer–lipid hybrid Ns system as a combination cancer treatment strategy. The formulation was composed of LPs-encapsulated CUR and albumin–PTX hybrid Ns (CUR-LPs-APNs) to evaluate its anticancer effect in MCF-7 and B16F10 carcinoma cell lines. The results showed the size of this system was approximately 250 nm with a high %EE for drugs poorly soluble. The drugs were loaded into polymer–lipid hybrid Ns (Table 4.2). This high entrapment of both drugs was attributed to the hydrophobicity of drugs and the ability of the lipid bilayer to incorporate into the core. This formulation delivered prolonged amounts of the drug at the time, 35% and nearly 20% of the CUR and Paclitaxel, respectively. The combination of PTX and CUR was significantly more cytotoxic than the individual drugs. Importantly, dual-drug-loaded nanocarriers exhibited a superior cytotoxic effect than a cocktail combination at a lower dose. These results showed that such co-loaded delivery systems could serve as a promising therapeutic approach to improve clinical outcomes against various malignancies. Other researchers as Jose et al. (2018) encapsulated CUR in DOTAP-based cationic LPs and then complexed it with STAT3 siRNA. This formulation was prepared by a thin film hydration method and complexed with STAT3 siRNA. The average particle size of CUR-loaded LPs with negatively charged siRNA was 192.6±9.0 nm and the other characteristic is in Table 4.2. The combination of CUR and STAT3 siRNA resulted in greater B16F10 cancer cell growth inhibition compared with individual agents. The in vivo efficacy studies were performed in the mouse model of melanoma skin cancer, the co-administration of CUR and STAT3 siRNA was effective in controlling the melanoma tumor progression compared with either liposomal CUR or STAT3 siRNA alone. Delivered CUR and signal transducer and activator of transcription 3 (STAT3) siRNA liposomal nanocomplex using iontophoresis to treat skin cancer.

In the same year, Sun et al. (2017) who developed a novel LP system functionalized with HA. This nanoformulation showed an average particle size, PDI, and ζ potential of 236 nm, 0.232, and 36.8 mV, respectively. Regarding CUR encapsulation, the %DL and %EE showed values of 13.2% and 66%, respectively, observed in Table 4.2. This novel CUR-LP modified with hyaluronan exhibited high potential in inhibiting acute myeloid leukemia (AML) cell. Moreover, the system played

TABLE 4.2
Summary of Curcumin-Loaded Liposomes Approach for Cancer Therapy Currently Reported

Type of system	Size (nm)/PDI	PZ (mV)	%EE	%DL	Release results	Cell lines evaluated	*In Vivo* model	References
CUR-liposome and albumin–PTX hybrid nanoparticles	200/0.200	+7	CUR: 99% PTX: 99.9%	CUR: 5.9% PTX: 9.09%	CUR: 50% PTX: 30% for 48 h	Human breast cancer cell line (MCF 7) and mouse melanoma cell line (B16F10)	–	Ruttala and Ko (2015)
CUR-Cationic liposome-siRNA complex	192/0.242	+56.48	86.8%	10%	–	Mouse melanoma cell line (B16F10)	C57BL/6 Mouse model of melanoma skin cancer	Jose et al. (2017)
CUR-Liposome functionalized with acid hyaluronic	236/0.232	–36.8	66%	13.2%	66% for 48 h	Leukemia cell lines (K562, KG-1, and MV4-11)	BALB/c mice and severe combined immunodeficient NPG mice	Sun et al. (2017)
CP and CUR -liposomes	294.6/0.119	+10.8	CUR: 93.4% CP: 99.5%	–	CP: 68.2% CUR: 74.1% for 24 h	Hepatocellular carcinoma cell line (HepG2)	H22 and HepG2 xenograft tumor mouse model	Cheng et al. (2018)
CUR -tetraether liposomes	200–220/0.29–0.17	–22.3/–5.89	83.5%	91.4%	–	Ovarian carcinoma cell line (SK-OV-3)	–	Duse et al. (2018)
CUR-liposomes using Eudragit S100 coating	36–70/N/A	–	98%	–	100% after 200 min	Human colon carcinoma cell line (Caco-2, clone TC7)	–	De Leo et al. (2018)
CUR-Liposomal dry powder inhalers	94.65/ N/A	–	–	–	–	Human bronchial epithelial cells lines (BEAS-2B) and (anti-A549)	Rat primary lung cancer models	Zhang et al. (2018)
CUR-Folate-conjugated liposomal	130.44/0.14	–25.10	4.15%	–	pH: 5.5 80% at 50 h	Non-tumorigenic epithelial (MCF-12A) and breast adenocarcinoma (MDA-MB-231) cell lines	–	Jadia et al. (2018)
CUR -CD133 aptamer A15 liposomes	86.6/0.21	–26.2	92.3%	–	60% for 40 h	Human prostate adenocarcinoma cell line (DU145)	Mice transplanted with DU145 cell line	Ma et al. (2019)
Cur-tetraether liposomes	149/0.46	–37.89	74%	1.5%	–	Lung carcinoma cell line (A549)	–	Lehmann et al. (2021)

Abbreviations: CUR, curcumin; PTX, paclitaxel; CP, cisplatin; ZP, Zeta potential; N/A, not available.

a critical role in downregulation of DNMT1 expression in AML, leading to DNA hypomethylation and reactivation of tumor suppressor genes such as miR-223. It was noteworthy that HA-CUR-LPs significantly prolonged the survival of xenogeneic AML mice. The above finding suggested that the Ns system HA-CUR-LPs may provide a new treatment method, either alone or in combination with other clinically effective agents, for AML patients. Recently, Cheng et al. (2018) formulated cisplatin (CP)/CUR-loaded LPs as a combination treatment strategy for hepatocellular carcinoma. In this study, CP and CUR co-loaded LPs were prepared by a reverse microemulsion and film dispersion method and their average particle size 294.6±14.8 nm with uniform size distribution (Table 4.2). *In vitro* study showed that the nano sized CP/CUR-LPs could synchronously release both drugs to achieve the synergistic effect against hepatocellular carcinoma cells based on the optimal ratio (1:8) of both drugs. Compared with free drug or encapsulated mono-drug therapy, the system LPs demonstrated higher antitumor activity *in vitro* against HepG2 cells with the IC_{50} of 0.62 μM. In addition, CP/CUR-LPs also increased intracellular ROS level during the HCC cells treatment. In vivo studies results showed improved antitumor activity with reduced adverse effects on both mouse hepatoma H22 and human HCC HepG2 xenografted models compared to single formulations. Duse et al. (2018) have developed tetraether lipid liposomal formulations with CUR where have upon excitation with light at a specific wavelength, CUR produces ROS in presence of oxygen, thereby exhibiting a cytotoxic effect toward the surrounding tissues, giving a total control on the onset of therapy. The results confirmed the formation of stable nanometer range liposomal vesicles (200–220 nm) containing CUR which is elsewise safe for normal cells, has apoptotic effect on tumor cells upon irradiation therefore selectively inhibiting tumor growth. Since production of cytotoxic ROS occurs only after irradiation of the intracellular photosensitizer. On the other hand, this year Saha and Pandey (2021) optimized CUR-loaded LPs for colonic drug delivery using Eudragit S100 coating as strategy to facilitate release of CUR in the gastrointestinal tract such targeting is to formulate LPs with a polymer coating. CUR was encapsulated into small unilamellar vesicles (SUVs) by the micelle-to-vesicle transition method in a simple and organic solvent-free way. CUR-loaded LPs were coated with Eudragit S100 by a fast and easily scalable pH-driven method. In particular, CUR-loaded LPs displayed size lower than 100 nm, EE of 98% (See Table 4.2). Such a coating was effective in protecting the LPs, which retained high CUR antioxidant activity after dissolution of the polymer shell in alkaline conditions. Results demonstrated that the adopted strategy used is suitable for the development of lipid systems for colonic delivery of hydrophobic bioactive molecules such as CUR.

Zhang et al. (2018), conducted studies on the treatment of lung cancer using liposomal CUR dry powder inhaler, whereby, the CUR-loaded LPs were freeze-dried to form the dry powder formulation. The uptake of CUR-LPs by human lung cancer (A549) cells was markedly greater and faster than free CUR. The high cytotoxicity on A549 cells and the low cytotoxicity of CUR-LPs on normal human bronchial (BEAS-2B) epithelial cells yielded a high selection index partly due to increased cell apoptosis. Besides, Lehmann et al. (2021) worked in the improvement of pulmonary photodynamic therapy with nebulization of CUR-loaded tetraether LPs. *In vitro* irradiation experiments on A549 revealed an excellent cytotoxic response of the nebulized LPs in which the stabilization of the lipid bilayer was the determining factor. Jadia et al. (2018) develop a folate-conjugated liposomal CUR system. The characterization of these LPs is expressed in Table 4.2. *In vitro* assays observed increased uptake of LPs compared to non-malignant cells, and also showed greater cytotoxicity resulting from treatments with free CUR and liposomal CUR (folate surface-modified and non-folate liposome types). The imaging and killing results indicated that the liposomal CUR formulations do not alter the therapeutic efficacy and selectivity of CUR to provide anticancer benefits. These results highlight the importance of folate-targeted LPs and its role in delivery of CUR for cancer treatment. In 2019, Ma et al. (2019) characterized CUR-loaded CD133 aptamer A15 LPs, the specification is in Table 4.2. Initial drug burst release was observed within 2 h and then, the drug was continuously released over 48 h. *In vitro* results showed that blank LPs had lower cellular cytotoxicity. LPs of CUR with or without A15 modified exhibited a similar trend of cellular cytotoxicity at the same concentration. With the extension of incubation time, A15-CUR-LPs showed a greater inhibitory effect on cells. Cell internalization in DU145 cells was higher for A15-CUR-LPs than others. An in vivo study using DU145 prostate carcinoma bearing mice showed that A15-CUR-LPs reduced tumor growth more than other forms of CUR.

4.4.3 Polymeric Nanoparticles in the Cancer Treatment

Polymeric nanoparticles (PNs) have sizes close to 100–500 nm and are extensively studied as carriers of promising colloidal drug delivery to overcome the limitations and stability problems of different drugs. These systems can incorporate active ingredients of a different nature with a relatively high drug load, have the ability to release the drug at a controlled and sustained rate at the target site, and protect the entity and activity of the active component during the transport from the systemic circulation. In recent years, different dosage forms of CUR nanoparticles for the treatment of cancer have been researched, such as poly (N-vinyl caprolactam) (PNVCL), poly (lactide-co-glycolide) (PLGA) and chitosan nanoparticles (CS), administered by different routes (Sun et al. 2012).

One of the most current and innovative biopolymers researched in the field of cancer treatment, is a PNVCL, a biocompatible polymer with thermoresponsive nature and complexation ability. Rejinold and cols. have developed CUR-loaded thermoresponsive chitosan-g-PNVCL nanoparticles (TRC-NPs) by a simple ionic cross-linking method as a cancer drug carrier. CUR-loaded TRC-NPs showed a nanometric size around 220 nm with a very high loading efficiency for all drug concentrations (Table 4.3). *In vitro* release studies were performed at two different temperatures (above and below LCST) to evaluate the nature of thermal response of the nanoparticle. After 72 h experiment, at above LCST 40% CUR has been released from the system, while at below LCST, only a 5% drug release is observed. The release profile observed above the LCST was around 10%–20% in the first 10 h, followed by a slower and almost constant continuous release. The system would follow a swelling-controlled release mechanism, especially in this initial period of release at the above LCST. The cytotoxicity studies evidenced that CUR-loaded TRC-NPs showed specific toxicity to cancer cells (PC3 and L929 cell lines) above their LCST. The hemolysis test showed that CUR-loaded and naked CRT-NPs are not hemolytic in the concentration range 0.002–10 mg/mL. Apoptosis assays showed that the cell death is concentration dependent and confirmed to be mediated by mitochondria. Beside, a differential sensitivity of CUR to normal and cancer cells was observed, evidencing that CRT-NPs loaded with CUR could be potential treatment for cancer (Sanoj Rejinold et al. 2011).

On the other hand, PLGA is a biodegradable and biocompatible polymer widely used and studied as a drug delivery system. Several different types of PLGA NPs for CUR encapsulation have been explored for cancer treatment. Tsai et al. have developed nanoparticulate PLGA systems to encapsulate CUR (CUR-NPs) by the high-pressure emulsification-solvent evaporation method (Tsai et al. 2011). CUR-NPs achieved a drug EE of around 50% and sustained release profile (see Table 4.3). Distribution studies evidenced that CUR could be detected in diverse organs, including liver, heart, spleen, lung, kidney, and brain. *In vivo* tests showed that CUR and CUR-NPs could cross the blood–brain barrier (BBB) and enter brain tissue, concentrating mainly on the hippocampus. After intravenous administration of CUR-NPs, the half-life ($t_{1/2}$) of CUR was significantly increased in the cerebral cortex (from 2.32 to 19.9) min and hippocampus (from 7.56 to 16.7) min. Additionally, CUR-NPs significantly prolonged retention time of CUR in the cerebral cortex (increased by 96%) and the hippocampus (increased by 83%). Otherwise, Prabhuraj and cols studied the combination of PLGA-CUR-loaded NPs coated with polyethylene glycol (PEG) and conjugated with different targeting agents, such as folic acid (FA), HA, or transferrin (Tf), with the help of amide coupling through EDC/NHS. Characterization studies showed spherical PLGA nanoparticles with a uniform PEG coating, ensuring a potentially long circulation time and helping to improve the solubility of the hydrophobic drug. The DL percentage decreased from PLGA-CUR to PLGA-CUR-PEG and PLGA-CUR-PEG-ligand NPs for FA, HA, and Tf respectively (Table 4.3). *In vitro* drug release showed that, maximum percentage of CUR was released at an acidic pH 6.0 of cancer cells (62.67%),

TABLE 4.3
Summary of Curcumin-Loaded Polymeric Formulations Approach for Cancer Therapy Currently Reported

Type of System	Size (nm)/ PDI	PZ (mv)	%EE	%DL	Release Results	Cell lines Evaluated	*In Vivo* Model	References
CUR-CS-g-PNVCL nanoparticles	180–220	N/A	~95%	N/A	10%–20% in about 10 h at above LCST	Prostate cancer cell line (PC3) and normal cell line (L929)	–	Sanoj Rejinold et al. (2011)
CUR- PLGA nanoparticles	163/0.053	−12.5	46.9%	N/A	59% at the first 12 h. After 6 days, 89.7%.	–	Male Sprague-Dawley rats	Tsai et al (2011)
PLGA-CUR-nanoparticles coated with PEG and different targeting moieties	200–275/ 0.189–0.297	N/A	PLGA-Cur 66.76%, PLGA-Cur-PEG 57.04% and PLGA-Cur-PEG-ligand 41.02% FA, 47.2% HA and 42.76% Tf	PLGA-Cur 3.5%, PLGA-Cur-PEG, 3%, PLGA-Cur-PEG-ligand, 2.15% (FA), 2.48% (HA), and 2.25%(Tf)	At 192 h PLGA-Cur 62.67% (pH 6.0) and 12.62% (pH 7.4); PLGA-Cur-PEG 96.9% (pH 6.0) and 72.5% (pH 7.4).	Breast cancer cell line (MDA-MB-231) and normal cell line (L929)	–	Prabhuraj et al. (2020)
ICA- PLGA and CUR- PLGA nanoparticles	30–250	–	74%	N/A	49% at 24 h. Cumulative drug release was 93% at 96 h	Human colorectal adenocarcinoma cell line (SW480)	–	Akhtar et al. (2020)
CUR- CS-GA nanoparticles	136/N/A	+50	95%	3.57%	76% at 8 h	Human colorectal adenocarcinoma cell line (HT29) and a human colon carcinoma cell line (HCT116)	–	Udompornmongkol and Chiang, (2015)

(Continued)

TABLE 4.3 (*Continued*)
Summary of Curcumin-Loaded Polymeric Formulations Approach for Cancer Therapy Currently Reported

Type of System	Size (nm)/ PDI	PZ (mv)	%EE	%DL	Release Results	Cell lines Evaluated	*In Vivo* Model	References
CUR- CS nanoparticles	652/N/A	–	51.7%	–	–	Breast cancer cell line (4T1)	BALB/c mice subjected to subcutaneous injection of 4T1 tumor cells	Kamalabadi-Farahani et al. (2018)
CUR- core-shell gold and CS nanoparticles	36.9/0.27		79%	11%	At 24 h ~30% (pH=7.4) and ~50% (pH=5.3)	Human breast adenocarcinoma cell line (MCF-7) and human hepatocellular carcinoma cell line (Huh7)	–	Amanlou et al. (2019)
CUR-hybrid nanoparticles of vanillin-CS	140–180	–	98.3% at 0.1, 0.75 and 1.0 for the drug to chitosan-vanillin, CFNP to CS-vanillin and TPP to CS-vanillin ratios, respectively	–	At 24 h (pH 1.2) chitosan with CFNP 97.1% and chitosan- vanillin with CFNP 78.3%. For pH 7.4, chitosan with CFNP 81.7% and chitosan-vanillin with CFNP 59.4%	Human breast adenocarcinoma cell line (MCF-7)	–	Kamaraj et al. (2018)
CUR-CS–alginate–sodium tripolyphosphate nanoparticles	<50/N/A	–	70%	–	29% at 2 h and a controlled release of 53% at 24 h	Human cervical epithelioid carcinoma cell line (HeLa)	–	Ahmadi et al. (2017)

Abbreviation: ZP, zeta potential; CUR, curcumin; PEG, polyethylene glycol; CS, chitosan; PNVCL, poly (N-vinyl caprolactam); PLGA, poly (lactide-co-glycolide); GA, gum arabic; ICA, Indole analog of CUR; TPP, tripolyphosphate; N/A, not available.

compared to physiological pH 7.4 (12.62%) of normal healthy cells, due to hydrolysis of PLGA in an acidic medium. *In vitro* experiments on MDA-MB-231 cells showed at 80 μM drug concentration, the viability gradually decreased from 92% to 41%, 32%, 39%, 19% and 8% for free CUR in PBS, to PLGA-CUR, PLGA-CUR-PEG, PLGA-CUR-PEG-ligands for Tf, HA and FA, respectively, demonstrating that PLGA can increase the solubility of CUR. Furthermore, PLGA-CUR-PEG-ligand NPs (with HA or FA) achieved higher cytotoxicity on MDA-MB-231 breast cancer cells, because of higher internalization of CUR through receptor-mediated endocytosis. This mechanism led to enhanced aggressive and metastatic breast cancer cells killing with targeted NPs (Prabhuraj et al. 2020). In addition, a research group explored the anticancer effect of the Indole analog of CUR (ICA) compared with CUR loading in PLGA NPs. In this study, ICA-loaded PLGA NPs (ICA-NPs) and CUR-loaded PLGA NPs (CUR-NPs) were synthesized by solvent evaporation technique for colorectal cancer (CRC) treatment. The optimized method achieves a percentage yield close to 74% encapsulated into the PLGA polymer matrix (see Table 4.3). NPs release the drug in a sustained manner over a prolonged period of time preserving the drug from degradation in wide ranges of pH environments. Cytotoxicity *in vitro* studies in human colon cancer cell line SW480 CRC suggested that drug-loaded NPs inhibits cell proliferation in a dose-dependent manner. The NPs treatment against SW480 cancer cell line triggered nuclear fragmentation, cell cycle blockade, inhibition of apoptosis and metastatic biomarkers. *In vitro* assays demonstrated that CUR-NPs and ICA-NPs induce cell cycle arrest followed trigger apoptosis by altering the Bax/Bcl-2 ratio and simultaneously have the potential to inhibit metastasis by modulating one of the different tissue inhibitors of metalloproteases found in humans (TIMP-2). Besides, they attenuate the metastatic potential of SW480, as it interferes with the expression of key metastatic markers (especially MMP-9) in the human colon cancer cell line. Finally, ICA-NPs and CUR-NPs treated SW480 cell line expresses lower levels of COX-2, enzyme that participates in the inflammation innate defense, associated with cancers and irregular growths in the intestinal tract (Sufi et al. 2020).

Another biopolymer most widely applied to prepare nanoformulation applied in cancer treatment is chitosan (CS) a linear polysaccharide obtained by the deacetylation of chitin and it consists of glucosamine and N-acetyl glucosamine units. CS has exhibited excellent physicochemical properties to be widely used as a drug carrier system and administered by various routes, such as biodegradable, biocompatible, mucoadhesive, non-toxic, film-forming, and low cost (Dev et al. 2010; Felt, Buri, and Gurny 1998; Prabaharan and Mano 2005). One of the most notable characteristics of this polymer is the amino group on the CS backbone can be protonated in acidic media, making CS the only cationic natural polysaccharide to develop nanomaterials as carriers. Whereas, CS is recognized as safe (GRAS) and approved by the United States FDA for dietary use, tissue engineering, and drug delivery (Kim, Tator, and Shoichet 2011). For these reasons, CUR-loaded CS-based nanocarriers are widely studied because they can improve the bioavailability and anticancer efficacy of CUR by passive or active targeting to cancer tissues. CS backbone has active sites that allow grafting different ligands on the surface, resulting in the targeted delivery to diseased organs. In addition, it is a pH-sensitive biopolymer that changes the molecular hydrophilic-hydrophobic balance and causes the disturbance of its nanostructure in contact to an acidic environment (pH 4–6). This property is very important in cancer treatment, as generally the environment that surrounds the tumor cells is acidic, this way could achieve the controlled release of CUR targeting in tumors to reduce harmful effects on normal cells (Hu and Luo 2021).

Udompornmongkol and cols. developed NPs formulated from the combination of chitosan with gum arabic (CS/GA-NPs), both natural polysaccharides. CS/GA-NPs loaded with CUR (CUR-NPs) were obtained by an emulsification-solvent diffusion method to enhance the effects of CUR against colorectal cancer (CCR). The CUR-NPs formed were positively charged with a particle size less than 200 nm and showed high CUR %EE (see Table 4.3). *In vitro* release studies indicated that CUR-NPs in simulated gastric and intestinal fluid only released 2% CUR, providing good evidence that NPs could reach the large intestine without being degraded by the upper parts of the digestive system. Furthermore, CUR-NPs awarded anti-colorectal cancer activities in HCT116 (human colorectal carcinoma cells) and HT29 (human colorectal adenocarcinoma cells) cell lines. Cell viability was significantly reduced when exposed to free CUR and CUR-NPs (1–50 mg/mL) in a time- and dose-dependent way. At 48 h, the IC50 of F-CUR and CUR-NPs were 6.945 and 2.329 mg/mL, respectively. From these results it can be inferred that colon cancer cells have had a higher absorption of CUR from NPs and more easily induce cell apoptosis, suggesting that they may be beneficial in CCR therapy (Udompornmongkol and Chiang 2015). Moreover, another research group evaluated the antitumor effect of co-delivery of CUR-loaded CS nanoparticles and tumor necrosis factor-related apoptosis-inducing ligand (TRAIL)-placental-derived mesenchymal stem cells (PDMSC) in a mice model of triple-negative breast cancer (TNBC). The TRAIL is an anti-cancer protein that induces apoptosis of tumor cells and has been considered as a promising neo-adjuvant therapy in different types of cancer. In this study, PDMSCs were genetically engineered to provide a delivery vehicle for TRAIL at the tumor site. CUR-loaded CS nanoparticles were used to make cancer cells sensitive to TRAIL treatment by producing ROS and increasing the apoptotic effect. The nanoparticles were synthesized by ionotropic gelation method using different concentrations of CS, tripolyphosphate (TPP) solution and CUR (Table 4.3). The results of both *in vitro* studies in breast cancer cell line 4T1 and mouse models of TNBC indicated that the combination of TRAIL-PDMSCs with CUR-CS NPs effectively induces apoptosis in tumor cells and significantly inhibits tumor growth *in vivo*, reducing tumor weight and volume to 0.34 and 0.44 as compared to single treatment and control (Kamalabadi-Farahani et al. 2018).

The combination of CS with different compounds has been widely studied in diverses publications. In this case, Amanlou and cols. have investigated the synthesis of CUR-loaded core-shell nanogels composed of gold NPs and CS (gold/CS nanogels). Gold NPs were obtained by reducing AuCl4 in the presence of TPP and CS. The particle size of the CUR-loaded gold/CS nanogels was less than 50 nm with high %DL and %EE of CUR (Table 4.3). The release of CUR from gold/CS nanogels in acidic media (pH 5.3) was twofold greater than in a neutral medium (pH 7.4). This behavior could be attributed to the acidic pH sensitivity of the carrier, preferably by the presence of CS, suggesting a high anticancer potency of the CUR-loaded nanostructures under the acidic tumor microenvironment. Cell viability assays showed better cytotoxic effects of CUR-loaded CS/gold nanogels compared to free CUR in MCF7 and Huh7 cell lines. The IC50 of nanostructures against MCF-7 cells was significantly lower than free CUR, with values of 12.6 and 28.2 μg/mL, respectively. CUR encapsulation in nanogels significantly increased cellular absorption (Amanlou et al. 2019).

Likewise, CUR-loaded hybrid NPs of vanillin-CS coated with superparamagnetic calcium ferrite (CFNPs) have been developed. The functionally modified vanillin-CS was obtained by the Schiff base reaction to enhance the hydrophobic drug EE and CFNPs were added to improve biocompatibility. The vanillin-chitosan-CFNPs, hybrid nanoparticle, was fabricated by ionic gelation method (Table 4.3). *In vitro* release studies were performed at various pH, initial DL concentrations, and magnetic fields. The release profile for all formulations showed the better-sustained release of CUR on simulated gastric fluid (pH 1.2) and simulated body fluid (pH 7.4). In most cases, the release mechanism follows non-Fickian diffusion which can be explained due to the more porous nature of the material or higher swelling ability in the release medium. Besides, an increase in rate constant was observed directly proportional to the initial DL concentration of the nanocarrier. Finally, the diffusional rate of the drug was faster with the aid of CFNPs in the presence of the magnetic field. Cytotoxicity assays evidenced that CUR-loaded chitosan-vanillin with CFNPs shows high cell toxicity (IC50 of 125 μg/mL) than the pure CS nanoparticles which show only 9% cell toxicity to MCF 7 cells (IC50 of 60 μg/mL). The combination of CFNPs, CUR and modified chitosan-vanillin has a synergistic and sustained effect on the release of CUR, greatly improving the cellular toxicity against the cancerous cell line (Kamaraj et al. 2018).

Furthermore, Ahmadi et al have found that the combination of CS-alginate (ALG)-sodium tripolyphosphate (STPP) forming CUR-loaded NPs (CS/ALG/STPP-NPs) suppressed proliferation and promoted the induction of apoptosis in human cervical epithelioid carcinoma cancer cells.

CS/ALG/STPP-NPs were synthesized by a novel ultrasonic-assisted method and showed an average size around 50 nm with high EE of the drug. The *in vitro* release profile showed an initial burst release followed by a drug sustained-release at 24 h of study. This effect could be due to the binding of CUR on the surface of NPs (Table 4.2). *In vitro* cytotoxicity assay evidenced that CUR-loaded CS/ALG/STPP-NPs displayed significant antitumor activity compared with the free CUR. After incubation with NPs loaded with CUR at a concentration of 50 µg/mL for 48 h, there was a significant increase in the expression of the apoptotic Bax gene and a decrease in the expression of the Bcl2 gene in human cervical epithelioid carcinoma (HeLa) cell line, compared to those treated with free CUR. The study suggests that CUR-loaded CS/ALG/STPP-Nps promove the induction of apoptosis and suppress the proliferation of HeLa cells (Ahmadi et al. 2017).

4.4.4 Nanoemulsions

Nanoemulsions (NEs), are lipid-based nanocarriers, have been intensively investigated as carriers for CUR suitable for being administered by various routes. Generally, NEs have a small droplet size between 20 and 200 nm and characterized by excellent long-term stability, solubilization capacity for hydrophobic drugs and biopharmaceutical advantages to enhance targeting and attenuate side effects of active molecules (Singh et al. 2017; Jiang, Liao, and Charcosset 2020). These systems can be obtained from different emulsifying agents such as biopolymers (i.e. proteins and polysaccharides), oils (i.e. essential oils and facts), surfactants (i.e.Tween-80), and phospholipids. Given these properties, NEs are particularly suitable for lipophilic compounds such as CUR due to improving their solubility and bioavailability (D'Angelo et al. 2021). Therefore, several researchers have studied the possibility of using NE as nanocarriers for CUR administration. Nikolic and cols. developed low-energy NEs stabilized by polysorbate 80 and soybean lecithin as vehicles for CUR delivery. CUR-NEs at concentrations of 1, 2 and 3 mg/mL presented particle size values <150 nm (Table 4.4; Nikolic et al. 2018). Physicochemical studies demonstrated that CUR is located at the oil-water interface of the NE, which could provide adequate stability to interact with biomolecules, keeping itself protected from external degradation. EPR antioxidant activity tests exhibited a very fast activity of CUR-loaded NE and demonstrated that the encapsulation of CUR in a developed nanocarrier protected it from degradation of external factors and preserved its activity. Cytotoxicity tests demonstrated that CUR-loaded NE significantly decreased cell survival of the HeLa cell line (IC_{50} of 22.89 µg/mL) and the human malignant melanoma Fem-x cell line (IC_{50} of 37.87 µg/mL). While free CUR presented notably lower IC_{50} values of 7.77 and 20.64 µg/mL for HeLa and Fem-x cell lines, respectively. Finally, the safety profile for CUR-NE increased compared to free CUR toward normal lung fibroblast (MRC-5) cell lines. IC_{50} values were 67.72 and 26.97 µg/mL for CUR-NE and free CUR, respectively (Nikolic et al. 2020). In another study, curcumin-loaded NEs o/w (CUR-NEs) were developed by the spontaneous nanoemulsification method using Miglyol 812 (as oil), Epikuron 145 V (as surfactant), ethanol, and acetone for cancer treatment. The NEs were obtained using safe excipients with particle sizes around 200 nm and a high CUR encapsulation yield (Table 4.4). Cytotoxicity studies evaluated cell viability after treatment with CUR-loaded NE in various cancer cell lines, such as colorectal adenocarcinoma (HT29-ATCC), human gastric adenocarcinoma (AGS), mammary gland adenocarcinoma (MDA-MB-231), murine melanoma (B16F10), and HT29-ATCC-derived cells with elevated metastatic potential (HT29-US). In crescent order, the IC_{50} values observed were 24, 26.2, 75.7 and 84.6 µM for AGS, MDA-MB-231, HT29-US and HT29-ATCC cells, respectively, indicating that the nanoparticulate system significantly reduced proliferation of the cancer cells. *In vivo* studies in C57BL/6 mice evidenced that a single dose of CUR-NEs applied topically to the area of primary tumors was enough to completely prevent reincident tumor growth and spontaneous lung metastasis, in comparison with the untreated animals which showed a 70% reincidence and tumor metastasis (Guerrero et al. 2018). Likewise, Machado and cols. developed CUR-NE (CUR-NE) formulation using soy phospholipids (Lipoid S100) and Poloxamer-188 (anionic surfactant), as a potential natural photosensitizing drug in photodynamic therapy (PDT) in breast cancer model cells.

TABLE 4.4
Summary of Curcumin-Loaded Nanoemulsions Approach for Cancer Therapy Currently Reported

Type of System	Size (nm)/PDI	PZ (mV)	%EE	%DL	Release Results	Cell Lines Evaluated	*In Vivo* Model	References
CUR-nanoemulsion with MCT, polysorbate 80 and soybean lecithin	<150/0.156–0.175	–30	N/A	N/A	At 6h, 10% for three formulations	Human cervical epithelioid carcinoma cell line (HeLa) and the human malignant melanoma cell line (Fem-x)	–	Nikolic et al. (2020, 2018)
CUR-nanoemulsion with Miglyol 812 and epikuron 145 V	195–217/<0.2	–30 to –36	95%	2.1%	–	Human gastric adenocarcinoma (AGS), colorectal adenocarcinoma (HT29-ATCC), cells derived from HT29- ATCC with elevated metastatic potential (HT29-US), human mammary gland adenocarcinoma (MDA-MB-231), and murine melanoma (B16F10)	C57BL/6 mice	Guerrero, et al (2018)
CUR-nanoemulsion with MCT, natural soy phospholipids, and poloxamer 188	<200/0.179	–46.3	N/A	N/A	–	Breast cancer cells lines (MCF-7)	–	Machado et al. (2019)
CUR-nanoemulsion with black pepper oil and Tween 80	25/0.286	–9.61	N/A	N/A	–	Fast-growing mouse neuroblastoma cell line (Neuro2A)	–	Moghaddasi et al. (2018)

Abbreviations: ZP: zeta potential, CUR: curcumin; MCT: medium-chain triglycerides, N/A: not available.

CUR-NE formulations were prepared by the interfacial pre-polymer deposition and spontaneous nanoemulsification method. The empty NE and CUR-loaded (CUR-NE) nanoemulsions showed similar physicochemical characteristics with a mean particle size around 200 nm and a monodisperse distribution in the colloidal medium (See Table 4.4). *In vitro* results showed that the application of CUR-NE associated with PDT resulted in a high phototoxic effect (cell viability reduced to 57%), and an increased ROS generation in MCF-7 breast cancer cell line. Besides, an increased level of caspase 3/7 activity in this cell model was observed, indicating that this therapy triggers mechanisms that lead to cell death (Machado et al. 2019). In another research, CUR was formulated into o/w NE (Nano-CUR) obtained with black pepper oil (BP oil) as an oil phase and Tween 80 as the surfactant at 1:9 ratio, by a modified microemulsion technique. Nano-CUR formulations showed a mean particle size close to 25 nm with negative potential electrokinetic (Table 4.4). Results demonstrated that Nano-CUR has higher antioxidant activity (~10%) in comparison to free CUR, which can be attributed to the reduced particle size, incorporating a large surface area, and the greater exposure of OH groups of CUR molecules to reduce free radicals. *In vitro* studies evidenced cytotoxicity in the Neuro2A cells (a fast-growing mouse neuroblastoma cell line) treated with Nano-CUR up to a concentration of 15.63 μg/mL (Moghaddasi et al. 2018).

4.4.5 Solid Lipid Nanoparticles

Solid lipid nanoparticles (SLNs) consist of a solid lipid core, in which lipophilic therapeutics can be efficiently entrapped, protected from chemical degradation, and controlled drug release.

SLNs proved attractive as carriers for controlled drug delivery (Wang et al. 2016) and there is evidence to support that SLNs can improve the bioavailability of poorly hydrophilic and lipophilic drugs. Li et al. (2009) and Luo et al. (2006) have improved the bioavailability of quercetin and vinpocetine. Furthermore, SLNs show low cytotoxicity to mammalian cells, demonstrating an acceptable tolerance to the body (Lin et al. 2017). Some advantage of SLNs over polymeric Ns is based on the lipid matrix, reported to efficiently improve pharmacokinetic profile (Fundarò et al. 2000; Zara et al. 2002), targeted drug delivery to tumor sites and efficacy, while minimizing systemic side effects (Jain et al. 2008; Reddy et al. 2006), decreasing the potential for acute, and chronic toxicity (Dudhipala and Veerabrahma 2016).

Rahman, Ramanathan, and Sankar (2014) described the development of SLNs of CUR by high-speed homogenization technique. Stearic acid was used as a lipid, tween 80 as a surfactant, and various co-surfactants were used for the preparation of SLN.

The particle sizes were in the range of 80–200 nm, and your specifications are in Table 4.5.

In vitro studies utilized the IMR 32 cell lines (Human Neuroblastoma cell line) in MTT assay to determine the cytotoxicity of CUR-SLNs. IC_{50} was found to be 129.33 and 60.08 μg/mL for CUR and CUR-SLNs, respectively. The authors described that these results are indicative that CUR uptake by neuroblastoma cells was increased sixtyfold through SLNs compared to plain CUR. Wang et al. (2018) developed CUR-SLNs by the emulsification evaporation-low temperature solidification method, using biocompatible and biodegradable materials. The optimized formulation had a well-defined spherical shape; a mean particle size of about 40 nm with a negative surface charge -25.3 ± 1.3 mV (Table 4.5). The CUR-SLNs showed a high uptake efficiency and a stronger cytotoxicity against SKBR3 cells, compared to cells treated by free CUR. After 48 h incubation, the IC50 value of CUR and CUR-SLNs on SKBR3 cells was 28.42 and 18.78 μM, respectively. Moreover, they observed that CUR-SLNs could inhibit cell migration. In recent years, paclitaxel (PTX) has been used to treat ovarian and breast carcinoma alone or in combination with other antineoplastic agents, such as cisplatin or carboplatin, with promising results (Miao et al. 2013). Taking advantage of the recognized activity of PTX, (Das et al. 2020) developed PTX-SLNs and CUR-SLNs by emulsion evaporation–solidification procedure, whose mean particle size and PDI were found to be within desirable limits, influenced by the percentage of lipid used (see Table 4.5). From the *in vitro* release studies, both formulations showed an initial burst release followed by a controlled

TABLE 4.5
Summary of Curcumin-Loaded Solid Lipid Nanoparticles Formulations Approach for Cancer Therapy Currently Reported

Type of System	Size (nm)/PDI	PZ (mV)	%EE	%DL	Release Results	Cell Lines Evaluated	*In Vivo* Model	References
CUR-SLNs by high-speed homogenization technique	80–200/0.451	N/A	58–85	78.12–93.33	50% of the SLN in the first 2h and 92% at 48h	Human Neuroblastoma cells line (IMR 32)	–	Rahman et al. (2014)
CUR-SLNs by the emulsification evaporation-low temperature solidification method	40/N/A	−25.3±1.3	72.47	23.38	–	Human breast cancer cells line (SKBR3)	–	Wang et al. (2018)
PTX-SLNs and CUR-SLNs	PTX: 136.6/0.211 CUR: 312.6/0.441	PTX: -25.4 CUR:-48.4	PTX-SLN: 97 CUR-SLN: 95.4	PTX-SLN: 3.6 CUR-SLN: 3.5	Initial burst release followed by a controlled release up to 48h		Toxicity studies in mice on their kidney and liver	Das et al. (2020)
CUR-SLNs CUR-TPGS	CUR-SLN: 125.2/0.268 CUR-TPGS: 285/0.283	CUR-SLN: −19.4 CUR-TPGS: 21.2	N/A	N/A	–	Hodgkin lymphoma cells line (L-540)	Hodgkin's lymphoma heterotopic xenograft model in SCID mice	Guorgui et al. (2018)
CUR-SLNs, with or without chitosan coating	175–190/<0.2	N/A	70–75	N/A	–	Human breast cancer cell line (MCF-7) Human Triple-Negative Breast Cancer Cell Line (TNBC MDA-MB-231) Murine mammary cancer JC cells (syngeneic with balb/C mice)	JC cells orthotopically implanted in female immunocompetent balb/C mice	Abd-Ellatef et al. (2020)

(*Continued*)

TABLE 4.5 (*Continued*)
Summary of Curcumin-Loaded Solid Lipid Nanoparticles Formulations Approach for Cancer Therapy Currently Reported

Type of System	Size (nm)/PDI	PZ (mV)	%EE	%DL	Release Results	Cell Lines Evaluated	*In Vivo* Model	References
(CUR+PIP)-SLNs	130.8/0.152	–20	CUR: 87.4 PIP:14.7	CUR: 19.56 µg/mg PIP: 3.26 µg/mg	A burst release of ~36% during the first 2h and 71.5% of CUR at 96h, and 89% of PIP in 24h	Paclitaxel-resistant human ovarian carcinoma cell line (A2780/Taxol)	–	Tang et al. (2017)
Liquid and solid lipids to encapsulate CUR into SLNs	150–200/ 0.187–0.276	–15 to –21.8	47.16–96.79	0.47–0.74	An initial burst release ~40% in the first 12h and, up to 70% over the next 72h.	Human breast cancer cell line (MCF-7)	Male Sprague-Dawley rats (280–300g)	Sun et al. (2013)
CUR- and RSV-SLNs	100–200	CUR-RSV-gelucire 50/13: –37.48 CUR-RSV-gelucire 50/13-HPβCD: -36.43	N/A	N/A	CUR-RSV-gelucire 50/13: 93.6 (at the end of 9h) CUR-RSV-gelucire 50/13-HPβCD: 104.64 (at the end of 9h)	Human colon cancer cell line (HCT 116)	–	Gumireddy et al. (2019)
CUR-SLNs by emulsification and low temperature solidification method	56.2	–26.2	N/A	37		Human adenocarcinoma cell line (A549)	–	Jiang et al. (2017)
SLNs were prepared using PEGylated Tefose 1500 (mixed PEG-6 stearate and PEG-32 stearate) with different surfactants.	167.60–520.00/ 0.323–1 SLNs, containing Tween 80:167.60/0.323	-17.92 to -35.90 SLNs, containing Tween 80: –26.91	35.51–40.02 SLNs, containing Tween 80: 40.01	N/A	SLNs, containing Tween 80: 75.02 and 98.25 at 24 and 48h	Human epidermoid squamous cell carcinoma cell line (A431)	Male Mus musculus Albino mice	Abdel Fadeel, Kamel, and Fadel (2020)

Abbreviations: PDI, polydispersity index; PZ, zeta potential; DL, drug loading; EE, encapsulation efficiency; CUR, curcumin; SLN, solid lipid nanoparticles; PTX, paclitaxel; TPGS, tocopheryl polyethyleÞne glycol succinate; PIP, piperine; PEG, polyethylene glycol; RSV, resveratrol; N/A, not available.

release up to 48 h. Toxicity studies in mice on their kidney and liver, hematological and biochemical parameters showed near normal values indicating that the SLNs formulation was highly tolerable in animals. The authors emphasize that in the mice group treated with the physical mixture of two formulations of PTX-SLNs and CUR-SLNs shows a better anticancer property than PTX-SLNs formulation alone, which both formulations may be a great potential for future nanoparticle-based medicine for the treatment of cancer.

Besides SLNs, other formulations have been studied. One of them was the formulation of CUR d-α-Tocopheryl PEG 1,000 succinate-stabilized CUR (CUR-TPGS) (Rachmawati et al. 2017, 2016) that provides a better *in vivo* kinetic profile than CUR. Beyond that its use in formulating CUR, a formulation containing paclitaxel formulated in TPGS has shown improved oral bioavailability in rats by about six-fold (Varma and Panchagnula 2005). Therefore, SLN and TPGS appear ideal formulations to enhance CUR's bioavailability *in vivo*. Guorgui et al. (2018) assessed whether formulating CUR in SLNs (CUR-SLNs) and in TPGS (CUR-TPGS) could improve the pharmacokinetics of CUR and enhance its anticancer effect in an animal model of Hodgkin's lymphoma (HL). CUR-SLNs and TPGS -CUR have a mean diameter below 300 nm (Table 4.5). Importantly, both CUR-SLNs and CUR-TPGS increased the level of CUR in the plasma, although the CUR when formulated in SLNs presents a better pharmacokinetic profile than CUR-TPGS and CUR alone.

An important point highlighted by the authors was the detection of CUR in the tumor, which was higher in the CUR-SLNs group, and the fact that tumor levels correlated inversely with tumor volume. These results suggested that the increased efficacy of CUR-SLNs may be closely related to enhanced CUR delivery to L-540 tumors (Goel, Jhurani, and Aggarwal 2008).

A critical obstacle to successful cancer treatment is multidrug resistance (MDR). Plasma membrane-associated adenosine triphosphate (ATP)-binding cassette (ABC) drug efflux transporters, such as the P-glycoprotein (Pgp) transporter, are the main thing responsible for MDR phenotypes. CUR is a Pgp inhibitor, but, as mentioned above, it is poorly soluble and bioavailable. To overcome these limitations, (Abd-Ellatef et al. 2020) described the efficacy and safety of CUR, loaded in biocompatible SLNs, with or without CS coating. Both CUR-loaded SLNs were five to ten-fold more effective than free CUR in increasing the intracellular retention and toxicity of doxorubicin in Pgp-expressing TNBC. CUR-loaded SLNs also effectively rescued the sensitivity to doxorubicin against drug-resistant TNBC tumors, without signs of systemic toxicity. To achieve the same chemosensitizing efficacy, free CUR had to be used at 25–50 μM, a range of concentrations that elicits cytotoxicity *in vitro* and is difficult to reach in vivo. From the analysis of the results, the authors suggested that the combination therapy, based on CUR-SLNs and doxorubicin, is an effective and safe approach to overcome the Pgp-mediated chemoresistance in TNBC.

Continuing research related to MDR and directing attention to the overexpression of Pgp, a key factor in the development of MDR, (Tang et al. 2017) investigated a novel co-delivery system of SLNs with TPGS and Brij 78® to enable CUR and Piperine (PIP) to overcome tumor drug resistance.

PIP has been reported to enhance the bioavailability of CUR both in preclinical studies and in studies on human volunteers (Shoba et al. 1998). PIP inhibits Pgp-mediated efflux in Caco-2 cells and CYP3A4-mediated drug metabolism (Bhardwaj et al. 2002). TPGS, a water-soluble succinate ester of vitamin E, can inhibit Pgp-mediated drug efflux and increase the oral bioavailability of anticancer drugs (Guo et al. 2013).

(CUR+PIP)-SLNs had a relatively high negative ζ and exhibited a mean particle diameter of ~130.8 nm (Table 4.5). Moreover, a smooth sphere morphology and uniform size distribution were observed by TEM.

The efficacy of overcoming MDR in tumor cells, cytotoxicity assay and cell uptake experiments of (CUR+PIP)-SLNs were conducted on the paclitaxel-resistant human ovarian carcinoma cell line (A2780/Taxol). The authors described that cell inhibition following treatment with CUR-SLNs or (CUR+PIP)-SLNs was higher than that observed following treatment with free CUR or (CUR-PIP) solutions ($P<0.01$). In addition, they demonstrated that blank SLNs, free Brij® or TPGS did not apparently inhibit cell growth, while that the SLNs with TPGS and Brij 78® could sensitize MDR cells. The authors emphasize that although the PIP load in (CUR-PIP)-SLNs (3.26±0.05 μg/mg) was not very high, the improvement in cytotoxicity exerted by CUR was remarkable.

Sun et al. (2013) used both liquid and solid lipids to encapsulate CUR into SLNs and thus improve the aqueous dispersibility and stability of CUR, prolonging its anticancer activity and cellular uptake and enhancing its bioavailability. CUR-SLNs were prepared by hot high-pressure homogenization combined with a high-shear dispersion method, using liquid lipid Sefsol-218® or Dynasan 114® as the lipid oil phase blended with solid lipids.

The results showed that CUR-SLNs have a near spherical shape with a size distributed between 150 and 200 nm under the TEM. Mixing suitable amounts of the liquid lipid Sefsol-218® reduced the particle size and stabilized the SLNs, but it slightly reduced the EE from 66% to 53%. The concentration of the surfactant Pluronic F68® significantly impacted particle size and EE in the CUR-SLNs (Table 4.5). These CUR-SLNs showed prolonged inhibition (at 72 h) of cell growth and cellular uptake in cancer cells, which differs from unformulated free CUR. Relative bioavailability was significantly increased after intravenous administration in rats.

The main disadvantage of these CUR-SLNs formulations was the low physical stability, observing precipitation of CUR after a few days.

Another formulation was developed by Gumireddy et al. (2019) to overcome the solubility problem of CUR by means of SLN. CUR- and resveratrol-containing SLNs were successfully synthesized by using gelucire 50/13 and gelucire 50/13- and Hydroxypropyl β-cyclodextrin (HPβCD) with a particle size in a desirable range of 100–200 nm (See Table 4.5).

SLNs made by gelucire 50/13 and gelucire 50/13- HPβCD were found to be physically stable with respect to particle size range over a period of 21 and 7 days, respectively. MTT assay results showed anticancer activity for CUR-resveratrol-gelucire 50/13- HPβCD (9.9 μM) and CUR-resveratrol-gelucire 50/13 (6.9 μM); IC_{50} values lower than the IC_{50} values found in the literature (Jayaprakasha, Chidambara Murthy, and Patil 2016).

It is important to highlight that CUR is a promising photosensitizer that has recently been discovered and could be used in PDT. Jiang et al. (2017) developed a SLNs that could act as an effective carrier to improve the photocytotoxicity of CUR for cancer treatment. CUR-SLNs were synthesized using emulsification and low temperature solidification methods and studied on the non-small cell lung cancer (A549) cells for PDT. The authors described that this novel system (see Table 4.5) dramatically increased intracellular shuttling of CUR and its accumulation in mitochondria, enhanced the hydrolytic stability of CUR, and increased light-induced ROS formation when CUR was exposed to 430 nm light. The results revealed that CUR-SLNs induced ~2.27-fold toxicity higher than free CUR at a low concentration of 15 μM under light excitation, stocking more cell cycles at G2/M phase. Western blot analysis revealed that CUR-SLNs increased the expression of caspase-3, caspase-9 proteins and promoted the ratio of Bax/Bcl-2.CUR-SLNs. Abdel Fadeel, Kamel, and Fadel (2020) also designed SLNs loaded with CUR to target skin cancer by PDT. SLNs were prepared using PEGylated Tefose 1,500 (mixed PEG-6 stearate and PEG-32 stearate) with different surfactants. SLNs, containing Tween 80, had the smallest particle size (167.60±15.12 nm) with uniform distribution (see Table 4.5) The dark and photo-cytotoxicity studies on a human skin cancer cell line (A431) revealed that, at all tested concentrations, the viability of cells treated with SLNs, containing Tween 80, was significantly lower than those treated by CUR suspension and, it decreased upon irradiation by blue light (410 nm). Results demonstrated that the PEGylated SLNs, containing Tween 80, could increase the penetration and deposition of CUR into skin layers, intensifying and extending its effect to reach deeper skin layers, in addition to its stabilizing effect.

4.5 CONCLUSION

Curcumin is a member of the ginger family, having many interesting pharmacological properties in the field of medicine. Despite its attractive and diverse pharmacological properties, it is less in use for therapeutic purposes due to its (1) less/poor aqueous solubility, (2) low oral bioavailability,

(3) limited physicochemical stability, (4) poor absorption from GIT, (5) rapid metabolism, and (6) short biological half-life. Curcumin has water solubility as low as 11 ng/mL, which causes low oral bioavailability. Different conjugates were prepared with the objective to increase the stability and solubility of curcumin in the water phase.

ACKNOWLEDGMENTS

The authors thank Consejo Nacional de Investigaciones Científicas y Técnicas, Universidad Nacional de Villa María, Universidad Nacional de Córdoba, Centro de Excelencia en Productos y Procesos de Córdoba and ARVO Foundation Collaborative Research Fellowship.

REFERENCES

Abd-Ellatef, Gamal Eldein Fathy, Elena Gazzano, Daniela Chirio, Ahmed Ragab Hamed, Dimas Carolina Belisario, Carlo Zuddas, Elena Peira, et al. 2020. "Curcumin-Loaded Solid Lipid Nanoparticles Bypass p-Glycoprotein Mediated Doxorubicin Resistance in Triple Negative Breast Cancer Cells." *Pharmaceutics* 12 (2). https://doi.org/10.3390/pharmaceutics12020096.

Abdel Fadeel, Doaa A., Rabab Kamel, and Maha Fadel. 2020. "PEGylated Lipid Nanocarrier for Enhancing Photodynamic Therapy of Skin Carcinoma Using Curcumin: In-Vitro/in-Vivo Studies and Histopathological Examination." *Scientific Reports* 10 (1): 1–10. https://doi.org/10.1038/s41598-020-67349-z.

Ahmadi, Fatemeh, Maryam Ghasemi-Kasman, Shahram Ghasemi, Maryam Gholamitabar Tabari, Roghayeh Pourbagher, Sohrab Kazemi, and Ali Alinejad-Mir. 2017. "Induction of Apoptosis in HeLa Cancer Cells by an Ultrasonic-Mediated Synthesis of Curcumin-Loaded Chitosan-Alginate-STPP Nanoparticles." *International Journal of Nanomedicine* 12 (November): 8545–56. https://doi.org/10.2147/IJN.S146516.

Akbar, Muhammad Usman, Khalid Mahmood Zia, Ahsan Nazir, Jamshed Iqbal, Syeda Abida Ejaz, and Muhammad Sajid Hamid Akash. 2018. "Pluronic-Based Mixed Polymeric Micelles Enhance the Therapeutic Potential of Curcumin." *AAPS PharmSciTech* 19 (6): 2719–39. https://doi.org/10.1208/s12249-018-1098-9.

Akbari, Abolfazl, Meghdad Sedaghat, Javad Heshmati, Seidamir Pasha Tabaeian, Sadegh Dehghani, Ana Beatriz Pizarro, Zahra Rostami, and Shahram Agah. 2021. "Molecular Mechanisms Underlying Curcumin-Mediated MicroRNA Regulation in Carcinogenesis; Focused on Gastrointestinal Cancers." *Biomedicine and Pharmacotherapy* 141: 111849. https://doi.org/10.1016/j.biopha.2021.111849.

Akhtar, B., F. Muhammad, B. Aslam, M. K. Saleemi, and A. Sharif. 2020. "Pharmacokinetic Profile of Chitosan Modified Poly Lactic Co-Glycolic Acid Biodegradable Nanoparticles Following Oral Delivery of Gentamicin in Rabbits." *International Journal of Biological Macromolecules* 164: 1493–500.

Amanlou, N., M. Parsa, K. Rostamizadeh, S. Sadighian, and F. Moghaddam. 2019. "Enhanced Cytotoxic Activity of Curcumin on Cancer Cell Lines by Incorporating into Gold/Chitosan Nanogels." *Materials Chemistry and Physics* 226 (March): 151–57. https://doi.org/10.1016/j.matchemphys.2018.12.089.

American Cancer Society. 2015. *Global Cancer Facts &* Figures 3rd Edition.

Araya-Sibaja, Andrea M., Norma J. Salazar-López, Krissia Wilhelm Romero, José R. Vega-Baudrit, J. Abraham Domínguez-Avila, Carlos A.Velázquez Contreras, Ramón E. Robles-Zepeda, Mirtha Navarro-Hoyos, and Gustavo A. González-Aguilar. 2021. "Use of Nanosystems to Improve the Anticancer Effects of Curcumin." *Beilstein Journal of Nanotechnology. Beilstein-Institut Zur Forderung der Chemischen Wissenschaften* 12: 1047–62. https://doi.org/10.3762/BJNANO.12.78.

Bhardwaj, Rajinder K., Hartmut Glaeser, Laurent Becquemont, Ulrich Klotz, Suresh K. Gupta, and Martin F. Fromm. 2002. "Piperine, a Major Constituent of Black Pepper, Inhibits Human P-Glycoprotein and CYP3A4." *Journal of Pharmacology and Experimental Therapeutics* 302 (2): 645–50. https://doi.org/10.1124/jpet.102.034728.

Bordoloi, Devivasha, Nand K. Roy, Javadi Monisha, Ganesan Padmavathi, and Ajaikumar B. Kunnumakkara. 2016. "Multi-Targeted Agents in Cancer Cell Chemosensitization: What We Learnt from Curcumin Thus Far." *Recent Patents on Anticancer Drug Discovery* 11 (1): 67–97. https://doi.org/10.2174/1574892810666151020101706.

Cabral, Horacio, Yoko Matsumoto, Kazue Mizuno, Qixian Chen, Mami Murakami, Miwako Kimura, Y. Terada, et al. 2011. "Accumulation of Sub-100 Nm Polymeric Micelles in Poorly Permeable Tumours Depends on Size." *Nature Nanotechnology* 6 (12): 815–23. https://doi.org/10.1038/nnano.2011.166.

Cheng, Yao, Pengxuan Zhao, Shuangping Wu, Tan Yang, Yan Chen, Xiaojuan Zhang, Chuanchuan He, et al. 2018. "Cisplatin and Curcumin Co-Loaded Nano-Liposomes for the Treatment of Hepatocellular Carcinoma." *International Journal of Pharmaceutics* 545 (1–2): 261–73. https://doi.org/10.1016/j.ijpharm.2018.05.007.

Cui, Wenqiang, Adnane Aouidate, Shouguo Wang, Qiuliyang Yu, Yanhua Li, and Shuguang Yuan. 2020. "Discovering Anti-Cancer Drugs via Computational Methods." *Frontiers in Pharmacology*. Frontiers Media S.A. https://doi.org/10.3389/fphar.2020.00733.

D'Angelo, Natália A., Mariana A. Noronha, Isabelle S. Kurnik, Mayra C.C. Câmara, Jorge M. Vieira, Luís Abrunhosa, Joana T. Martins, et al. 2021. "Curcumin Encapsulation in Nanostructures for Cancer Therapy: A 10-Year Overview." *International Journal of Pharmaceutics*. Elsevier B.V. https://doi.org/10.1016/j.ijpharm.2021.120534.

Das, Piu, Sanjit Kumar Roy, Arjit Guha, and Ketousetuo Kuotsu. 2020. "Fabrication and Comparative Evaluation of Curcumin and Paclitaxel-Loaded Solid Lipid Nanoparticle; the Pathway of Effective Cancer Therapy." *International Journal of Pharmaceutical Sciences and Research* 11 (3): 1110–20. https://doi.org/10.13040/IJPSR.0975-8232.11(3).1110-20.

De Leo, V., F. Milano, E. Mancini, R. Comparelli, L. Giotta, A. Nacci, F. Longobardi, et al. 2018. Encapsulation of Curcumin-Loaded Liposomes for Colonic Drug Delivery in a pH-Responsive Polymer Cluster Using a pH-Driven and Organic Solvent-Free Process. *Molecules* 23 (4): 739. https://doi.org/10.3390/molecules23040739.

Dev, Ashish, Jithin C. Mohan, V. Sreeja, H. Tamura, G. R. Patzke, F. Hussain, S. Weyeneth, S. V. Nair, and R. Jayakumar. 2010. "Novel Carboxymethyl Chitin Nanoparticles for Cancer Drug Delivery Applications." *Carbohydrate Polymers* 79 (4): 1073–79. https://doi.org/10.1016/j.carbpol.2009.10.038.

Dewangan, Rukmani, A. K. Sharma, Naveen Kumar, S. K. Maiti, Himani Singh, A. K. Gangwar, Sameer Shrivastava, Sonal, and Amit Kumar. 2012. "In-Vitro Biocompatibility Determination of Bladder Acellular Matrix Graft." *Trends in Biomaterials and Artificial Organs* 25 (4): 161–71.

Dudhipala, Narendar, and Kishan Veerabrahma. 2016. "Candesartan Cilexetil Loaded Solid Lipid Nanoparticles for Oral Delivery: Characterization, Pharmacokinetic and Pharmacodynamic Evaluation." *Drug Delivery* 23 (2): 395–404. https://doi.org/10.3109/10717544.2014.914986.

Duse, Lili, Shashank Reddy Pinnapireddy, Boris Strehlow, Jarmila Jedelská, and Udo Bakowsky. 2018. "Low Level LED Photodynamic Therapy Using Curcumin Loaded Tetraether Liposomes." *European Journal of Pharmaceutics and Biopharmaceutics* 126: 233–41. https://doi.org/10.1016/j.ejpb.2017.10.005.

Etheridge, Michael L., Stephen A. Campbell, Arthur G. Erdman, Christy L. Haynes, Susan M. Wolf, and Jeffrey McCullough. 2013. "The Big Picture on Nanomedicine: The State of Investigational and Approved Nanomedicine Products." *Nanomedicine: Nanotechnology, Biology, and Medicine* 9 (1): 1–14. https://doi.org/10.1016/j.nano.2012.05.013.

Fadus, Matthew C., Cecilia Lau, Jai Bikhchandani, and Henry T. Lynch. 2017. "Curcumin: An Age-Old Anti-Inflammatory and Anti-Neoplastic Agent." *Journal of Traditional and Complementary Medicine*. National Taiwan University. https://doi.org/10.1016/j.jtcme.2016.08.002.

Felt, Olivia, Pierre Buri, and Robert Gurny. 1998. "Chitosan: A Unique Polysaccharide for Drug Delivery." *Drug Development and Industrial Pharmacy* 24 (11): 979–93. https://doi.org/10.3109/03639049809089942.

Feng, Xia, Caixia Wang, Baorui Lin, and Fang Xu. 2006. "Methoxy Poly(Ethylene Glycol)-Conjugated Linoleic Acid Polymeric Micelles for Paclitaxel Delivery." *Colloid Journal* 68 (6): 779–83. https://doi.org/10.1134/S1061933X06060160.

Fundarò, Anna, Roberta Cavalli, Alessandro Bargoni, Daniela Vighetto, Gian Paolo Zara, and Maria Rosa Gasco. 2000. "Non-Stealth and Stealth Solid Lipid Nanoparticles (SLN) Carrying Doxorubicin: Pharmacokinetics and Tissue Distribution after i.v. Administration to Rats." *Pharmacological Research* 42 (4): 337–43. https://doi.org/10.1006/phrs.2000.0695.

Giordano, Antonio, and Giuseppina Tommonaro. 2019. "Curcumin and Cancer." *Nutrients*. MDPI AG. https://doi.org/10.3390/nu11102376.

Goel, Ajay, and Bharat B. Aggarwal. 2010. "Curcumin, the Golden Spice from Indian Saffron, Is a Chemosensitizer and Radiosensitizer for Tumors and Chemoprotector and Radioprotector for Normal Organs." *Nutrition and Cancer*. https://doi.org/10.1080/01635581.2010.509835.

Goel, Ajay, Sonia Jhurani, and Bharat B. Aggarwal. 2008. "Multi-Targeted Therapy by Curcumin: How Spicy Is It?" *Molecular Nutrition and Food Research* 52 (9): 1010–30. https://doi.org/10.1002/mnfr.200700354.

Gou, Maling, Ke Men, Huashan Shi, Mingli Xiang, Juan Zhang, Jia Song, Jianlin Long, et al. 2011. "Curcumin-Loaded Biodegradable Polymeric Micelles for Colon Cancer Therapy *in Vitro* and in Vivo." *Nanoscale* 3 (4): 1558–67. https://doi.org/10.1039/c0nr00758g.

Guerrero, Simón, Mariela Inostroza-Riquelme, Pamela Contreras-Orellana, Victor Diaz-Garcia, Pablo Lara, Andrea Vivanco-Palma, Areli Cárdenas, et al. 2018. "Curcumin-Loaded Nanoemulsion: A New Safe and Effective Formulation to Prevent Tumor Reincidence and Metastasis." *Nanoscale* 10 (47): 22612–22. https://doi.org/10.1039/c8nr06173d.

Gumireddy, Ashwini, Ryann Christman, Dunesh Kumari, Amit Tiwari, E. Jeffrey North, and Harsh Chauhan. 2019. "Preparation, Characterization, and In Vitro Evaluation of Curcumin- and Resveratrol-Loaded Solid Lipid Nanoparticles." *AAPS PharmSciTech* 20 (4). https://doi.org/10.1208/s12249-019-1349-4.

Guo, Yuanyuan, Jun Luo, Songwei Tan, Ben Oketch Otieno, and Zhiping Zhang. 2013. "The Applications of Vitamin e TPGS in Drug Delivery." *European Journal of Pharmaceutical Sciences* 49 (2): 175–86. https://doi.org/10.1016/j.ejps.2013.02.006.

Guorgui, Jacob, Ruixue Wang, George Mattheolabakis, and Gerardo G. Mackenzie. 2018. "Curcumin Formulated in Solid Lipid Nanoparticles Has Enhanced Efficacy in Hodgkin's Lymphoma in Mice." *Archives of Biochemistry and Biophysics* 648: 12–19. https://doi.org/10.1016/j.abb.2018.04.012.

Gupta, Subash C., Gorkem Kismali, and Bharat B. Aggarwal. 2013. "Curcumin, a Component of Turmeric: From Farm to Pharmacy." *BioFactors*. https://doi.org/10.1002/biof.1079.

Gupta, Subash C., Sridevi Patchva, and Bharat B. Aggarwal. 2013. "Therapeutic Roles of Curcumin: Lessons Learned from Clinical Trials." *AAPS Journal*. https://doi.org/10.1208/s12248-012-9432-8.

Gupta, Subash C., Bokyung Sung, Ji Hye Kim, Sahdeo Prasad, Shiyou Li, and Bharat B. Aggarwal. 2013. "Multitargeting by Turmeric, the Golden Spice: From Kitchen to Clinic." *Molecular Nutrition and Food Research*. https://doi.org/10.1002/mnfr.201100741.

Hanahan, Douglas, and Robert A Weinberg. 2000. "The Hallmarks of Cancer Review Evolve Progressively from Normalcy via a Series of Pre." *Cell* 100 (1): 57–70. https://doi.org/10.1016/s0092-8674(00)81683-9.

Hanahan, Douglas, and Robert A. Weinberg. 2011. "Hallmarks of Cancer: The next Generation." *Cell*. https://doi.org/10.1016/j.cell.2011.02.013.

Hassaninasab, Azam, Yoshiteru Hashimoto, Kaori Tomita-Yokotani, and Michihiko Kobayashi. 2011. "Discovery of the Curcumin Metabolic Pathway Involving a Unique Enzyme in an Intestinal Microorganism." *Proceedings of the National Academy of Sciences of the United States of America* 108 (16): 6615–20. https://doi.org/10.1073/pnas.1016217108.

Heger, Michal, Rowan F. van Golen, Mans Broekgaarden, and Martin C. Michel. 2014. "The Molecular Basis for the Pharmacokinetics and Pharmacodynamics of Curcumin and Its Metabolites in Relation to Cancers." *Pharmacological Reviews*. https://doi.org/10.1124/pr.110.004044.

Hewlings, Susan J., and Douglas S. Kalman. 2017. "Curcumin: A Review of Its Effects on Human Health." *Foods*. MDPI AG. https://doi.org/10.3390/foods6100092.

Hoelder, Swen, Paul A. Clarke, and Paul Workman. 2012. "Discovery of Small Molecule Cancer Drugs: Successes, Challenges and Opportunities." *Molecular Oncology*. John Wiley and Sons Ltd. https://doi.org/10.1016/j.molonc.2012.02.004.

Hu, Qiaobin, and Yangchao Luo. 2021. "Chitosan-Based Nanocarriers for Encapsulation and Delivery of Curcumin: A Review." *International Journal of Biological Macromolecules*. Elsevier B.V. https://doi.org/10.1016/j.ijbiomac.2021.02.216.

Hu, Yuzhu, Yihong He, Jianrui Ji, Songping Zheng, and Yongzhong Cheng. 2020. "Tumor Targeted Curcumin Delivery by Folate-Modified MPEG-PCL Self-Assembly Micelles for Colorectal Cancer Therapy." *International Journal of Nanomedicine* 15: 1239–52. https://doi.org/10.2147/IJN.S232777.

Huang, Min, Jin Jian Lu, and Jian Ding. 2021. "Natural Products in Cancer Therapy: Past, Present and Future." *Natural Products and Bioprospecting* 11 (1): 5–13. https://doi.org/10.1007/s13659-020-00293-7.

Imai, Kohzoh, and Akinori Takaoka. 2006. "Comparing Antibody and Small-Molecule Therapies for Cancer." *Nature Reviews Cancer*. https://doi.org/10.1038/nrc1913.

Jadia, Rahul, Janel Kydd, Brandon Piel, and Prakash Rai. 2018. "Liposomes Aid Curcumin's Combat with Cancer in a Breast Tumor Model." *Oncomedicine* 3: 94–109. https://doi.org/10.7150/oncm.27938.

Jain, Sanjay K., Akash Chaurasiya, Yashwant Gupta, Anekant Jain, Pradeep Dagur, Beenu Joshi, and Vishwa M. Katoch. 2008. "Development and Characterization of 5-FU Bearing Ferritin Appended Solid Lipid Nanoparticles for Tumour Targeting." *Journal of Microencapsulation* 25 (5): 289–97. https://doi.org/10.1080/02652040701799598.

Jayaprakasha, Guddadarangavvanahally K., Kotamballi N. Chidambara Murthy, and Bhimanagouda S. Patil. 2016. "Enhanced Colon Cancer Chemoprevention of Curcumin by Nanoencapsulation with Whey Protein." *European Journal of Pharmacology* 789: 291–300. https://doi.org/10.1016/j.ejphar.2016.07.017.

Jiang, Shan, Rongrong Zhu, Xiaolie He, Jiao Wang, Mei Wang, Yechang Qian, and Shilong Wang. 2017. "Enhanced Photocytotoxicity of Curcumin Delivered by Solid Lipid Nanoparticles." *International Journal of Nanomedicine* 12: 167–78. https://doi.org/10.2147/IJN.S123107.

Jiang, Tian, Wei Liao, and Catherine Charcosset. 2020. "Recent Advances in Encapsulation of Curcumin in Nanoemulsions: A Review of Encapsulation Technologies, Bioaccessibility and Applications." *Food Research International*. Elsevier Ltd. https://doi.org/10.1016/j.foodres.2020.109035.

Jose, A., S. Labala, and V. V. Venuganti. 2017. "Co-Delivery of Curcumin and STAT3 siRNA Using Deformable Cationic Liposomes to Treat Skin Cancer." *Journal of Drug Targeting* 25 (4): 330–41. https://doi.org/10.1080/1061186X.2016.1258567.

Jose, Anup, Suman Labala, Kunal Manoj Ninave, Sudeep Kumar Gade, and Venkata Vamsi Krishna Venuganti. 2018. "Effective Skin Cancer Treatment by Topical Co-Delivery of Curcumin and STAT3 SiRNA Using Cationic Liposomes." *AAPS PharmSciTech* 19 (1): 166–75. https://doi.org/10.1208/s12249-017-0833-y.

Kamalabadi-Farahani, Mohammad, Mohammad Vasei, Naser Ahmadbeigi, Somayeh Ebrahimi-Barough, Masoud Soleimani, and Reza Roozafzoon. 2018. "Anti-Tumour Effects of TRAIL-Expressing Human Placental Derived Mesenchymal Stem Cells with Curcumin-Loaded Chitosan Nanoparticles in a Mice Model of Triple Negative Breast Cancer." *Artificial Cells, Nanomedicine and Biotechnology* 46 (sup3): S1011–21. https://doi.org/10.1080/21691401.2018.1527345.

Kamaraj, Sriram, Uma Maheswari Palanisamy, Meera Sheriffa Begum Kadhar Mohamed, Arthanareeswaran Gangasalam, Gover Antoniraj Maria, and Ruckmani Kandasamy. 2018. "Curcumin Drug Delivery by Vanillin-Chitosan Coated with Calcium Ferrite Hybrid Nanoparticles as Carrier." *European Journal of Pharmaceutical Sciences* 116 (April): 48–60. https://doi.org/10.1016/j.ejps.2018.01.023.

Kamble, Sumedh, Pegah Varamini, Markus Müllner, Théophile Pelras, and Ramin Rohanizadeh. 2020. "Bisphosphonate-Functionalized Micelles for Targeted Delivery of Curcumin to Metastatic Bone Cancer." *Pharmaceutical Development and Technology* 25 (9): 1118–26. https://doi.org/10.1080/10837450.2020.1798458.

Kemp, Jessica A., and Young Jik Kwon. 2021. "Cancer Nanotechnology: Current Status and Perspectives." *Nano Convergence* 8 (1): 34. https://doi.org/10.1186/s40580-021-00282-7.

Kim, Howard, Charles H. Tator, and Molly S. Shoichet. 2011. "Chitosan Implants in the Rat Spinal Cord: Biocompatibility and Biodegradation." *Journal of Biomedical Materials Research - Part A* 97 A (4): 395–404. https://doi.org/10.1002/jbm.a.33070.

Kocaadam, Betül, and Nevin Şanlier. 2017. "Curcumin, an Active Component of Turmeric (Curcuma Longa), and Its Effects on Health." *Critical Reviews in Food Science and Nutrition* 57 (13): 2889–95. https://doi.org/10.1080/10408398.2015.1077195.

Kotha, Raghavendhar R., and Devanand L. Luthria. 2019. "Curcumin: Biological, Pharmaceutical, Nutraceutical, and Analytical Aspects." *Molecules*. MDPI AG. https://doi.org/10.3390/molecules24162930.

Kulkarni, Vitthal S., and Charles Shaw. 2016. "Surfactants, Lipids, and Surface Chemistry." In *Essential Chemistry for Formulators of Semisolid and Liquid Dosages*, 5–19. https://doi.org/10.1016/b978-0-12-801024-2.00002-9.

Kunnumakkara, Ajaikumar B., Devivasha Bordoloi, Choudhary Harsha, Kishore Banik, Subash C. Gupta, and Bharat B. Aggarwal. 2017. "Curcumin Mediates Anticancer Effects by Modulating Multiple Cell Signaling Pathways." *Clinical Science*. Portland Press Ltd. https://doi.org/10.1042/CS20160935.

Kunnumakkara, Ajaikumar B., Devivasha Bordoloi, Ganesan Padmavathi, Javadi Monisha, Nand Kishor Roy, Sahdeo Prasad, and Bharat B. Aggarwal. 2016. "Curcumin, the Golden Nutraceutical: Multitargeting for Multiple Chronic Diseases." *British Pharmacological Society*. https://doi.org/10.1111/bph.13621.

Kurita, T., and Y. Makino. 2013. "Novel Curcumin Oral Delivery Systems." *Anticancer Research* 33 (7): 2807–21.

Kurnik, Isabelle S., Natália A. D'Angelo, Priscila G. Mazzola, Marlus Chorilli, Daniel T. Kamei, Jorge F.B. Pereira, António A. Vicente, and André M. Lopes. 2021. "Polymeric Micelles Using Cholinium-Based Ionic Liquids for the Encapsulation and Release of Hydrophobic Drug Molecules." *Biomaterials Science* 9 (6): 2183–96. https://doi.org/10.1039/d0bm01884h.

Kuttan, Ramadasan, P. Bhanumathy, K. Nirmala, and M. C. George. 1985. "Potential Anticancer Activity of Turmeric (Curclima LongA)." *CancerLetters* 29 (2): 197–202.

Lachowicz, Dorota, Alicja Karabasz, Monika Bzowska, Michał Szuwarzyński, Anna Karewicz, and Maria Nowakowska. 2019. "Blood-Compatible, Stable Micelles of Sodium Alginate - Curcumin Bioconjugate for Anti-Cancer Applications." *European Polymer Journal* 113: 208–19. https://doi.org/10.1016/j.eurpolymj.2019.01.058.

Lampe, V., and J. Milobedzka. 1913. "Studien Über Curcumin." *Berichte der deutschen chemischen Gesellschaft* 46: 2235–40.

Lehmann, Jennifer, Michael R. Agel, Konrad H. Engelhardt, Shashank R. Pinnapireddy, Sabine Agel, Lili Duse, Eduard Preis, Matthias Wojcik, and Udo Bakowsky. 2021. "Improvement of Pulmonary Photodynamic Therapy: Nebulisation of Curcumin-Loaded Tetraether Liposomes." *Pharmaceutics* 13 (8). https://doi.org/10.3390/pharmaceutics13081243.

Lestari, Maria L.A.D., and Gunawan Indrayanto. 2014. "Curcumin." In *Profiles of Drug Substances, Excipients and Related Methodology*, 39:113–204. Academic Press Inc. https://doi.org/10.1016/B978-0-12-800173-8.00003-9.

Li, Hou Li, Xiao Bin Zhao, Yu Kun Ma, Guang Xi Zhai, Ling Bing Li, and Hong Xiang Lou. 2009. "Enhancement of Gastrointestinal Absorption of Quercetin by Solid Lipid Nanoparticles." *Journal of Controlled Release* 133 (3): 238–44. https://doi.org/10.1016/j.jconrel.2008.10.002.

Li, Xingyi, Tongke Chen, Lu Xu, Zhaoliang Zhang, Lingli Li, and Hao Chen. 2014. "Preparation of Curcumin Micelles and the in Vitro and in Vivo Evaluation for Cancer Therapy." *Journal of Biomedical Nanotechnology* 10 (8): 1458–68. https://doi.org/10.1166/jbn.2014.1840.

Lin, Chih Hung, Chun Han Chen, Zih Chan Lin, and Jia You Fang. 2017. "Recent Advances in Oral Delivery of Drugs and Bioactive Natural Products Using Solid Lipid Nanoparticles as the Carriers." *Journal of Food and Drug Analysis* 25 (2): 219–34. https://doi.org/10.1016/j.jfda.2017.02.001.

Lin, Xiaokun, Lihua Wang, Liqian Zhao, Zheng Zhu, Tongzuan Chen, Shinuo Chen, Yecheng Tao, et al. 2020. "Curcumin Micelles Suppress Gastric Tumor Cell Growth by Upregulating ROS Generation, Disrupting Redox Equilibrium and Affecting Mitochondrial Bioenergetics." *Food and Function* 11 (5): 4146–59. https://doi.org/10.1039/d0fo00260g.

Liu, Lei, Lu Sun, Qinjie Wu, Wenhao Guo, Ling Li, Yishan Chen, Yuchen Li, Changyang Gong, Zhiyong Qian, and Yuquan Wei. 2013. "Curcumin Loaded Polymeric Micelles Inhibit Breast Tumor Growth and Spontaneous Pulmonary Metastasis." *International Journal of Pharmaceutics* 443 (1–2): 175–82. https://doi.org/10.1016/j.ijpharm.2012.12.032.

López-Lázaro, Miguel. 2008. "Anticancer and Carcinogenic Properties of Curcumin: Considerations for Its Clinical Development as a Cancer Chemopreventive and Chemotherapeutic Agent." *Molecular Nutrition and Food Research*. https://doi.org/10.1002/mnfr.200700238.

Luo, Yi Fan, Da Wei Chen, Li Xiang Ren, Xiu Li Zhao, and Jing Qin. 2006. "Solid Lipid Nanoparticles for Enhancing Vinpocetine's Oral Bioavailability." *Journal of Controlled Release* 114 (1): 53–59. https://doi.org/10.1016/j.jconrel.2006.05.010.

Ma, Qi, Wei Qian, Wei Tao, Yanling Zhou, and Boxin Xue. 2019. "Delivery of Curcumin Nanoliposomes Using Surface Modified with CD133 Aptamers for Prostate Cancer." *Drug Design, Development and Therapy* 13: 4021–33. https://doi.org/10.2147/DDDT.S210949.

Machado, Francielly Cristina, Renata Prandini Adum de Matos, Fernando Lucas Primo, Antonio Claudio Tedesco, Paula Rahal, and Marília Freitas Calmon. 2019. "Effect of Curcumin-Nanoemulsion Associated with Photodynamic Therapy in Breast Adenocarcinoma Cell Line." *Bioorganic and Medicinal Chemistry* 27 (9): 1882–90. https://doi.org/10.1016/j.bmc.2019.03.044.

Mansouri, Kamran, Shna Rasoulpoor, Alireza Daneshkhah, Soroush Abolfathi, Nader Salari, Masoud Mohammadi, Shabnam Rasoulpoor, and Shervin Shabani. 2020. "Clinical Effects of Curcumin in Enhancing Cancer Therapy: A Systematic Review." *BMC Cancer*. BioMed Central Ltd. https://doi.org/10.1186/s12885-020-07256-8.

Marqus, Susan, Elena Pirogova, and Terrence J. Piva. 2017. "Evaluation of the Use of Therapeutic Peptides for Cancer Treatment." *Journal of Biomedical Science* 24 (1). https://doi.org/10.1186/s12929-017-0328-x.

Miao, Jing, Yong Zhong Du, Hong Yuan, Xing Guo Zhang, and Fu Qiang Hu. 2013. "Drug Resistance Reversal Activity of Anticancer Drug Loaded Solid Lipid Nanoparticles in Multi-Drug Resistant Cancer Cells." *Colloids and Surfaces B: Biointerfaces* 110: 74–80. https://doi.org/10.1016/j.colsurfb.2013.03.037.

Milobedzka, J., S. Kostanecki, and V, Lampe. 1910. "Zur Kenntnis Des Curcumins." *Ber Deut Chem Ges* 43: 2163–70.

Moghaddasi, Fariba, Mohammad Reza Housaindokht, Majid Darroudi, Mohammad Reza Bozorgmehr, and Abbas Sadeghi. 2018. "Synthesis of Nano Curcumin Using Black Pepper Oil by O/W Nanoemulsion Technique and Investigation of Their Biological Activities." *LWT* 92 (June): 92–100. https://doi.org/10.1016/j.lwt.2018.02.023.

Nelson, Kathryn M., Jayme L. Dahlin, Jonathan Bisson, James Graham, Guido F. Pauli, and Michael A. Walters. 2017. "The Essential Medicinal Chemistry of Curcumin." *Journal of Medicinal Chemistry* 60 (5): 1620–37. https://doi.org/10.1021/acs.jmedchem.6b00975.

Nikolic, Ines, Dominique Jasmin Lunter, Danijela Randjelovic, Ana Zugic, Vanja Tadic, Bojan Markovic, Nebojsa Cekic, et al. 2018. "Curcumin-Loaded Low-Energy Nanoemulsions as a Prototype of Multifunctional Vehicles for Different Administration Routes: Physicochemical and in Vitro Peculiarities Important for Dermal Application." *International Journal of Pharmaceutics* 550 (1–2): 333–46. https://doi.org/10.1016/j.ijpharm.2018.08.060.

Nikolic, Ines, Evgenia Mitsou, Ana Damjanovic, Vassiliki Papadimitriou, Jelena Antic-Stankovic, Boban Stanojevic, Aristotelis Xenakis, and Snezana Savic. 2020. "Curcumin-Loaded Low-Energy Nanoemulsions: Linking EPR Spectroscopy-Analysed Microstructure and Antioxidant Potential with in Vitro Evaluated Biological Activity." *Journal of Molecular Liquids* 301 (March). https://doi.org/10.1016/j.molliq.2020.112479.

Prabaharan, M., and J. F. Mano. 2005. "Chitosan-Based Particles as Controlled Drug Delivery Systems." *Drug Delivery: Journal of Delivery and Targeting of Therapeutic Agents* 12 (1): 41–57. https://doi.org/10.1080/10717540590889781.

Prabhuraj, R. S., Kartik Bomb, Rohit Srivastava, and Rajdip Bandyopadhyaya. 2020. "Selection of Superior Targeting Ligands Using PEGylated PLGA Nanoparticles for Delivery of Curcumin in the Treatment of Triple-Negative Breast Cancer Cells." *Journal of Drug Delivery Science and Technology* 57 (June). https://doi.org/10.1016/j.jddst.2020.101722.

Prasad, Sahdeo, Subash C. Gupta, Amit K. Tyagi, and Bharat B. Aggarwal. 2014. "Curcumin, a Component of Golden Spice: From Bedside to Bench and Back." *Biotechnology Advances*. Elsevier Inc. https://doi.org/10.1016/j.biotechadv.2014.04.004.

Priyadarsini, Kavirayani Indira. 2014. "The Chemistry of Curcumin: From Extraction to Therapeutic Agent." *Molecules*. MDPI AG. https://doi.org/10.3390/molecules191220091.

"PubChem Compound Summary for CID 969516, Curcumin." n.d. Curcumin | $C_{21}H_{20}O_6$ | CID 969516 - PubChem (nih.gov).

Rachmawati, Heni, Aditya Trias Pradana, Dewi Safitri, and I. Ketut Adnyana. 2017. "Multiple Functions of D-α-Tocopherol Polyethylene Glycol 1000 Succinate (TPGS) as Curcumin Nanoparticle Stabilizer: In Vivo Kinetic Profile and Anti-Ulcerative Colitis Analysis in Animal Model." *Pharmaceutics* 9 (3): 1–13. https://doi.org/10.3390/pharmaceutics9030024.

Rachmawati, Heni, Dewi Safitri, Aditya Trias Pradana, and I. Ketut Adnyana. 2016. "TPGS-Stabilized Curcumin Nanoparticles Exhibit Superior Effect on Carrageenan-Induced Inflammation in Wistar Rat." *Pharmaceutics* 8 (3): 1–13. https://doi.org/10.3390/pharmaceutics8030024.

Rahman, Mohamed Habibur, Muthiah Ramanathan, and Veintramuthu Sankar. 2014. "Preparation, Characterization and in Vitro Cytotoxicity Assay of Curcumin Loaded Solid Lipid Nanoparticle in IMR32 Neuroblastoma Cell Line." *Pakistan Journal of Pharmaceutical Sciences* 27 (5): 1281–85.

Reddy, Lakkireddy Harivardhan, Jawahar Singh Adhikari, Bilikere Srinirasa Rao Dwarakanath, Rakesh Kumar Sharma, and Rayasa Ramachandra Murthy. 2006. "Tumoricidal Effects of Etoposide Incorporated into Solid Lipid Nanoparticles after Intraperitoneal Administration in Dalton's Lymphoma Bearing Mice." *AAPS Journal* 8 (2): 254–62. https://doi.org/10.1208/aapsj080229.

Ruttala, Hima Bindu, and Young Tag Ko. 2015. "Liposomal Co-Delivery of Curcumin and Albumin/Paclitaxel Nanoparticle for Enhanced Synergistic Antitumor Efficacy." *Colloids and Surfaces B: Biointerfaces* 128: 419–26. https://doi.org/10.1016/j.colsurfb.2015.02.040.

Sabet, Saman, Ali Rashidinejad, Laurence D. Melton, and Duncan J. McGillivray. 2021. "Recent Advances to Improve Curcumin Oral Bioavailability." *Trends in Food Science and Technology*. Elsevier Ltd. https://doi.org/10.1016/j.tifs.2021.02.006.

Saha, P., and M. M. Pandey. 2021. "A New Fluorescence-Based Method for Rapid and Specific Quantification of Rotigotine in Chitosan Nanoparticles." *Spectrochimica Acta Part A: Molecular and Biomolecular Spectroscopy*, 267 (Part 2): 120555.

Saha, S., A. Adhikary, P. Bhattacharyya, T. Das, and G. Sa. n.d. "Death by Design- Where Curcumin Sensitizes Drug-Resistant Tumours." *Anticancer Research* 32: 2567–84.

Sak, Katrin. 2020. "Radiosensitizing Potential of Curcumin in Different Cancer Models." *Nutrition and Cancer*. Routledge. https://doi.org/10.1080/01635581.2019.1681480.

Salehi, Bahare, Zorica Stojanović-Radić, Jelena Matejić, Mehdi Sharifi-Rad, Nanjangud V. Anil Kumar, Natália Martins, and Javad Sharifi-Rad. 2019. "The Therapeutic Potential of Curcumin: A Review of Clinical Trials." *European Journal of Medicinal Chemistry*. Elsevier Masson s.r.l. https://doi.org/10.1016/j.ejmech.2018.12.016.

Sanoj Rejinold, N., M. Muthunarayanan, V. V. Divyarani, P. R. Sreerekha, K. P. Chennazhi, S. V. Nair, H. Tamura, and R. Jayakumar. 2011. "Curcumin-Loaded Biocompatible Thermoresponsive Polymeric Nanoparticles for Cancer Drug Delivery." *Journal of Colloid and Interface Science* 360 (1): 39–51. https://doi.org/10.1016/j.jcis.2011.04.006.

Sarker, Satyajit D., Lutfun Nahar, Anca Miron, and Mingquan Guo. 2020. *Anticancer Natural Products. Annual Reports in Medicinal Chemistry*. 1st ed. Vol. 55. Elsevier Inc. https://doi.org/10.1016/bs.armc.2020.02.001.

Shaikh, Siraj, Javed Shaikh, Yusufi Sadia Naba, Kailas Doke, Khursheed Ahmed, and Mujahid Yusufi. 2021. "Curcumin: Reclaiming the Lost Ground against Cancer Resistance." *Cancer Drug Resistance*. https://doi.org/10.20517/cdr.2020.92.

Shanmugam, Muthu K., Grishma Rane, Madhu Mathi Kanchi, Frank Arfuso, Arunachalam Chinnathambi, M. E. Zayed, Sulaiman Ali Alharbi, Benny K.H. Tan, Alan Prem Kumar, and Gautam Sethi. 2015. "The Multifaceted Role of Curcumin in Cancer Prevention and Treatment." *Molecules*. MDPI AG. https://doi.org/10.3390/molecules20022728.

Shehzad, Adeeb, Munibah Qureshi, Muhammad Nabeel Anwar, and Young Sup Lee. 2017. "Multifunctional Curcumin Mediate Multitherapeutic Effects." *Journal of Food Science* 82 (9): 2006–15. https://doi.org/10.1111/1750-3841.13793.

Shimizu, Kana, Masafumi Funamoto, Yoichi Sunagawa, Satoshi Shimizu, Yasufumi Katanasaka, Yusuke Miyazaki, Hiromichi Wada, Koji Hasegawa, and Tatsuya Morimoto. 2019. "Anti-Inflammatory Action of Curcumin and Its Use in the Treatment of Lifestyle-Related Diseases." *European Cardiology Review* 14 (2): 117–22. https://doi.org/10.15420/ecr.2019.17.2.

Shoba, Guido, David Joy, Thangam Joseph, M. Majeed, R. Rajendran, and P. S.S.R. Srinivas. 1998. "Influence of Piperine on the Pharmacokinetics of Curcumin in Animals and Human Volunteers." *Planta Medica* 64 (4): 353–56. https://doi.org/10.1055/s-2006-957450.

Singh, Yuvraj, Jaya Gopal Meher, Kavit Raval, Farooq Ali Khan, Mohini Chaurasia, Nitin K. Jain, and Manish K. Chourasia. 2017. "Nanoemulsion: Concepts, Development and Applications in Drug Delivery." *Journal of Controlled Release* 252: 28–49. https://doi.org/10.1016/j.jconrel.2017.03.008.

Soudamini, K.K., and R. Kuttan. 1988. "Cytotoxic and Tumour Reducing Properties of Curcumin." *The Indian Journal of Pharmacology* 20: 95–101.

Srinivasan, K. R. 1953. "A Chromatographic Study of the Curcuminoids in Curcuma Longa, L." *Journal of Pharmacy and Pharmacology* 5 (1): 448–57. https://doi.org/10.1111/j.2042-7158.1953.tb14007.x.

Stewart, Bernard W., and Christopher P. Wild. 2014. *World Cancer Report 2014*, edited by Bernard W. Stewart and Christopher P. Wild. Lyon: International Agency for Research on Cancer, World Health Organization. ISBN 978-92-832-0443-5.

Sufi, Shamim Akhtar, Muddasarul Hoda, Sankar Pajaniradje, Victor Mukherjee, Selvaraj Mohane Coumar, and Rukkumani Rajagopalan. 2020. "Enhanced Drug Retention, Sustained Release, and Anti-Cancer Potential of Curcumin and Indole-Curcumin Analog-Loaded Polysorbate 80-Stabilizied PLGA Nanoparticles in Colon Cancer Cell Line SW480." *International Journal of Pharmaceutics* 588 (October). https://doi.org/10.1016/j.ijpharm.2020.119738.

Sun, Dan, Jian Kang Zhou, Linshu Zhao, Zhe Yu Zheng, Jiao Li, Wenchen Pu, Shaoyang Liu, et al. 2017. "Novel Curcumin Liposome Modified with Hyaluronan Targeting CD44 Plays an Anti-Leukemic Role in Acute Myeloid Leukemia in Vitro and in Vivo." *ACS Applied Materials and Interfaces* 9 (20): 16857–68. https://doi.org/10.1021/acsami.7b02863.

Sun, Jiabei, Chao Bi, Hok Man Chan, Shaoping Sun, Qingwen Zhang, and Ying Zheng. 2013. "Curcumin-Loaded Solid Lipid Nanoparticles Have Prolonged in Vitro Antitumour Activity, Cellular Uptake and Improved in Vivo Bioavailability." *Colloids and Surfaces B: Biointerfaces* 111: 367–75. https://doi.org/10.1016/j.colsurfb.2013.06.032.

Sun, Min, Xun Su, Buyun Ding, Xiuli He, Xiuju Liu, Aihua Yu, Hongxiang Lou, and Guangxi Zhai. 2012. "Advances in Nanotechnology-Based Delivery Systems for Curcumin." *Nanomedicine*. https://doi.org/10.2217/nnm.12.80.

Tan, Bee Ling, and Mohd Esa Norhaizan. 2019. "Curcumin Combination Chemotherapy: The Implication and Efficacy in Cancer." *Molecules*. MDPI AG. https://doi.org/10.3390/molecules24142527.

Tang, Jingling, Hongyu Ji, Jinmei Ren, Mengting Li, Nannan Zheng, and Linhua Wu. 2017. "Solid Lipid Nanoparticles with TPGS and Brij 78: A Co-Delivery Vehicle of Curcumin and Piperine for Reversing P-Glycoprotein-Mediated Multidrug Resistance in Vitro." *Oncology Letters* 13 (1): 389–95. https://doi.org/10.3892/ol.2016.5421.

Tian, Cihui, Sajid Asghar, Yurui Xu, Zhipeng Chen, Jingwei Zhang, Qineng Ping, and Yanyu Xiao. 2018. "Tween 80-Modified Hyaluronic Acid-Ss-Curcumin Micelles for Targeting Glioma: Synthesis, Characterization and Their in Vitro Evaluation." *International Journal of Biological Macromolecules* 120: 2579–88. https://doi.org/10.1016/j.ijbiomac.2018.09.034.

Tomeh, Mhd Anas, Roja Hadianamrei, and Xiubo Zhao. 2019. "A Review of Curcumin and Its Derivatives as Anticancer Agents." *International Journal of Molecular Sciences*. MDPI AG. https://doi.org/10.3390/ijms20051033.

Tsai, Yin Meng, Chao Feng Chien, Lie Chwen Lin, and Tung Hu Tsai. 2011. "Curcumin and Its Nano-Formulation: The Kinetics of Tissue Distribution and Blood-Brain Barrier Penetration." *International Journal of Pharmaceutics* 416 (1): 331–38. https://doi.org/10.1016/j.ijpharm.2011.06.030.

Udompornmongkol, Panisa, and Been Huang Chiang. 2015. "Curcumin-Loaded Polymeric Nanoparticles for Enhanced Anti-Colorectal Cancer Applications." *Journal of Biomaterials Applications* 30 (5): 537–46. https://doi.org/10.1177/0885328215594479.

Varma, Manthena V.S., and Ramesh Panchagnula. 2005. "Enhanced Oral Paclitaxel Absorption with Vitamin E-TPGS: Effect on Solubility and Permeability in Vitro, in Situ and in Vivo." *European Journal of Pharmaceutical Sciences* 25 (4–5): 445–53. https://doi.org/10.1016/j.ejps.2005.04.003.

Verma, Vivek. 2016. "Relationship and Interactions of Curcumin with Radiation Therapy." *World Journal of Clinical Oncology*. Baishideng Publishing Group Co., Limited. https://doi.org/10.5306/wjco.v7.i3.275.

Vogel A., Jr. 1842. "No Title." *Journal de Pharma. et de Chemie* 3: 20.

Vogel Pelletier. 1815. "No Title." *Journal de Pharmacie.* I:289.

Wahlström, Bo, and G. Blennow. 1978. "A Study on the Fate of Curcumin in the Rat." *Acta Pharmacologica et Toxicologica* 43 (2): 86–92. https://doi.org/10.1111/j.1600-0773.1978.tb02240.x.

Wang, Man, Shuai Jiang, Li Zhou, Fei Yu, Han Ding, Peifeng Li, Meng Zhou, and Kun Wang. 2019. "Potential Mechanisms of Action of Curcumin for Cancer Prevention: Focus on Cellular Signaling Pathways and MiRNAs." *International Journal of Biological Sciences*. NLM (Medline). https://doi.org/10.7150/ijbs.33710.

Wang, Taoran, Qiaobin Hu, Mingyong Zhou, Yan Xia, Mu Ping Nieh, and Yangchao Luo. 2016. "Development of 'All Natural' Layer-by-Layer Redispersible Solid Lipid Nanoparticles by Nano Spray Drying Technology." *European Journal of Pharmaceutics and Biopharmaceutics* 107: 273–85. https://doi.org/10.1016/j.ejpb.2016.07.022.

Wang, Wenrui, Tiantian Chen, Henan Xu, Baihui Ren, Xiaodan Cheng, Rongrong Qi, Haibo Liu, et al. 2018. "Curcumin-Loaded Solid Lipid Nanoparticles Enhanced Anticancer Efficiency in Breast Cancer." *Molecules* 23 (7): 1–13. https://doi.org/10.3390/molecules23071578.

Wicki, Andreas, Dominik Witzigmann, Vimalkumar Balasubramanian, and Jörg Huwyler. 2015. "Nanomedicine in Cancer Therapy: Challenges, Opportunities, and Clinical Applications." *Journal of Controlled Release*. Elsevier B.V. https://doi.org/10.1016/j.jconrel.2014.12.030.

Zara, Gian Paolo, Roberta Cavalli, Alessandro Bargoni, Anna Fundaro, Daniela Vighetto, and Maria Rosa Gasco. 2002. "Intravenous Administration to Rabbits of Non-Stealth and Stealth Doxorubicin-Loaded Solid Lipid Nanoparticles at Increasing Concentrations of Stealth Agent: Pharmacokinetics and Distribution of Doxorubicin in Brain and Other Tissues." *Journal of Drug Targeting* 10 (4): 327–35. https://doi.org/10.1080/10611860290031868.

Zhang, Tongtong, Yanming Chen, Yuanyuan Ge, Yuzhen Hu, Miao Li, and Yiguang Jin. 2018. "Inhalation Treatment of Primary Lung Cancer Using Liposomal Curcumin Dry Powder Inhalers." *Acta Pharmaceutica Sinica B* 8 (3): 440–48. https://doi.org/10.1016/j.apsb.2018.03.004.

Zhao, Guangfu, Yan Sun, and Xiaoyan Dong. 2020. "Zwitterionic Polymer Micelles with Dual Conjugation of Doxorubicin and Curcumin: Synergistically Enhanced Efficacy against Multidrug-Resistant Tumor Cells." *Langmuir* 36 (9): 2383–95. https://doi.org/10.1021/acs.langmuir.9b03722.

Zheng, Songping, Xiang Gao, Xiaoxiao Liu, Ting Yu, Tianying Zheng, Yi Wang, and Chao You. 2016. "Biodegradable Micelles Enhance the Antiglioma Activity of Curcumin in Vitro and in Vivo." *International Journal of Nanomedicine* 11: 2721–36. https://doi.org/10.2147/IJN.S102450.

Zhong, Lei, Yueshan Li, Liang Xiong, Wenjing Wang, Ming Wu, Ting Yuan, Wei Yang, et al. 2021. "Small Molecules in Targeted Cancer Therapy: Advances, Challenges, and Future Perspectives." *Signal Transduction and Targeted Therapy*. Springer Nature. https://doi.org/10.1038/s41392-021-00572-w.

Zhou, Siying, Sijie Zhang, Hongyu Shen, Wei Chen, Hanzi Xu, Xiu Chen, Dawei Sun, Shanliang Zhong, Jianhua Zhao, and Jinhai Tang. 2017. "Curcumin Inhibits Cancer Progression through Regulating Expression of MicroRNAs." *Tumor Biology*. SAGE Publications Ltd. https://doi.org/10.1177/1010428317691680.

Zhu, Julia, Katherine Z. Sanidad, Elvira Sukamtoh, and Guodong Zhang. 2017. "Potential Roles of Chemical Degradation in the Biological Activities of Curcumin." *Food and Function* 8 (3): 907–14. https://doi.org/10.1039/c6fo01770c.

5 Lipid-Based Carriers with Phytoconstituents for Cancer Therapy

V. Sankar, B. Shanmuganathan Alias Viswesh, M. Karthikeyan, K. Umaa, S. Ramesh, S. Malathi, and A. Justin

5.1 INTRODUCTION

Phytomedicines have been widely used since ancient times around the world due to their excellent therapeutic effects against many diseases (Sajid et al., 2019). The phytomedicines are still today used as a formulation to deliver the active constituents and a combination of the Indian system of medicine and formulations is still used in the management of many diseases.

In *Ayurveda*, the natural elements were used to treat the root cause of the disease (Parasuraman et al., 2014). The therapeutic effectiveness of medicinal plants basically depends on the delivery of active compounds to the site. So that the phytochemical, phytopharmacological, and pharmaceutical sciences have enabled the development of delivering medicinal plant products to improve biological activities.

Though the active constituents of plant extracts (flavonoids, tannins, and terpenoids) are highly soluble in water it has less absorption and less permeability through the lipid membranes of the cells. Generally, the plant-derived active constituents have less bioavailability and efficacy due to their high molecular size and poor absorption (Silva et al.,2013).

The development of nanotechnology-based formulations with herbal medicine might be helpful to deliver the active principle of the herbal extract with the required dose to minimize side effects and potentiate the activity of herbal medicines.

5.2 NANOCARRIER SYSTEMS AND THEIR APPLICATIONS

5.2.1 Nano suspension with Phytoconstituents

Nanosuspensions are drug delivery systems that contain a pure therapeutic agent and stabilizer (steric polymers such as polyethylene glycol (PEG) with a typical size range of 10–100 nm) causing the emergence of particles with a size <1 μm. It can be an appropriate choice to solve low bioavailability by increasing drug surface area that comes into contact with biological media, low, poor pharmacokinetics of insoluble drugs, and drug aqueous solubility. Drug nanosuspensions emerged as one solution to delivering such hydrophobic drugs.

The most widely used stabilizers, surfactants, and polymers, in nanosuspensions include Tween® 80, polyvinyl alcohols (PVA), pluronics (F68 and F127), polyethylene glycols (PEG), poloxamer-188, D-α-tocopherol polyethylene glycol 1,000 succinates or vitamin E-TPGS, etc.

These nanoparticles are capable of accumulating in targeted areas such as cancer tissues and infarct zones with minimal damage to healthy tissues.

Nanosuspensions are often prepared by commercially available methods such as precipitation ultrasonication technique, solvent precipitation method, high-pressure homogenization, media milling, emulsification, and melt emulsification.

DOI: 10.1201/9781003292692-5

5.2.1.1 Application of Nanosuspensions

**Nanosuspensions are used in oral, parenteral, ocular, and pulmonary drug delivery systems.

Oral administration is the first patient choice because of painless and noninvasive administration. In addition, oral formulations have several advantages for the pharmaceutical industry such as easy manufacturing, short production time, and reasonable production cost.

Oleanolic acid, which has many applications such as hepato-protective, antitumor, antibacterial, anti-inflammatory, and antiulcer effects, has low aqueous solubility which results in erratic pharmacokinetics after oral administration. Applying oleanolic acid in the form of nanosuspension increases the dissolution rate to about 90% in the first 20 min compared to just 15% for micronized drug powder (Yadollahi et al., 2014).

A comparative study of atovaquonein the form of micronized particles and nanosuspensions showed that the latter decreased infectivity from 40% to 15 (Yadollahi et al., 2014). In another example, buparvaquone nanosuspensions reduced infection from 2.0 to 1.02 and micronized particles only to 1.47 (Yadollahi et al., 2014).

5.2.1.2 Formulation of Nanosuspensions by Different Techniques

DOX nanosuspensions were prepared by the solvent precipitation method, which is one of the bottom-up techniques (Ibrahim et al., 2018). In brief, 50 mg of DOX was dissolved in 5 mL (chloroform). The drug solution in the organic solvent was dropped slowly to 20 mL of an aqueous solution containing a specific concentration of the stabilizer and stirred at 800 rpm. The dispersion was allowed to stir for 4 h to ensure complete removal of the organic solvent.

The resulting nanosuspension was then characterized for its particle size and zeta potential. The prepared DOX nanosuspension was freeze-dried at −60°C and the vacuum pressure was less than 1 mbar and stored in tightly closed containers protected from light until further studies.

Naringenin (NRG) nanosuspensions were formulated by the sonoprecipitation method. NRG was dissolved in ethanol (30 mg/mL) and the polymer was added to distilled water to prepare solvent (organic phase) and antisolvent system (aqueous phase), respectively. In brief, 1 mL of the organic phase was quickly injected using a syringe (22 needle gauze) into 10 mL of aqueous phase under magnetic stirring at 1,000 rpm for 5 min which led to the precipitation of drug particles.

Samples were immediately subjected to sonication using the ultrasonic processor for a time period of 8 min under ice-cold conditions (3°C–5°C). After sonication, stirring was continued for 3 h to allow evaporation of the solvent. For long-term storage of the final product, the freshly prepared nanosuspensions were stored in a refrigerator for 12 h to maintain cold conditions followed by chilling using dry ice (–75°C) and subsequently freeze-dried at –50°C for 12 h, followed by a secondary drying phase at 20°C for 4 h. The schematic representation for the preparation of nanosuspensions is shown in Figure 5.1.

Cyadox (CYX) nanosuspension was produced by recrystallization based on acid–base neutralization combined with high-pressure homogenization. Briefly, 10 g CYX was dissolved in 85 mL hot mixed solution containing 15 mmol NaOH and 2 g stabilizers (poloxamer 407, poloxamer 188, sodium alginate, Pluronic F-68, polyvinyl pyrrolidone or polyvinyl alcohol).

After the dissolution of the drug, 15 mL 4°C 1.0 mol/L HCL solution was quickly poured into the mixed solution under stirring by using a thermostat magnetic stirrer to recrystallize the cyadox based on acid–base neutralization. The recrystallized cyadox passed through a high-pressure homogenizer for 3 cycles at 300 bar and finally 6 cycles at 1,200 bar to obtain uniform small particle size. The nanosuspensions were collected into glass vials, labeled, and used for subsequent tests.

5.2.1.3 Nanosuspension Characterization

5.2.1.3.1 Particle Charge (Zeta Potential)

Particle charge plays an important role in ensuring stable nanosuspensions. The electric charge on a particle surface provides electrostatic repulsion between the drug nanoparticles and in this way

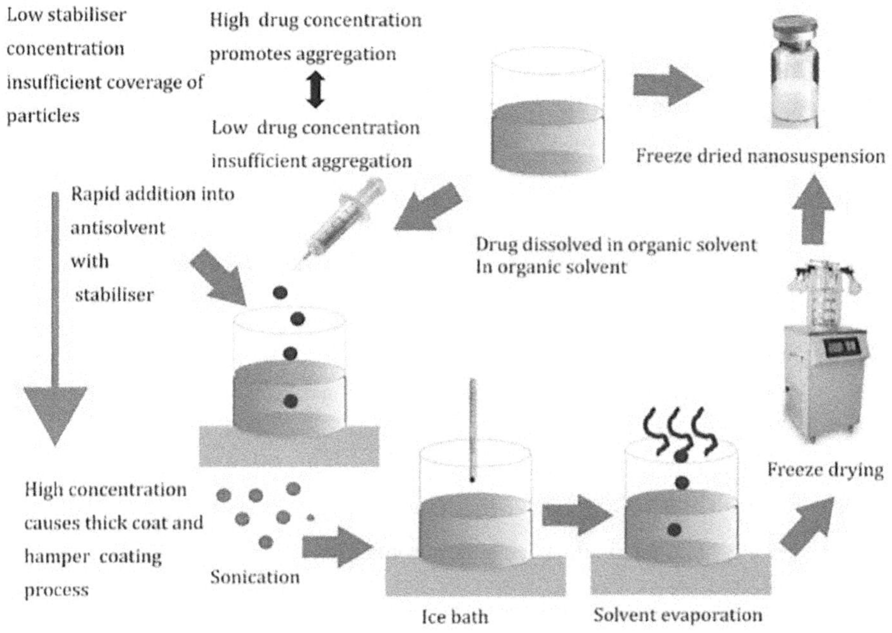

FIGURE 5.1 Schematic presentation for preparation of nanosuspensions by sonoprecipitation method; double arrows show the optimization steps in the process. (Scientific Image and Illustration Tool adopted: www.biorender.com.)

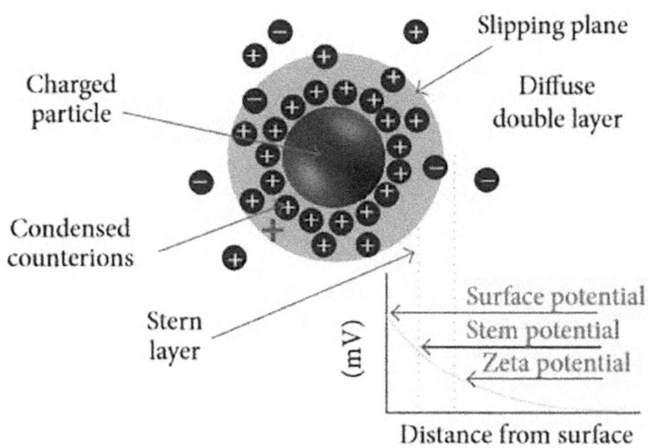

FIGURE 5.2 Electric double layer around a charged particle. (Scientific Image and Illustration Tool adopted: www.biorender.com.)

prevents particles from aggregation and precipitation. The schematic in Figure 5.2 provides an illustration of the electric double layer around a charged particle. The double layer consists of a stern layer and a diffusion layer of opposite ions.

The electric potential at the shear plane is known as the zeta potential. It is considered that a minimum zeta potential of ±30 mV is required to ensure pure electrostatic stabilization. When electrostatic stabilization is combined with steric stabilization (by using appropriate polymers), a zeta potential of ±20 mV could be sufficient to prevent drug particles from aggregation and precipitation.

Steric stabilization is defined as stabilization caused by the adsorbed and hydrated polymer layers on the dispersed particle.

The zeta potential can also be measured by applying an ultrasound wave which induces the so-called electroacoustic phenomena.

5.2.1.3.2 Stability

Precipitation of particles is another phenomenon that should be taken into account when considering the stability of nano suspensions. According to Stoke's law, decreasing particle size, reducing the density difference of solid phase, and increasing the viscosity of the medium decrease the precipitation velocity.

Common stabilizers used to formulate nanosuspensions include polysorbates, povidones, poloxamer, lecithin, polyoleate, and cellulose polymers. A mixture of surfactants and polymers has been found to be beneficial for the long-term stabilization of nanosuspensions. Nano suspensions loaded with phytocompounds are cited in Table 5.1.

5.2.2 Nanoemulsion with Phytoconstituents

Nanoemulsions are colloidal particulates of submicrometer size range varies from 10 to 1,000 nm acting as carriers of drug molecules.

It is a thermodynamically unstable system, which can be stabilized by the presence of an emulsifying agent (emulgent or emulsifier). The dispersed phase is also known as an internal phase or the discontinuous phase while the outer phase is called the dispersion medium, external phase, or continuous phase. The emulsifying agent is also known as intermediate or interphase. The term 'nanoemulsion' also refers to a mini emulsion which is fine oil/water or water/oil dispersion shown in Figure 5.3 stabilized by an interfacial film of surfactant molecule having droplet size range 20–600 nm.

Because of its small size, nanoemulsions are transparent. There are three types of nanoemulsion which can be formed:

- Oil-in-water nanoemulsion in which oil is dispersed in the continuous aqueous phase,
- Water in oil nanoemulsion in which water droplets are dispersed in the continuous oil phase,
- Bi-continuous nanoemulsions.

5.2.2.1 Advantages of Nano Emulsion

- It may be used as a substitute for liposomes and vesicles.
- It improves the bioavailability of drugs.
- It is non-toxic and non-irritant in nature.
- It has improved physical stability.

TABLE 5.1
Phytocompounds Loaded in Nanosuspension

	Naringenin Nanosuspension (Mohammed et al., 2021)	Docetaxel Nanaosuspension (Ibrahim et al., 2018)
Biological source	Flavonoids in citrus fruits	Analogue of taxol
Method of preparation	Sonoprecipitation method	Solvent precipitation
XRD	Detects crystallinity of drug upon precipitation from antisolvent	Nanoparticles are stabilized by 5% HPMC and 5% SPC
Zeta potential	Measured by dynamic light scattering in Malvern^R Zetasizer	Measured by Zetasizer

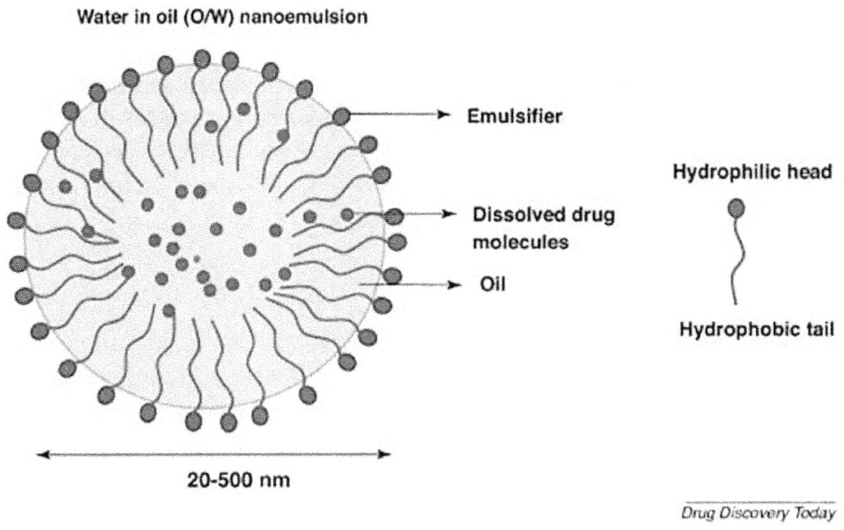

FIGURE 5.3 Water in oil-type nanoemulsion. (Scientific Image and Illustration Tool adopted: www.biorender.com.)

- Nanoemulsions have small-sized droplets having greater surface area providing greater absorption.
- It can be formulated in a variety of formulations such as foams, creams, liquids, and sprays.
- It provides better uptake of oil-soluble supplements in cell culture technology.
- It helps to solubilize lipophilic drugs.
- Helpful in taste masking.
- Less amount of energy is required.

5.2.2.2 Preparation of Piplartine Nano Emulsion (Fofaria et al., 2016)

The first type of oil-in-water nanoemulsion was formulated by the self-emulsification method. Based on the solubility assay, the surfactant mix (Smix) was screened by using different weight ratios of surfactant (Tween 80 or Solutol HS) and cosurfactant (PEG 400 or Transcutol).

The oil and the surfactant mixture (premix) were then mixed at a weight ratio of 9:1, 8:2, 7:3, 6:4, 5:5, 4:6, 3:7, 2:8, and 1:9. The oil and surfactant mixture formed 10% of the total volume. The remaining 90% was water which was added in a drop-wise manner with constant stirring. Upon addition of water, formulations were visually inspected as transparent, translucent nanoemulsions, milky emulsions, or phase separation.

The second type of oil-in-water nanoemulsion was prepared by homogenization–sonication technique using a surfactant mix of Tween 80 and DSPE-PEG 5000. The oil phase used was either 10% (−) or 20% (+). The surfactant used was 2% (−) or 1% (+)Tween 80. The cosurfactant was either not used (−) or used (+).

Based on this, eight different types of formulations were prepared. The oil phase was prepared by heating Capmul PG-8 at 50°C with or without PL (2.5 mg/mL). Once the drug was dissolved, DSPE-PEG 5000 (1% w/v) was dissolved in the oil phase. The aqueous phase was taken separately and prepared by dissolving Tween 80 in deionized water (87% w/v) at 50°C.After heating the oil and water phase for 5 min, the water phase was gradually added to the oil phase with constant and rapid stirring. The coarse emulsion was then homogenized at 9,500 rpm for 10 min using a homogenizer followed by sonication for 120 min using a bath sonicator to yield nanoemulsion.

5.2.2.3 Preparation of Baicalin Nanoemulsion (El-Gogary et al., 2019)

Baicalin was first dissolved in PEG 400 and mixed with soy-lecithin, Tween 80, and IPM, after which the required amount of water(1:2:1.5:3.75:8.25, w/w) was added and the solution stirred to obtain a clear and transparent liquid, ie, baicalin-loaded nanoemulsion (BAN-1) (Table 5.2).

The resulting nanoemulsions were filled with nitrogen gas and tightly sealed and stored at room temperature. In addition, baicalin was dissolved in the final nanoemulsion formulations to obtain baicalin-loaded nanoemulsions (BAN-2).

5.2.2.4 Characterization of Nanoemulsion

5.2.2.4.1 Changes in Nanoemulsions Solubility after Dilution

To investigate the solubilization in the diluted nanoemulsions, it was diluted 30 times with normal saline at 37°C. The samples (100 μL) were withdrawn at regular intervals over 24 h and centrifuged at 10,000 rpm for 10 min to remove the precipitated drug if any. The concentration of the drug in nanoemulsions was determined by using HPLC after appropriate dilution.

5.2.2.4.2 Stability

A stable nanoemulsion is characterized by the absence of the internal phase, absence of creaming, absence of deterioration by microorganisms, and maintenance of elegance with respect to appearance, color, odor, and consistency.

5.2.2.5 Miscellaneous Instability

Emulsions may deteriorate under extremely high or low temperatures or in the presence of light. Hence emulsions are usually packed in air-tight, colored containers and stored at moderate temperature.

5.2.3 Liposomes with Phytoconstituents

Liposomes are small artificial vesicles of spherical shape with particle sizes ranging from 30 nm to several micrometers that can be created from cholesterol and natural non-toxic phospholipids. Due to their size and hydrophobic and hydrophilic character (besides biocompatibility), furthermore, the choice of bilayer components determines the 'rigidity' or 'fluidity' and the charge of the bilayer (Figure 5.4).

For instance, unsaturated phosphatidylcholine species from natural sources (egg or soybean phosphatidylcholine) give much more permeable and less stable bilayers, whereas the saturated phospholipids with long acyl chains (for example, dipolmitoyl phosphatidylcholine) form a rigid, but impermeable bilayer.

Phospholipid bilayer membranes can transport aqueous or lipid drugs, depending on the nature of those drugs. Because lipids are amphipathic (both hydrophobic and hydrophilic) in aqueous media.

TABLE 5.2
Phytocompounds Loaded in Nanoemulsion

	Baicalin Nanoemulsion	Piplartine Nanoemulsion
Biological source	Flavone glucuronide derived from dried roots of *Scutellarea baicalensis*	Alkaloid found in black pepper
Method of preparation	Nanoemulsification	Self-emulsification
Droplet size determination	Measured using Malvern zeta sizer	Using Malvern zeta sizer
In vitro release study	Using the dialysis bag method	Using USP-II apparatus (paddle type)

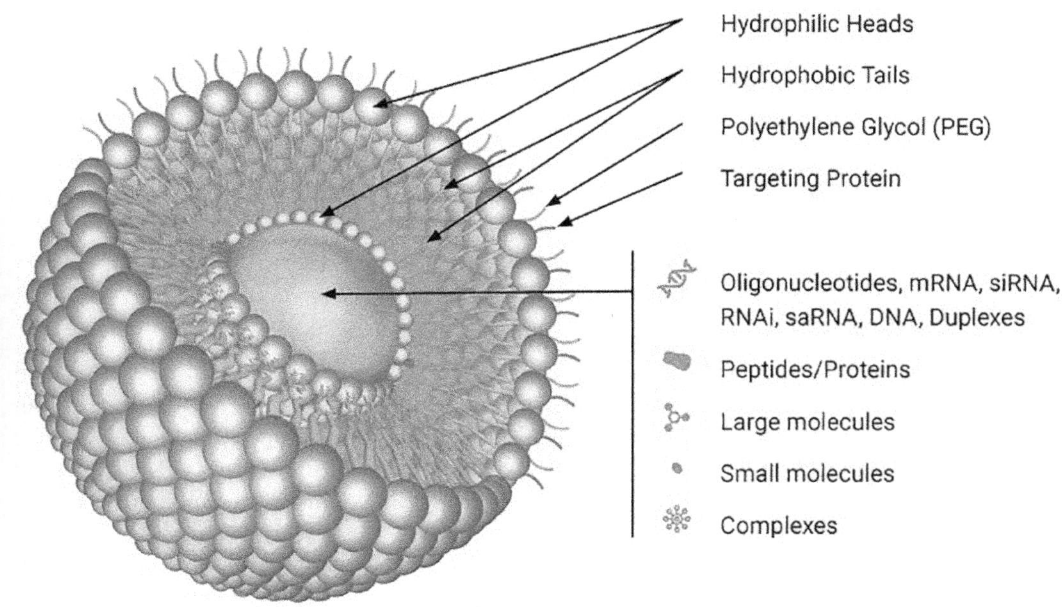

FIGURE 5.4 Bilayer liposome. (Scientific Image and Illustration Tool adopted: www.biorender.com.)

5.2.3.1 Advantages of Liposome

- Liposomes increased the efficacy and therapeutic index of the drug.
- Liposome increased stability via encapsulation.
- Liposomes are non-toxic, flexible, biocompatible, completely biodegradable, and nonimmunogenic for systemic and non-systemic administrations.
- Liposomes reduce the toxicity of the encapsulated agent (Taxol).
- Liposomes help reduce the exposure of sensitive tissues to toxic drugs.
- Flexibility to couple with site-specific ligands to achieve active targeting.

5.2.3.2 Disadvantages of Liposomes

- Low solubility,
- Short half-life,
- Sometimes phospholipid undergoes oxidation and hydrolysis-like reaction,
- Leakage and fusion of encapsulated drug/molecules,
- Production cost is high.

5.2.3.3 Preparation of Apigenin-Loaded Liposomes (Banerjee et al., 2016)

The apigenin nanoparticles were produced by the LAP technique, followed by freeze-drying.

Apigenin-loaded liposomes were prepared by using different molar ratios of (1:1, 1:2, and 1:3) based on earlier reported techniques. Briefly, accurately weighed amounts of apigenin and phospholipid were placed into a 100 mL round-bottom flask, and dissolved in 20 mL of a mixture of 1,4-dioxane:methanol (14:6). The reaction temperature of the reflux was controlled at 40°C, 50°C, or 60°C using a water bath for 2 h. The flask contents were then concentrated to obtain a dry residue.

This dried material was dissolved in a 50 mL mixture of chloroform: methanol (45:5). The solution was then concentrated to 2–3 mL and poured into 100 mL of n-hexane. The resulting product was precipitated and filtered. Subsequently, the liposome was dried completely under vacuum at 40°C. The dried product was transferred to a light-protected (amber) glass vial purged with nitrogen, and stored at room temperature (Table 5.3).

TABLE 5.3
Phytocompounds Loaded in Liposome

	Celastrol	Apigenin
Biological source	Derived from herb *Tripterygium wilfordii*	Flavone occurs in various fruits and vegetables
Method of preparation	Extrusion method	Lipid film hydration method
Cell viability assay	Measured using cell counting kit-8	Measured using culture technique

5.2.3.4 Preparation of Daidzein Long-Circulating Nanoliposome (Wang et al., 2019)

Daidzein long-circulating nanoliposome (DLCL) was prepared by thin film evaporation-sonication method described briefly as follows: soybean phosphatidylcholine, cholesterol, DSPE-mPEG 2000, and daidzein were dissolved in a round-bottomed flask with 10 mL chloroform-methanol (1:4, v/v) mixture. Under the conditions of vacuum and 40°C (water bath), the mixture was dried to form a thin film in the rotary evaporation apparatus, and then hydrated with 20 mL ultrapure water by sonication (80 w) for 24 min in an ice bath.

The liposomal suspension was extruded three times by filtering through 0.45 and 0.22 μm microporous membranes in turn. The prepared DLCL solution was stored at 4°C. For long-term preservation, 3% sucrose was added (used as a lyoprotectant) to the DLCL suspension and freeze-drying preservation at −20°C.

5.2.3.5 Preparation of Celastrol-Loaded Liposome (Wolfram et al., 2014)

The two compositions of the celastrol liposomes were as follows: (1) DOPC, cholesterol, and DSPE mPEG 2000 (7:4:3 molar ratio) and (2) DSPC, cholesterol, and DSPE mPEG 2000 (6:3:1 molar ratio).

Celastrol was added at a concentration of 1 mg/40 mg of phospholipids. The components were subsequently placed in a round-bottom flask and dissolved using a mixture of methanol and chloroform (1:2 v/v). The lipid film was made with a Rotavapor (55°C, 50 rpm).

The formulations were hydrated with phosphate-buffered saline (PBS, 1 mL/20 mg phospholipids) and subsequently heated in a 60°C water bath for 3 min and mixed with a vortex mixer for 3 min. The hydration process was repeated three times.

The formulations were then extruded through a series of polycarbonate membranes to increase homogeneity (pore size: 400, 200, and 100 nm) using an extrusion device (Table 5.3).

5.2.3.6 Liposomal Characterization

5.2.3.6.1 Serum Stability

Add 250 mL of the liposome formulation to 1 mL of serum solution (70% fetal bovine serum (FBS) and 30% PBS). The resulting mixture was incubated at 37°C and stirred with an electronic stirrer (700 rpm), to mimic the conditions of blood circulation.

Samples were taken at regular intervals and diluted in MilliQ water (1:50 v/v) for dynamic light scattering analysis of size and polydispersity, using a Zetasizer.

5.2.4 Solid Lipid Nanoparticles in Phytoconstituents

Solid lipid nanoparticles (SLNs) are introduced as an alternative carrier system to traditional colloidal carriers such as—emulsions, liposomes, and polymeric micro-and nanoparticles. Nanoparticles made from solid lipids are attracting major attention as novel colloidal drug carrier for intravenous applications as they have been proposed as an alternative particulate carrier system.

SLN are submicron colloidal carriers ranging from 50 to 1,000 nm, which are composed of physiological lipid, dispersed in water or in aqueous surfactant solution. SLN offer unique properties such as small size, large surface area, high drug loading and the interaction of phases at the interface and are attractive for their potential to improve performance of pharmaceuticals.

SLNs are composed of physiologically tolerated lipid components which are in solid state at room temperature.

Advantages of SLN:

- Control and/or target drug release.
- Excellent biocompatibility and improve stability of pharmaceuticals.
- High and enhanced drug content and easy to scale up and sterilize.
- Better control over the release kinetics of encapsulated compounds.
- Enhanced bioavailability of entrapped bioactive compounds.
- Chemical protection of labile incorporated compounds.
- Much easier to manufacture than biopolymeric nanoparticles.
- No special solvent is required.
- Conventional emulsion manufacturing methods applicable.
- Raw materials essential the same as in emulsions.
- Very high long-term stability.
- Can be subjected to commercial sterilization procedures.

Disadvantages of SLN:

- Particle growth.
- Unpredictable gelation tendency.
- Unexpected dynamics of polymeric transitions.

5.2.4.1 Preparation of Epigallocatechin-3-Gallate-Loaded SLN: (Ramesh and Mandal, 2019)

5.2.4.1.1 Fabrication of SLN

To fabricate SLN, glycerol monostearate (GMS) was used as a source of lipid, while polyoxyethylene stearate (S-40) and poloxamer 188 (F-68) were used as surfactant and cosurfactant, respectively. Briefly, polyoxyethylene stearate (S-40, 700 mg) and poloxamer 188 (F-68, 300 mg) were mixed in 5 mL of water and melted at 70°C temperature in a water bath followed by the addition of 50 mg of GMS with continuous stirring until it became transparent.

Epigallocatechin-3-Gallate(EGCG)-loaded SLN (was prepared by adding 5 mg of EGCG to this mixture and stirring constantly which formed thermodynamically stable water/oil microemulsion). The water/oil microemulsion was dispersed in 50 mL ice-cold aqueous solution of 0.8% S-40 and sonicated for 1.5 h and stored at 4°C for further use.

5.2.4.1.2 Lyophilization of SLN

Before lyophilization, the prepared formulation was centrifuged to remove unentrapped EGCG and stored in glass vials at −80°C for 24 h. The frozen nanoformulation was further lyophilized to obtain free-flowing freeze-dried product.

5.2.4.2 Preparation of Aloe-Emodin SLNs (Chen et al., 2014)

High-pressure homogenization (HPH) was utilized for preparation of aloe-emodin SLNs (AE-SLNs) (Li et al., 2006). Pharmaceutical lipid material and drug were precisely weighed and dissolved in a water bath at 70°C–80°C.

Meanwhile, a certain amount of surfactant was dissolved in water at the same temperature and slowly added to the melted lipid mixture with mechanical stirring at a dropping speed of 1 mL/min.

TABLE 5.4
Phytocompounds Loaded in Solid Lipid Nanoparticle

	Genistein SLN	Aloe-Emodin SLN
Biological source	Iso-flavone found in Leguminosae species	Anthraquinone and isomer of emodin from aloe latex
Method of preparation	Melt dispersion technique	High-pressure homogenization
In vitro dissolution study	USP-II type used	The dialysis bag method is used
Stability	HPLC analysis	Differential scanning calorimetry

Subsequently, it was continuously stirred for 20 min and then the pre-emulsion was passed through a high-pressure homogenizer. The AE-SLNs suspensions were thus obtained and stored in a refrigerator at 4°C. Besides, lyophilized AE-SLNs were also obtained for further physical characterizations (Table 5.4).

5.2.4.3 Preparation of Genistein-SLNs: (Kim et al., 2017)

Genistein-SLNs were prepared using a hot homogenization method. The same amount of genistein and glyceryl palmitostearate (GPS) was melted in a water bath maintained at 90°C. The surfactants for the lipid phase, L-α-lecithin, and vitamin E-TPGS were also dissolved inconcentrations of 5% (w/w) and 0.05% (w/v), respectively in the melted mixture. Then, 10 mL distilled water containing poloxamer 188 (2%, w/w), as a surfactant, was heated under the same conditions.

The organic phase was subsequently added to the aqueous phase, and the mixture was sonicated using a probe-type sonicator. The resulting emulsion was processed using a high-pressure homogenizer at 15,000 psi and 10–20 cycles to produce genistein-SLNs, which were cooled at the ambient temperature and stored in a refrigerator at 4°C until used.

To prepare a suspension of genistein, 300 mg of genistein powder was sieved to a particle size range of 150–200 μm and added to 10 mL distilled water containing 0.2% (w/v) Na-CMC, a suspending agent, followed by vigorous vortexing.

5.2.4.4 Characterization of SLNs

The size distribution of the SLNs was evaluated by analyzing images captured using an optical microscope.

5.2.5 Nanostructured Lipid Carriers

Nanostructured lipid carriers (NLCs) are lipid-based formulations that have been broadly studied as drug delivery systems (Figure 5.5). They have a solid matrix at room temperature and are considered superior to many other traditional lipid-based nanocarriers such as nanoemulsions, liposomes, and SLNs due to their enhanced physical stability, improved drug loading capacity and biocompatibility (Figure 5.6).

Advantages of NLCs:

- Better physical stability.
- Ease of preparation and scale-up.
- Increased dispersability in an aqueous medium.
- High entrapment of lipophilic drugs and hydrophilic drugs.
- Controlled particle size.
- An advanced and efficient carrier system in particular for lipophilic substances.
- Increase of skin occlusion.

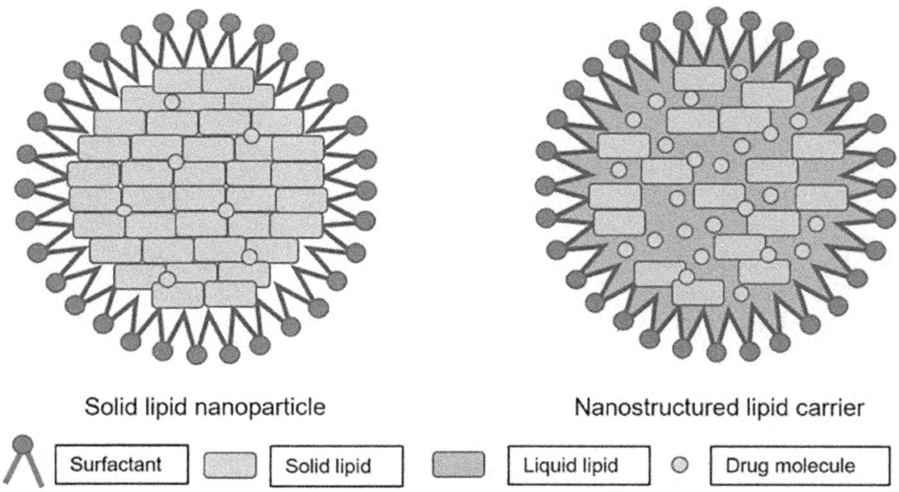

FIGURE 5.5 Components of nanostructured lipid carrier.

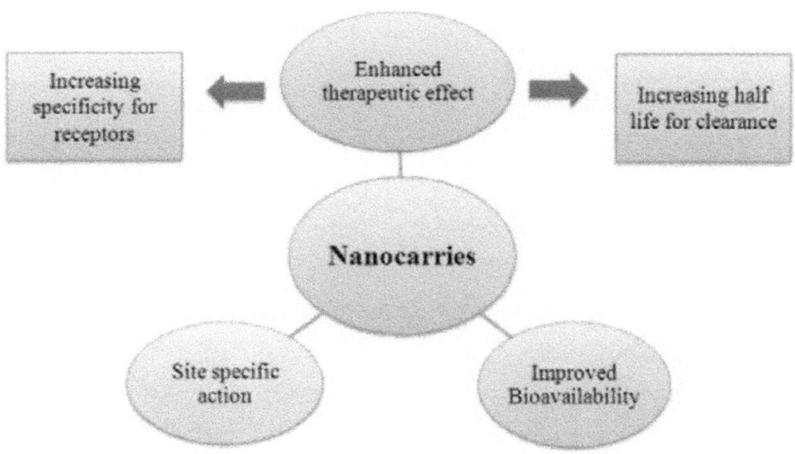

FIGURE 5.6 Merits of nanocarrier. (Scientific Image and Illustration Tool adopted: www.biorender.com.)

- Extended release of the drug.
- One of the carriers of choice for topically applied drugs because their lipid components have an approved status of excipients used in commercially available topical cosmetic or pharmaceutical preparations.
- The small size of the lipid particles ensures close contact to the stratum corneum thus enhancing drug penetration into the mucosa or skin.
- Improve benefit/risk ratio.
- Increase skin hydration and elasticity.
- These carriers are highly efficient systems due to their solid lipid matrixes, which are also generally recognized as safe or have a regulatory accepted status.

Limitations with lipid nanoparticles:

Despite the great potential of NLCs in targeted delivery, they face certain limitations like:

- Cytotoxic effects related to the nature of the matrix and concentration,
- Irritative and sensitizing action of some surfactants,

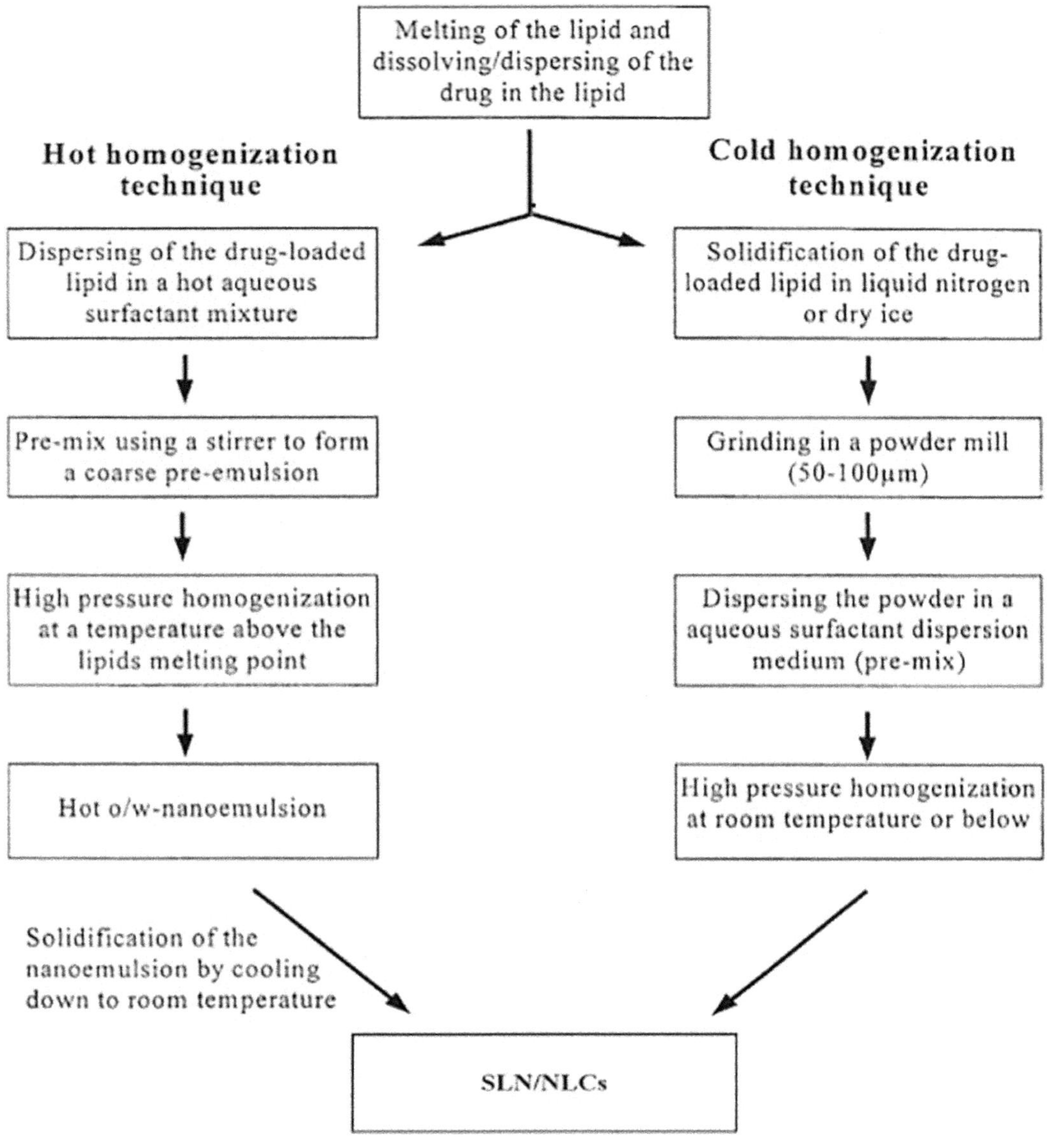

FIGURE 5.7 Preparation methods of solid lipid nanoparticle and nano lipid carrier. (Scientific Image and Illustration Tool adopted: www.biorender.com.)

- Application and efficiency in the case of protein and peptide drugs and gene delivery systems still need to be better exploited,
- Lack of sufficient preclinical and clinical studies with these nanoparticles in case of bone repair (Figure 5.7).

Preparation of NLCs:

Fabrication of Quercetin and linseed oil co-loaded NLCs (Huang et al., 2017)

The quercetin and linseed oil co-loaded NLCs were produced by using melting emulsification together with the HPH technique. The concentrations of linseed oil (liquid lipid), GMS (solid lipid),

P6 (surfactant), Tween 80 (surfactant), and 1,1-propylene glycol (cosurfactant) were employed as 6%, 4%, 5%, 5% and 2% (w/w) respectively for the fabrication of NLCs.

The concentrations of quercetin were 0.1%, 0.2%, and 0.3% (w/w) separately. Briefly, certain amounts of GMS, P6, Tween 80, and 1, 1-propylene glycol were mixed together and heated up to 70°C (more than 5°C above the melting point of GMS). After the above components were melted, linseed oil and quercetin precisely weighted were added to the mixture and stirred at 70°C until all the crystals disappeared.

Subsequently, the distilled water kept at the same temperature was slowly injected into the mixture of oils and surfactants with stirring (500 rpm) for 30 min. The resultant pre-emulsion was further subjected to a preheated high-pressure homogenizer for 6 cycles at 800 bar. Due to the melting point of GMS being around 60°C, boiling distilled water was utilized to preheat the high-pressure homogenizer to avoid any precipitation during the homogenization.

The high-pressure homogenizer consists of a charging hopper, a homogenization chamber, and a manual press valve. When the pre-emulsion passed through the charging hopper into the homogenization chamber, the increasing homogenization pressure raised the intensity of cavitations, turbulence, and hydraulic shear forces inside the homogenization chamber, and then smaller particles of NLC were achieved. Finally, the fabricated NLC was cooled down at room temperature, and then quercetin and linseed oil co-loaded NLC was acquired

5.2.5.1 Characterization of Nanolipid Carriers

5.2.5.1.1 Rheological Analysis

Rheological measurements of NLC were determined by a rotational rheometer equipped with a cone-and-plate test geometry. The steady shear rheological measurement of NLC was performed at a constant frequency of 1 Hz with a shear rate ranging from 1 to 100 Hz.

An oscillation stress sweep test was carried out to determine the linear viscoelastic region for the following experiment. Subsequently, an oscillation frequency test was conducted over a frequency range of 0.01–10 Hz at selected stress under the linear viscoelastic region (0.5%). All tests were carried out at 25°C.

5.2.5.2 Common Characterization Techniques for Lipid Carriers

5.2.5.2.1 Particle Size and Zeta Potential

The particle size and polydispersity index (PDI) of lipid carriers were measured by the DLS technique at a fixed scattering angle of 90° and temperature of 25°C. Prior to the measurement, NLC was diluted with distilled water to obtain a suitable scattering intensity. The zeta potential of NLC was also assessed by determining the droplet electrophoretic light scattering using the DLS technique at 25°C.

5.2.5.2.2 Morphology

The morphology of lipid carriers was investigated by TEM. A sample diluted with distilled water(1:50)was dropped onto a film-coated copper grid. Subsequently, the sample was stained with a 2% w/v solution of phosphor tungstic acid or with 1% uranyl acetate for 10 min. The overflowing fluid was wiped off with filter paper and the sample was allowed to dry at room temperature before being captured under the TEM.

5.2.5.3 Drug Entrapment Efficiency

Nano lipid carriers were purified using ultracentrifugation (45,000 rpm, 4°C, 1h) to separate the particles from the supernatant (has unentrapped drug). The remaining pellet was then solubilized with solvent to release the drug from the nanocarriers, allowing the drug conc. to be determined by creating a standard curve in solvent The absorbance was read with a UV-spectrophotometer at a suitable wavelength.

5.2.5.4 Differential Scanning Calorimetry

Samples were accurately weighed (approximately 2 mg) and sealed in aluminum pans. Nitrogen was used as flush gas. The scan was recorded using a Thermal Analyzer from 40 to 300°C at a heating rate of 10°C/min, using an Al_2O_3 pan as a reference and the physical status of SLNs was evaluated.

5.2.5.4.1 X-ray Diffraction

The X-ray diffraction (XRD) analysis was performed by an X-ray diffractometer. Samples were prepared by lyophilization to evaporate the water. All samples were scanned from 50° to 60°with a scan step size of 0.02° and speed of 0.15 sec/step. The X-ray source used was a Cu anode.

5.3 CONCLUSION

Currently, nanotechnology has a wide range of applications in pharmaceutical industries and research. Typically nanocarriers for drug delivery have potential in many aspects than conventional drug delivery systems.

The growing applications of nanotechnology in the medical field urge researchers to find the impact of nanoparticles on human body organs, tissues, and the immune system. The surface modifications of the nanocarrier system enable the modulation of the immune system either by evading the immune system to prevent allergic reactions or by enhancing the immunogenic response.

Various nanocarrier systems are currently being developed to improve drug delivery, especially to combat cancer, diabetes, and brain targeting. Among the developed carrier systems, lipid-based nanoparticles are one of the promising drug delivery candidates, based on drug loading, biocompatibility, and prolonged release. Polymeric nanocarriers also been have widely studied and used in targeted drug delivery due to their accessibility, stability, biocompatibility, biodegradability, lack of toxicity, safety, and cost-effectiveness.

The mechanism of drug delivery by nanotechnology is not still well-explored, however, various types of nanocarriers have been in use still it requires a clear idea to deliver the drug safely and conveniently.

REFERENCES

Banerjee, Kacoli, Shubhadeep Banerjee, and Mahitosh Mandal. Enhanced chemotherapeutic efficacy of apigenin liposomes in colorectal cancer based on flavone-membrane interactions, 2017, *Journal of colloid and interface science*, 491: 98–110. https://doi.org/10.1016/j.jcis.2016.12.025

Ruie Chen, Shengpeng Wang, Jinming Zhang, Meiwan Chen, Yitao Wang, 2014, Aloe-emodin loaded solid lipid nanoparticles: formulation design and in vitro anti-cancer study. *Drug Delivery*, https://doi.org/10.3109/10717544.2014.882446

Riham El-Gogary, Maha Nasr, Abd-Allah Hend, Abdel-Mottaleb Mona, 2020, Nanoparticulate systems for wound healing. In Nanopharmaceuticals Jan 1, pp. 73–90. Elsevier. http://dx.doi.org/10.1016/B978-0-12-817778-5.00004-X.

Neel M. Fofaria, Hussaini Syed Sha Qhatta, Xinli Liu, Sanjay K. Srivastava, 2016, Nanoemulsion formulations for anti-cancer agent piplartine– characterization, toxicological, pharmacokinetics and efficacy studies. *International Journal of Pharmaceutics*, https://doi.org/10.1016/j.ijpharm.2015.11.045.

Juan Huang, Qiang Wang, Tong Li, Nan Xia, Qiang Xia, 2017, Nanostructured lipidcarrier as a strategy for encapsulation of Quercitin and linseed oil: preparation and in vitro characterization studies. *Journal of Food Engineering*, https://doi.org/10.1016/j.jfoodeng.2017.07.002

Mohamed A. Ibrahim., Gamal A. Shazly, Fadilah S. Aleanizy, Fulwah Y. Alqahtani, Gehan M. Elosaily, 2018, Formulation and evaluation of docetaxel nanosuspensions: in-vitro evaluation and cytotoxicity. *Saudi Pharmaceutical Journal*, https://doi.org/10.1016/j.jsps.2018.07.018

Jeong Tae Kim, Sonia Barua, Hyeongmin Kim, Seong-Chul Hong, Seung-Yup Yoo, Hyojin Jeon, Yeongjin Cho, Sangwon Gil, Kyungsoo Oh, Jaehwi Lee, 2017, Absorption study of genistein using solid lipid microparticles and nanoparticles: control of oral bioavailability by particle sizes, *Biomolecules & Therapeutics*, https://doi.org/10.4062/biomolther.2017.095

Ying-chao Li, Lei Dong, Ai Jia, Xin-ming Chang, Hui Xue, 2006, Preparation of solid lipid nanoparticles loaded with traditional Chinese medicine by high-pressure homogenization. Nan Fang yi ke da xue xue bao, *Journal of Southern Medical University*. May 1;26(5):541–4. PMID: 16762843.

Bhia Mohammed, Mahzad Motallebi , Banafshe Abadi B, Atefeh Zarepour, Pereira-Silva Miguel, Saremnejad Farinaz, Santos Ana Claudia paiva santos, Zarrabi Ali, Melero Ana, Seid Mahdi Jafari and Mehdi shakibaei, 2021, Naringenin nano-delivery systems and their therapeutic applications. *Pharmaceutics*, 13, 291.

Subramani Parasuraman, Gan Siaw Thing, and Sokkalingam Arumugam Dhanaraj, 2014, Polyherbal formulation: concept of ayurveda. *Pharmacognosy Reviews*, 8(16):73–80. https://doi.org/10.4103/0973-7847.134229

Nithya Ramesh, Abulkalam Azad Mandal, 2019, Pharmacokinetic, toxicokinetic, and bioavailability studies of epigallocatechin-3-gallate loaded solid lipid nanoparticle in rat model. *Drug Development and Industrial Pharmacy*, https://doi.org/10.1080/03639045.2019.1634091

Mohammad Sajid, Swaranjit Singh Cameotra, Mohd Sajjad Ahmad Khan, Iqbal Ahmad, 2019, Nanoparticle-based delivery of phytomedicines: challenges and opportunities. *New Look to Phytomedicine*, 597–623.

Amélia M. Silva, Carlos Martins-Gomes, Eliana B. Souto, and Joana F. Fangueiro, 2013, Delivery of phytochemicals in nanoparticles: concepts and challenges. *Current Pharmaceutical Design*, 19(41):7165–7187

Qiao Wang, Wenjin Liu, Hong Liu,Yong Chen, 2019, Preparation and pharmacokinetic study of daidzein long-circulating liposomes. *Nano Scale Research Letters*, https://doi.org/10.1186/s11671-019-3164-y

Joy Wolfram, Krishna Suri, Yi Huang, Roberto Molinaaro, Carlotta Borsoi, Bronwyn Scott, Kathryn Boom, Donatella Paolino, Massimo Fresta, Jianghua Wang, Mauro Ferran, Christian Celia,Haifa Shen, 2014, Evaluation of anti-cancer activity of Celasterol liposomes in prostate cancer cells. *Journal of Microencapsulation*, https://doi.org/10.3109/02652048.2013.879932

Roya Yadollahi, Krasimir Vasile, Spomenka Simovic, 2014, Nanosuspension technologies for delivery of poorly soluble drugs. *Hindawi Publishing Corporation Journal of Nanomaterials*, https://doi.org/10.1155/2015/216375

6 Application of Phytodrug Delivery in Various Therapeutic Applications for Neurodegenerative Diseases

Abhishek Wahi, Riya Nagpal, Anusha Thota, Sushma Dev, and Priti Jain

6.1 INTRODUCTION

Neurodegenerative disorders (NDs) are a group of neurological disorders with heterogeneous clinical and pathological traits, which are indicated by a disordered functioning of the nervous system. The various NDs include Parkinson's disease (PD), Alzheimer's disease (AD), spinal cord injury (SCI), Huntington's disease, and amyotrophic lateral sclerosis (ALS). NDs are a major cause of death worldwide, notably in the elderly. Numerous NDs, including amyloidosis, -synucleinopathies, tauopathies, and TDP-43 proteinopathies, are diverse in their clinical manifestations and underlying physiological circumstances while frequently developing in conjunction with other disorders (Figure 6.1). The major volume of NDs is categorized into familial and sporadic forms, based on the family history of the disease condition. Furthermore, NDs are frequently observed due to misfolding of proteins in glial and neural cells (Erkkinen, Kim, and Geschwind 2018; Kovacs 2016). Conventionally, NDs are characterized as discrete clinical disorders because they attack different regions of the brain and experience distinct symptoms. Still, they persist in several common symptoms at the cellular, molecular, and genetic levels. These symptoms include the aggregation and spread of misfolded proteins, vulnerability of particular neurons, vascular dysfunction, activation of the immune system finally loss of synapses, and brain damage. Modified mechanisms at the cellular level correspond between various NDs, and they include synaptic toxicity, mitochondrial defects, autophagic–lysosomal dysfunctions, and other mechanisms such as the liquid-phase transitions and stress granule pathways (Ochneva et al. 2022). Ultimately, for all ND, the most implicated factor is microglial cells are being reported for neurodegeneration, more specifically, microglial responses shall be shifted from the role of neuroprotective to a deleterious role ("Focus on Neurodegenerative Disease" 2018; Sweeney et al. 2018).

6.2 OBSTACLE IN DRUG DELIVERY TO CNS

6.2.1 Blood–Brain Barrier

The CNS has built a multitude of barriers to safeguard the brain from neurotoxins, circulating blood cells, and viruses. The permeability of the blood–brain barrier (BBB), blood–cerebrospinal fluid (CSF), blood–spinal cord, and blood–retinal barriers all varies. The BBB is the most comprehensive of these (Figure 6.2). A discontinuous layer of pericytes and tightly connected brain endothelial cells make up the majority of its structure (Cipolla 2009). Endothelial cells in the BBB are covered with microvessels and capillaries inside the brain and are densely linked via connection

DOI: 10.1201/9781003292692-6

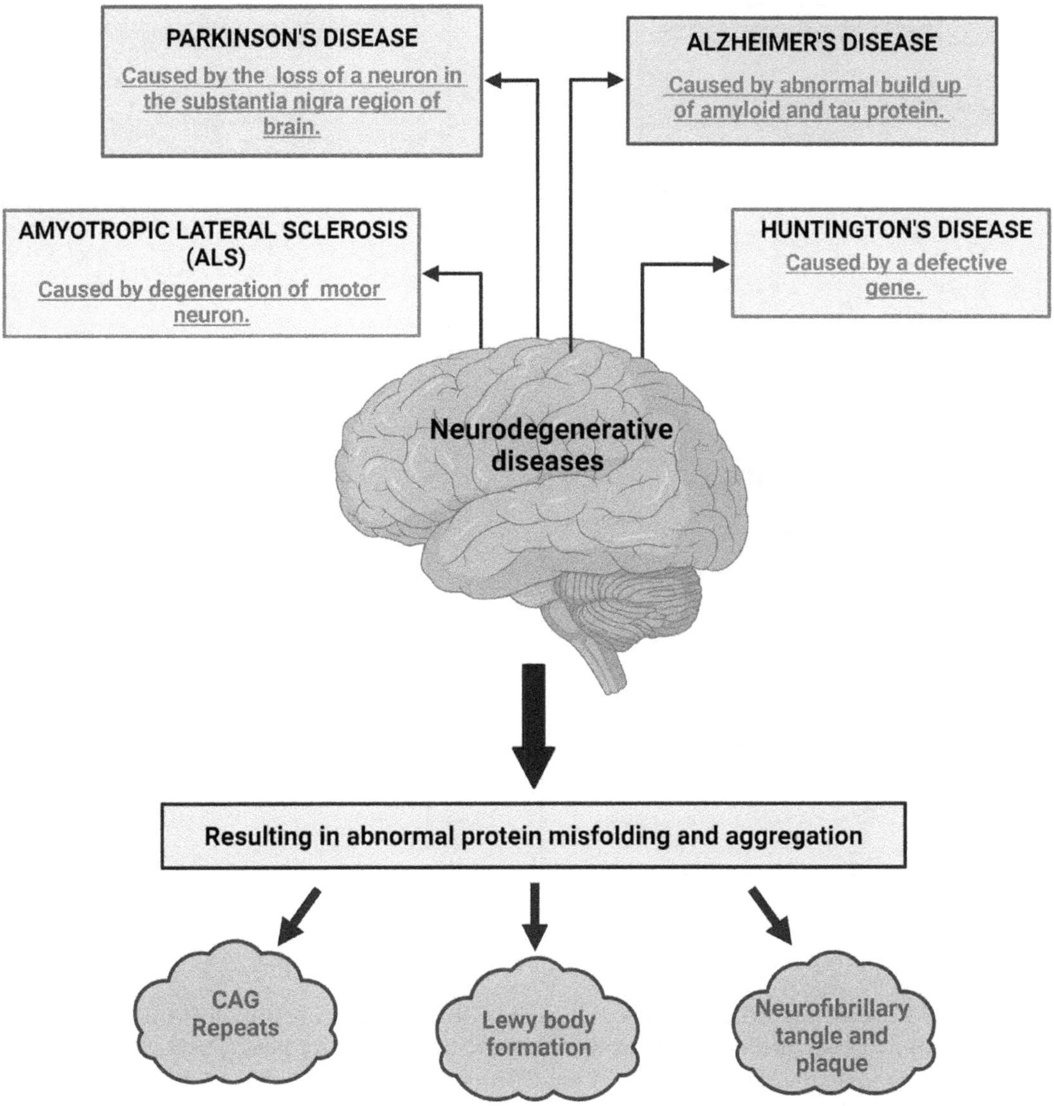

FIGURE 6.1 General pathogenesis of neurodegenerative diseases (NDs).

points such as adherens junctions and tight junctions, enabling minimal gaps between the cells, and have fewer endocytic vesicles and fenestrae than other organ capillaries (Teleanu et al. 2018; Furtado et al. 2018).

Although endothelial cells are tightly coupled by a strong cohesive system, the BBB allows for the selective transit of cells and tiny substances to the brain extracellular matrix, astrocytes, pericytes, and microglial cells surround brain capillary endothelial cells and enclosed by basal lamina, which is made up of heparin sulfate, type IV collagen, laminin, and fibronectin. Drugs delivered systemically must pass through BBB which protects the CNS from hazardous elements while also providing CNS cells with required nutrients and removing CNS waste materials (Tajes et al. 2014). According to reports, 98% of neuroprotective pharmacological entities are unable to penetrate this BBB, which prevents the medicine from having enough bioavailability to effectively treat ND.

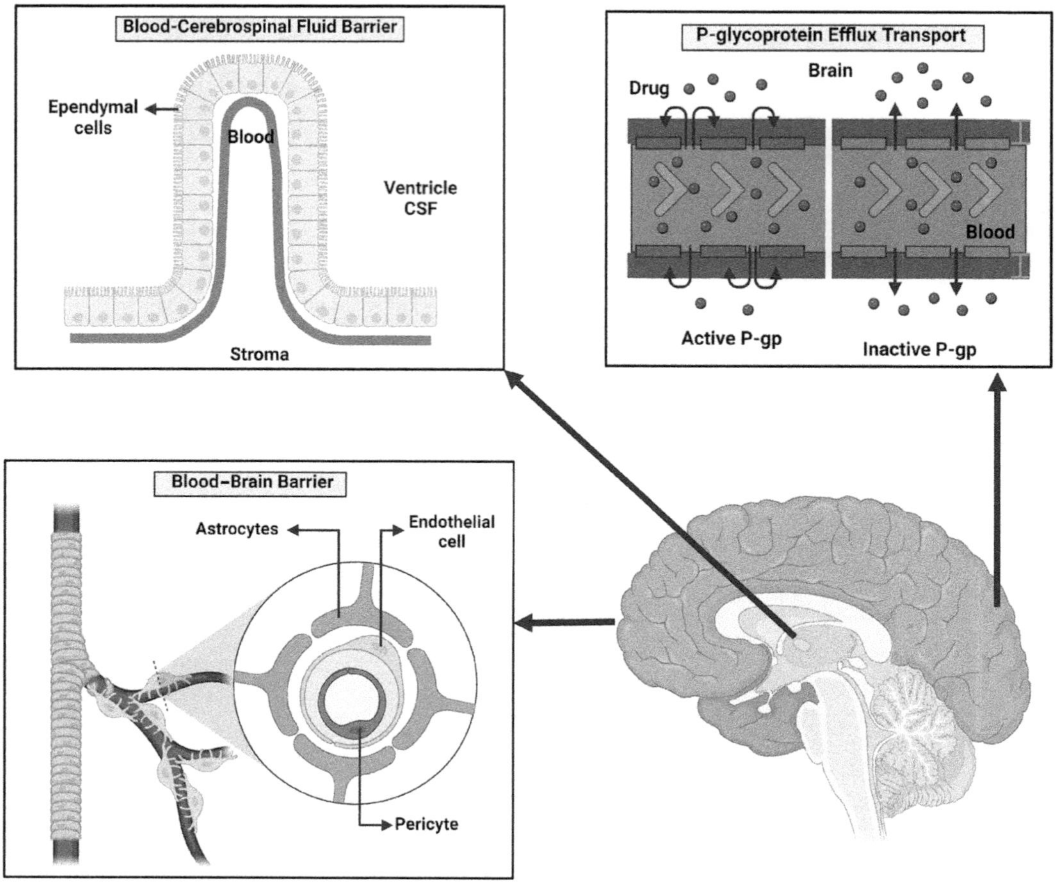

FIGURE 6.2 Major limitations to drug transport to the brain.

As a result, there is less drug dispersion over the BBB, which limits the utilization of several popular administration techniques including the oral and intravenous routes (Pandey, Sharma, and Gupta 2016). Despite the tight junctions of the BBB restricting the substantial passive transfer of comparatively tiny hydrophilic molecules from the blood to the brain, numerous transport systems (influx systems) capable of eliciting the BBB facilitate the entrance of vital hydrophilic and lipophilic substances such as glucose and amino acids.

6.2.2 Blood–Cerebrospinal Barrier

After the BBB, a drug molecule must cross through the blood–cerebrospinal fluid barrier (BCFB), which is composed of choroid plexus epithelial cells (Figure 6.2; Sarkar et al. 2017). The choroid plexus epithelium (CPE) is a secretory epithelium with distinct cellular transport networks (Damkier, Brown, and Praetorius 2013). Similar to the BBB, this epithelial barrier separates blood from CSF and is compacted by tight junctions among adjoining cells on the interstitial fluid surface. An uninterrupted band of tight junction proteins links the BBB-forming cells to the central nervous system and blocks the paracellular gap. It's a system made up of detoxifying enzymes and multi-specific efflux transport proteins that work together to keep potentially hazardous substances out of the CNS (Strazielle and Ghersi-Egea 2016).

6.2.3 Multidrug Resistance Proteins

Research findings from immunocytochemistry, molecular biology, transport studies, and biochemistry, revealed that cerebral endothelial cells contain a variety of metabolic enzymes such as alkaline phosphatase, cytochrome P450, and glutathione transferases enzymes as well as ATP-dependent efflux transport proteins like P-glycoproteins (P-gp) and non-uniformly arranged multidrug-resistance proteins which create a barrier to Central nervous system medication delivery (van Assema et al. 2012). P-gp, a plasma membrane-associated protein, is a vital ATP-dependent efflux carrier (Figure 6.2) at the BBB produced by the expression of the multidrug resistance protein (MDR1) gene (Xu et al. 2012).

6.2.4 Other Challenges

The treatment options currently available for NDs are limited and many therapy strategies mitigate the symptoms but not the underlying physiological condition. There are many challenges usually encountered during the therapy of these conditions. The main problem is the permeability of the BBB, which resists drug delivery to reach a therapeutic dose at the desired tissue of the brain. Hence, the urge is to formulate bioactive materials that can reach the brain site by crossing BBB (Sudhakar and Richardson 2019). Other challenges include safety concerns and long-term safety issues while treating the patients. Lastly, it is very important to diagnose the disease condition at an early stage, because many of them do not experience symptoms at the early stages of disease progression (Budd Haeberlein and Harris 2015; Poovaiah et al. 2018).

A promising technology to increase brain drug delivery is pronounced by gene therapy. The gene therapy done by altering or modifying gene expression using altered viral vectors, which are derived from adeno-associated viruses (AAV), plays a prominent role in restoration, neuroprotection, and alteration of the underlying mechanisms. Furthermore, the innovation of a convection-enhanced delivery (CED) in the 19th century and further upgraded to magnetic resonance imaging (MRI)-coupled with CED in current years, showed a route map for monitoring of the drug given by infusion through co-administration of the contrast agent used for MRI to have a designed action at the required anatomical site. Clinical trials for PD and AD mainly use this technique for treatment (LeWitt et al. 2011; Mittermeyer et al. 2012; Warren Olanow et al. 2015; Rafii et al. 2018).

Currently, many researchers are focusing on protein misfolding. The ultimate goal is to prevent the production of the proteins that are tangled in the NDs, prevent protein aggregation, inhibit or remove their spreading, and alleviate their disease progression and toxicity. Consequently, synthetic agents were developed that are capable of preventing aberrant chemical modulators of autophagy, protein aggregation, and antibodies to avoid protein misfolding. Stem cell therapy is a new technique for treating NDs that has emerged as an alternative treatment option in recent years. Stem cells have the capability to regrow themselves rapidly and able to produce all types of cells. Currently, a wide range of cell lines have been adopted for the renewal of stem cells. Mesenchymal stem cells (MSCs) are considered appropriate due to their high inherence, high number, and low immunogenicity. Besides this, they can be easily identified and separated into glial cells and neurons, when cultured under the same conditions. MSCs can be extracted from many regions, however, the most widely used MSCs are from the bone marrow (BMSCs), umbilical cord (UC-MSCs), and adipose tissue (ASCs) (De Gioia et al. 2020; Furno, Mannino, and Giuffrida 2018; Furno et al. 2013). Despite many attempts, MSCs have encountered many challenges in *in vivo* models. Such issues can be minimized by developing biomaterials in both *in vivo* and *in vitro* models. Biomaterial holds promising results for the development of 3D systems and for the treatment to enhance the stem cell effect in the therapy of brain injuries and disorders. Certainly, bioengineered scaffolds have an excellent ability to mimic the extracellular matrix (ECM) environment. Finally, it has good control over cellular behavior via functionalization, which recommends scaffolds to be combined with cells in the area of neural regeneration (Volkman and Offen 2017; Gattazzo, Urciuolo, and Bonaldo 2014; Carradori et al. 2017).

Apart from these, medicinal plants like *Ginkgo Biloba*, curcumin, *Withania somnifera*, *Bacopa monnieri,* and Wolfberry are also being studied extensively to treat NDs in preclinical and clinical studies. It has been explored that a natural medicinal agent in the nanosize range has the same activity as in their native state (Griffin et al. 2017). Nanodrug delivery increases the drug's bioavailability, allowing it to target cells and tissues more precisely. To carry medicinal plants for drug delivery in the treatment of NDs, nanoparticles, polymeric nanomicelles, complex polymers nanocrystals, and nanofibers are being developed and have opened a new platform of research for the delivery of phytoconstituents across BBB.

6.3 COMMONLY USED MEDICINAL PLANTS AND PHYTOCHEMICALS FOR THE TREATMENT OF NDS

6.3.1 *Curcuma longa*

Turmeric (*Curcuma longa*) is a tropical herbaceous rhizomatous plant in the Zingiberaceae, ginger family. It's native to regions of the Indian subcontinent and Southeast Asia. Ayurveda recommends turmeric for a wide range of medical conditions including infections, respiratory problems, edema, and rheumatism (Darvesh et al. 2012). Turmeric possesses antilipidogenic, anticarcinogenic, cardioprotective, and neuroprotective effects, according to anecdotal evidence (Aggarwal et al. 2007). The primary phytoconstituents present in curcumin are the polyphenols called Curcuminoids. Curcumin in commercial form is essentially a blend of three types of curcuminoids; Curcumin, demethoxycurcumin, and bisdemethoxycurcumin (Preeti et al. 2020). Curcumin is also known to show antioxidant, for example, curcumin-mediated lipid peroxidation inhibition has been reported in a number of experimental settings, with curcumin's antioxidant efficacy being 10 times that of vitamin E and anti-inflammatory properties, like curcuminoids show inhibitory effects against various inflammatory cytokines and enzymes. Curcumin's neuroprotective properties are due to its multifaceted biochemical ability to influence a variety of molecular events (Darvesh et al. 2012). Curcumin is reported to reduce inflammation by downregulation of pro- and anti-inflammatory mediators like cyclooxygenase-2, IL-8, and tumor necrosis factor-alpha (TNF-α). It has also been considered to be effective against Ab and other protein aggregation, making it a good candidate for treating NDs, particularly AD (Zhang et al. 2019). In AD, curcumin tends to boost macrophage functioning as it aids in removing amyloid plaque. A study done by Zhang et al. (2006) reported that curcumin-treated macrophages had a larger absorption and ingestion of plaque. Furthermore, an *in vivo* study done on male albino rats revealed that curcumin is an effective drug candidate to treat depression. Curcumin was found to attenuate olfactory bulbectomy-induced behavioral deficits such as immobility time, step-down passive avoidance deficits, and a rise in the level of open-area activity. Moreover, the levels of 3,4-dihydroxyphenylacetic acid, 5-hydroxyindoleacetic acid, noradrenaline, and serotonin released in excess in depression, in male albino rats' hippocampus were reversed after prolonged use of curcumin during the course of treatment. Curcumin aids in the normalization of 5-hydroxyindoleacetic acid, dopamine, and noradrenaline levels in the frontal lobe of the rat's brain (Chang et al. 2016). Nonetheless, researches also prove that curcumin could also be used therapeutically in the treatment of PD. For example, in MES23.5 cells treated with 6-OHDA, Wang, Du, et al. (2009) discovered that curcumin has antioxidant and anti-inflammatory properties. Curcumin also prevented the production of α-synuclein fibrils and aggregation in PD (Pandey et al. 2008). Though, there are outstanding therapeutic qualities of curcumin but still it has inadequate pharmacokinetics, with low absorption, negligible bioavailability, and rapid elimination from the body. To address these shortcomings, several nanoformulations have been developed to improve Curcumin's pharmacokinetics and bioavailability (Bollimpelli et al. 2016).

6.3.2 Ashwagandha

Ashwagandha (*Withania somnifera*) is an Ayurvedic herb from the Solanaceae family. It is commonly known as the Indian winter cherry and is widely recognized in India for its therapeutic action. Ashwagandha is among the most important herbal drugs in Ayurvedic medicine, having been used for over 3,000 years in managing stress, and energy rejuvenation, as well as in enhancing cognitive performance (Pratte et al. 2014). Antiepileptic, antidepressant, anticoagulant, anti-inflammatory, antiarthritic, antioxidant, antipyretic, and antidiabetic properties have been illustrated by this medicinal herb, as well as it also possesses certain soothing effects such as relieving pain, regeneration, and growth-promoting attributes (Subbaraju et al. 2006). The bioactive constituents of Ashwagandha, such as withanolide A, withanolide IV, and withanolide VI, tend to reconstruct pre- and postsynapses, as well as to contribute to the restoration of neuronal dendrites and axons. *W. somnifera* extracts were shown to have antistress activity when assessed on Wistar rats by Bhattacharya and Muruganandam (2003).

6.3.3 *Ginkgo biloba*

Ginkgo biloba is thought to have originated in the remote alpine valleys of eastern China. *G. biloba* extract (GBE) has a medicinal property obtained from dried leaves of *G. biloba* that is used to improve blood flow and to cure memory loss, which would include AD (Oken, Storzbach, and Kaye 1998). GBE is known to consist of an active component which is well-known for its antioxidant qualities and its ability to prevent blood platelet aggregation. Additionally, the extract contains antioxidant, anti-inflammatory, wound-healing, radical-scavenging, and neuroprotective activities against NDs like AD and PD (Yuan et al. 2017; Zeng et al. 2018). Moreover, in the case of brain damage, *G. biloba* can be administered as a neuroprotective and an antidepressant medicine. The *G. biloba* plant has been reported to have beneficial effects on AD, including memory enhancement, dementia of vascular origin, cognitive disorders, and its antioxidative actions in combination with other medications that increase their potency or lessen their psychiatric side effects (Werneke, Turner, and Priebe 2006). It is beneficial in the treatment of depression, psychosis, anxiety, and schizophrenia. It improves cerebral blood flow to treat issues including anxiety, hearing issues, stress, poor memory, poor attention, thinking, social behavior, and dementia in AD that are brought on by a lack of brain blood flow (Noorbala and Akhondzadeh 2006).

6.3.4 Ginseng

Ginseng is an herb in the Araliaceae family. It has been used as a medication in Asia for millennia, particularly in China, Japan as well as in Korea (Baker et al. 1995). Ginsenosides are steroid-like compounds having a four-ring hydrophobic framework with sugar moieties linked to one other through glycosidic bonds. The amount and position of sugar moieties, as well as the diversity of the sugar components, may influence the unique action of each ginsenoside. Distinct pharmacological action has been reported for every ginsenoside, so in the same tissue, a particular ginsenoside might have numerous effects (Chen, Chiou, and Zhang 2008). According to the study by Nah, Park, and McCleskey (1995), ginseng does have the ability to inhibit voltage-dependent Ca^{2+} channels via a receptor linked to a toxin-sensitive G protein. Also, Rg1, one of the bioactive components of *Panax ginseng*, might even be a potential neuroprotective medication because it lowers elevated levels of iron, lowers divalent metal transport 1 (DMT1) expression, and raises expression of ferroportin 1 in the C57BL/6 mice that have been MPTP-treated due to which dopamine, as well as its metabolites, are increased in the striatum, and TH expression is continued to increase in the SN (Wang, Xu, et al. 2009). In addition to Rg1, Re, and Rd have neuroprotective action. It was reported that Rd reduces NO creation and PGE2 synthesis, which slows the loss of dendritic processes, In mesencephalic primary cultures that are treated with LPS and Re, raised expression of bcl-2 protein and bcl-2 mRNA was observed while reduced expressions of iNOS, bax, and bax mRNA were seen. It also weakens the cleavage of caspase-3 (Xu et al. 2005). Microglia are CNS-resident innate immune cells that respond to a range of stimuli, including infection, traumatic brain injury, toxic metabolites, and autoimmune. It was reported that in LPS-stimulated microglia, ginsenoside Rh1 suppressed the expression of pro-inflammatory cytokines, iNOS and COX-2, whereas boosted the anti-inflammatory cytokine IL-10 expression. Inhibition of HO-1 diminished these effects (Jung et al. 2010). Furthermore, In MPTP and MPP+PDs models, decreased locomotor dysfunction as well as altered tyrosine hydroxylase-positive cell loss in the substantia nigra was observed when ginseng extract G115 was administered orally (Van Kampen et al. 2003). In AD, there is a loss of ACh in the brain. Ginsenoside Rg5 monitors memory deficits and neurodegeneration in the brains of streptozotocin-induced rats with impaired memory (Chu et al. 2014). Other studies also revealed that *P. ginseng* and its constituents may have the potential to delay the progression of PD, AD, stroke, and other NDs.

6.3.5 *Bacopa monnieri*

B. monnieri, also widely recognized as Brahmi, is well-known in Ayurvedic medicine mostly for its medicinal properties (Chaudhari et al. 2017). It contains the bioactive ingredient Bacoside A which is known to show neuropharmacological properties. It also contains various phytoconstituents such as apigenin, alkaloid brahmine, and nicotine (Deepak and Amit 2004; Anand et al. 2014). It has the ability to regenerate nerve cells as well as improve memory which makes it a useful medication in NDs. Extract of *B. monnieri* (EBm) promotes mechanisms of free radical scavenging and helps to protect cells of the hippocampus, striatum, and prefrontal cortex from neurotoxicity and DNA damage associated with AD. Detoxification and adhering of free radical scavenging metallic ions, as well as intensifying antioxidant functionalities, are all implicated in the neuroprotection provided by bacosides (Russo et al. 2003). It also inhibits LOX activity, thus reducing the generation of scavenging ROS, lipid peroxides, and divalent metals and reducing the creation of lipid peroxides (Limpeanchob et al. 2008). It was reported that similar to donepezil, rivastigmine, and galantamine, EBm administration protects cholinergic neurons and inhibits anticholinesterase activity. It also lowers beta-amyloid accumulation in the hippocampus region and stress-induced hippocampal damage (Uabundit et al. 2010).

6.3.6 Wolfberry

The fruit of *Lycium barbarum*, Wolfberry, has long been utilized in Chinese medicine as an antiaging remedy and as a culinary element (Chang and So 2008). The extract of wolfberry was found to have neuroprotective properties by decreasing Aβ-induced neurotoxicity (Shi et al. 2017). Ho and his coworkers explored how wolfberries affected neurological impairment brought on by high plasma levels of homocysteine (Hcy). *L. barbarum* extract has been used to cleave tau as well as decrease tau phosphorylation, which is brought on by Hcy (Ho et al. 2010). This study revealed that LB is effective in preventing cell death which has been induced by glutamate and phosphorylation of the c-jan N-terminal kinase (JNK) (Ho et al. 2009). *L. barbarum* is effective at preventing secondary disintegration of retinal ganglion cells and also lowers p-ERK and p-JNK levels. The fruitless wolfberry–sprout extract (FWE) was recently found to have antioxidant properties as well as enhanced learning abilities and also tends to strengthen the memory caused in chronic mild unforeseen stress rats (Mocan et al. 2014; Liu et al. 2018).

6.3.7 Quercetin

Quercetin is a bioflavonoid present in a variety of fruits, vegetables, and herbal origin oils, and it has been shown to have neuroprotective and anti-inflammatory properties (Amanzadeh et al. 2019). The primary characteristic of quercetin is the presence of many hydroxyl (-OH) groups in its structure. These groups have the potential to attach to reactive oxygen species (ROS) and maintain cell viability (Dhaouadi et al. 2009). As a result of its ability to cure neurodegenerative diseases by regulating the oxidative content of cells or interacting with proteins in pathways that support brain cell survival, quercetin has also been developed as an anti-neurodegeneration drug. As a result, quercetin is now viewed as a crucial therapeutic substance for treating neurodegenerative diseases, which may represent a substantial advancement in the treatment of these conditions (Sharma et al. 2016).

6.3.8 *Terminalia chebula*

T. chebula (Combretaceae) is a herbal medicine with a variety of therapeutic characteristics and is known to contain numerous biochemicals such as phenolic compounds, hydrolyzable tannins, and flavonoids, making it a promising medicinal choice (Afshari, Sadeghnia, and Mollazadeh 2016). In traditional systems of medicines like Unani, Ayurveda, and homeopathy, *T. chebula* has been used to treat geriatric illnesses and improve memory and cognitive brain functions (Saleem et al. 2002). *T. chebula* demonstrated anxiolytic activity and is comparable to the conventional medication diazepam. *T. chebula* shows antioxidant properties comparable to scavenging free radicals and has significant pharmacological properties related to dementia therapy. *T. chebula* leaf extract also protects nerve cells from ischemia and increases the mortality rate of microglia cells via lipopolysaccharide (Lee et al. 2017).

6.3.9 *Tinospora cordifolia*

Tinospora cordifolia (*T. cordifolia*) is a member of the Menispermaceae family, which includes the giloe plant. Steroids, alkaloids, diterpenoid lactones, aliphatic acids, and glycosides, are among the chemical compounds isolated from the plant. The memory-improving effects of *T. cordifolia* are related to immunological stimulation and increased acetylcholine generation (Nayak et al. 2018). It boosts glutathione levels and the copper–zinc gene expressions including superoxide dismutase (SOD) and gamma-glutamyl-cysteine ligase, which contribute significantly to brain damage during hypoxia and ischemia. *T. cordifolia* also boosts the brain's dopamine levels. *T. cordifolia* has thus been proven to protect against neurodegenerative alterations and to improve cognition, learning, and memory.

6.3.10 *Allium sativum*

Garlic, or *A. sativum*, is a member of the Amaryllidaceae family and it has been utilized for therapeutic purposes since time immemorial. It has been used to cure hyperglycemia, dementia, high blood pressure, diabetes, atherosclerosis, thrombosis, and malignancy. One of the key components present in mature garlic extract (AGE), S-allyl cysteine (SAC), has been thoroughly explored. The antioxidant activity of SAC is both indirect and direct. It reduces lipid peroxidation and DNA fragmentation while also lowering oxidation and nitration. Moreover, SAC provides protection to dopamine levels in Parkinsonian models like 1-methyl-4-phenyl pyridinium (MPP) and 6-hydroxydopamine (6-OHDA), prevents oxidative harm, and promotes lipid peroxidation (Powell, Dillon, and Matson 2018). *A. sativum* AS is deemed safe for human use as an ingredient in regular diets and has a wide range of positive applications (Majewski 2014).

6.3.11 *Coriandrum sativum* L.

Coriander (*Coriandrum sativum* L.) is a parsley-like evergreen shrub belonging to the family Apiaceae. In Persian, this plant is known as Geshniz. *C. sativum* is a Mediterranean native that is widely cultivated across the world. In the neurological system, this medicinal herb elevates antioxidant concentrations, lowers oxidant levels, and suppresses the action of acetylcholinesterase. This plant is also neuroprotective because it produces fewer pro-inflammatory markers (Khazdair et al. 2019). Coriander is also a great source of triglycerides like petroselinic acid and a plethora of essential oils (EOs), which are required for brain development and functioning (Sahib et al. 2013). Furthermore, the *Coriandrum sativum* extract was observed to decrease the amount of cerebral infarct, calcium levels, and lipid peroxidation (LPO), as well as enhance total protein concentration and the levels of the enzymes CAT, SOD, and GSH. The leaves of *C. sativum* have also been reported to lessen the memory loss brought on by scopolamine and diazepam (Owoeye et al. 2019).

6.3.12 Resveratrol

Resveratrol, chemically 3,4',5-Trihydroxystilbene, is a form of natural polyphenol from the class of stilbene, abundantly found in grapes, raspberries, blueberries, and mulberries which provides a number of health benefits (Vidavalur et al. 2006). It is known to have antidiabetic, anti-apoptotic, anti-inflammatory, antiviral, and antioxidative action (Berardi et al. 2009; Baarine et al. 2011). According to Dasgupta and Milbrandt (2007), resveratrol has a neuroprotective effect as it stimulates AMP kinase and thus affects neuronal homeostasis. Feng et al. (2002) investigated the effect of immune modulation at low doses of resveratrol and concluded that low doses of resveratrol enhanced cell-mediated immune response by triggering cytokine production and altering macrophage activity. It was also reported that resveratrol also protected neurons from cell death induced by β-amyloid (Feng et al. 2013).

6.3.13 *Acorus calamus*

A. calamus, also referred to as sweet flag, is a member of the Acoraceae family that functions as a rejuvenator of the central nervous system and has positive effects on the brain as it helps in improving memory, cognitive behavior, and ability (Sharma et al. 2020). The predominance of α- and β-asarone is found in the *A. calamus*, and β-asarone inhibits beta-amyloid-induced neuronal apoptosis in the hippocampus by decreasing the expression of Bcl-w and Bcl-2, which leads to caspase-3 activation and phosphorylation of c-Jun N-terminal kinase (JNK). *A. calamus* also exhibited an inhibitory effect on AChE with an IC_{50} value of 188 g/mL, in addition to the above-mentioned benefits (Bandelow, Michaelis, and Wedekind 2017). It also has beneficial benefits in terms of enhancing dopaminergic nerve activity, and so functions as a neuroprotective for PD. The oral administration of various *A. calamus* fractions, particularly the aqueous fraction, managed to prevent memory impairment and strain by controlling oxidative stress and inflammation (Esfandiari et al. 2018).

6.3.14 *Galanthus nivalis*

G. nivalis, widely recognized as a snowdrop, is a member of the Amaryllidaceae family. Galantamine, a tertiary isoquinoline alkaloid, is the chief component found in *G. nivalis* bulbs and flowers. The medication is an AChE inhibitor that is both competitive and specialized in nature. It can stimulate nicotinic receptors, which improves sensory perception and memory (Wattmo et al. 2013). The drug also modifies receptors of nicotinic acetylcholine on cholinergic neurons, specifically of the α-7 and α-34 subtypes, this boosts the release of acetylcholine (Prvulovic, Hampel, and Pantel 2010).

6.3.15 *Centella asiatica*

The Apiaceae (Umbelliferae) family member *C. asiatica*, which has been used for a long time in the Indian Ayurvedic medicinal system, has been found to have neuroprotective benefits. It is utilized as an alternative medicine therapy for brain functioning in the aging population (Golla and Tirupathi 2017). *C. asiatica*'s primary active constituents are saponins (also known as triterpenoids), which contain a trisaccharide moiety linked to aglyconeasiatic acid, madecassoside, and madasiatic acid.

Other components include brahmoside and brahminoside, which may be engaged throughout the CNS action (Gohil, Patel, and Gajjar 2010). *C. asiatica* has significant antioxidant activity, is competent in radical scavenging, decreases ferric ions, as well as increases glutathione-S-transferase activity, which revives glutathione levels. *C. asiatica* reduces Aβ accumulation in the central nervous system (Chen et al. 2016). Asiaticoside's amelioration of the colchicine-induced decline in AChE activity and suppression of nitric oxide-induced neurodegenerative diseases may further contribute to the possibility that *C. asiatica* has therapeutic benefits in NDs (Kumar et al. 2006).

6.3.16 Piperine

Piperine (1-piperoylpiperidine), is a nitrogenous-smelling naturally occurring substance in the fruits of black pepper, long pepper, and other piper species (Chonpathompikunlert, Wattanathorn, and Muchimapura 2010). It is known to show various biological and medicinal actions like antioxidant, anti-inflammatory, anticancer action, and cardiovascular protection (Ren and Zuo 2019). Furthermore, many preclinical studies have demonstrated that it can be a potential therapeutic agent in CNS disorders such as AD, PD, and epilepsy. Piperine's neuroprotective action may also be linked to the reduction of neuronal network synchronization and Ca^{2+} overloading in hippocampal neuronal cells as well as inhibition of presynaptic glutamic acid release, according to *in vitro* investigations (Fu, Sun, and Zuo 2010; Hsieh, Chang, and Wang 2019). Piperine's anti-inflammatory and antioxidant capabilities may be associated with its neuroprotective action against PD, as indicators like tumor necrosis factor (TNF) and inducible nitric oxide synthase (iNOS) as well as oxidative stress were restored in an animal model of PD (Correia et al. 2015).

6.4 POSSIBLE DRUG DELIVERY SYSTEMS TO OVERCOME BBB FOR TREATMENT OF NDS

As shown in Figure 6.3, various drug delivery systems depending on the make-up of the carrier material can be employed for the drug delivery to CNS like liposomes, polymers, emulsions, solid-lipid NPs, dendrimers, and gold nanoparticles. In this article, we go over applications for nanodrug delivery in treating various NDs.

6.4.1 Liposomal Drug Delivery System for Neurodegenerative Diseases

Liposomes are synthetic spherical lipid vesicles that can be composed of cholesterol with one or more concentric phospholipid bilayer membranes (Nsairat et al. 2022). The structure is surrounded by an aqueous layer that can be used to encapsulate active ingredients. There are different types of methods well established in liposomal preparations; which may include: Multilamellar vesicles (MLVs): liposomes are formed spontaneously by gentle shaking thereby producing big MLVs. Generally, these liposomes consist of multiple lipid bilayers extended to 14 layers, where every layer is divided by a hydrophilic layer. MLVs are a heterogeneous mixture that can be anywhere between 500 and 5,000 nm in size. Large unilamellar vesicles (LUVs) or small unilamellar vesicles (SUVs) can result from homogenizing MLVs. One lipid layer with a diameter of 25–100 nm makes up the structure of the liposomes. LUVs are large unilamellar vesicles that have a single layer of lipid and have a diameter between 200 and 800 nm (Spuch and Navarro 2011).

The most often used technology is stealth liposome which is based on liposome systems used for active drug delivery. This technique was developed by simple alteration of the liposome membrane surface, a method that was accomplished by modifying hydrophilic polymer conjugates.

6.4.1.1 Neuropharmacological Liposomal Agent

Liposomes are widely used as nanocarriers because of their much higher carrying capacity. Since, many years, this drug delivery system has shown excellent drug release patterns in the brain, as the drug is mainly entrapped in vesicles, thereby preventing rapid degradation or elimination.

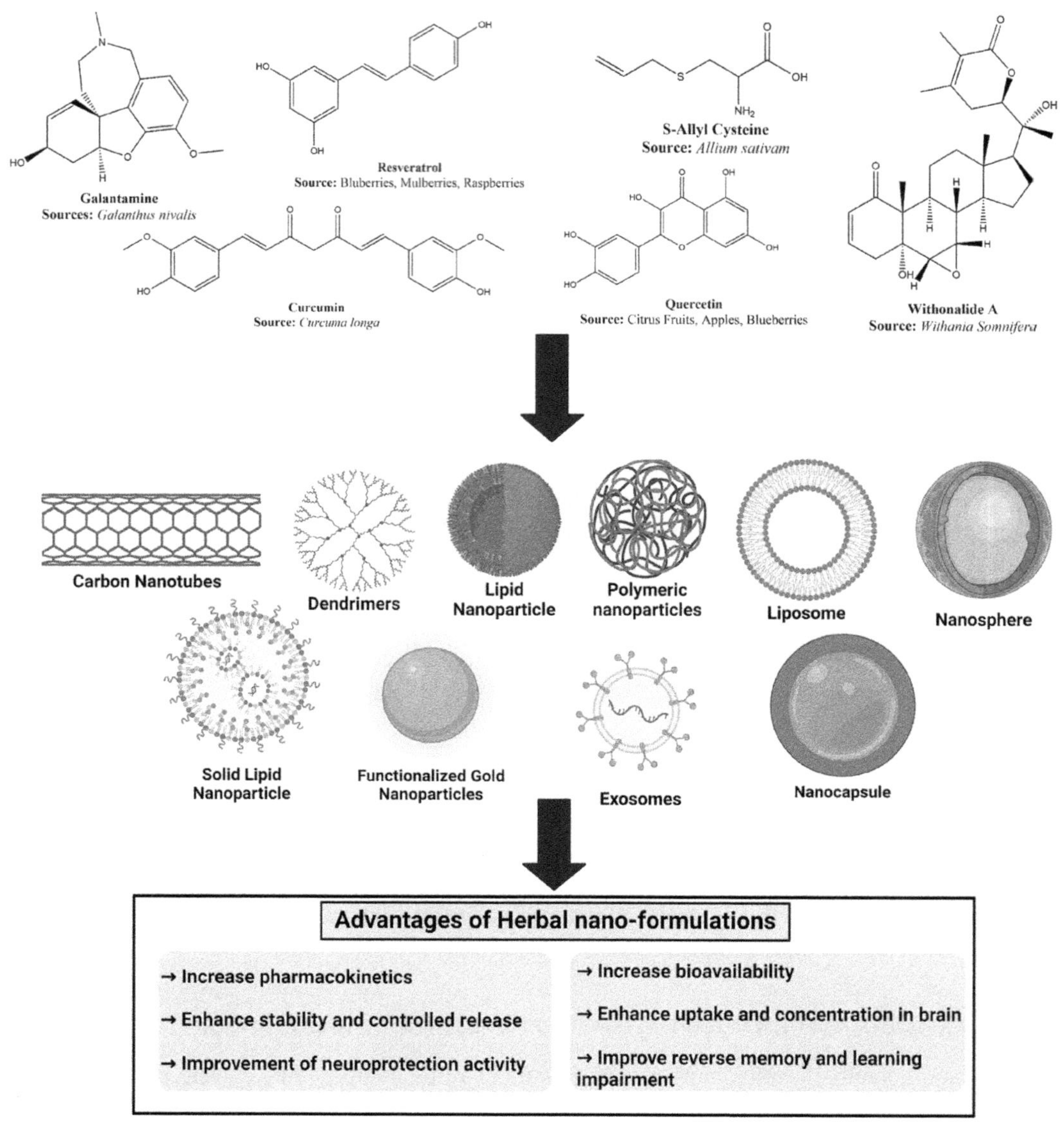

FIGURE 6.3 Nanoformulation of phytomedicines for the treatment of neurodegenerative diseases.

Additionally, it also promotes penetration through the BBB (Vieira and Gamarra 2016). Besides this, liposomes don't produce negative biological effects that normally encounters when a foreign substance enters the system. The pre-treatment and requisite formulation targets the brain or close proximities of the brain, therefore these are designed to be nontoxic, noncarcinogenic, nonthrombogenic, nonimmunogenic, and biodegradable. Nanodrug delivery systems like polymers, liposomes, and other nanoparticles can be employed as effective treatment options as to upgrade neurodegeneration therapies. Among these therapy options, the liposomal drug delivery system will have a considerable impact on neurology. Liposomal drug formulations progress from various design strategies for the enhancement of drug distribution over unbound drugs. Formulations based on the reticuloendothelial systems markedly decrease systemic exposure because of increased levels of unbound drugs without causing ease in penetrating the brain. Liposomal preparations need constant upgradation in formulation by using various lipids and lead molecules. For instance, liposomes without cholesterol and high-density lipoprotein may lead to degradation of the liposomal drug system, as well as liposomes with less concentration of cholesterol, and high-density lipoprotein can cause the contents to leak out. The development of novel delivery systems for NDs becomes one of the biggest challenges in medical care, which is a need of the hour. The advancement in liposome drug delivery systems can give a solution to bridle these challenges for early diagnosis and for further treating such neurodegenerative illnesses like AD and PD.

6.4.2 Nanophytosome-Based Drug Delivery System

Phytosomes are the lipid-based vesicular systems, in which herbal drugs are generally attached to phospholipids. The rapid advancement of nanophytomedicines (nano size of phytomedicine), polymers attached with phytomedicines, and altering the surface characteristics of herbal drugs, results in enhancement of the permeability, solubility, and lastly the bioavailability of herbal preparations.[1] Structurally, phytosomes differ from liposomes, as liposomes have their active ingredient encapsulated in the hydrophilic vesicles or within the layers, while in phytosomes, active moieties are part of the membrane itself. Many studies have evidenced that phytosomes can enhance the absorption of drugs through oral topical routes, consequently, they can improve bioavailability, thereby small doses can produce a desirable therapeutic effect. Therefore, it produces remarkable therapeutic benefits (Fernandes et al. 2021).

6.4.3 Nanoformulations

6.4.3.1 Nanocapsules, Nanospheres, and Polymeric Nanoparticles

Since nanoparticles have very great drug load capacity, the formulation can support and protect the drug's active substances from deterioration. As a result, it enhances the likelihood of penetration power of the drug to reach the region of the brain. Because of their comprehensive structures and specific features, they could even elude immune cells, facilitating drug delivery to the BBB. Generally, Nanospheres are tightly packed crosslinked particles generated through microemulsion polymerization, whereas nanocapsules are formed through a polymeric envelope encircled by a lipid-filled casing. In regard to nanospheres, a study discovered that when carboxylated polystyrene nanospheres were administered intravenously, they extravasated into the brain during ischemia-induced stress, which partially breached the BBB which significantly shows that nanospheres can be used as efficient CNS drug delivery system (Kreuter 2001). Polymeric nanoparticles (PNPs) seem to be colloidal particles that contain a drug encapsulated inside a biopolymer-based carrier such as PLA, PLGA, and PGA, having sizes in order of 1–1,000 nm (Chopra, Kumar, and Singh 2021). It was reported that the capacity of PLGA NPs to traverse the BBB was increased by surface modification, allowing for better tissue targeting siRNA and a few drugs also. This method also makes it easier to deliver medications to their intended targets in a controlled and/or sustained manner (Shakeri et al. 2020; Calzoni et al. 2019).

6.4.3.2 Polymeric Nanosuspensions and Nanogels

Nanogels are nanosized formulations of hydrogels, that are either ionic-monomers, non-ionic or copolymerized such as PEG, PEI, or/and Pluronic. Such networks inflate in water and include molecules that are oppositely charged such as siRNA, oligonucleotides, tiny molecular mass medicines, and DNA proteins, through ionic interactions (Vinogradov, Batrakova, and Kabanov 1999). The particle size of the gels ranges from 20 to 200 nm. This system has a drug loading capacity of 40%–60%. The various research studies revealed that the structures of nanogels may increase the brain loading capacity and reduce the spleen and liver holding capacity of oligonucleotides. Individual nanogel particles collapsed due to the solubility of PEG chains and thus forms are stable dispersions that do not undergo phase separation. Nanosuspensions are described as crystalline drug particles that are prepared by complexing of non-ionic surfactants or lipids (Jahangir et al. 2020).

6.4.3.3 Nanofibers

Nanofibers can be prepared by using distinct polymers and thus it has discrete physical properties and application capacity. Nanofibers are fibers with a size range within the nanometers and these are produced by electrospinning (Mazza et al. 2013). These have been made to resemble the fibrous makeup of brain tissue. Since it has a wide surface area and surface volume, it is the appropriate bioactive matrix for binding cells, loading drugs, and characterizing to offer reinforced biophysical and biochemical features to aid in tissue regeneration. Most studies on neurite regeneration and synaptic formation, however, focused on the differentiation of stem cells into dopaminergic neurons for the treatment of PD, while very few studies on the treatment of neurodegenerative diseases were reported (Sivandzade and Cucullo 2021; Wei et al. 2023). A principal characteristic that nanofibers should have, that is, strict alignment, resulting in promoting the regrowth of axons. Nanofibers can be created using synthetic materials, but they can also start as natural materials. These are stuffed with pharmaceuticals, demonstrated resistance to swelling, enhanced stiffness, and lysozyme breakdown, which leads to better alignment and cell culture proliferation. Nanofibers are safer than carbon nanotubes, in which, the risk of pollution of air is lower, comparatively. Moreover, neural prosthetics are designed and produced by the nanofibers (Bordoni et al. 2020).

6.4.3.4 Carbon Nanotubes

Carbon nanotubes (CNTs) are considered a promising tool for non-chemical methods for the regeneration of neural tissue. They have excellent properties to improve systemic circulation, by improving the drug loading capacity, stability, retention effect, permeability, and availability at desired sites. These structures may also control the rate at which drugs are released, making it easier to monitor drugs and image how they work. Its adaptability to a variety of stimuli, including pH, temperature, pressure, chemicals, magnetic fields, and electric fields, has also been demonstrated (Jamali et al. 2017). One of the most amazing methods for neurological therapy is the use of nanostructures made of carbon. CNTs have a complicated cylindrical nanostructure built of carbon allotropes. To reduce their electrical stimulation, they are frequently employed. These can be used to upgrade neuro-engineering for purposes such as neural differentiation, neuroprotection, interface, regeneration, and stimulation by acting as supports or by fusing with other biodegradable materials. CNTs are graphite sheets that cruise into cylindrical tubes in a size range of 0.4–2 nm, and with long lengths size ranges from nanometers to micrometers. These are available as single-walled multi-walled CNTs, based on their geometry. Because CNTs contain graphite, they may aid in conducting electrical stimuli, making brain tissue engineering easier. To encourage electrical conduction in inert materials, CNTs are typically added to biomaterials as a supplement. The fundamental unit for tissue engineering is to produce conductivity without external stimuli is easy *in vitro*, while it becomes hard to accomplish *in vivo* (Guo et al. 2017).

6.4.4 Hydrogels

Hydrogels are crosslinked hydrophilic molecules that exist in three-dimensional, water-insoluble polymer crosslinks that clench together by physical or chemical bonds. In physical gels, the network is obtained by secondary forces or molecular entanglements, and these gels can be reversible by alteration of physical conditions like pH or temperature. In the case of chemical gels, the network is cross-linked by the polymers and these can be charged or uncharged. The hydrogels' hydrophilic properties, porosity, and smooth consistency allow them to convey nutrients, oxygen, and other soluble components to more closely mimic tissues than other biomaterials. A wide range of hydrogels have been used, either alone or in a combination at different ratios in the field of neuro-regeneration and degeneration. Natural hydrogels like gelatin, collagen, xyloglucan, alginate sodium, and hyaluronic acid (HA) while synthetic hydrogels like polyethylene glycol and polyacrylamide are widely used in NDs. Hydrogel can be widely used presently for the delivery of drugs in treating NDs. In general, cells can be enclosed in 3D-hydrogel structures and then implanted in the brain. Alternatively, the cells can be grafted in a pre-gel state, which then transforms into a gel directly in the brain tissue, allowing for more precise drug delivery to the implanted cells' localized tissue. The 3-D structures may help to grow cells rapidly in a permissive condition in which all other requirements are. Additionally, the structure also resists the inflammatory cells and other adverse elements present in the host tissue. Many research studies mainly focused on various drug delivery methods in preventing and treating AD. Peptide-amphiphile hydrogels can release the neuroprotective compound and antioxidant and hydrogels prepared from xanthan and gellan gum can release the active ingredient. Furthermore, many research studies evidenced that the encapsulation of neural stem cells (NSCs) in various types of hydrogels may increase the life span and cognitive capacity of stem cells (Rajput et al. 2018). In the case of PD, the authors mainly focused on the release of dopamine. Chitosan or gelatin and dextran or gelatin hydrogels help to release dopamine delivery, both types had showed excellent release of the drug. Besides this, it also revealed that there is an improvement in behavioral and motor changes in PD mice after the therapy (Spuch et al. 2010; Ren et al. 2017; Senthilkumar et al. 2007).

6.4.5 Polymeric Micelles

Polymeric micelles are a form of nanomaterial made from the self-assembly of amphiphilic molecules in water at a specific concentration known as critical micelle concentration (Niu, Chen, and Gao 2019). The polymeric micelles frequently comprised an amphiphilic block copolymer and a center core–shell of polymer blocks that are hydrophobic in nature (e.g., poly (D,L-lactide), poly (caprolactone), poly (propylene glycol) (PPG), and many more) and a shell of hydrophilic polymer blocks (regularly PEG) (Gilmore et al. 2008). The fluctuation in size and dimensions of polymeric micelles have been observed from 10 to 100 nm. Their center can consolidate high levels (up to 20%wt–30%wt) of poorly soluble medications forestalling untimely medication delivery and drug degradation. The shell of micelles gets stabilized by dispersion and protects the drug from serum proteins and untargeted cells. In the wake of arriving at target cells, the drug is let out of the micelle through diffusion. Such nanomaterials are diversified and can carry variable solutes (both hydrophilic and hydrophobic drugs). Nanomicelles are benefitted due to their controllable size and shape, and high encapsulation effectiveness and capacity to load drugs. By focusing on conveyance, they are being developed as theranostics (drug carrier as well as diagnostic imaging tool) (Torchilin 2004).

6.4.6 Exosomes

Exosomes are one of the regular endogenous nanocarriers having a size range from 30 to 150 nm and have a unique and complex lipid bilayer structure, which is rumored a "floating jug" in the living body. Exosomes are secreted by an assortment of cells such as B cells, macrophages, T cells, and dendritic cells in the endosomal compartment (Niu, Chen, and Gao 2019). These exosomes because highly versatile advantages like nanoscopic size, high bioavailability, capacity to communicate with surrounding cells, both fundamentally and locally, immunogenicity, competence to transcend biological barriers like the BBB, the impressive potential for tissue-targeted delivery, epitome for carrying different unstable biological molecules like lipids, chemicals, proteins, and qualities, thus have counted the exosomes as reasonable promising carriers for further developing the drug carriers for therapy of various NDs (Sarko and McKinney 2017). Exosomes are assumed to play double parts by either helping with spreading the hazardous proteins like prions, a-synuclein, and tau, in this manner, speeding up the movement of the disease; or because of their capacity in shipping cell elements across the BBB, and thereby decreasing the risk of NDs (Jan et al. 2017). Alvarez-Erviti et al. (2011) engineered dendritic cell-derived exosomes to present with nervous system-specific protein (rabies viral glycoprotein) and contain siRNA that specifically targeted neurons, microglia, and oligodendrocytes in the mice brains to achieve a knockdown of BACE1, a therapeutic target for AD. This shows that exosomes have the potential to be used in AD.

6.4.7 Dendrimers

Dendrimers are polymers having a polymeric hyperbranched nearly spherical configuration and a three-dimensional framework which are extensively used during delivery of drugs (Du et al. 2013). They can be conjugated with pro-drugs or imaging agents or formed noncovalently with biological agents like DNA. As a result, they can be utilized as drug delivery vehicles, molecular imaging, or therapeutics. Various research studies have been conducted in recent years which have highlighted the significance of dendrimers as drug carriers in NDs. Srinageshwar et al. effectively replaced surface amine groups with biocompatible hydroxyl (-OH) groups to create poly-amidoamine (PAMAM) dendrimers. Additionally, they showed that the created dendrimers led to the effective migration and delivery of drug candidates across the BBB with minimum invasiveness in mouse models (Srinageshwar et al. 2017).

6.5 TREATMENT OF NEURODEGENERATIVE AILMENTS USING NANOFORMULATIONS OF PHYTOCONSTITUENTS/HERBAL EXTRACTS

Various recent studies on nanocarriers of phytoconstituents and herbal extracts are listed out in Table 6.1. Curcumin is a biologically diverse phenolic compound originating from the family of plants Curcuma longa, it has a unique chemical framework that allows it to have therapeutic potential and effects. Cur may act on distinct pharmacological and biological sites, like growth factors, gene transcription factors, and cytokines (Bengmark 2006; Wang, Li, et al. 2009). It can control inflammation by suppressing anti-inflammatory mediators such as interleukin 8 (IL-8 cyclooxygenase-2 (COX-2), and TNF-α. Additionally, it has proven that it is effective on Aβ and supplementary protein aggregation, making it favorable to enhance NDs, more specifically AD (Del Prado-Audelo et al. 2019). Regardless of its excellent therapeutic benefits, it produced adequate pharmacodynamic, pharmacokinetic, and bioavailability parameters. To bridle such limitation, various NPs were developed such as nanoliposomes, nanocapsules, and nanomicelles to upgrade the bioavailability, pharmacokinetics, and pharmacodynamic properties of the drug (Bollimpelli et al. 2016). Also, there are some drawbacks to resveratrol such as poor water solubility, low bioavailability, and rapid metabolism. It was discovered that the resveratrol-loaded mesoporous nanoselenium (MSe-Res/Fc-b-CD/Bor) delivery system improved memory deficits, reduced oxidative stress, and suppressed Aβ aggregation, in another investigation (Sun et al. 2019). Lu et al. (2009) reported that resveratrol-loaded polymeric micelles were effective in *in vitro* inhibition of damages caused by induction of Aβ by decreasing oxidative stress and apoptosis. Also, the damage to the spinal cord in rats, chitosan-coated PLGA NPs of resveratrol lowered inflammatory cytokines, increased IL-10, improved neuroprotection, and improved functional recovery (Wang et al. 2019). Etman et al. formulated a microemulsion of piperine that had good efficiency, improved clinical results, and enhanced PIP delivery to the brain in patients with AD (Etman et al. 2018). Kuo et al. formulated ApoE-QC-RA-PA liposomes which have been proven to traverse the BBB to re-establish Aβ_{1-42}'s neurotoxicity in an AD model (Kuo, Chen, and Rajesh 2018). The same formulation showed reduced peroxidation of lipids, production of Aβ plaques, and, activity of acetylcholinesterase enzyme. EGCG is a product of the esterification of gallic acid and epigallocatechin, the most common catechin in tea. Due to their interactions with key proteins including transthyretin, Aβ, huntingtin, andα-synuclein, these chemicals may have positive effects on NDs due to their antioxidant potencies (Singh, Mandal, and Khan 2016). Thus, Smith and his developed nanolipidic EGCG which has its EGCG bioavailability and neuronal α-secretase activity in AD and HIV-associated dementia mice models (Smith et al. 2010). Furthermore, thymoquinone which is a bioactive lipophilic molecule has shown various pharmacological effects in NDs and cognitive diseases. Ismail and his co-workers developed a nanoemulsion of thymquinone which controlled the activity of β- and γ-secretase and increased the breakdown of Aβ plaques (Ismail et al. 2017). Kundu et. al. formulated glyceryl monooleate (GMO) nanoparticles co-encapsulated with piperine

TABLE 6.1
Various Studies on Nanotechnology-Based Phytodrug Delivery Systems

S. No	Active Constituent	Delivery System Formulated	*In-Vivo* Model/*In-Vitro* Cell Line Used	Disease under Study	Molecular Targets	Inference	References
1	Curcumin	Dendrimers	Neuro-2a cell line	PD	α-Synuclein aggregation	Developed formulation increased the solubility and stability of curcumin. It inhibits abnormal aggregation of α-synuclein, which adds to CUR's proven therapeutic benefits for PD. Along with this, it reduced the oxidative stress.	Igartúa et al. (2023)
2	The root extract of *Paeonia moutan*	Gold nanoparticles	Murine microglial BV2 cells	PD	Inflammatory cytokines & interleukins, and oxidative stress system	It displayed anti-inflammatory, anti-oxidant activity. Furthermore, in the *in vivo* model, it reduces neural inflammation and enhances motor coordination.	Xue et al. (2019)
3	Quercetin	Nanocrystals	6-hydroxydopamine (6-OHDA)-induced Parkinson-like model in male rats	PD	Dopaminergic and Oxidative stress	Quercetin nanocrystals improved the bioavailability of quercetin and it also reversed the hippocampal oxidative stress and memory impairment.	Ghaffari, Hajizadeh Moghaddam, and Zare (2018)
4	Resveratrol/ Grape extract	Solid lipid nanoparticles (SLNs)	Human brain-like endothelial cells	AD	BBB transcytosis and Amyloid cascades	The extract actively reduced fibrillar development, with the extract having stronger action. Also, the SLNs demonstrated greater transcytosis and encapsulation efficiency.	Loureiro et al. (2017)
5	Piperine	Polymeric nanoparticles	Colchicine-induced AD Wistar rats	AD	Cholinergic and oxidative stress systems	Piperine NPs effectively reduced the behavioral impairment in the AD rat model as compared to its free form via inhibiting AChE and the oxidative stress environment.	Elnaggar et al. (2015)

(Continued)

TABLE 6.1 (*Continued*)
Various Studies on Nanotechnology-Based Phytodrug Delivery Systems

S. No	Active Constituent	Delivery System Formulated	*In-Vivo* Model/*In-Vitro* Cell Line Used	Disease under Study	Molecular Targets	Inference	References
6	Galantamine	Polymeric nanoparticles	Scopolamine-induced AD rats	AD	Cholinergic system and notch signaling	Improved cholinergic transmission, exploratory behavior, and spatial memory in rats. The histopathological deterioration was recovered after Notch signaling and beta-amyloid formation were blocked. Prepared nanoparticles also improved the delivery of galantamine to the brain and reduced its elimination.	El-Ganainy et al. (2021)
7	Anthocyanins	Polymeric gold nanoparticles	HT22 neuronal cells and $A\beta_{1-42}$ Mouse Model of AD	AD	BACE-1 and Amyloid precursor protein	Prevented neurodegeneration and apoptosis in mice given injections of $A\beta_{1-42}$. It also protected neurons from $A\beta_{1-42}$ synaptic dysfunction and improved memory power.	Ali et al. (2017)
8	Berberine	Carbon nanotubes	β-amyloid induced AD in Wistar rats	AD	Cholinergic systems and BBB transcytosis,	In comparison to its free form, berberine-loaded carbon nanotubes coated with phospholipids and polysorbates significantly improved memory impairment and reduced AChE activity in AD rats.	Lohan et al. (2017)
9	Thymoquinone	SLNs	Albino male rats	HD	Neuro-inflammation	Thymoquinone's solubility, bioavailability, and absorption were all improved by SLNs. Additionally, it improved the drug payload and sustained release capability. Thymoquinone in SLNs therefore prevents 3-Nitropropionic acid-induced inflammation and degeneration.	Ramachandran and Thangarajan (2018)
10	Sumac	Phytosomes	Streptozotocin (STZ)-induced rat model of AD	AD	Oxidative stress	The developed formulation enhanced the bioavailability of sumac across BBB and reduced the oxidative stress and anxiety-like behaviors that STZ causes in the rat model of AD.	Hajizadeh Moghaddam et al. (2020)

(*Continued*)

TABLE 6.1 (*Continued*)
Various Studies on Nanotechnology-Based Phytodrug Delivery Systems

S. No	Active Constituent	Delivery System Formulated	*In-Vivo* Model/*In-Vitro* Cell Line Used	Disease under Study	Molecular Targets	Inference	References
11	Resveratrol	SLNs	hCMEC/D3 cell line	ALS	Inflammation, and Oxidative stress	Resveratrol is delivered to the brain very effectively while being protected from degradation and crossing the BBB resulting in increased bioavailability.	Neves, Queiroz, and Reis (2016)
12	Curcumin and fish oil	Spongosome and cubosome lipid NPs	The human neuroblastoma SHSY5Y cells	HD	Oxidative stress	A significant decrease in the ROS accumulation was observed. Additionally, it offered protection against ROS buildup and H_2O_2-induced cell death.	Rakotoarisoa et al. (2019)
13	Epigallocatechin gallate	Carbonized polymer dots	SH-SY5Y cells, *C. elegans* mutant CL2006 and wild-type N2 models	AD	Oxidative stress and Aβ fibrils	Inhibition of β-sheet-rich Aβ-aggregate formation, the quick disassembly of mature Aβ-40 fibrils, the detection of plaques created by A-aggregation, the scavenging of Aβ-induced intracellular ROS, and the extension of the life span of the *in-vivo* models under research may all be accomplished with the aid of artificially created drug delivery systems.	Lin et al. (2023)
14	Artichoke (*Cynara cardunculus* L)	Chitosan-coated,artichoke-loaded SLNs	Swiss albino STZ-induced mice model of AD	AD	Oxidative stress and amyloidogenic pathway	It improved cognitive functions and spatial memory. Moreover, it provided neuroprotection, reduced neural inflammation, and decreased Aβ and Tau levels.	El-Nashar et al. (2022)
15	*Cannabis sativa* extract	Nanoliposomes	Male IRC mice	Epilepsy	Oxidative stress	Prepared formulation improved solubility, and delivery of *Cannabis* extract to the CNS and thus enhanced bioavailability.	Mihailova et al. (2022)

and curcumin inhibited S-protein oligomerization and fibril formation, reduced rotenone-induced toxicity, reduced oxidative stress by reducing glutathione depletion induced by rotenone, inhibited apoptosis by reducing the ratio of Bcl-2 to Bax, and started the autophagic pathway *in-vitro*. Additionally, *in-vivo* experiments on PD animal models showed that these nanoparticles could pass the BBB, restore the motor coordination impairment caused by rotenone, and stop the degradation of dopaminergic neurons (Kundu et al. 2016). Machado et al. developed retinoic acid-loaded nanoparticles and observed effective reduction in microglial activation in N9 microglial cells and suppressed nitric oxide release and boosted arginase-1 and IL-4 production (Machado-Pereira et al. 2017). This study indicates that retinoic acid-loaded nanoparticles can be used as a promising therapeutic agent for brain inflammatory diseases.

6.6 CHALLENGES IN HARNESSING NANOTECH FOR NEURODEGENERATIVE DISORDERS

The 21st century has seen extensive research in the field of nanotechnology. Due to the requirements and uses of nanosystems in various pharmaceutical sciences domains, it has attracted public interest (Kumari et al. 2022). Because of its size, form, and structural characteristics, nanosystems have a number of advantages over conventional systems. It does, however, have a few really delicate issues.

There isn't a single treatment approach that can be utilized to treat all patients because each person's DNA, neurological functioning, and features are different. These neurodegenerative illnesses must be diagnosed and treated individually for each patient, which is a difficult undertaking that hinders the creation of an all-encompassing strategy or method of treatment. If the administration is not done correctly, the NPs used in this process could possibly be absorbed by the blood or other body parts instead of the desired target area (Kim, Lee, and Chen 2013). Recently, the researchers have brought this issue to light. The medication will not be absorbed or used to treat the intended area for neurodegenerative illnesses such as AD and PD if it is not administered correctly. Some have stated that the therapy has not yet been demonstrated to be 100% effective because just a little amount of the drug actually gets to where it is needed (Rahman et al. 2015). Nanosystems can interact with biological systems thanks to their distinct physicochemical features, which may have unforeseen and unwelcome results. Consequently, a thorough examination of nanobio-interactions is required to guarantee the appropriate and secure usage of nanomaterials. Furthermore, another important problem that needs more research is the creation of nanoparticles that target particular brain cells. Targeting particular brain cells could increase its potential therapeutic value in the setting of NDs, such as dopaminergic neurons (the main target in PD), neural stem cells (neuronal repair), or microglia (neuroinflammation). Microglial activation is a key element in NP neurotoxicity as it participates in the neurodegenerative pathogenic process in the vast majority of CNS illnesses. The same factors that require research into the neurotoxicity of a more conventional treatment system apply to the neurotoxicity of nanoparticles (Rajput et al. 2023). Therefore, extensive research on the neurotoxicity of nanotechnology-based phytodrug delivery systems is required.

6.7 CONCLUSION

The frequency of NDs is increasing along with an increase in longevity globally, necessitating novel therapeutic approaches to improve both the symptoms and quality of life of individuals with such diseases. Neurodegenerative ailments have similar cellular and subcellular characteristics, as well as the same molecular signaling pathways that can lead to inflammation, necroptosis, and apoptosis. Degradation of protein, oxidative stress, tissue damage, lifestyle exposures, mitochondrial abnormalities, hereditary history, and aberrant accumulation of protein in neurons are among the most widely researched environmental variables that cause NDs. The CNS is carefully protected by a number of barriers. So, in order for a proper medicine to reach the CNS, it must basically cross the BBB. In this context, herbal therapy, in general, offers intriguing alternatives to existing neurodegenerative disease treatments. Medicinal herbs and their phytoconstituents, such as *A. sativum*, *A. calamus*, resveratrol, *G. nivalis*, *W. somnifera*, *G. biloba*, and *C. longa*, are used to cure or suppress neurological illnesses and relieve brain abnormalities in *in vivo* or preclinical studies. Nevertheless, the broad range of approaches tends to reflect the difficulties involved in transporting therapeutic and imaging agents across the BBB. However, the pharmacokinetic features of herbal medicine/natural substances severely limit their potential as their therapeutic applications are restricted by specific characteristics like limited selectivity, low solubility, diminished absorption, and poor bioavailability. Further factors that contribute to their low plasma levels include chemical breakdown, rapid metabolism, and rapid clearance. Hence herbal medication has been included in several drug delivery formulations to circumvent these constraints.

In recent years, nanotechnology has been considered a leading new field for medication delivery. These nanoformulations can target drugs to specific cells and lower required doses. Nanosystems regulated by plants, potentially enhance the pharmacokinetic properties as well as the bioactivity of bioactive agents in the CNS, boost drug–brain penetration, and promote dispersion or avoid the development of complexes inside the brain. By enhancing the bioavailability, specificity, cellular absorption, and efficacy of phytoconstituent nanoformulations (micelle, nanoparticles, solid lipid nanoparticles, liposomes, etc.), the pharmacokinetic constraints faced by phytoconstituents can be controlled. Furthermore, employing natural herbs in the nanosize range as a therapeutic agent has been shown to have similar activity as in their raw state. Nanocarriers based on phytochemicals provide a number of significant benefits, including being environmentally acceptable, less hazardous, safe, affordable, easily scalable, and generating nanoparticles of precise size. Therefore, it is anticipated that these recent developments in the novel delivery mechanism for plant-based natural medicines will change the therapeutic strategies used to manage NDs.

However, a number of problems still need to be solved, including the systemic toxicity of utilizing NCs in the brain. The *in-vitro* and *in vivo* approaches must be validated to move these nanoformulations from laboratory testing to clinical trials. Establishing precise guidelines for the quality, efficacy, and safety assessment of these products as well as conducting sufficient clinical trials to guarantee the occurrence of few adverse effects—long-term evaluations are also required for the clinical effectiveness of nanosystems in the field of neurological care—are essential if these nanoformulations are to be properly marketed. Furthermore, to succeed commercially in the market, cost-effectiveness in the scale-up of such nanoformulations must be taken into account. Once the unresolved problems with the toxicity of NCs are further assessed, the future of nanocarrier-based therapies against NDs appears bright and advanced.

Funding: The authors declare that no funds, grants, or other support were received during the preparation of this manuscript.

Competing interests: The authors declare that they have no known competing financial interests or personal relationships that could have appeared to influence the work reported in this paper.

REFERENCES

Afshari, Amir R, Hamid R Sadeghnia, and Hamid Mollazadeh. 2016. "A Review on Potential Mechanisms of *Terminalia Chebula* in Alzheimer's Disease." edited by Berend Olivier. *Advances in Pharmacological Sciences*. Hindawi Publishing Corporation: 8964849. doi:10.1155/2016/8964849.

Aggarwal, Bharat B, Chitra Sundaram, Nikita Malani, and Haruyo Ichikawa. 2007. "Curcumin: The Indian Solid Gold." edited by B. B. Aggarwal, Y. J. Surh, and S. Shishodia. *Advances in Experimental Medicine and Biology*, Vol. 595. Springer: 1–75. doi:10.1007/978-0-387-46401-5_1.

Ali, Tahir, Min Ju Kim, Shafiq Ur Rehman, Ashfaq Ahmad, and Myeong Ok Kim. 2017. "Anthocyanin-Loaded PEG-Gold Nanoparticles Enhanced the Neuroprotection of Anthocyanins in an Aβ(1-42) Mouse Model of Alzheimer's Disease." *Molecular Neurobiology* 54 (8): 6490–506. doi:10.1007/s12035-016-0136-4.

Alvarez-Erviti, Lydia, Yiqi Seow, HaiFang Yin, Corinne Betts, Samira Lakhal, and Matthew J A Wood. 2011. "Delivery of SiRNA to the Mouse Brain by Systemic Injection of Targeted Exosomes." *Nature Biotechnology* 29 (4): 341–45. doi:10.1038/nbt.1807.

Amanzadeh, Elnaz, Abolghasem Esmaeili, Soheila Rahgozar, and Maryam Nourbakhshnia. 2019. "Application of Quercetin in Neurological Disorders: From Nutrition to Nanomedicine." *Reviews in the Neurosciences* 30 (5): 555–72. doi:10.1515/revneuro-2018-0080.

Anand, T, K Bhanu Prakash, M D Pandareesh, and Farhath Khanum. 2014. "Development of Bacoside Enriched Date Syrup Juice and Its Evaluation for Physical Endurance." *Journal of Food Science and Technology* 51 (12): 4026–32. doi:10.1007/s13197-013-0955-5.

Baarine, Mauhamad, Sijo Joseph Thandapilly, Xavier Lieben Louis, Frédéric Mazué, Liping Yu, Dominique Delmas, Thomas Netticadan, Gérard Lizard, and Norbert Latruffe. 2011. "Pro-Apoptotic versus Anti-Apoptotic Properties of Dietary Resveratrol on Tumoral and Normal Cardiac Cells." *Genes & Nutrition* 6 (2): 161–69. doi:10.1007/s12263-011-0232-z.

Baker, Joseph T, Robert P Borris, Brad Carté, Geoffrey A Cordell, Djaja D Soejarto, Gordon M Cragg, Mahabir P Gupta, Maurice M Iwu, Domingo R Madulid, and Varro E Tyler. 1995. "Natural Product Drug Discovery and Development: New Perspectives on International Collaboration." *Journal of Natural Products* 58 (9): 1325–57. doi:10.1021/np50123a003.

Bandelow, Borwin, Sophie Michaelis, and Dirk Wedekind. 2017. "Treatment of Anxiety Disorders." *Dialogues in Clinical Neuroscience* 19 (2): 93–107. doi:10.31887/DCNS.2017.19.2/bbandelow.

Bengmark, Stig. 2006. "Curcumin, an Atoxic Antioxidant and Natural NFkappaB, Cyclooxygenase-2, Lipooxygenase, and Inducible Nitric Oxide Synthase Inhibitor: A Shield against Acute and Chronic Diseases." *JPEN. Journal of Parenteral and Enteral Nutrition* 30 (1): 45–51. doi:10.1177/014860710603000145.

Berardi, Valerio, Francesca Ricci, Mauro Castelli, Gaspare Galati, and Gianfranco Risuleo. 2009. "Resveratrol Exhibits a Strong Cytotoxic Activity in Cultured Cells and Has an Antiviral Action against Polyomavirus: Potential Clinical Use." *Journal of Experimental & Clinical Cancer Research* 28 (1): 96. doi:10.1186/1756-9966-28-96.

Bhattacharya, S K, and A V Muruganandam. 2003. "Adaptogenic Activity of Withania Somnifera: An Experimental Study Using a Rat Model of Chronic Stress." *Pharmacology, Biochemistry, and Behavior* 75 (3): 547–55. doi:10.1016/s0091-3057(03)00110-2.

Bollimpelli, V Satish, Prashant Kumar, Sonali Kumari, and Anand K Kondapi. 2016. "Neuroprotective Effect of Curcumin-Loaded Lactoferrin Nano Particles against Rotenone Induced Neurotoxicity." *Neurochemistry International* 95 (May): 37–45. doi:10.1016/j.neuint.2016.01.006.

Bordoni, Matteo, Eveljn Scarian, Federica Rey, Stella Gagliardi, Stephana Carelli, Orietta Pansarasa, and Cristina Cereda. 2020. "Biomaterials in Neurodegenerative Disorders: A Promising Therapeutic Approach." *International Journal of Molecular Sciences* 21 (9). doi:10.3390/ijms21093243.

Budd Haeberlein, S L, and T J R Harris. 2015. "Promising Targets for the Treatment of Neurodegenerative Diseases." *Clinical Pharmacology and Therapeutics* 98 (5): 492–501. doi:10.1002/cpt.195.

Calzoni, Eleonora, Alessio Cesaretti, Alice Polchi, Alessandro Di Michele, Brunella Tancini, and Carla Emiliani. 2019. "Biocompatible Polymer Nanoparticles for Drug Delivery Applications in Cancer and Neurodegenerative Disorder Therapies." *Journal of Functional Biomaterials* 10 (1). doi:10.3390/jfb10010004.

Carradori, Dario, Joel Eyer, Patrick Saulnier, Véronique Préat, and Anne des Rieux. 2017. "The Therapeutic Contribution of Nanomedicine to Treat Neurodegenerative Diseases via Neural Stem Cell Differentiation." *Biomaterials* 123 (April): 77–91. doi:10.1016/j.biomaterials.2017.01.032.

Chang, Raymond Chuen-Chung, and Kwok-Fai So. 2008. "Use of Anti-Aging Herbal Medicine, Lycium Barbarum, against Aging-Associated Diseases. What Do We Know So Far?" *Cellular and Molecular Neurobiology* 28 (5): 643–52. doi:10.1007/s10571-007-9181-x.

Chang, Xue-Run, Li Wang, Jing Li, and Dian-Shui Wu. 2016. "Analysis of Anti-Depressant Potential of Curcumin against Depression Induced Male Albino Wistar Rats." *Brain Research* 1642 (July): 219–25. doi:10.1016/j.brainres.2016.03.010.

Chaudhari, Kaustubh S, Nishant R Tiwari, Rakesh R Tiwari, and Rohan S Sharma. 2017. "Neurocognitive Effect of Nootropic Drug Brahmi (Bacopa Monnieri) in Alzheimer's Disease." *Annals of Neurosciences* 24 (2): 111–22. doi:10.1159/000475900.

Chen, Chieh-fu, Wen-fei Chiou, and Jun-tian Zhang. 2008. "Comparison of the Pharmacological Effects of Panax Ginseng and Panax Quinquefolium." *Acta Pharmacologica Sinica* 29 (9): 1103–8. doi:10.1111/j.1745-7254.2008.00868.x.

Chen, Chien-Li, Wen-Hao Tsai, Chun-Jen Chen, and Tzu-Ming Pan. 2016. "Centella Asiatica Extract Protects against Amyloid β(1-40)-Induced Neurotoxicity in Neuronal Cells by Activating the Antioxidative Defence System." *Journal of Traditional and Complementary Medicine* 6 (4): 362–69. doi:10.1016/j.jtcme.2015.07.002.

Chonpathompikunlert, Pennapa, Jintanaporn Wattanathorn, and Supaporn Muchimapura. 2010. "Piperine, the Main Alkaloid of Thai Black Pepper, Protects against Neurodegeneration and Cognitive Impairment in Animal Model of Cognitive Deficit like Condition of Alzheimer's Disease." *Food and Chemical Toxicology : An International Journal Published for the British Industrial Biological Research Association* 48 (3): 798–802. doi:10.1016/j.fct.2009.12.009.

Chopra, Hitesh, Sandeep Kumar, and Inderbir Singh. 2021. "Biopolymer-Based Scaffolds for Tissue Engineering Applications." *Current Drug Targets* 22 (3): 282–95. doi:10.2174/13894501219992011 02140408.

Chu, Shenghui, Junfei Gu, Liang Feng, Jiping Liu, Minghua Zhang, Xiaobin Jia, Min Liu, and Danian Yao. 2014. "Ginsenoside Rg5 Improves Cognitive Dysfunction and Beta-Amyloid Deposition in STZ-Induced Memory Impaired Rats via Attenuating Neuroinflammatory Responses." *International Immunopharmacology* 19 (2): 317–26. doi:10.1016/j.intimp.2014.01.018.

Cipolla, Marilyn J. 2009. *The Cerebral Circulation.* San Rafael (CA). doi:10.4199/C00005ED1V01 Y200912ISP002.

Correia, Alyne, Cruz AAP, Aquino ATRD, Diniz JRG, Santana KBF, Cidade PIM, Peixoto JD, et al. 2015. "Neuroprotective Effects of Piperine, an Alkaloid from the Piper Genus, on the Parkinson's Disease Model in Rats." *Journal of Neurology and Therapeutics* 1 (November): 1–8. doi:10.14312/2397-1304.2015-1.

Damkier, Helle H, Peter D Brown, and Jeppe Praetorius. 2013. "Cerebrospinal Fluid Secretion by the Choroid Plexus." *Physiological Reviews* 93 (4): 1847–92. doi:10.1152/physrev.00004.2013.

Darvesh, Altaf S, Richard T Carroll, Anupam Bishayee, Nicholas A Novotny, Werner J Geldenhuys, and Cornelis J Van der Schyf. 2012. "Curcumin and Neurodegenerative Diseases: A Perspective." *Expert Opinion on Investigational Drugs* 21 (8): 1123–40. doi:10.1517/13543784.2012.693479.

Dasgupta, Biplab, and Jeffrey Milbrandt. 2007. "Resveratrol Stimulates AMP Kinase Activity in Neurons." *Proceedings of the National Academy of Sciences of the United States of America* 104 (17): 7217–22. doi:10.1073/pnas.0610068104.

Deepak, M, and A Amit. 2004. "The Need for Establishing Identities of 'Bacoside A and B', the Putative Major Bioactive Saponins of Indian Medicinal Plant Bacopa Monnieri." *Phytomedicine* 11 (2): 264–68. doi:10.1078/0944-7113-00351.

De Gioia, Roberta, Fabio Biella, Gaia Citterio, Federica Rizzo, Elena Abati, Monica Nizzardo, Nereo Bresolin, Giacomo Pietro Comi, and Stefania Corti. 2020. "Neural Stem Cell Transplantation for Neurodegenerative Diseases." *International Journal of Molecular Sciences* 21 (9). doi:10.3390/ijms21093103.

Del Prado-Audelo, María L, Isaac H Caballero-Florán, Jorge A Meza-Toledo, Néstor Mendoza-Muñoz, Maykel González-Torres, Benjamín Florán, Hernán Cortés, and Gerardo Leyva-Gómez. 2019. "Formulations of Curcumin Nanoparticles for Brain Diseases." *Biomolecules* 9 (2). doi:10.3390/biom9020056.

Dhaouadi, Z, M Nsangou, N Garrab, E H Anouar, K Marakchi, and S Lahmar. 2009. "DFT Study of the Reaction of Quercetin with ·O2- and ·OH Radicals." *Journal of Molecular Structure: THEOCHEM* 904 (1): 35–42. doi:10.1016/j.theochem.2009.02.034.

Du, Xin, Bingyang Shi, Ji Liang, Jingxu Bi, Sheng Dai, and Shi Zhang Qiao. 2013. "Developing Functionalized Dendrimer-like Silica Nanoparticles with Hierarchical Pores as Advanced Delivery Nanocarriers." *Advanced Materials (Deerfield Beach, Fla.)* 25 (41): 5981–85. doi:10.1002/adma.201302189.

El-Ganainy, Samar O, Mennatallah A Gowayed, Mahmoud Agami, Passant Mohamed, Marwa Belal, Ragwa M Farid, and Amira S Hanafy. 2021. "Galantamine Nanoparticles Outperform Oral Galantamine in an Alzheimer's Rat Model: Pharmacokinetics and Pharmacodynamics." *Nanomedicine* 16 (15): 1281–96. doi:10.2217/nnm-2021-0051.

Elnaggar, Yosra S R, Samar M Etman, Doaa A Abdelmonsif, and Ossama Y Abdallah. 2015. "Intranasal Piperine-Loaded Chitosan Nanoparticles as Brain-Targeted Therapy in Alzheimer's Disease: Optimization, Biological Efficacy, and Potential Toxicity." *Journal of Pharmaceutical Sciences* 104 (10): 3544–56. doi:10.1002/jps.24557.

El-Nashar, Heba A S, Haidy Abbas, Mariam Zewail, Mohamed H Noureldin, Mai M Ali, Marium M Shamaa, Mohamed A Khattab, and Nehal Ibrahim. 2022. "Neuroprotective Effect of Artichoke-Based Nanoformulation in Sporadic Alzheimer’s Disease Mouse Model: Focus on Antioxidant, Anti-Inflammatory, and Amyloidogenic Pathways." *Pharmaceuticals*. doi:10.3390/ph15101202.

Erkkinen, Michael G, Mee-Ohk Kim, and Michael D Geschwind. 2018. "Clinical Neurology and Epidemiology of the Major Neurodegenerative Diseases." *Cold Spring Harbor Perspectives in Biology* 10 (4). doi:10.1101/cshperspect.a033118.

Esfandiari, Ebrahim, Mustafa Ghanadian, Bahman Rashidi, Amir Mokhtarian, and Amir M Vatankhah. 2018. "The Effects of Acorus Calamus L. in Preventing Memory Loss, Anxiety, and Oxidative Stress on Lipopolysaccharide-Induced Neuroinflammation Rat Models." *International Journal of Preventive Medicine* 9 (October): 85. doi:10.4103/ijpvm.IJPVM_75_18.

Etman, Samar M, Yosra S R Elnaggar, Doaa A Abdelmonsif, and Ossama Y Abdallah. 2018. "Oral Brain-Targeted Microemulsion for Enhanced Piperine Delivery in Alzheimer's Disease Therapy: In Vitro Appraisal, In Vivo Activity, and Nanotoxicity." *AAPS PharmSciTech* 19 (8): 3698–711. doi:10.1208/s12249-018-1180-3.

Feng, Xiaowen, Nan Liang, Dexiao Zhu, Qing Gao, Lei Peng, Haiman Dong, Qingwei Yue, et al. 2013. "Resveratrol Inhibits β-Amyloid-Induced Neuronal Apoptosis through Regulation of SIRT1-ROCK1 Signaling Pathway." *PloS One* 8 (3): e59888. doi:10.1371/journal.pone.0059888.

Feng, Yong-Hong, Wen-Liang Zhou, Qing-Li Wu, Xiao-Yu Li, Wei-Min Zhao, and Jian-Ping Zou. 2002. "Low Dose of Resveratrol Enhanced Immune Response of Mice." *Acta Pharmacologica Sinica* 23 (10): 893–97.

Fernandes, Filipe, Mónica Dias-Teixeira, Cristina Delerue-Matos, and Clara Grosso. 2021. "Critical Review of Lipid-Based Nanoparticles as Carriers of Neuroprotective Drugs and Extracts." *Nanomaterials (Basel, Switzerland)* 11 (3): 563. doi:10.3390/nano11030563.

"Focus on Neurodegenerative Disease." 2018. *Nature Neuroscience* 21 (10): 1293. doi:10.1038/s41593-018-0250-x.

Fu, Min, Zhao-Hui Sun, and Huan-Cong Zuo. 2010. "Neuroprotective Effect of Piperine on Primarily Cultured Hippocampal Neurons." *Biological & Pharmaceutical Bulletin* 33 (4): 598–603. doi:10.1248/bpb.33.598.

Furno, Debora Lo, Giuliana Mannino, and Rosario Giuffrida. 2018. "Functional Role of Mesenchymal Stem Cells in the Treatment of Chronic Neurodegenerative Diseases." *Journal of Cellular Physiology* 233 (5): 3982–99. doi:10.1002/jcp.26192.

Furno, Debora Lo, Rosalia Pellitteri, Adriana C E Graziano, Rosario Giuffrida, Carlo Vancheri, Elisa Gili, and Venera Cardile. 2013. "Differentiation of Human Adipose Stem Cells into Neural Phenotype by Neuroblastoma- or Olfactory Ensheathing Cells-Conditioned Medium." *Journal of Cellular Physiology* 228 (11): 2109–18. doi:10.1002/jcp.24386.

Furtado, Denzil, Mattias Björnmalm, Scott Ayton, Ashley I Bush, Kristian Kempe, and Frank Caruso. 2018. "Overcoming the Blood-Brain Barrier: The Role of Nanomaterials in Treating Neurological Diseases." Advanced Materials (Deerfield Beach, Fla.) 30 (46): e1801362. doi:10.1002/adma.201801362.

Gattazzo, Francesca, Anna Urciuolo, and Paolo Bonaldo. 2014. "Extracellular Matrix: A Dynamic Microenvironment for Stem Cell Niche." *Biochimica et Biophysica Acta* 1840 (8): 2506–19. doi:10.1016/j.bbagen.2014.01.010.

Ghaffari, Fatemeh, Akbar Hajizadeh Moghaddam, and Mahboobeh Zare. 2018. "Neuroprotective Effect of Quercetin Nanocrystal in a 6-Hydroxydopamine Model of Parkinson Disease: Biochemical and Behavioral Evidence." *Basic and Clinical Neuroscience* 9 (5): 317–24. doi:10.32598/bcn.9.5.317.

Gilmore, Jamie L, Xiang Yi, Lingdong Quan, and Alexander V Kabanov. 2008. "Novel Nanomaterials for Clinical Neuroscience." *Journal of Neuroimmune Pharmacology : The Official Journal of the Society on NeuroImmune Pharmacology* 3 (2): 83–94. doi:10.1007/s11481-007-9099-6.

Gohil, Kashmira J, Jagruti A Patel, and Anuradha K Gajjar. 2010. "Pharmacological Review on Centella Asiatica: A Potential Herbal Cure-All." *Indian Journal of Pharmaceutical Sciences* 72 (5): 546–56. doi:10.4103/0250-474X.78519.

Golla, Padmavathi, and Hemapriya Tirupathi. 2017. "To Evaluate and Compare Antidepressant Activity of Centella Asiatica in Mice by Using Forced Swimming Test." *International Journal of Basic & Clinical Pharmacology* 5 (5): 2017–20.

Griffin, Sharoon, Muhammad Irfan Masood, Muhammad Jawad Nasim, Muhammad Sarfraz, Azubuike Peter Ebokaiwe, Karl-Herbert Schäfer, Cornelia M Keck, and Claus Jacob. 2017. "Natural Nanoparticles: A Particular Matter Inspired by Nature." *Antioxidants (Basel, Switzerland)* 7 (1). doi:10.3390/antiox7010003.

Guo, Qing, Xian-Tao Shen, Yuan-Yuan Li, and Shun-Qing Xu. 2017. "Carbon Nanotubes-Based Drug Delivery to Cancer and Brain." *Current Medical Science* 37 (5): 635–41. doi:10.1007/s11596-017-1783-z.

Hajizadeh Moghaddam, A, M Mosavi mehr, V Hasantabar, and S Khanjani Jelodar. 2020. "Antioxidant and Anti-Anxiety Effects of Sumac Nano-Phytosome in Rat Model of Alzheimer's Disease: An Experimental Study." *RUMS_JOURNAL* 19 (7): 647–60. doi:10.29252/jrums.19.7.647.

Ho, Yuen-Shan, Man-Shan Yu, Xi-Fei Yang, Kwok-Fai So, Wai-Hung Yuen, and Raymond Chuen-Chung Chang. 2010. "Neuroprotective Effects of Polysaccharides from Wolfberry, the Fruits of Lycium Barbarum, against Homocysteine-Induced Toxicity in Rat Cortical Neurons." *Journal of Alzheimer's Disease : JAD* 19 (3): 813–27. doi:10.3233/JAD-2010-1280.

Ho, Yuen-Shan, Man-Shan Yu, Suet-Yi Yik, Kwok-Fai So, Wai-Hung Yuen, and Raymond Chuen-Chung Chang. 2009. "Polysaccharides from Wolfberry Antagonizes Glutamate Excitotoxicity in Rat Cortical Neurons." *Cellular and Molecular Neurobiology* 29 (8): 1233–44. doi:10.1007/s10571-009-9419-x.

Hsieh, Ting Yang, Yi Chang, and Su Jane Wang. 2019. "Piperine-Mediated Suppression of Voltage-Dependent Ca(2+) Influx and Glutamate Release in Rat Hippocampal Nerve Terminals Involves 5HT(1A) Receptors and G Protein Bγ Activation." *Food & Function* 10 (5): 2720–28. doi:10.1039/c8fo02189a.

Igartúa, Daniela E, Florencia González-Lizárraga, Carolina S Martinez, Silvia del V Alonso, César L Ávila, Rosana Chehín, Nadia S Chiaramoni, and M Jimena Prieto. 2023. "PAMAM Dendrimers of Generation 4.5 Loaded with Curcumin Interfere with α-Synuclein Aggregation." *OpenNano* 11: 100140. doi:https://doi.org/10.1016/j.onano.2023.100140.

Ismail, Norsharina, Maznah Ismail, Nur Hanisah Azmi, Muhammad Firdaus Abu Bakar, Zhang Yida, Maizaton Atmadini Abdullah, and Hamidon Basri. 2017. "Thymoquinone-Rich Fraction Nanoemulsion (TQRFNE) Decreases Aβ40 and Aβ42 Levels by Modulating APP Processing, up-Regulating IDE and LRP1, and down-Regulating BACE1 and RAGE in Response to High Fat/Cholesterol Diet-Induced Rats." *Biomedicine & Pharmacotherapy* 95: 780–88. doi:https://doi.org/10.1016/j.biopha.2017.08.074.

Jahangir, Mohammed Asadullah, Chettupalli Anand, Abdul Muheem, Sadaf Jamal Gilani, Mohamad Taleuzzaman, Ameeduzzafar Zafar, Mohammed Jafar, Surajpal Verma, and Md Abul Barkat. 2020. "Nano Phytomedicine Based Delivery System for CNS Disease." *Current Drug Metabolism* 21 (9): 661–73. doi:10.2174/1389200221666200523161003.

Jamali, Simin, Hossein Mostafavi, Ghasem Barati, Mehdi Eskandari, and Samad Nadri. 2017. "Differentiation of Mesenchymal Stem Cells -Derived Trabecular Meshwork into Dopaminergic Neuron-like Cells on Nanofibrous Scaffolds." *Biologicals : Journal of the International Association of Biological Standardization* 50 (November): 49–54. doi:10.1016/j.biologicals.2017.09.002.

Jan, Arif Tasleem, Mudasir A Malik, Safikur Rahman, Hye R Yeo, Eun J Lee, Tasduq S Abdullah, and Inho Choi. 2017. "Perspective Insights of Exosomes in Neurodegenerative Diseases: A Critical Appraisal." *Frontiers in Aging Neuroscience* 9 (September): 317. doi:10.3389/fnagi.2017.00317.

Jung, Ji-Sun, Jin A Shin, Eun-Mi Park, Jung-Eun Lee, Young-Sook Kang, Sung-Won Min, Dong-Hyun Kim, Jin-Won Hyun, Chan-Young Shin, and Hee-Sun Kim. 2010. "Anti-Inflammatory Mechanism of Ginsenoside Rh1 in Lipopolysaccharide-Stimulated Microglia: Critical Role of the Protein Kinase A Pathway and Hemeoxygenase-1 Expression." *Journal of Neurochemistry* 115 (6): 1668–80. doi:10.1111/j.1471-4159.2010.07075.x.

Khazdair, Mohammad Reza, Akbar Anaeigoudari, Milad Hashemzehi, and Reza Mohebbati. 2019. "Neuroprotective Potency of Some Spice Herbs, a Literature Review." *Journal of Traditional and Complementary Medicine* 9 (2): 98–105. doi:10.1016/j.jtcme.2018.01.002.

Kim, Tae Hyung, Seulki Lee, and Xiaoyuan Chen. 2013. "Nanotheranostics for Personalized Medicine." *Expert Review of Molecular Diagnostics* 13 (3): 257–69. doi:10.1586/erm.13.15.

Kovacs, Gabor G. 2016. "Molecular Pathological Classification of Neurodegenerative Diseases: Turning towards Precision Medicine." *International Journal of Molecular Sciences* 17 (2). doi:10.3390/ijms17020189.

Kreuter, J. 2001. "Nanoparticulate Systems for Brain Delivery of Drugs." *Advanced Drug Delivery Reviews* 47 (1): 65–81. doi:10.1016/s0169-409x(00)00122-8.

Kumar, Anil, Neha Seghal, Satyanaryana Venketeshwara Padi, and Pattipati Sreenivaslu Naidu. 2006. "Differential Effects of Cyclooxygenase Inhibitors on Intracerebroventricular Colchicine-Induced Dysfunction and Oxidative Stress in Rats." *European Journal of Pharmacology* 551 (1–3): 58–66. doi:10.1016/j.ejphar.2006.08.076.

Kumari, Amrita, Neha Raina, Abhishek Wahi, Khang W Goh, Pratibha Sharma, Riya Nagpal, Atul Jain, Long C Ming, and Madhu Gupta. 2022. "Wound-Healing Effects of Curcumin and Its Nanoformulations: A Comprehensive Review." *Pharmaceutics*. doi:10.3390/pharmaceutics14112288.

Kundu, Paromita, Manasi Das, Kalpalata Tripathy, and Sanjeeb K Sahoo. 2016. "Delivery of Dual Drug Loaded Lipid Based Nanoparticles across the Blood-Brain Barrier Impart Enhanced Neuroprotection in a Rotenone Induced Mouse Model of Parkinson's Disease." *ACS Chemical Neuroscience* 7 (12): 1658–70. doi:10.1021/acschemneuro.6b00207.

Kuo, Yung-Chih, I- Yin Chen, and Rajendiran Rajesh. 2018. "Use of Functionalized Liposomes Loaded with Antioxidants to Permeate the Blood-Brain Barrier and Inhibit β-Amyloid-Induced Neurodegeneration in the Brain." *Journal of the Taiwan Institute of Chemical Engineers* 87: 1–14. doi:10.1016/j.jtice.2018.03.001.

Lee, Dong Young, Hyun Woo Kim, Heejung Yang, and Sang Hyun Sung. 2017. "Hydrolyzable Tannins from the Fruits of Terminalia Chebula Retz and Their α-Glucosidase Inhibitory Activities." *Phytochemistry* 137: 109–16. doi:10.1016/j.phytochem.2017.02.006.

LeWitt, Peter A, Ali R Rezai, Maureen A Leehey, Steven G Ojemann, Alice W Flaherty, Emad N Eskandar, Sandra K Kostyk, et al. 2011. "AAV2-GAD Gene Therapy for Advanced Parkinson's Disease: A Double-Blind, Sham-Surgery Controlled, Randomised Trial." *The Lancet. Neurology* 10 (4): 309–19. doi:10.1016/S1474-4422(11)70039-4.

Limpeanchob, Nanteetip, Somkiet Jaipan, Saisunee Rattanakaruna, Watoo Phrompittayarat, and Kornkanok Ingkaninan. 2008. "Neuroprotective Effect of Bacopa Monnieri on Beta-Amyloid-Induced Cell Death in Primary Cortical Culture." *Journal of Ethnopharmacology* 120 (1): 112–17. doi:10.1016/j.jep.2008.07.039.

Lin, Xiaoding, Wei Liu, Xiaoyan Dong, and Yan Sun. 2023. "Epigallocatechin Gallate-Derived Carbonized Polymer Dots: A Multifunctional Scavenger Targeting Alzheimer's β-Amyloid Plaques." *Acta Biomaterialia* 157: 524–37. doi:https://doi.org/10.1016/j.actbio.2022.11.063.

Liu, Shu-Ying, Shuai Lu, Xiao-Lin Yu, Shi-Gao Yang, Wen Liu, Xiang-Meng Liu, Shao-Wei Wang, et al. 2018. "Fruitless Wolfberry-Sprout Extract Rescued Cognitive Deficits and Attenuated Neuropathology in Alzheimer's Disease Transgenic Mice." *Current Alzheimer Research* 15 (9): 856–68. doi:10.2174/1567205015666180404160625.

Lohan, Shikha, Kaisar Raza, S K Mehta, Gurjit Kaur Bhatti, Sumant Saini, and Bhupinder Singh. 2017. "Anti-Alzheimer's Potential of Berberine Using Surface Decorated Multi-Walled Carbon Nanotubes: A Preclinical Evidence." *International Journal of Pharmaceutics* 530 (1–2): 263–78. doi:10.1016/j.ijpharm.2017.07.080.

Loureiro, Joana A, Stephanie Andrade, Ana Duarte, Ana Rute Neves, Joana Fontes Queiroz, Cláudia Nunes, Emmanuel Sevin, et al. 2017. "Resveratrol and Grape Extract-Loaded Solid Lipid Nanoparticles for the Treatment of Alzheimer's Disease." *Molecules (Basel, Switzerland)* 22 (2). doi:10.3390/molecules22020277.

Lu, Xiaowei, Chenbo Ji, Huae Xu, Xiaolin Li, Haixia Ding, Min Ye, Zhenshu Zhu, et al. 2009. "Resveratrol-Loaded Polymeric Micelles Protect Cells from Abeta-Induced Oxidative Stress." *International Journal of Pharmaceutics* 375 (1–2): 89–96. doi:10.1016/j.ijpharm.2009.03.021.

Machado-Pereira, Marta, Tiago Santos, Lino Ferreira, Liliana Bernardino, and Raquel Ferreira. 2017. "Anti-Inflammatory Strategy for M2 Microglial Polarization Using Retinoic Acid-Loaded Nanoparticles." edited by Mirella Giovarelli. *Mediators of Inflammation*. Hindawi: 6742427. doi:10.1155/2017/6742427.

Majewski, Michał. 2014. "Allium Sativum: Facts and Myths Regarding Human Health." *Roczniki Panstwowego Zakladu Higieny* 65 (1): 1–8.

Mazza, Mariarosa, Rebecca Notman, Jamshed Anwar, Alison Rodger, Matthew Hicks, Gary Parkinson, Dave McCarthy, et al. 2013. "Nanofiber-Based Delivery of Therapeutic Peptides to the Brain." *ACS Nano* 7 (2): 1016–26. doi:10.1021/nn305193d.

Mihailova, Ljubica, Jana Tchekalarova, Dushko Shalabalija, Nikola Geskovski, Veronika Stoilkovska Gjorgievska, Gjoshe Stefkov, Paoleta Krasteva, Maja Simonoska Crcarevska, and Marija Glavas Dodov. 2022. "Lipid Nano-Carriers Loaded with Cannabis Sativa Extract for Epilepsy Treatment - in Vitro Characterization and in Vivo Efficacy Studies." *Journal of Pharmaceutical Sciences* 111 (12): 3384–96. doi:10.1016/j.xphs.2022.09.012.

Mittermeyer, Gabriele, Chadwick W Christine, Kathryn H Rosenbluth, Suzanne L Baker, Philip Starr, Paul Larson, Paul L Kaplan, John Forsayeth, Michael J Aminoff, and Krystof S Bankiewicz. 2012. "Long-Term Evaluation of a Phase 1 Study of AADC Gene Therapy for Parkinson's Disease." *Human Gene Therapy* 23 (4): 377–81. doi:10.1089/hum.2011.220.

Mocan, Andrei, Laurian Vlase, Dan Cristian Vodnar, Cristina Bischin, Daniela Hanganu, Ana-Maria Gheldiu, Radu Oprean, Radu Silaghi-Dumitrescu, and Gianina Crişan. 2014. "Polyphenolic Content, Antioxidant and Antimicrobial Activities of Lycium Barbarum L. and Lycium Chinense Mill. Leaves." *Molecules (Basel, Switzerland)* 19 (7): 10056–73. doi:10.3390/molecules190710056.

Nah, S Y, H J Park, and E W McCleskey. 1995. "A Trace Component of Ginseng That Inhibits Ca2+ Channels through a Pertussis Toxin-Sensitive G Protein." *Proceedings of the National Academy of Sciences of the United States of America* 92 (19): 8739–43. doi:10.1073/pnas.92.19.8739.

Nayak, Puravi, Prashant Tiwari, Shakti Prusty, and Pratap Sahu. 2018. "Phytochemistry and Pharmacology of Tinospora Cordifolia: A Review." *Systematic Reviews in Pharmacy* 9 (July): 70–78. doi:10.5530/srp.2018.1.14.

Neves, Ana Rute, Joana Fontes Queiroz, and Salette Reis. 2016. "Brain-Targeted Delivery of Resveratrol Using Solid Lipid Nanoparticles Functionalized with Apolipoprotein E." *Journal of Nanobiotechnology* 14 (1): 27. doi:10.1186/s12951-016-0177-x.

Niu, Xiaoqian, Jiejian Chen, and Jianqing Gao. 2019. "Nanocarriers as a Powerful Vehicle to Overcome Blood-Brain Barrier in Treating Neurodegenerative Diseases: Focus on Recent Advances." *Asian Journal of Pharmaceutical Sciences* 14 (5): 480–96. doi:https://doi.org/10.1016/j.ajps.2018.09.005.

Noorbala, Ahmad-Ali, and Shahin Akhondzadeh. 2006. "Attention-Deficit/Hyperactivity Disorder: Etiology and Pharmacotherapy." *Archives of Iranian Medicine* 9 (4): 374–80.

Nsairat, Hamdi, Dima Khater, Usama Sayed, Fadwa Odeh, Abeer Al Bawab, and Walhan Alshaer. 2022. "Liposomes: Structure, Composition, Types, and Clinical Applications." *Heliyon* 8 (5): e09394. doi:10.1016/j.heliyon.2022.e09394.

Ochneva, Aleksandra, Yana Zorkina, Olga Abramova, Olga Pavlova, Valeriya Ushakova, Anna Morozova, Eugene Zubkov, Konstantin Pavlov, Olga Gurina, and Vladimir Chekhonin. 2022. "Protein Misfolding and Aggregation in the Brain: Common Pathogenetic Pathways in Neurodegenerative and Mental Disorders." *International Journal of Molecular Sciences* 23 (22). doi:10.3390/ijms232214498.

Oken, B S, D M Storzbach, and J A Kaye. 1998. "The Efficacy of Ginkgo Biloba on Cognitive Function in Alzheimer Disease." *Archives of Neurology* 55 (11): 1409–15. doi:10.1001/archneur.55.11.1409.

Owoeye, O, F I Obazie, Foluso Ayobami Atiba, and Adefolarin Obanisola Malomo. 2019. "Comparative Neuroprotective Effect of Celosia Argentea Linn. and Vitamin E on Mercury-Induced Oxidative and Histological Parameters of Rat Brain." *Nigerian Journal of Physiological Sciences : Official Publication of the Physiological Society of Nigeria* 34 (2): 167–75.

Pandey, Neeraj, Jeffrey Strider, William C Nolan, Sherry X Yan, and James E Galvin. 2008. "Curcumin Inhibits Aggregation of Alpha-Synuclein." *Acta Neuropathologica* 115 (4): 479–89. doi:10.1007/s00401-007-0332-4.

Pandey, Pawan Kumar, Ashok Kumar Sharma, and Umesh Gupta. 2016. "Blood Brain Barrier: An Overview on Strategies in Drug Delivery, Realistic in Vitro Modeling and in Vivo Live Tracking." *Tissue Barriers* 4 (1): e1129476. doi:10.1080/21688370.2015.1129476.

Poovaiah, N, Z Davoudi, H Peng, B Schlichtmann, S Mallapragada, B Narasimhan, and Q Wang. 2018. "Treatment of Neurodegenerative Disorders through the Blood-Brain Barrier Using Nanocarriers." *Nanoscale* 10 (36): 16962–83. doi:10.1039/c8nr04073g.

Powell, Chadwick R, Kearsley M Dillon, and John B Matson. 2018. "A Review of Hydrogen Sulfide (H(2)S) Donors: Chemistry and Potential Therapeutic Applications." *Biochemical Pharmacology* 149 (March): 110–23. doi:10.1016/j.bcp.2017.11.014.

Pratte, Morgan A, Kaushal B Nanavati, Virginia Young, and Christopher P Morley. 2014. "An Alternative Treatment for Anxiety: A Systematic Review of Human Trial Results Reported for the Ayurvedic Herb Ashwagandha (Withania Somnifera)." *Journal of Alternative and Complementary Medicine (New York, N.Y.)* 20 (12): 901–8. doi:10.1089/acm.2014.0177.

Preeti, Rathaur, Raja Waseem, P.W. Ramteke., and A Suchit. 2020. "Turmeric - The Golden Spice of Life." *Turmeric* 3 (07): 21–34. doi:10.1201/9781420006322-7.

Prvulovic, David, Harald Hampel, and Johannes Pantel. 2010. "Galantamine for Alzheimer's Disease." *Expert Opinion on Drug Metabolism & Toxicology* 6 (3): 345–54. doi:10.1517/17425251003592137.

Rafii, Michael S, Mark H Tuszynski, Ronald G Thomas, David Barba, James B Brewer, Robert A Rissman, Joao Siffert, and Paul S Aisen. 2018. "Adeno-Associated Viral Vector (Serotype 2)-Nerve Growth Factor for Patients with Alzheimer Disease: A Randomized Clinical Trial." *JAMA Neurology* 75 (7): 834–41. doi:10.1001/jamaneurol.2018.0233.

Rahman, Mahfoozur, Sohail Akhter, Mohammad Zaki Ahmad, Javed Ahmad, Richard T Addo, Farhan Jalees Ahmad, and Chantal Pichon. 2015. "Emerging Advances in Cancer Nanotheranostics with Graphene Nanocomposites: Opportunities and Challenges." *Nanomedicine* 10 (15): 2405–22. doi:10.2217/nnm.15.68.

Rajput, Amarjitsing, Aditi Bariya, Ahmed Allam, Sarah Othman, and Shital B Butani. 2018. "In Situ Nanostructured Hydrogel of Resveratrol for Brain Targeting: In Vitro-in Vivo Characterization." *Drug Delivery and Translational Research* 8 (5): 1460–70. doi:10.1007/s13346-018-0540-6.

Rajput, Mithun Singh, Nilesh Prakash Nirmal, Viral Patel, Purnima Dey Sarkar, and Manan Raval. 2023. "Chapter 27- Phytoconstituents-Based Nanoformulations for Neurodegenerative Disorders." In, edited by Siddhartha Pati, Tanmay Sarkar, and Dibyajit Lahiri. *Recent Frontiers of Phytochemicals*, 463–82. Elsevier. doi:10.1016/B978-0-443-19143-5.00003-7.

Rakotoarisoa, Miora, Borislav Angelov, Vasil M Garamus, and Angelina Angelova. 2019. "Curcumin- and Fish Oil-Loaded Spongosome and Cubosome Nanoparticles with Neuroprotective Potential against H2O2-Induced Oxidative Stress in Differentiated Human SH-SY5Y Cells." *ACS Omega* 4 (2): 3061–73. doi:10.1021/acsomega.8b03101.

Ramachandran, Surekha, and Sumathi Thangarajan. 2018. "Thymoquinone Loaded Solid Lipid Nanoparticles Counteracts 3-Nitropropionic Acid Induced Motor Impairments and Neuroinflammation in Rat Model of Huntington's Disease." *Metabolic Brain Disease* 33 (5): 1459–70. doi:10.1007/s11011-018-0252-0.

Ren, Tianjing, and Zhong Zuo. 2019. "Role of Piperine in CNS Diseases: Pharmacodynamics, Pharmacokinetics and Drug Interactions." *Expert Opinion on Drug Metabolism & Toxicology* 15 (10): 849–67. doi:10.1080/17425255.2019.1672658.

Ren, Yizhuo, Xin Zhao, Xiaofeng Liang, Peter X Ma, and Baolin Guo. 2017. "Injectable Hydrogel Based on Quaternized Chitosan, Gelatin and Dopamine as Localized Drug Delivery System to Treat Parkinson's Disease." *International Journal of Biological Macromolecules* 105 (Pt 1): 1079–87. doi:10.1016/j.ijbiomac.2017.07.130.

Russo, Alessandra, Francesca Borrelli, Agata Campisi, Rosaria Acquaviva, Giuseppina Raciti, and Angelo Vanella. 2003. "Nitric Oxide-Related Toxicity in Cultured Astrocytes: Effect of Bacopa Monniera." *Life Sciences* 73 (12): 1517–26. doi:10.1016/s0024-3205(03)00476-4.

Sahib, Najla Gooda, Farooq Anwar, Anwarul-Hassan Gilani, Azizah Abdul Hamid, Nazamid Saari, and Khalid M Alkharfy. 2013. "Coriander (Coriandrum Sativum L.): A Potential Source of High-Value Components for Functional Foods and Nutraceuticals--A Review." *Phytotherapy Research : PTR* 27 (10): 1439–56. doi:10.1002/ptr.4897.

Saleem, Ammar, Michael Husheem, Pirkko Härkönen, and Kalevi Pihlaja. 2002. "Inhibition of Cancer Cell Growth by Crude Extract and the Phenolics of Terminalia Chebula Retz. Fruit." *Journal of Ethnopharmacology* 81 (3): 327–36. doi:10.1016/s0378-8741(02)00099-5.

Sarkar, Alika, Ismat Fatima, Qazi Mohammad Sajid Jamal, Usman Sayeed, Mohammad Kalim A Khan, Salman Akhtar, Mohammad Amjad Kamal, Alvina Farooqui, and Mohammad Haris Siddiqui. 2017. "Nanoparticles as a Carrier System for Drug Delivery Across Blood Brain Barrier." *Current Drug Metabolism* 18 (2): 129–37. doi:10.2174/1389200218666170113125132.

Sarko, Diana K, and Cindy E McKinney. 2017. "Exosomes: Origins and Therapeutic Potential for Neurodegenerative Disease." *Frontiers in Neuroscience*. https://www.frontiersin.org/article/10.3389/fnins.2017.00082.

Senthilkumar, Karuppagounder S, Karuppagounder S Saravanan, Goutam Chandra, Kizhakke M Sindhu, A Jayakrishnan, and Kochupurackal P Mohanakumar. 2007. "Unilateral Implantation of Dopamine-Loaded Biodegradable Hydrogel in the Striatum Attenuates Motor Abnormalities in the 6-Hydroxydopamine Model of Hemi-Parkinsonism." *Behavioural Brain Research* 184 (1): 11–18. doi:10.1016/j.bbr.2007.06.025.

Shakeri, Shahryar, Milad Ashrafizadeh, Ali Zarrabi, Rasoul Roghanian, Elham G Afshar, Abbas Pardakhty, Reza Mohammadinejad, Anuj Kumar, and Vijay K Thakur. 2020. "Multifunctional Polymeric Nanoplatforms for Brain Diseases Diagnosis, Therapy and Theranostics." *Biomedicines*. doi:10.3390/biomedicines8010013.

Sharma, D R, W Y Wani, A Sunkaria, R J Kandimalla, R K Sharma, D Verma, A Bal, and K D Gill. 2016. "Quercetin Attenuates Neuronal Death against Aluminum-Induced Neurodegeneration in the Rat Hippocampus." *Neuroscience* 324 (June): 163–76. doi:10.1016/j.neuroscience.2016.02.055.

Sharma, Vineet, Rohit Sharma, DevNath Singh Gautam, Kamil Kuca, Eugenie Nepovimova, and Natália Martins. 2020. "Role of Vacha (Acorus Calamus Linn.) in Neurological and Metabolic Disorders: Evidence from Ethnopharmacology, Phytochemistry, Pharmacology and Clinical Study." *Journal of Clinical Medicine* 9 (4): 1176. doi:10.3390/jcm9041176.

Shi, Zhongshan, Lihui Zhu, Tingting Li, Xiaoya Tang, Yonghui Xiang, Xinjia Han, Luoxing Xia, et al. 2017. "Neuroprotective Mechanisms of Lycium Barbarum Polysaccharides Against Ischemic Insults by Regulating NR2B and NR2A Containing NMDA Receptor Signaling Pathways." *Frontiers in Cellular Neuroscience*. https://www.frontiersin.org/article/10.3389/fncel.2017.00288.

Singh, Neha Atulkumar, Abul Kalam Azad Mandal, and Zaved Ahmed Khan. 2016. "Potential Neuroprotective Properties of Epigallocatechin-3-Gallate (EGCG)." *Nutrition Journal* 15 (1): 60. doi:10.1186/s12937-016-0179-4.

Sivandzade, Farzane, and Luca Cucullo. 2021. "Regenerative Stem Cell Therapy for Neurodegenerative Diseases: An Overview." *International Journal of Molecular Sciences* 22 (4). doi:10.3390/ijms22042153.

Smith, Adam, Brian Giunta, Paula C Bickford, Michael Fountain, Jun Tan, and R Douglas Shytle. 2010. "Nanolipidic Particles Improve the Bioavailability and Alpha-Secretase Inducing Ability of Epigallocatechin-3-Gallate (EGCG) for the Treatment of Alzheimer's Disease." *International Journal of Pharmaceutics* 389 (1–2): 207–12. doi:10.1016/j.ijpharm.2010.01.012.

Spuch, Carlos, Desiree Antequera, Aitziber Portero, Gorka Orive, Rosa Ma Hernández, Jose A Molina, Felix Bermejo-Pareja, José L Pedraz, and Eva Carro. 2010. "The Effect of Encapsulated VEGF-Secreting Cells on Brain Amyloid Load and Behavioral Impairment in a Mouse Model of Alzheimer's Disease." *Biomaterials* 31 (21): 5608–18. doi:10.1016/j.biomaterials.2010.03.042.

Spuch, Carlos, and Carmen Navarro. 2011. "Liposomes for Targeted Delivery of Active Agents against Neurodegenerative Diseases (Alzheimer's Disease and Parkinson's Disease)." edited by Sophia Antimisiaris. *Journal of Drug Delivery*. Hindawi Publishing Corporation: 469679. doi:10.1155/2011/469679.

Srinageshwar, Bhairavi, Sarah Peruzzaro, Melissa Andrews, Kayla Johnson, Allison Hietpas, Brittany Clark, Crystal McGuire, et al. 2017. "PAMAM Dendrimers Cross the Blood-Brain Barrier When Administered through the Carotid Artery in C57BL/6J Mice." *International Journal of Molecular Sciences* 18 (3). doi:10.3390/ijms18030628.

Strazielle, Nathalie, and Jean-François Ghersi-Egea. 2016. "Potential Pathways for CNS Drug Delivery Across the Blood-Cerebrospinal Fluid Barrier." *Current Pharmaceutical Design* 22 (35): 5463–76. doi:10.2174/1381612822666160726112115.

Subbaraju, Gottumukkala V, Mulabagal Vanisree, Chirravuri V Rao, Chillara Sivaramakrishna, Pratha Sridhar, Bolleddula Jayaprakasam, and Muraleedharan G Nair. 2006. "Ashwagandhanolide, a Bioactive Dimeric Thiowithanolide Isolated from the Roots of Withania Somnifera." *Journal of Natural Products* 69 (12): 1790–92. doi:10.1021/np060147p.

Sudhakar, Vivek, and R Mark Richardson. 2019. "Gene Therapy for Neurodegenerative Diseases." *Neurotherapeutics : The Journal of the American Society for Experimental NeuroTherapeutics* 16 (1): 166–75. doi:10.1007/s13311-018-00694-0.

Sun, Jing, Chunfang Wei, Yanan Liu, Wenjie Xie, Mengmeng Xu, Hui Zhou, and Jie Liu. 2019. "Progressive Release of Mesoporous Nano-Selenium Delivery System for the Multi-Channel Synergistic Treatment of Alzheimer's Disease." *Biomaterials* 197 (March): 417–31. doi:10.1016/j.biomaterials.2018.12.027.

Sweeney, Melanie D, Kassandra Kisler, Axel Montagne, Arthur W Toga, and Berislav V Zlokovic. 2018. "The Role of Brain Vasculature in Neurodegenerative Disorders." *Nature Neuroscience* 21 (10): 1318–31. doi:10.1038/s41593-018-0234-x.

Tajes, Marta, Eva Ramos-Fernández, Xian Weng-Jiang, Mònica Bosch-Morató, Biuse Guivernau, Abel Eraso-Pichot, Bertrán Salvador, Xavier Fernàndez-Busquets, Jaume Roquer, and Francisco J Muñoz. 2014. "The Blood-Brain Barrier: Structure, Function and Therapeutic Approaches to Cross It." *Molecular Membrane Biology* 31 (5): 152–67. doi:10.3109/09687688.2014.937468.

Teleanu, Daniel Mihai, Cristina Chircov, Alexandru Mihai Grumezescu, Adrian Volceanov, and Raluca Ioana Teleanu. 2018. "Blood-Brain Delivery Methods Using Nanotechnology." *Pharmaceutics* 10 (4). doi:10.3390/pharmaceutics10040269.

Torchilin, V P. 2004. "Targeted Polymeric Micelles for Delivery of Poorly Soluble Drugs." *Cellular and Molecular Life Sciences : CMLS* 61 (19–20): 2549–59. doi:10.1007/s00018-004-4153-5.

Uabundit, Nongnut, Jintanaporn Wattanathorn, Supaporn Mucimapura, and Kornkanok Ingkaninan. 2010. "Cognitive Enhancement and Neuroprotective Effects of Bacopa Monnieri in Alzheimer's Disease Model." *Journal of Ethnopharmacology* 127 (1): 26–31. doi:10.1016/j.jep.2009.09.056.

van Assema, Daniëlle M E, Mark Lubberink, Ronald Boellaard, Robert C Schuit, Albert D Windhorst, Philip Scheltens, Adriaan A Lammertsma, and Bart N M van Berckel. 2012. "P-Glycoprotein Function at the Blood-Brain Barrier: Effects of Age and Gender." *Molecular Imaging and Biology* 14 (6): 771–76. doi:10.1007/s11307-012-0556-0.

Van Kampen, Jackalina, Harold Robertson, Theo Hagg, and Robert Drobitch. 2003. "Neuroprotective Actions of the Ginseng Extract G115 in Two Rodent Models of Parkinson's Disease." *Experimental Neurology* 184 (1): 521–29. doi:10.1016/j.expneurol.2003.08.002.

Vidavalur, Ramesh, Hajime Otani, Pawan K Singal, and Nilanjana Maulik. 2006. "Significance of Wine and Resveratrol in Cardiovascular Disease: French Paradox Revisited." *Experimental and Clinical Cardiology* 11 (3): 217–25.

Vieira, Débora B, and Lionel F Gamarra. 2016. "Getting into the Brain: Liposome-Based Strategies for Effective Drug Delivery across the Blood-Brain Barrier." *International Journal of Nanomedicine* 11 (October): 5381–5414. doi:10.2147/IJN.S117210.

Vinogradov, Serguei, Elena Batrakova, and Alexander Kabanov. 1999. "Poly(Ethylene Glycol)-Polyethyleneimine NanoGelTM Particles: Novel Drug Delivery Systems for Antisense Oligonucleotides." *Colloids and Surfaces B: Biointerfaces* 16 (1): 291–304. doi:https://doi.org/10.1016/S0927-7765(99)00080-6.

Volkman, Rotem, and Daniel Offen. 2017. "Concise Review: Mesenchymal Stem Cells in Neurodegenerative Diseases." *Stem Cells (Dayton, Ohio)* 35 (8): 1867–80. doi:10.1002/stem.2651.

Wang, Jun, Xi-Xun Du, Hong Jiang, and Jun-Xia Xie. 2009. "Curcumin Attenuates 6-Hydroxydopamine-Induced Cytotoxicity by Anti-Oxidation and Nuclear Factor-Kappa B Modulation in MES23.5 Cells." *Biochemical Pharmacology* 78 (2): 178–83. doi:10.1016/j.bcp.2009.03.031.

Wang, Jun, Hua-Min Xu, Hai-Dong Yang, Xi-Xun Du, Hong Jiang, and Jun-Xia Xie. 2009. "Rg1 Reduces Nigral Iron Levels of MPTP-Treated C57BL6 Mice by Regulating Certain Iron Transport Proteins." *Neurochemistry International* 54 (1): 43–48. doi:10.1016/j.neuint.2008.10.003.

Wang, Shao-Ling, Ying Li, Ying Wen, Yan-Feng Chen, Li-Xin Na, Song-Tao Li, and Chang-Hao Sun. 2009. "Curcumin, a Potential Inhibitor of up-Regulation of TNF-Alpha and IL-6 Induced by Palmitate in 3T3-L1 Adipocytes through NF-KappaB and JNK Pathway." *Biomedical and Environmental Sciences : BES* 22 (1): 32–39. doi:10.1016/S0895-3988(09)60019-2.

Wang, Xiaowu, Gang Li, Peng Zhang, Wei Li, and Xijing He. 2019. "Surface Engineering of Resveratrol to Improve Neuro-Protection and Functional Recovery after Spinal Cord Injury in Rat." *Journal of Drug Delivery Science and Technology* 49: 89–96. doi:https://doi.org/10.1016/j.jddst.2018.10.016.

Warren Olanow, C, Raymond T Bartus, Tiffany L Baumann, Stewart Factor, Nicholas Boulis, Mark Stacy, Dennis A Turner, et al. 2015. "Gene Delivery of Neurturin to Putamen and Substantia Nigra in Parkinson Disease: A Double-Blind, Randomized, Controlled Trial." *Annals of Neurology* 78 (2): 248–57. doi:10.1002/ana.24436.

Wattmo, Carina, Erik Jedenius, Kaj Blennow, and Asa K Wallin. 2013. "Dose and Plasma Concentration of Galantamine in Alzheimer's Disease - Clinical Application." *Alzheimer's Research & Therapy* 5 (1): 2. doi:10.1186/alzrt156.

Wei, Min, Zhaofei Yang, Song Li, and Weidong Le. 2023. "Nanotherapeutic and Stem Cell Therapeutic Strategies in Neurodegenerative Diseases: A Promising Therapeutic Approach." *International Journal of Nanomedicine* 18: 611–26. doi:10.2147/IJN.S395010.

Werneke, Ursula, Trevor Turner, and Stefan Priebe. 2006. "Complementary Medicines in Psychiatry: Review of Effectiveness and Safety." *The British Journal of Psychiatry : The Journal of Mental Science* 188 (February): 109–21. doi:10.1192/bjp.188.2.109.

Xu, Bei-Bei, Chun-Qing Liu, Xi Gao, Wan-Qin Zhang, Shi-Wei Wang, and Ying-Lin Cao. 2005. "Possible Mechanisms of the Protection of Ginsenoside Re against MPTP-Induced Apoptosis in Substantia Nigra Neurons of Parkinson's Disease Mouse Model." *Journal of Asian Natural Products Research* 7 (3): 215–24. doi:10.1080/10286020410001690172.

Xu, Dan-Hua, Miao Yan, Huan-De Li, Ping-Fei Fang, and Yan-Wen Liu. 2012. "Molecular and Cellular Pharmacology." *European Journal of Pharmacology* 690 (1–3): 68–76. doi:10.1016/j.ejphar.2012.06.032.

Xue, Jinwei, Tongtong Liu, Yongdan Liu, Ye Jiang, Vidya Devanatha Desikan Seshadri, Surapaneni Krishna Mohan, and Li Ling. 2019. "Neuroprotective Effect of Biosynthesised Gold Nanoparticles Synthesised from Root Extract of Paeonia Moutan against Parkinson Disease - In Vitro & In Vivo Model." *Journal of Photochemistry and Photobiology B: Biology* 200: 111635. doi:10.1016/j.jphotobiol.2019.111635.

Yuan, Qiuju, Chong-Wen Wang, Jun Shi, and Zhi-Xiu Lin. 2017. "Effects of Ginkgo Biloba on Dementia: An Overview of Systematic Reviews." *Journal of Ethnopharmacology* 195 (January): 1–9. doi:10.1016/j.jep.2016.12.005.

Zeng, Kuan, Mengzhu Li, Jichang Hu, Yacoubou Abdoul Razak Mahaman, Jian Bao, Fang Huang, Yiyuan Xia, et al. 2018. "Ginkgo Biloba Extract EGb761 Attenuates Hyperhomocysteinemia-Induced AD Like Tau Hyperphosphorylation and Cognitive Impairment in Rats." *Current Alzheimer Research* 15 (1): 89–99. doi:10.2174/1567205014666170829102135.

Zhang, Laura, Milan Fiala, John Cashman, James Sayre, Araceli Espinosa, Michelle Mahanian, Justin Zaghi, et al. 2006. "Curcuminoids Enhance Amyloid-Beta Uptake by Macrophages of Alzheimer's Disease Patients." *Journal of Alzheimer's Disease : JAD* 10 (1): 1–7. doi:10.3233/jad-2006-10101.

Zhang, Mengjun, Xinlong Zang, Mingyue Wang, Zhen Li, Mingxi Qiao, Haiyang Hu, and Dawei Chen. 2019. "Exosome-Based Nanocarriers as Bio-Inspired and Versatile Vehicles for Drug Delivery: Recent Advances and Challenges." *Journal of Materials Chemistry B* 7 (15): 2421–33. doi:10.1039/C9TB00170K.

7 New Generation Nanocarriers for Ocular Delivery of Resveratrol
A Multiple Beneficial

Agustina Alaimo, Cecilia Tettamanti, Oscar E. Perez, Daniel A. Allemandi, and Daniela A. Quinteros

7.1 INTRODUCTION

Nanotechnology is currently a forward-looking bet given its potential to significantly impact various biological disciplines. In particular, the branch of nanomedicine has emerged as a result of the accelerated development of the field of nanotechnology and its application in this area (Gorantla et al. 2020; Weng et al. 2017). Its application is expected to lead to the expansion of improved devices and delivery systems to protect agents from degradation or avoid an immune response in the body by promoting their controlled release. The introduction of nanotechnology in the pharmacological and biomedical fields has triggered a revolution in terms of its potential application as therapeutic agents (Wu, Wang, and Li 2020; Bisht et al. 2018). Another important goal in nanomedicine is to design and develop NP to improve the penetration and delivery of resveratrol (RSV) or other drugs in tissues that are difficult to access. This is the case with the unique anatomy of the eye, which is characterized by the presence of barriers that protect the anterior and posterior segments. These anatomical mechanisms protecting the eye provide a natural resistance. For this reason, the agent currently used to treat eye disease must be applied frequently to compensate for the loss of efficacy of administration. This limitation poses a particularly significant challenge in the field of experimental ophthalmology. Thus, nanomedicine is expected to offer the possibility of overcoming ocular barriers and enabling more efficient drug delivery, especially in the posterior segment (Dinte et al. 2020; Gorantla et al. 2020; Harikumar and Sonia 2011). On the other hand, nanotechnology offers great potential to improve all traditional ocular drug delivery systems. In this context, it is possible to design nanoparticles (NPs) in order to improve relevant properties (bioavailability, solubility, permeability, stability, and therapeutic efficacy) of drugs of ophthalmologic interest such as RSV, a non-flavonoid polyphenolic compound obtained from natural resources and shown to have several beneficial properties for human health (DiSanto et al. 2021).

According to sales expectative, it is expected the ophthalmologic medicine market generate ~US$35 billion in sales by early 2025, with a CAGR of approximately 4.7% between 2018 and 2024, indicating the enormous reach of ocular products, which is driving scientists to develop new drug platforms (Gorantla et al. 2020).

In summary, when the reader delves into this book chapter, they will first find introductory information on complex ocular anatomy, which in itself justifies why their pharmacological approach is a currently unresolved challenge. Second, the reader will find a chemical disruption in RSV. One will be able to understand why RSV is known as the "miracle molecule" of wine, but it will also reveal its physicochemical limitations. Toward the end of the text, the reader is provided with an exhaustive summary of scientific reports in which NP is made of various materials. The NPs were developed for two purposes: to improve the delivery pathways in the eye and to protect against RSV.

DOI: 10.1201/9781003292692-7

7.2 HUMAN EYE: ANATOMY AND PHYSIOLOGY "AT THE GLANCE"

Fibrous tunica: consists of the sclera and cornea, which together form a semi-elastic and protective structure that determines with great precision the optical geometry of the visual apparatus (Wang et al. 2017; Ansari and Nadeem 2016; Ludwig, Jessu, and Czyz 2021; Forrester et al. 2016):

The eyeball is covered and protected by the sclera and since is an opaque membrane; it prevents light from passing through. It is about 1.1 mm thick in the posterior region near the optic nerve insertion and tapers toward the equator, while it reaches a thickness of 0.4 mm in the areas where the rectus muscles are located. The sclera is a viscoelastic structure consisting of dense connective tissue with some elastic fibers, which enables the regulation of intraocular pressure and contains collagen which makes up most of the dry weight of the sclera.

The cornea is a transparent tissue located in the anterior part of the collagenous wall of the eyeball. It consists of a tissue without blood vessels but with abundant nerves originating from the ophthalmic nerve. Its properties allow light to pass through and form an optical image on the retina, a process explained in the next chapter. The cornea forms a dome-shaped shield that protrudes straight out from the spherical shape of the eyeball. The degree of its curvature varies in different people as well as in the same person throughout life, being more pronounced in youth than in older age. The cornea has a particular shape since it is slightly larger horizontally than vertically. In the adult, the cornea diameter is about 11 mm and a thickness of about 0.5 mm in the central part and increases toward the periphery. Also, the cornea consists of a multilayered structure in which 5 main layers are distinguished, from the surface to the interior:

a. The anterior surface of the cornea is covered by an epithelial membrane which is about 50 µm thick. It is a stratified layer with cells that are normally not keratinized. This epithelium has the same functions as any other, such as cleaning, maintaining renewal functions, and acting as a barrier to the outside world. The corneal epithelium is protected by a tear film that keeps it safe from toxic, chemical, or foreign components. This film consists of an outer lipid layer and a deep aqueous mucosa that interact with the epithelial cells to allow the release of the tear film with each blink.
b. The corneal epithelium is separated from the stroma by the anterior elastic lamina or Bowman's membrane. It is a homogeneous acellular layer with a dense mass of collagen fibers that intertwine to form a network that stabilizes the connection between the epithelium and stroma. It is a sheet that is resistant to trauma but lacks repair capacity.
c. The stroma is a compact, firm, dense, and fibrous layer that occupies most of the cornea. It contains about 85% water and between 200 and 250 overlapping collagen flakes that run parallel to the surface. Their dimensions, which are copious minor than the light wavelength and the regularity of their spacing, are the factors that determine the transparent character of the cornea.
d. Descemet's membrane is extremely thin, between 10 and 15 µm wide, whose thickness increases over the years and it is located between the stroma and the endothelium. It is believed to have protective functions.
e. The endothelium is responsible for nutrient supply toward the cornea from the anterior chamber which contains the aqueous humor. This process allows the dynamic aqueous balance, which is critical to the transparency of the cornea. It is worth noting that an excess of fluids could alter the physiology of corneal fibers and lead to the appearance of light scattering.

Vascular tunic: constitutes a compact and coherent structure compounded by the iris, ciliary body, and choroid, being this latter highly vascular and 0.1–0.15 mm thick. Its color depends on the pigmentation of the stroma, but it is generally dark brown and covers almost five-sixths of the back of the eye. It is also penetrated by the optic nerve, where it is firmly fused to the sclera. The blood flow through the choroid is very high and plays a crucial role in regulating the temperature of the eye and maintaining intraocular pressure. On the other hand, anatomically choroid is compounded by different layers which perform different functions. One of these is to provide a smooth surface for the pigment cells and receptors of the retina to align properly. The ciliary body joins the choroid at the back and the iris at the front. It is a triangular structure that forms a complete ring around the eyeball and consists of the ciliary muscle and the ciliary processes. It is brown and has a folded anterior part surrounding the base of the iris and another smooth annular part. On the inside, it has a toothed profile where it merges with the choroid and retina. The ciliary processes, which are longitudinal ridges, are responsible for the production of aqueous humor. The ciliary processes are the origin of the zonular fibers that enter the lens and hold it in its anatomical position. So one of the functions of the ciliary body is to support the lens and allow it to accommodate. Like the cornea, the lens of the eye is a transparent structure, a property maintained by the regularity of the elastic fibers of the cells that make it up. The ciliary muscle involves the smooth muscle fibers that are arranged circularly, longitudinally, and radially. When it is relaxed, the tension of the ligament of the lens increases, and the lens assumes a flatter curvature and allows good focusing on distant objects. When the muscle contracts, the lens takes on a round shape and increases the ability to focus on close objects. The vitreous chamber occupies the eyeball from the lens to the retina. This chamber is filled with a transparent and gelatinous substance called the vitreous humor. Unlike aqueous humor, this substance cannot be replaced. The iris is an adaptable diaphragm with a central opening, the pupil, which controls the amount of light entering the eye. The pupil diameter varies between 1 and 8 mm, although it can reach a higher value under the influence of drugs. It is a structure that is visible through the cornea and gives the eye its color. The main purpose of the iris, on the other hand, is to regulate the quantity of light that reaches the eye and to control the opening of the pupil to different intensities of ambient brightness. It has two muscles that perform opposite movements: the sphincter, which serves to contract the pupil, and the dilator (Forrester et al. 2016).

Nerve tunica: The retina borders the inside of the back of the eyeball and is where the image is produced. The retina is protected by the surrounding sclera and cornea and remains in place. Its average thickness is 250 μm and it consists of ten layers: Photoreceptors under a layer of neurons connected to them. The receptor layer lies deep in the retina, which means that the photon of light must first pass through most of the layers of the retina before it reaches the receptors. The receptors absorb the photons and convert them into neuronal signals, which are then processed by the neuronal cell network. The pigment epithelium of the retina is the outermost layer and serves both as a primary metabolic support for the receptor cells and as a source of brightness reduction for incoming photons. Although the neurons are located in front of the photoreceptor cells, they do not prevent incoming light from being absorbed by the receptors because the neurons are transparent, and the photoreceptors are light-sensitive (Gupta et al. 2015; Forrester et al. 2016).

Photoreceptors convert light into action potentials, distinguish wavelengths that allow us to discriminate colors, and operate over a very wide range of light intensities, from very bright to very dark. Both types of photoreceptors, rods, and cones, are named essentially for their shape. The rods are longer than the cones and are cylindrical at one end, while the cones have a pointed end. The rods are sensitive to low levels of luminosity, especially in dim light, and are particularly useful for night vision. The cones do not respond to low light, but they are extremely sensitive to bright light. Therefore, the cones are responsible for color vision and the ability of humans to see fine details. Photoreceptors convert light energy first into chemical energy and then into neuronal activity. When light hits a photoreceptor, a series of chemical reactions are triggered, resulting in a change in membrane potential (Lamb, Collin, and Pugh 2007; Forrester et al. 2016).

Human vision is clearer in the center of the visual field than in the peripheral areas because there are a greater number of photoreceptors in the center of the retina. The most visually important area of the retina is concentrated in an area called the macula, which is located right in the middle of the posterior pole of the eye. It only covers about 5 mm in diameter. Within the macula, there is a smaller area called the fovea where light from outside is concentrated. It is in this fovea area that there are only cones, which are immersed in a depression due to the absence of the other layers. The fovea is the place where the retina has the greatest potential for visual acuity. On the way from the photoreceptors to the brain, three types of cells are found: horizontal, amacrine, and bipolar.

The amacrine cells in particular connect the photoreceptors with the bipolar cells or with the second neuronal layer. This second layer is formed by the retinal ganglion cells, whose axons converge in a bundle toward the optic disc and leave the eye to form the optic nerve. Several important structures are distinguished in the posterior region of the retina, the optic nerve part. The optic disc is the region where the fibers of the retinal neurons form the optic nerve, which is connected to the brain. It is lighter in color than the adjacent retina and the exit of the central retinal artery and its branches in the vessels that supply the retina can be seen on its surface. It is also called the retinal dot because it lacks light-sensitive photoreceptors (Gupta et al. 2015; Binder 2004).

7.3 OCULAR DRUG DELIVERY: "THE GREAT CHALLENGE"

The pharmaceutical industry oriented to the ophthalmology area presents great challenges when it comes to administering drugs efficiently. The limitations to overcome are precisely the natural anatomical and physiological barriers, which have evolved due to the protective microenvironment they provide against foreign particles and substances (Nettey et al. 2016). A brief and introductory mention of the most used ocular administration routes in the ophthalmic field is described below (Figure 7.1):

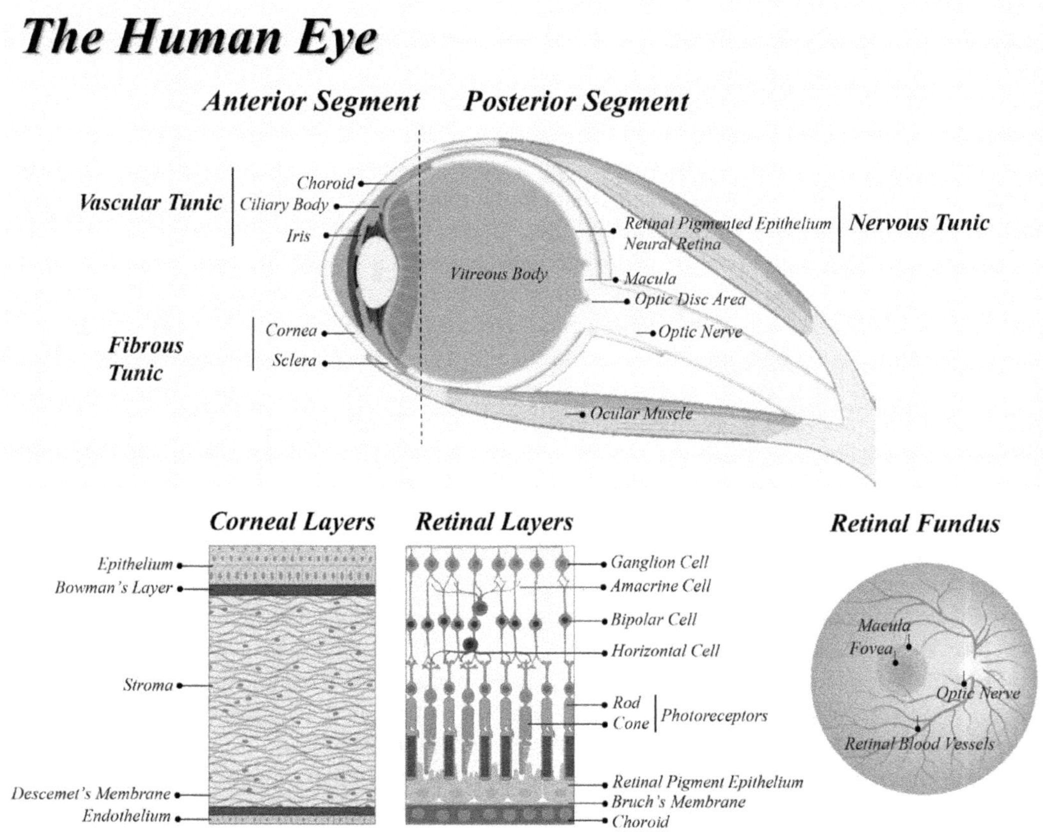

FIGURE 7.1 Diagram of the anatomy of the human eye (upper panel). The anatomical structure of the retina and cornea (lower panel).

Topical route: is the most frequent method for ocular drug delivery and more often for anterior segment conditions, e.g., conjunctivitis treatment. This administration pathway must face the cornea, conjunctiva, and sclera, which are the main barriers to the penetration of the active principle into the eye. The corneal epithelium, through its tight junctions, makes it difficult for water-soluble substances to pass into the stroma. On the contrary, lipophilic substances can cross the corneal epithelium but the passage at the level of the stroma is reduced. The conjunctival epithelium also constitutes a barrier to drug penetration, although to a lesser extent than the corneal epithelium. However, the conjunctival vasculature will absorb a substantial portion of the drug that has passed through the conjunctival epithelium. The most anterior portion of the sclera partially allows the permeation of the active principle, achieving access to the anterior part of the humor vitreous (Almeida et al. 2015).

The posterior segment of the eye diseases involve devastating vision consequences and affect large numbers of people around the world, e.g. age-related macular degeneration (AMD), diabetic retinopathy (DR), glaucoma, and uveitis, among others. Unfortunately for posterior segment pathologies, the use of drug-containing eye drops, ointments, and/or gels is impractical. Therefore, the difficulty of drug access into the eyeball and the defense mechanisms of the eye, lead to the use of other routes of administration (Wang et al. 2017; Yasin, Thakur, and Rupenthal 2018; Peng, Tang, and Zhou 2017):

a. Systemic administration: this route is employed for the treatment of posterior eye segment diseases. Also, the oral formulation would be suitable and tolerable for patients. Nevertheless, this pathway has poor access to the aqueous humor and the vitreous, due to the blood-water barrier of the ciliary processes and the internal and external blood–retinal barriers, preventing the drug from reaching the extravascular space of the retina and the vitreous. As a consequence, the doses required to reach therapeutic levels are too high, which implies a greater risk of generating unwanted adverse effects.
b. To carry out a more efficient drug delivery more efficient for vitreoretinal diseases, two main pathways have been established:
c. Intravitreal administration: consists of a drug administration route that involves the direct injection of a formulation into the vitreous humor. In this way, it is possible to achieve high drug concentrations in the vitreous and neural retina with fewer side effects than the systemic administration. Moreover, the blood–retinal barrier is made up of both epithelial and endothelial cells together with efflux transporters; all of them constitute a barrier to inner-retinal cells. Drugs of lower molecular weight are quickly eliminated and that is the reason why the frequency of administration must be increased. As a consequence, a higher probability of undesirable side effects may occur because of frequent injections, such as retinal detachment, retinal hemorrhage, and endophthalmitis. Even though the utility of this administration route cannot be debated, the invasiveness of the technique makes it difficult to consider as the ideal for drug delivery into the posterior segment
d. Subretinal administration: this route has more direct effects on target cells in the subretinal space than the intravitreal route. The subretinal space involves the space between the retinal pigment epithelium and the photoreceptors. For this reason, it is mainly used for gene and cell therapies, involving stem cells mainly in patients with vision-threatening disorders attributable to mutations in photoreceptor and/or retinal pigment epithelium genes, e.g. retinitis pigmentosa or AMD. Like intravitreal injections, subretinal injection leads to more frequent problems, such as vitreous hemorrhage, recurrence of submacular hemorrhage, retinal detachment, and development of choroidal neovascularization.
e. The drug administration into the posterior segment of the eye requires the search for less harmful routes of administration than intravitreal or subretinal injection. In this sense, periocular injections seem to provide another alternative for the posterior segment of the eye.
f. Periocular injections: these routes are primarily around the eyeball, either below the conjunctiva of the eye (subconjunctival), around the eyeball (peribulbar), or behind the eyeball

(retrobulbar). This route is ideal to achieve a drug release in high concentrations in the vitreous, since the sclera offers less resistance to the penetration of substances than the blood-aqueous and haemato-retinal barriers, being permeable even to large molecules. In addition, the periocular route provides a virtual space where the active principle is stored, as a reservoir, for its progressive release. Despite the advantages of periocular administration, this route is not without its challenges when it comes to delivering therapeutically effective drug concentrations to the focus site in the posterior segment. There are several physiological barriers, care interests, and issues that would hamper patient compliance.

7.4 RESVERATROL

7.4.1 Natural Polyphenols: Returning to Our Ancestral Medicine

Throughout the ages, humans have focused on disease, one of the leading reasons for mortality. Despite the advances of modern times, diseases continue to be a threat to health regardless of the economy of the countries, affecting both urban and rural areas and all ethnic groups on the planet (Pan et al. 2014; Rasouli, Farzaei, and Khodarahmi 2017). Phytochemicals are plant-derived bioactives that regulate metabolic events (Correia et al. 2012; Veeresham 2012; Watkins et al. 2015). Agreeing with human fossil data, the employment of plants as medicines goes back, at least, to the middle Paleolithic age (Fabricant and Farnsworth 2001). About 1,000 medicinal substances derived from plants are in the records of the ancient civilization of Mesopotamia 2,600 BC. Likewise, the Indian Ayurveda system, which is 5,000 years old, also offers natural drug molecules. Chinese literature mentioned the systematic use of herbal medicine from 2,500 years ago (Kumar and Goel 2019; Rasouli, Farzaei, and Khodarahmi 2017). In 400 BC, Hippocrates, the famous physician of ancient Greece, quoted "*Let food be your medicine and medicine be your food*" to emphasize the importance of nutrition (Chishti 1988; Rasouli, Farzaei, and Khodarahmi 2017). Hence, the significance of these findings exceeds the limit of remaining in an anthropogenic or archaeological treasure. Particularly, the value of nutrition has been captured by pharmaceutical research as a methodology of medicinal agents and has gained considerable interest in worldwide health debates (Yuan et al. 2016; Fabricant and Farnsworth 2001). In their many forms, natural components may have certain limitations, but they are still a valuable repository of human knowledge. Also, the revaluation of the traditional medicine of our ancestors is nowadays plenty in vogue (Tilburt and Kaptchuk 2008; Hano and Tungmunnithum 2020). According to the World Health Organization, an increasing world population is using herbs for their healthcare needs; however, it is still the case of the conventional modern drugs that are still chemically synthesized, and this development implies many risks, from high failure and economic cost to a high rate of adverse events (Pan et al. 2014). In sum, the therapeutic attempt to alternative compounds extracted from plants is an option for a rigorous search of brand-new drugs. Twenty years ago, Fabricant and Farnsworth (2001), claimed that plant exploitation as a source of therapeutic agents may have many strengths, since natural bioactive could be isolated for direct use as drugs, as templates for a generation of novel semi-synthetic entities with higher activity and/or lower toxicity, and/or as pharmacologic tools. Among the plant-derived bioactive molecules, polyphenols are included (Durazzo et al. 2019). In the last 20 years, we have witnessed a revolution in the investigation of polyphenolic compounds precisely because of their high potential for health benefits (Tufarelli et al. 2017). As indicated in the exhaustive review performed by DiSanto et al. (2021), polyphenols are plant secondary metabolites that constitute a family with at least 10,000 components. Polyphenols are classified into two large groups, flavonoids, and non-flavonoids, based on the presence of one or more hydroxyl groups attached to a benzene ring. The non-flavonoids are sub-classified into phenolic acids, tannins, xanthones, lignans, and stilbenes. In relation to the latter, the identified stilbenes include oxyresveratrol, piceid, gnetol, piceatannol, arachidin, rhapontigenin, pinostilbene, pterostilbene, isorhapontigenin, pinosylvin and RSV. Nowadays, the latter is the most popular bioactive under eye-researchers for its pharmacological and biomedical applications (Russo et al. 2021; Pannu and Bhatnagar 2019).

7.4.2 Chemical and Biological Properties of Resveratrol

RSV (3,4′,5-trihydroxystilbene) has been used in traditional medicine for more than 2,000 years. Its relevance lies in its easy access, which is why the scientific community and companies that explore food additives, cosmetics, and natural medicine have focused on this molecule. In 1940, RSV was initially isolated from the root of *Veratrum grandiflorum*, a flower frequently found in grassland areas of Europe and Asia, in 1940. The term "resveratrol" has its origin in the combination of the Latin prefix *Res*, meaning "which comes from," *veratr*, from the plant "*Veratrum*," and the suffix *ol*, for its alcohol chemical groups (Delmas et al. 2021). RSV is generally recognized as a safe (GRAS) phytochemical and is mainly found in red wine, grape berry skins, and seeds (Li et al. 2020). Its concentration in grapes fluctuates from 0.16 to 3.5 μg/g, meanwhile, dry grape skin contains about 24 μg/g. Cranberry raw juice comprises about 0.2 mg/L. Also, RSV ranges from μg/g (pistachios, peanuts) to ng/g (blueberries, bilberries). Remarkably, different varieties of red wines contain a considerably better quantity of this bioactive (0.1–14.3 mg/L) than white wines (0.1–2.1 mg/L) (Alaimo et al. 2020; DiSanto et al. 2021).

The chemical structure of RSV is identified in two isomeric forms, *cis*- and *trans*-RSV. The latter isoform undergoes to *cis* by isomerization when exposed to UV irradiation. The *trans* isomer is the dominant isoform in terms of its occurrence and diverse biological activities.

The relevance of RSV in plant physiology remains in its capacity to prevent the growth of pathogenic fungi. For this reason, RSV belongs to the so-called group of phytoalexins. In this sense, this property was determined in the RSV from the vine. The efficacy of the RSV against Botrytis cinerea fungus infection was due to the *trans-organic* isoform (Delmas et al. 2021). Also, this stilbene can inhibit the growth of some pathogenic microorganisms, such as gram-negative and gram-positive bacteria and other fungi (e.g., *Candida* species). Moreover, other bioactive effects, such as being antioxidant, anti-inflammatory, antimicrobial, anti-aging, antiangiogenic, neuroprotector, and anti-tumoral have also been reported; however, the most valuable property is its ability to act as a powerful detoxifying compound (Figure 7.2).

The antioxidant property of the RSV depends on the composition of the functional groups in the nuclear structure. Thus, the configuration, substitution, and total number of hydroxyl groups significantly influence several mechanisms of antioxidant activity, such as the capability to scavenge radicals and chelate metal ions. The hydroxyl group at the 4′ position is not the only key to

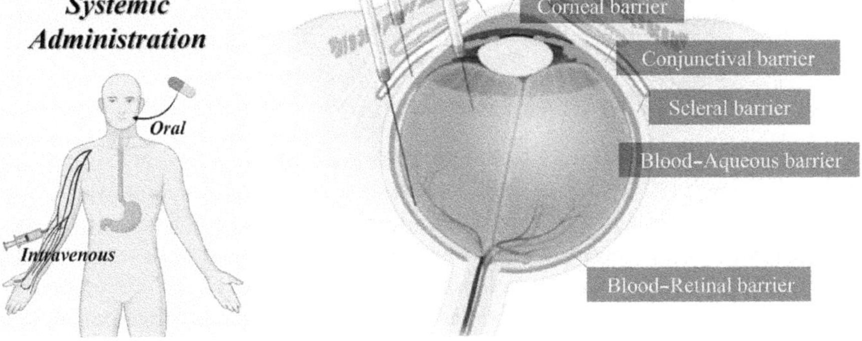

FIGURE 7.2 Systemic and ocular routes of drug administration.

antioxidant activity, but so are the 3-OH and 5-OH groups. The study of the antioxidant effect against hydroxyl (OH) and hydroperoxyl (•OOH) radicals in simulated aqueous media by using computational kinetic and density functional quantum chemistry methods revealed that RSV can act as an efficient •OOH, and also presumably •OOR, radical scavenger. In this way, RSV not only decreases the levels of reactive species but also acts as a promoter of the up-regulation of antioxidant enzymes (glutathione peroxidase, superoxide dismutase); therefore, the oxidized products of lipids and proteins decrease (Salehi et al. 2018; Pannu and Bhatnagar 2020). At a molecular biology scale, the favorable effects of RSV have been accredited to the triggering of the SIRT-1 / AMPK / PGC1-α axis, which leads to favoring mitochondrial oxidative phosphorylation system and thus promotes the activation of PGC1-α transcription factor. Therefore, mitochondrial protection is conferred, decreasing the production of mitochondrial reactive oxygen species (ROS). Regarding the anti-inflammatory activity, RSV acts by down-regulating cytokines such as TNF-α, IFN-γ, IL-6, IL-8, and IL-1β. Also, transcription factors and enzymes such as NF-κB and COX-2 are also targeted to reduce inflammation. Hence, RSV prevents cell death by regulating oxidative stress and inflammation. Additionally, RSV down-regulates pro-apoptotic Bcl-2-associated X protein (Bax), inhibits the intracellular increase of Ca^{2+} ion concentration as well as the main apoptotic executor, the caspase-3 (Pannu and Bhatnagar 2019).

Overall, RSV can restore cellular energetics and limit cell death which ultimately contributes to health. In this sense, it has been reported that RSV was effective against metabolic syndrome, obesity, cardiovascular, neurodegenerative, ocular, and age-related diseases, among others.

7.4.3 Evidence of Resveratrol Against Ocular Diseases

Several reports focused on the effects of RSV against environmental factors (e.g., acrolein present in cigarette smoke and car exhaust fumes, air pollution, LED lights, and UV light) and ocular disorders that lead to progressive vision loss and/or blindness (e.g., corneal wound healing, cataracts, glaucoma, uveitis, DR, AMD). The main biological actions in the eye include antioxidant, antiapoptotic, anticancer, anti-inflammatory, antiangiogenic, and vasodilator. At present, *in vitro* and *in vivo* experimental studies have presented evidence that RSV provides great benefits toward the treatment of various eye disorders (Sarubbo et al. 2017; Koushki et al. 2018; Delmas et al. 2021; Abu-Amero, Kondkar, and Chalam 2016). Notably, various effects and mechanisms suggested to be regulated by RSV have not yet been completely proven in humans. More investigations employing human clinical trials are necessary to verify the beneficial properties and mechanisms in order to establish its efficacy. In particular, Abu-Amero, Kondkar, and Chalam (2016) postulated that it is highly necessary to carry out more clinical trials at the ophthalmological level. This is because RSV is already available in dietary supplements and its generally favorable findings are readily used by nutritionally based companies. One year later, a study carried out in humans determined that RSV could be detected in ocular tissues after oral administration, although in low concentration. However, not everything is so unfortunate since the authors mentioned that the metabolites detected could play a beneficial role in the treatment of ocular diseases (Wang et al. 2017).

7.4.4 Physic-Chemical Limitations of Resveratrol

The definition of bioavailability refers to the proportion of nutrients that are digested, absorbed, and metabolized through the usual metabolic routes of assimilation. In humans, the RSV absorption is especially transepithelial when acquired orally. However, its bioavailability is low as a result of its rapid metabolism in the intestine and liver. In particular, when it is in contact with intestinal enterocytes, the RSV's phenolic groups are conjugated with sulfates, and due to glucuronidation, the molecule is not efficiently metabolized. Therefore, the nutritional and therapeutic value of RSV is lost. On the other hand, the intestinal microbiota hydrogenates the *trans* aliphatic double bonds of RSV in such a manner that a portion of the biologically active molecule is missed (Chung et al.

2020). As a consequence, very low levels arrive both intracellularly and in plasma. This is a complication effective concentration in blood and tissues is crucial. The bioavailability of RSV depends on age and sex; however, on average, after the oral consumption of 25–1,000 mg, the total metabolites of RSV reach a peak plasma concentration of 400–500 ng/mL comparable to ~2 μM (Alaimo et al. 2020; Delmas et al. 2021). Another of the main limitations of RSV is the poor aqueous solubility (<1 mg/mL) with rapid degradation upon exposure to light, oxygen, temperature, and oxidative enzymes. In recent years, with the advent of nanotechnology, studies have been carried out to obtain formulations that stabilize and protect RSV from degradation and improve both its solubility and bioavailability. In this way, it is also expected for the nanosystem to maintain a sustained release and to be able to transport RSV to specific locations in the eye, avoiding the natural barriers of the eye (described in Section 7.2) (Buosi et al. 2020; Pannu and Bhatnagar 2019). The RSV delivery can be through the systemic or ocular (local) pathways (Figure 7.3). Whatever, in order to overcome the aforementioned limitations, several encapsulation methods have been used in different materials to overcome. Thus, NP becomes the "Trojan horse" that allows an efficient RSV delivery by overcoming natural obstacles and preserving its biological function (Figure 7.4). Detailed information about the use of RSV in various ocular diseases and its possible employment in treatments will be provided in the next section.

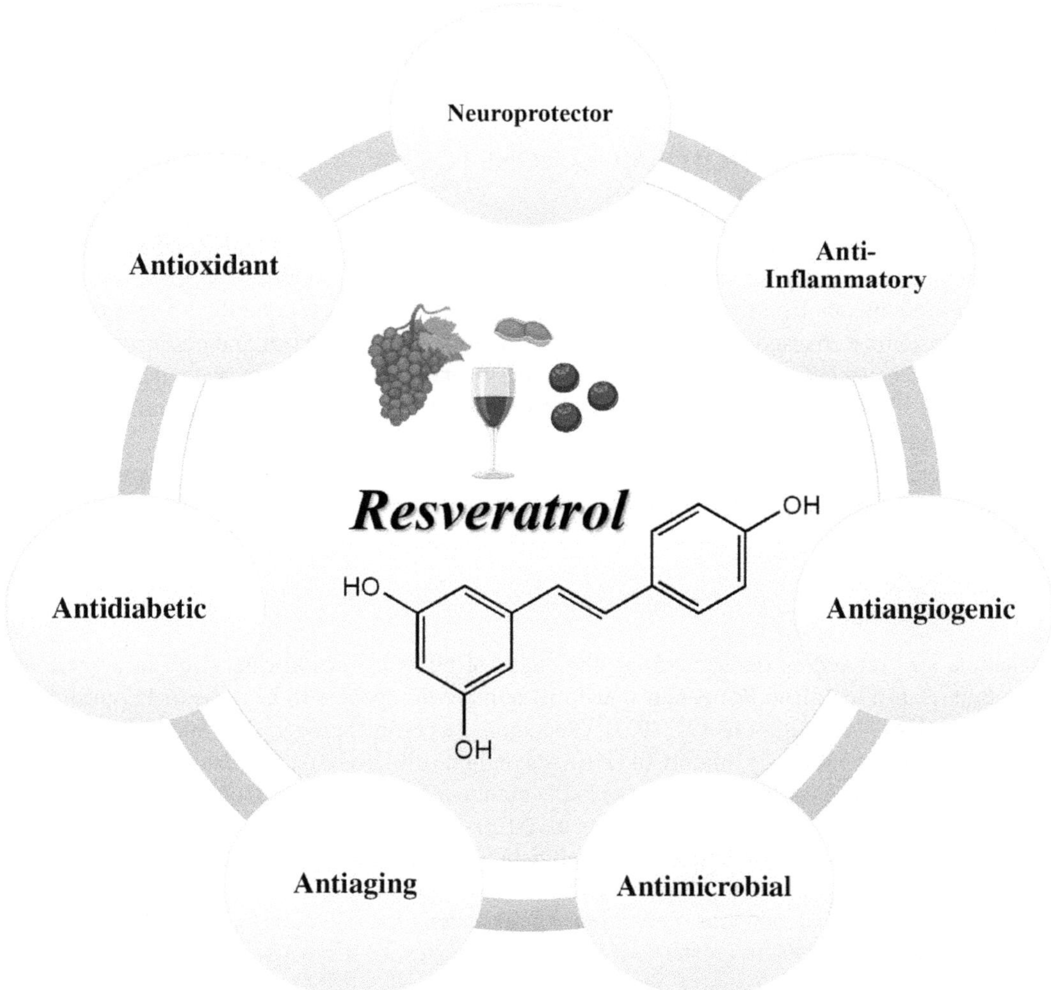

FIGURE 7.3 Biological activities and potential health benefits of RSV.

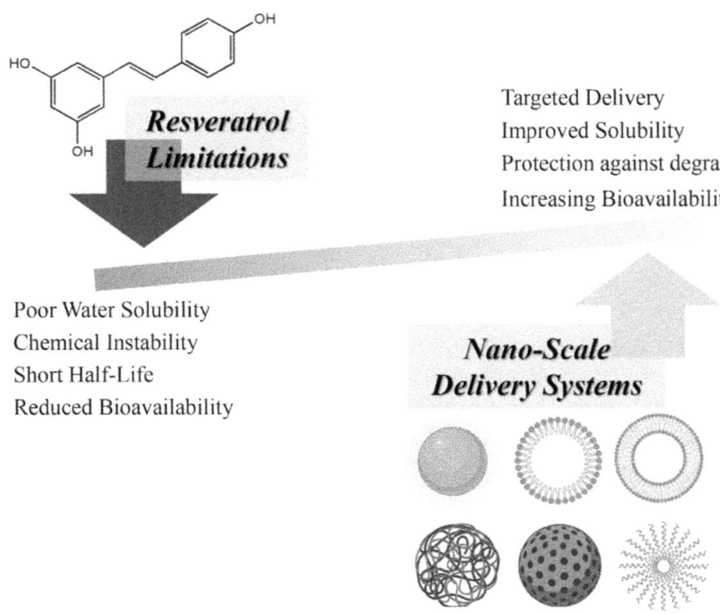

FIGURE 7.4 Schematic illustrating the RSV-associated limitations and how the nanoencapsulation and delivery could overcome them.

7.5 RESVERATROL NANOFORMULATIONS FOR OCULAR DELIVERY AND TREATMENTS

Nanotechnology has shown promising results by using NP-based formulations in order to encapsulate RSV. Such formulations have shown greater stability and bioavailability of RSV, as well as a reduction in its metabolism that corresponds to an increase in tissue availability. These nanotechnological systems were used for different ocular pathologies in the anterior and posterior segments. The focus of the present chapter book is the comparison of the obtained results from the RSV encapsulation of different types of generated NP. We discussed the NP size (nm), polydispersity index (PDI), ζ-potential, drug loading (DL), encapsulation efficiency (EE) as well as the RSV release profile from NP and the *in vitro*/*in vivo* studies performed. For this purpose, an extensive bibliographic search was carried out using the databases of available scientific bibliographical references. Notably, the first report was published in 2016. Those reports were summed up in Table 7.1.

7.5.1 Glaucoma

Glaucoma was ranked as the second leading cause of blindness worldwide. Currently, there are more than 76 million glaucoma patients, and this number is expected to increase to 112 million by 2040. (Wu et al. 2021; Adelson et al. 2021). Glaucoma is a chronic, progressive, and reversible optic neuropathy. In this pathology, mainly the retinal ganglion cells (RGC) undergo apoptosis as a function of time. As a result, there is a decrease in the number of axons derived from these cells with the consequent atrophy of the optic nerve (Wu et al. 2021).

The incorporation of RSV alone or in combination with riluzole in rats with a glaucoma model was proved to have a protective effect on RGC. In addition, it was found that if treatment was started immediately after model induction, rather than 3 weeks later, protection was higher. The combination of riluzole and RSV results in greater RGC protection, suggesting a synergy. (Pirhan et al. 2016).

In a study performed by Razali et al. (2015), it was proved that RSV was able to reduce intraocular pressure in a steroid-induced ocular hypertension model. The action mechanism was by activating the A1 Adenosine receptor, which increased matrix metalloproteinase (MMP-2) activity thus reducing trabecular meshwork extracellular matrix deposition.

TABLE 7.1
Resveratrol-Loaded Nanoparticles for Ocular Diseases Treatments

Formulation	Particle Size (nm) ζ Potential (mV)	Biological Model	Disease	Main Findings	References
PEG-CS NPs	129.8 N/A	Rabbits	Glaucoma	Non-irritating Reduces the intraocular pressure	Pandian et al. (2016) Natesan et al. (2017)
PMs-RSV-Rh6G	3579±0.2 +21.0±4.6	D407 cells	Diabetic Retinopathy	Non-cytotoxic Anti-VEGF and anti-inflammatory activity	Rugină et al. (2019)
RSV-AuNPs	20 N/A	Rats	Diabetic Retinopathy	Increases the retinal expression of PEDF Decreases the retinal expression of VEGF-1 Reduces retinal mRNA expressions of VEGF-1, TNFα, MCP-1, ICAM-1, and IL-6, IL-1β	Dong et al. (2019)
RSV-Ro5	500 N/A	CHO-K1 cells HRECs cells	Diabetic Retinopathy	Non-cytotoxic Increases cell proliferation	Gonzalez-Perez (2021)
Sol-RSV NMs	50.1±1.3 −3.5	Mouse HREC cells	Inflammatory and oxidative stress disease	High ocular tolerance Improves corneal wound healing Inhibition of anti-inflammation mediators Favors production of antioxidant factors	Li et al. (2020)
RSV-RMLCNs	163.3 +46.4	Rabbits	Anterior and Segment Eye Diseases	Improves sustained drug release on the eye surface	Saha et al. (2021)
Gelucire/HPßCD NPs	329.9±1.9 N/A	Bovine lenses	Cataract	Antioxidant activity	Vora et al. (2019)
ROLG	N/A N/A	Rats	Corneal neovascularization	Enhances drug permeability across corneal tissues Anti-VEGF activity	Li et al. (2021)
HCS-TPP NG	140 +32±2	ARPE-19 cells	AMD and related pathologies	Non-cytotoxic Intracellular drug delivery Resveratrol protection against UV light	Buosi et al. (2020)
PLGA-RSV NPs	102.7±2.8 −47.3±0.9	ARPE-19 cells	AMD	Non-cytotoxic Anti-VEGF activity	Bhatt et al. (2020)

Abbreviations: PEG-CS NPs, polyethylene glycols modified chitosan nanoparticles (NPs), resveratrol-loaded; PMs-RSV-Rh6G, polymeric microcapsules (Poly(allylamine hydrochloride) and Polystyrene sulfonate) with resveratrol, covered by rhodamine 6G (Rh6G); RSV-AuNPs, resveratrol coated gold nanoparticles; RSV-Ro5, nanoparticles of resveratrol and Ro5-3335 stabilized with polyvinyl alcohol; Sol-RSV NMs, polyvinyl caprolactampolyvinyl acetate–polyethylene glycol graft copolymer Soluplus® and resveratrol nanomicelles; RSV-RMLCNs: resveratrol-loaded mucoadhesive lecithin/chitosan nanoparticles; Gelucire/HPßCD NPs, lipid-cyclodextrin-based nanoparticles, polyphenol-loaded. ROLG, Resveratrol-loaded ocular lamellar crystalline gel; CS-TPP NG, nanogels (NG) based on high molecular weight chitosan (HCS) crosslinked with sodium tripolyphosphate (TPP); PLGA-RSV NPs, resveratrol poly (lactic-co-glycolic acid) (PLGA) nanoparticles; D407, Human Retinal Pigment Epithelial cells; CHO-K1, Chinese Hamster Ovary cell line; HREC, human retinal endothelial cell line; ARPE, human retinal pigment epithelial cell line; AMD, age-related macular degeneration; N/A, not available.

Nanotechnology was used in this opportunity to enhance the bioavailability of RSV for the glaucoma treatment. Pandian and co-authors reported polymeric nanoparticles with RSV-loaded-polyethylene glycols (PEG) modified chitosan (CS) NP (PEG-CS NPs). These nanoparticles were designed by ionic gelation and exhibited a particle size and EE of 129.8 nm and 91.9±1.1% respectively. As the concentration of PEG increases, the particle size and PDI of the systems increase, and EE and DL of RSV into NPs decrease. The drug release studies demonstrate an effect burst release of approx 45% followed by controlled release. Studies in vivo proved that the NPs were non-irritant, and these nanoparticles reduced the intraocular pressure by 4.3±0.5 mmHg up to 8 h in normotensive rabbits.

In conclusion, these systems deliver RSV efficiently to the ocular tissues and reduce the intraocular pressure for the treatment of glaucoma. (Pandian et al. 2017). In the same year, the same group of investigators studied concurrent administration of RSV and quercetin. The bioavailability of RSV was improved, and NP could reduce intraocular pressure efficiently (Natesan et al. 2017).

7.5.2 Diabetic Retinopathy

The World Health Organization estimates that more than 420 million persons have diabetes, and this number is projected to increase to 578 million by 2030 (Natesan et al. 2017). Is common in patients diabetic who have gradually lost vision and this is because they present DR. Is an eye condition that causes changes in the retina, more precisely in the blood vessels. When left untreated, DR can scar and damage the retina (Nentwich and Ulbig 2015; Al-Hussaini, Kittaneh, and Kilarkaje 2021). Two stages can be distinguished in DR, non-proliferative and proliferative. Non-proliferative DR is an early stage of disease, in which new blood vessels do not grow and damage is limited to the retina. As vessels become clogged, retinopathy increases from mild to severe. As the disease advances, proliferative retinopathy develops, this stage is in which the blood vessels close causing the appearance of new abnormal vessels that can leak blood into the vitreous. The scar tissue created can cause a retinal detachment. If these new vessels interfere with flow, there can be a buildup of pressure that can lead to glaucoma. (R. P. Singh et al. 2019; DeLos Ríos, Gonzalez Pérez, and Londoño López 2021). In this sense, higher sugar levels in blood characteristic of this pathology activate many specific genes related to vasoactive peptides and inflammatory cytokines, such as the vascular endothelial growth factor (VEGF) and interleukin 6 (IL-6) (Rugină et al. 2019) The currently available treatments are invasive and mainly target neovascularization through the use of intraocular injection with anti-inflammatory medicines such as corticosteroids or injection of anti-VEGF, laser treatment and surgery. The drugs used could cause unwanted side effects, and even their delivery is sometimes a disadvantage. For the purpose of reducing or eliminating the adverse effects, the development of novel vehicles might be very useful, focusing on drugs capable of blocking the VEGF action (Rugină et al. 2019). RSV has an inhibitory effect on hyperglycemia-induced retinal pigment epithelial cell inflammation. Precisely, RSV improved high blood glucose levels induced parameters., and the decrease of Gap junction intercellular communication, secretion of inflammation biomarkers IL-6 and IL-8, downregulation of connexin 43 (Cx43), and activation of transforming growth factor-β1 (TGF-β), protein kinase Cβ (PKCβ), and cyclooxygenase-2 (COX-2) (Losso, Truax, and Richard 2010). Moreover, RSV reduces retinal neurodegeneration and increases the thickness of the retina and the inner nuclear layer in diabetic rats. Moreover, RSV was shown to decrease various pro-inflammatory cytokines as well as TNF-α and apoptosis.

Rugină et al. (2019) obtained an RSV-delivery system using polymeric microcapsules based on porous $CaCO_3$ sacrificial templates. Briefly, $CaCO_3$ particles were first prepared, and then both RSV and dextran were added. After that, four alternative layers of polyelectrolytes Poly (allylamine hydrochloride), and Polystyrene sulfonate and covered by rhodamine 6G (Rh6G). Finally, the $CaCO_3$ core was dissolved with EDTA-obtained hollow PMs-RSV-Rh6G. The diameter of microcapsules was 3579±0.2 nm, the zeta potential was +21.0±4.6 mV and the EE resulted in 96.8±4.0%. For biological evaluation, the PMs-RSV-Rh6G microcapsules were incubated into a human retinal pigment epithelium (RPE) D407 cell line grown with different glucose concentrations. Viability studies were carried out by incubating WST-1 cell proliferation reagent with D407 cells in order to assay the mitochondrial activity as an indicator of viable cells. Also, It was reported that polymeric microcapsules RSV-Rh6G did not produce cell toxicity after 24h. On the other hand, the VEGF secretion and the levels of IL-6 were measured in the culture medium of D407 cells, in the presence or absence of a nanoparticle system. These measurements revealed that polymeric microcapsules RSV-Rh6Gs inhibited the VEGF increased levels and significantly reduced the IL-6 levels in the cell culture exposed at different glucose concentrations. Finally, this work shows that particles were internalized into D407 cells. A fluorescence lifetime imaging microscopy was used to study the localization of microcapsules into D407 cells. It was concluded that cells can engulf 3–4 microcapsules per cell, and confocal laser scanning microscopy was used to differentiate whether the microcapsules are just adherent to the outer membrane or incorporated by cells. The incorporation of microcapsules into cells was demonstrated by the redistribution of the actin cytoskeleton around polymeric microcapsule-RSV-Rh6G when the cells engulf the microcapsules. Therefore, these results indicate that PMs-RRES-Rh6G could be a potential intraocular RSV-delivery vehicle with anti-VEGF activity in the therapy of DR.

In a study performed by Dong et al. (2019), RSV-loaded gold NP (RSV-AuNPs) were synthesized through an eco-friendly process and their effect on diabetic rats was evaluated. In the generation of RSV-AuNPs, RSV was used as a stabilizing and reducing agent. The results reported that NPs were spherical and had a particle size of 20nm, with a face-centered cubic crystal structure. The experiments were carried out with diabetic rats, were administered orally with RSV-AuNPs for a period of 3months and several parameters were evaluated in investigational rats at the end of treatment. These studies demonstrated that rats treated with RSV-AuNPs did not significantly decrease in body weight, meanwhile, glucose levels decreased, and the levels of glycosylated hemoglobin were reduced when compared with diabetic untreated rats. When diabetic rats were treated with RSV-AuNPs, a decrease in retinal vascular permeability was observed in histological sections. Besides, in relation to the morphological variations in the retina, a greater number of vessels were observed in the different layers of the retina (ONL, INL, and GCL) in the diabetic rats which decreased when treated with RSV-AuNPs. On the other hand, the retinas of diabetic rats were found to be significantly thinner than non-diabetic rats. In rats treated with AuNPs, the retina was observed to be thicker, as well as the GCL and nerve fiber layer (NFL). Rats treated with RSV-AuNPs were evaluated on the gene expression of retinal pro-angiogenic factors, inflammatory cytokines, and chemokines. As is known, the retinal expression of VEGF proteins and mRNA levels increase in a diabetic model in rats, these values declined on treatment with RSV-AuNPs. In addition, the treatment with RSV-loaded AuNPs showed reduced phosphorylation of nuclear factor kappa B p65 and Extracellular signal-regulated kinase 1–2. In conclusion, the RSV-AuNPs have a protective effect redeveloping the balance among the inhibitors and stimulators of angiogenesis, and ameliorating retinal inflammation.

Gonzalez-Perez et al. (2021) developed NP formulation based on RSV and 7-chloro-1,3-dihidro-5-(1H-pirrol-2-yl)-2H-1,4-benzodiazepin-2-ona (Ro5). The authors evaluated the potential application in the treatment of DR (RSV-Ro5). Briefly, the nanoemulsion stabilized with polyvinyl alcohol was prepared by the phase inversion method. The RSV-Ro5 nanodrops were spherical, and stable over time, with a particle size lower than 500nm and a pH around 6. In addition, *in vitro* tests were carried out to evaluate cell cytotoxicity and proliferation. For the cytotoxicity tests, CHO-K1 cells were used with the formulations containing RSV, Ro5, and their combination, in different concentrations. The results showed that the formulations did not have serious cytotoxic effects on CHO-K1 cells. Proliferation assays were carried out with HREC cell lines using formulations containing RSV, Ro5, and RSV- Ro5. The results obtained showed an increase in the percentage of cell proliferation for all formulations. Even the RSV nanoemulsion was the one presenting the highest proliferation percentage.

7.5.3 Inflammatory and Oxidative Stress Disease

Several eye diseases involve inflammation and oxidative stress events. Thus, it contributes to the occurrence of many ocular diseases, such as dry eye syndrome, cataracts, autoimmune and inflammatory uveitis, and conjunctival and corneal diseases. RSV is known to have the free radical scavenging ability with anti-inflammatory effects.

In order to overcome the drawbacks of RSV applications to the eye, Li et al. (2020) developed a Soluplu® (polyvinyl caprolactam–polyvinyl acetate–polyethylene glycol graft copolymer) micelle ophthalmic solution containing RSV prepared via the film method. The formulation proved to have the highest EE (98.8 ± 1.9%) with a mean micelle size of 50.1 ± 1.3 nm and a PDI of 0.081 ± 0.002, with an $\zeta = -3.5$ mV. The samples were characterized by transmission electron microscopy (TEM), uniform particle sphere or quasi-sphere shape was observed. The ophthalmic solution showed good storage stability at 4°C, and the RSV remaining in micelles was 92.3 ± 4.1% after 12 weeks of storage. Furthermore, the chemical stability of RSV within micelles was significantly better than the free RSV solution. The Sol-RSV NMs for ophthalmic solution demonstrated high ocular tolerance and improved the *in vitro* passive and *in vivo* permeation, in relation to free RSV. Besides, *in vivo,* the inhibition of antioxidant factors as well as the anti-inflammation mediators and corneal wound healing were favored.

In order to remedy the problems related to the drug permanence time on the ocular surface and the bioactive bioavailability, Saha et al. (2021) designed RSV-loaded mucoadhesive lecithin/chitosan NP (RSV-MLCNs). The RSV-MLCNs were formed by ionic gelation and the resultant particles had the following characteristics: particle size of about 163.3 nm, a polydispersity of 0.254 with potential electrokinetic 46.4 mV, and EE of 97.03%. The release of RSV was 96.87 with the zero-order release kinetics. Moreover, several studies were performed to evaluate the mucoadhesive properties, stability, and ocular retention. In short, the results suggested that this RSV-MLCNs formulation presented optimal mucoadhesive properties and good stability for 30 days. Besides, RSV-MLCNs were retained on the ocular surface for a prolonged time, due to mucus layer interaction, ensuring the sustained drug release.

The cataract appears when the lens, due to aging, becomes opaque and loses its transparency. This causes objects to be out of focus and blurry images are created. It is the main cause of preventable blindness in the world and affects almost 8 out of 10 people over 75 years of age (A. Singh and Bodakhe 2020; Delmas et al. 2021).

In an experimental model of naphthalene-induced age-related cataracts in rats, RSV retarded lenticular opacity, restored antioxidants, protein contents, Ca^{2+} ATPase function, and decreased lipid peroxidation in the lenses of rats treated with RSV. (A. Singh and Bodakhe 2020). In human lens FHL 124 cell line, RSV significantly inhibits the TGFβ2-induced expression of the myofibroblast marker alpha-smooth muscle actin (α-SMA) and human capsular bags following simulated cataract surgery, demonstrating the capacity of RSV to avoid the epithelial-to-mesenchymal transition associated with posterior capsule opacification (Smith, Eldred, and Wormstone 2019).

Vora et al. (2019) formulated NP by combining Gelucire 50/13 and HPßCD to form a desired matrix to encapsulate RSV. The size of this formulation was 329.9 ± 1.9 nm which significantly decreased after freeze-drying. The PDI obtained was <0.3 and the EE was 72.2% ± 1.5%. The profile release of this polyphenol showed biphasic behavior, which suggests that first there was an effect burst release of the RSV and then there was a sustained release in the time.

7.5.4 Corneal Neovascularization

This disease is the presence of abnormal blood vessels in the cornea. They are an unequivocal sign of corneal suffering associated with numerous causes, especially occur as a result of inflammation, and infectious, degenerative, and traumatic disorders. The invasion of new blood vessels into the avascular cornea causes vision impairment and even blindness (Mahmoudzadeh et al. 2020).

RSV has a potent anti-VEGF effect; hence, it has been used successfully for the treatment of corneal chemical burns (Chen et al. 2013).

Li et al. (2021) evidenced that a gel formulation of RSV had higher retention on the ocular surface. These formulations are easily and safely administered and allow for enhanced drug permeability across corneal tissues, resulting in a high anti-VEGF effect with the consequent corneal NV suppression.

7.5.5 Age-Related Macular Degeneration

One of the top causes of irreversible blindness among those people over 50 years of age is AMD. Statistically, it is the third most predominant eye disease leading to loss of vision with a blindness occurrence of 8.7%. By 2040, it is predicted that around 288 million individuals worldwide will suffer AMD (Fleckenstein et al. 2021). Several environmental and genetic risk factors associated with the progression of the disease have been established, including advanced age, white race, heredity, metabolic disorder, smoking, ultraviolet (UV) radiation, and air pollution. (Mitchell et al. 2018). All these factors could initiate AMD through the generation of free radicals including superoxide anion (O_2), nitric oxide (NO•), and hydroxyl radical (OH•), which creates oxidative stress and inflammation in eye tissues. At present, it is known that inflammation and oxidative stress play an important role in the initiation and progression of AMD (Ruan, Jiang, and Gericke 2021; Dieguez et al. 2019; Diéguez et al. 2020; Alaimo et al. 2019).

The dry AMD affects around 85% of patients; this form progresses more slowly than the wet form and there is no currently approved medical treatment or cure. Also, in the dry AMD, small yellowish deposits, called drusen, accumulate on the retina, underneath the macula, causing it to degenerate across time. Hence, since accumulation of these drusen occurs, central vision is lost (Delmas et al. 2021; Rabin et al. 2013). Wet AMD is the form less common but much more serious, it happens when abnormal blood vessels grow under the retina. These new blood vessels may break or leak, causing fluid accumulation and/or hemorrhages, distortion, and even loss of vision. Currently, some therapies used for the treatment of wet AMD consist of the inhibition of the atypical growth of vessels. To this end, laser photocoagulation may be performed or VEGF inhibitors are used, but these therapies have many side effects. About dry AMD, no therapies have shown efficacy. Treatment focuses on delaying the loss of visual function (Fleckenstein et al. 2021)

In summary, key events have been detected in AMD like oxidative damage, reduced RPE activities, chronic inflammation, and enhanced apoptosis. Furthermore, neovascularization has a very relevant role in complications associated with this disease. Therefore, RSV could be useful in the treatment of this disease mostly due to its antioxidant capacity, anti-inflammatory effect, or through its antiangiogenic actions Delmas et al. (2021).

Buosi et al. (2020) developed nanogels (NG) synthesized from a high molecular weight CS and sodium tripolyphosphate (ionic gelation method) in order to entrap RSV. The characterization studies demonstrated that these NG presented a distribution size around 140 nm, a ζ-potential of 32±2 mV, and an EE of 59%±1%. AFM and TEM images proved that NG presented a rounded morphology. The performed experiments to measure photodegradation proved that CS protects RSV degradation induced by ultraviolet light. Biocompatibility tests were performed in human ARPE-19 by measuring the concentration of IL-6 and IL-8; these results revealed that NGs were neither cytotoxic nor inflammatory. By CLSM, authors confirmed that an endo-lysosomal escape of NG occurs, after cellular internalization, and this step is critical for an effective delivery system. In short, the authors predict that these nanogels could represent novel RVS carriers, with the possibility of its application in the treatment of ocular diseases.

Another group synthesized poly lactic-*co*-glycolic acid (PLGA) polymeric NP RSV-loaded (PLGA-RSV NPs). Bhatt et al. (2020) prepared O/W emulsion followed of solvent evaporation. The obtained NP showed a mean size of 102.7±2.8 nm with a ζ-potential of –47.3±0.9 mV. Moreover, the EE was 65.2%. *In vitro* drug release studies demonstrated that the PLGA-RES NPs formulation had an initial burst of drug release at 1 h and then NP showed sustained cumulative drug release with

83.4%±0.6% at the end of day 3. The cell viability studies, using ARPE-19 cell line, revealed that the obtained formulation was non-cytotoxic. Cellular uptake of NP by ARPE-19 cells was evaluated by using CLSM. NPs were observed in ARPE-19 cells after 2 h of incubation, indicating an efficient endocytosis uptake. In addition, the RSV encapsulated into PLGA-based NP exhibited effective inhibition of VEGF expression levels. Thus, these NP could improve the therapeutic potential of the RSV and may be used for neovascular AMD treatment.

7.6 CONCLUDING REMARKS

This chapter summarizes the current state of knowledge on the development of different strategies to overcome the physicochemical limitations of RSV and the physiological barriers of the eye. Encapsulation nanosystems have been designed and developed to avoid and/or reduce these barriers.

The specific properties and biological safety profiles for NP need further investigation in *in vitro* and in vivo models until clinical application. It is important to note that whether NP has desirable therapeutic and safety profiles depends on a variety of factors, including biodistribution pattern, RSV concentration, dose, shape, size, surface charge, and solubility. However, the available evidence (Table 7.1) allows a general conclusion with a positive prognosis. It is possible to confirm that RSV encapsulating NPs do not induce toxicity or inflammation in tissues or culture cells. Such effects were independent of the chemical nature of the encapsulating material. On the other hand, the ζ-potential is suitable for describing various properties of colloidal systems designed for drug or bioactive substance delivery, i.e., for nanomedicine applications. As a rule, the absolute values of ζ-potential of NP solutions are usually in the range of +100 to –100 mV. The magnitude of this value gives an indication of colloidal stability. ζ-potential values > +25 mV or < –25 mV indicate that the NPs have a high degree of stability due to electrostatic repulsion. At lower dispersion values, they tend to aggregate, coagulate, or flocculate due to van der Waals attraction between particles (Tekade 2019). As documented in Table 7.1, only five reports mentioned the ζ-potential values of the NPs. Four of these reports produced NP with a ζ-potential > +20 or < –20 mV, which is clearly critical for their optimal stability. However, the nanomicrobes generated with the commercial polymeric solubilizer (Soluplus®) do not fall under this rule as they have virtually no charge (–3.5 mV). In this case, this nanoformulation does not appear to be supported by electrostatic stabilization, suggesting that other forces determine the stability of the system. This could be due to hydrophobic interactions, steric overlap interactions, or van der Waals interactions (Pérez et al. 2014).

Although few reports have been published so far, most of them are related to ocular diseases of the posterior segment. Yet, it is the ocular segment that presents the greatest difficulties for drug access and administration. For this reason, it is the ocular segment that requires rapid attention to find a solution, because the way to achieve an optimal dose in this area is through frequent intravitreal injections. The adverse effects of these injections are discomfort or pain at the injection site, subconjunctival hemorrhage, transient elevation of intraocular pressure, and endophthalmitis. In this way, these effects contribute to patients discontinuing their treatment. In addition, the NP tested showed great potential for RSV transmission to key cellular targets depending on the disease being studied (e.g., RPE in AMD; RGCs in glaucoma). These findings were underscored by the positive results that were able to reverse changes in the inflammatory and pro-apoptotic molecular cascades tested in cell cultures and in pathological conditions in animals.

In summary, the range of nanomedicine applications in ophthalmology is growing. The use of nanoformulations would favor drug bioavailability, allow diffusion through anatomical barriers, and reduce adverse effects associated with the use of conventional ophthalmic drugs. In short, the use of nanotechnology in the field of ophthalmology would allow the benefits of various drugs and bioactives, such as RSV, to be optimized and their negative effects, such as cytotoxicity and low biodisponibility, to be reduced. This opens a large window into the spectrum of so-called personalized medicine, which will most likely require new studies with specific designs for populations with individual characteristics.

ACKNOWLEDGMENTS

The authors want to thank to Agencia Nacional de Promoción Científica y Tecnológica (PICTs 2017-1683, and 2015-3866), Universidad de Buenos Aires (20020190100297BA), Universidad Nacional de Cordoba, ARVO Foundation Collaborative Research Fellowship and CONICET of Argentina. Figures were created with Created with BioRender platform.

REFERENCES

Abu-Amero, Khaled K., Altaf A. Kondkar, and Kakarla V. Chalam. 2016. "Resveratrol and Ophthalmic Diseases." *Nutrients*. https://doi.org/10.3390/nu8040200.

Adelson, Jaimie D, Rupert R A Bourne, Paul Svitil Briant, Seth R Flaxman, Hugh R B Taylor, Jost B Jonas, Amir Aberhe Abdoli, et al. 2021. "Causes of Blindness and Vision Impairment in 2020 and Trends over 30 Years, and Prevalence of Avoidable Blindness in Relation to VISION 2020: The Right to Sight: An Analysis for the Global Burden of Disease Study." *Lancet Global Health* https://doi.org/10.1016/S2214-109X(20)30489-7.

Alaimo, Agustina, Guadalupe García Liñares, Juan Marco Bujjamer, Roxana Mayra Gorojod, Soledad Porte Alcon, Jimena Hebe Martínez, Alicia Baldessari, Hernán Edgardo Grecco, and Mónica Lidia Kotler. 2019. "Toxicity of Blue Led Light and A2E Is Associated to Mitochondrial Dynamics Impairment in ARPE-19 Cells: Implications for Age-Related Macular Degeneration." *Archives of Toxicology* 93 (5): 1401–15. https://doi.org/10.1007/s00204-019-02409-6.

Alaimo, Agustina, Mariana Carolina Di Santo, Ana Paula Domínguez Rubio, Gabriela Chaufan, Guadalupe García Liñares, and Oscar Edgardo Pérez. 2020. "Toxic Effects of A2E in Human ARPE-19 Cells Were Prevented by Resveratrol: A Potential Nutritional Bioactive for Age-Related Macular Degeneration Treatment." *Archives of Toxicology* 94 (2): 553–72. https://doi.org/10.1007/s00204-019-02637-w.

Al-Hussaini, Heba, Rawan Sharhabeel Kittaneh, and Narayana Kilarkaje. 2021. "Effects of Trans-Resveratrol on Type 1 Diabetes-Induced up-Regulation of Apoptosis and Mitogen-Activated Protein Kinase Signaling in Retinal Pigment Epithelium of Dark Agouti Rats." *European Journal of Pharmacology* 904 (August). https://doi.org/10.1016/J.EJPHAR.2021.174167.

Almeida, Hugo, Maria Helena Amaral, Paulo Lobao, Christian Frigerio, and Jose Manuel Sousa Lobo. 2015. "Nanoparticles in Ocular Drug Delivery Systems for Topical Administration: Promises and Challenges." *Current Pharmaceutical Design* 21 (36): 5212–24. https://doi.org/10.2174/1381612821666150923095155.

Ansari, Mohammad Wakeel, and Ahmed Nadeem. 2016. *Anatomy of the Orbit. Atlas of Ocular Anatomy*. Springer International Publishing. https://doi.org/10.1007/978-3-319-42781-2_1.

Bhatt, Priyanka, Gulimirerouzi Fnu, Deepak Bhatia, Amna Shahid, and Vijaykumar Sutariya. 2020. "Nanodelivery of Resveratrol-Loaded PLGA Nanoparticles for Age-Related Macular Degeneration." *AAPS PharmSciTech* 21 (8). https://doi.org/10.1208/S12249-020-01836-4.

Binder, Susanne. 2004. "The Macula: Diagnosis, Treatment and Future Trends," Springer International Publishing. https://doi.org/10.1007/978-3-7091-7985-7.

Bisht, Rohit, Abhirup Mandal, Jagdish K. Jaiswal, and Ilva D. Rupenthal. 2018. "Nanocarrier Mediated Retinal Drug Delivery: Overcoming Ocular Barriers to Treat Posterior Eye Diseases." *Wiley Interdisciplinary Reviews: Nanomedicine and Nanobiotechnology*. https://doi.org/10.1002/wnan.1473.

Buosi, F.S., A. Alaimo, M.C. Di Santo, F. Elías, G. García Liñares, S.L. Acebedo, M.A. Castañeda Cataña, et al. 2020. "Resveratrol Encapsulation in High Molecular Weight Chitosan-Based Nanogels for Applications in Ocular Treatments: Impact on Human ARPE-19 Culture Cells." *International Journal of Biological Macromolecules* 165: 804–21. https://doi.org/10.1016/j.ijbiomac.2020.09.234.

Chen, Liang Ke, Peng Fei Qiang, Qi Ping Xu, Yi Hua Zhao, Fang Dai, and Lu Zhang. 2013. "Trans-3,4,5,4′-Tetramethoxystilbene, a Resveratrol Analog, Potently Inhibits Angiogenesis *in vitro* and *in vivo*." *Acta Pharmacologica Sinica* 34 (9): 1174–82. https://doi.org/10.1038/APS.2013.60.

Chishti, Hakim G M. 1988. *The Traditional Healer's Handbook*. Inner Traditions Bear and Company.

Chung, Ill Min, Umadevi Subramanian, Prabhu Thirupathi, Baskar Venkidasamy, Ramkumar Samynathan, Baniekal Hiremath Gangadhar, Govindasamy Rajakumar, and M. Thiruvengadam. 2020. "Resveratrol Nanoparticles: A Promising Therapeutic Advancement over Native Resveratrol." *Processes* 8 (4): 1–30. https://doi.org/10.3390/PR8040458.

Correia, Roberta T.P., Kátia C. Borges, Maria F. Medeiros, and Maria I. Genovese. 2012. "Bioactive Compounds and Phenolic-Linked Functionality of Powdered Tropical Fruit Residues." *Food Science and Technology International* 18 (6): 539–47. https://doi.org/10.1177/1082013211433077.

Delmas, Dominique, Clarisse Cornebise, Flavie Courtaut, Jianbo Xiao, and Virginie Aires. 2021. "New Highlights of Resveratrol: A Review of Properties against Ocular Diseases." *International Journal of Molecular Sciences* 22 (3): 1–29. https://doi.org/10.3390/ijms22031295.

Diéguez, Hernán H., María F. González Fleitas, Marcos L. Aranda, Juan S. Calanni, María I. Keller Sarmiento, Mónica S. Chianelli, Agustina Alaimo, et al. 2020. "Melatonin Protects the Retina from Experimental Nonexudative Age-Related Macular Degeneration in Mice." *Journal of Pineal Research* 68 (4): 1–13. https://doi.org/10.1111/jpi.12643.

Dieguez, Hernán H., Horacio E. Romeo, Agustina Alaimo, María F. González Fleitas, Marcos L. Aranda, Ruth E. Rosenstein, and Damián Dorfman. 2019. "Oxidative Stress Damage Circumscribed to the Central Temporal Retinal Pigment Epithelium in Early Experimental Non-Exudative Age-Related Macular Degeneration." *Free Radical Biology and Medicine* 131: 72–80. https://doi.org/10.1016/j.freeradbiomed.2018.11.035.

Dinte, Elena, Oliviu Vostinaru, Ovidiu Samoila, Bogdan Sevastre, and Ede Bodoki. 2020. "Ophthalmic Nanosystems with Antioxidants for the Prevention and Treatment of Eye Diseases." *Coatings* 10 (1): 36. https://doi.org/10.3390/coatings10010036.

DiSanto, Mariana Carolina, Cecilia Luciana D'Antoni, Ana Paula Domínguez Rubio, Agustina Alaimo, and Oscar Edgardo Pérez. 2021. "Chitosan-Tripolyphosphate Nanoparticles Designed to Encapsulate Polyphenolic Compounds for Biomedical and Pharmaceutical Applications - A Review." *Biomed Pharmacotherapy* 142: 11970.

Dong, Yi, Guangming Wan, Panshi Yan, Cheng Qian, Fuzhen Li, and Guanghua Peng. 2019. "Fabrication of Resveratrol Coated Gold Nanoparticles and Investigation of Their Effect on Diabetic Retinopathy in Streptozotocin Induced Diabetic Rats." *Journal of Photochemistry and Photobiology. B, Biology* 195 (June): 51–57. https://doi.org/10.1016/J.JPHOTOBIOL.2019.04.012.

Durazzo, Alessandra, Massimo Lucarini, Eliana B. Souto, Carla Cicala, Elisabetta Caiazzo, Angelo A. Izzo, Ettore Novellino, and Antonello Santini. 2019. "Polyphenols: A Concise Overview on the Chemistry, Occurrence, and Human Health." *Phytotherapy Research* 33 (9): 2221–43. https://doi.org/10.1002/ptr.6419.

Fabricant, D. S., and Farnsworth, N. R. 2001. "The Value of Plants Used in Traditional Medicine for Drug Discovery." *Environmental Health Perspectives* 109 (SUPPL. 1): 69–75. https://doi.org/10.1289/ehp.01109s169.

Fleckenstein, Monika, Tiarnán D. L. Keenan, Robyn H. Guymer, Usha Chakravarthy, Steffen Schmitz-Valckenberg, Caroline C. Klaver, Wai T. Wong, and Emily Y. Chew. 2021. "Age-Related Macular Degeneration." *Nature Reviews Disease Primers* 7 (1): 1–25. https://doi.org/10.1038/s41572-021-00265-2.

Forrester, John V., Andrew D. Dick, Paul G. McMenamin, Fiona Roberts, and Eric Pearlman. 2016. *The Eye Basic Sciences in Practice.* 4th Edition. Elsevier. https://doi.org/10.1016/C2012-0-07682-X.

Gonzalez-Perez, Juliana, Sarah De Los Ríos-Sánchez, Marybel Montoya-Álvarez, Pedronel Araque-Marín, and Martha E. Londoño-López 2021. "Resveratrol/Ro5-3335 Nanoemulsion for Potential Application in Retinal Drug Delivery". In 2021 IEEE 2nd International Congress of Biomedical Engineering and Bioengineering, CI-IB and BI 2021. https://doi.org/10.1109/CI-IBBI54220.2021.962612.

Gorantla, Srividya, Vamshi Krishna Rapalli, Tejashree Waghule, Prem Prakash Singh, Sunil Kumar Dubey, Ranendra N. Saha, and Gautam Singhvi. 2020. "Nanocarriers for Ocular Drug Delivery: Current Status and Translational Opportunity." *RSC Advances* 10 (46): 27835–55. https://doi.org/10.1039/d0ra04971a.

Gupta, Mrinali Patel, Alexandra A. Herzlich, Theodor Sauer, and Chi Chao Chan. 2015. "Retinal Anatomy and Pathology." *Developments in Ophthalmology* 55: 7–17. https://doi.org/10.1159/000431128.

Hano, Christophe, and Duangjai Tungmunnithum. 2020. "Plant Polyphenols, More than Just Simple Natural Antioxidants: Oxidative Stress, Aging and Age-Related Diseases." *Medicines* 7 (5): 26. https://doi.org/10.3390/MEDICINES7050026.

Harikumar, S. L., and Arora, S. 2011. "Nanotechnological Approaches in Ophthalmic Delivery Systems." *International Journal of Drug Development and Research* 3 (4): 9–19.

Koushki, Mehdi, Nasrin Amiri-Dashatan, Nayebali Ahmadi, Hojjat Allah Abbaszadeh, and Mostafa Rezaei-Tavirani. 2018. "Resveratrol: A Miraculous Natural Compound for Diseases Treatment." *Food Science and Nutrition* 6 (8): 2473–90. https://doi.org/10.1002/fsn3.855.

Kumar, Naresh, and Nidhi Goel. 2019. "Phenolic Acids: Natural Versatile Molecules with Promising Therapeutic Applications." *Biotechnology Reports* 24: e00370. https://doi.org/10.1016/j.btre.2019.e00370.

Lamb, Trevor D, Shaun P Collin, and Edward N Pugh. 2007. "Evolution of the Vertebrate Eye: Opsins, Photoreceptors, Retina and Eye Cup NIH Public Access." *Nature Reviews Neuroscience*. https://doi.org/10.1038/nrn2283.

Li, Mengshuang, Ling Zhang, Rong Li, and Meixing Yan. 2020. "New Resveratrol Micelle Formulation for Ocular Delivery: Characterization and *in vitro/in vivo* Evaluation." *Drug Development and Industrial Pharmacy* 46 (12): 1960–70. https://doi.org/10.1080/03639045.2020.1828909.

Li, Minshu, Xiang Yu, Lin Zhu, Yiguang Jin, and Zhihong Wu. 2021. "Ocular Lamellar Crystalline Gels for Sustained Release and Enhanced Permeation of Resveratrol against Corneal Neovascularization." *Drug Delivery* 28 (1): 206–17. https://doi.org/10.1080/10717544.2021.1872739.

Losso, Jack N., Robert E. Truax, and Gerald Richard. 2010. "Trans-Resveratrol Inhibits Hyperglycemia-Induced Inflammation and Connexin Downregulation in Retinal Pigment Epithelial Cells." *Journal of Agricultural and Food Chemistry* 58 (14): 8246–52. https://doi.org/10.1021/JF1012067.

Ludwing, P.E., Jessu, R., Czyz, C.N. 2024. Book: Physiology, Eye. StatPearls Publishing. Bookshelf ID: NBK470322

Mahmoudzadeh, Raziyeh, Saeed Heidari-Keshel, Mohammad Mehrpour, Fahimeh Asadi Amoli, Leila Aghajanpour, and Alireza Lashay. 2020. "Comparison of Topical Pigment Epithelium-Derived Factor (PEDF) with Topical Bevacizumab for Accelerating the Regression of Corneal Neovascularization in an Experimental Model of Rabbit Corneal Angiogenesis." *Ocular Immunology and Inflammation*. https://doi.org/10.1080/09273948.2020.1751211.

Mitchell, Paul, Gerald Liew, Bamini Gopinath, and Tien Y. Wong. 2018. "Age-Related Macular Degeneration." *The Lancet*. https://doi.org/10.1016/S0140-6736(18)31550-2.

Natesan, Subramanian, Saravanakumar Pandian, Chandrasekar Ponnusamy, Rajaguru Palanichamy, Sivakumar Muthusamy, and Ruckmani Kandasamy. 2017. "Co-Encapsulated Resveratrol and Quercetin in Chitosan and Peg Modified Chitosan Nanoparticles: For Efficient Intra Ocular Pressure Reduction." *International Journal of Biological Macromolecules* 104 (Pt.B): 1837–45. https://doi.org/10.1016/j.ijbiomac.2017.04.117.

Nentwich, MM, and MW Ulbig. 2015. "Diabetic Retinopathy - Ocular Complications of Diabetes Mellitus." *World Journal of Diabetes* 6 (3): 489. https://doi.org/10.4239/WJD.V6.I3.489.

Nettey, Henry, Yvonne Darko, Oluyemisi A. Bamiro, and Richard T. Addo. 2016. "Ocular Barriers." In: *Ocular Drug Delivery: Advances, Challenges and Applications*, 27–36. Springer International Publishing. https://doi.org/10.1007/978-3-319-47691-9_3.

Pan, Si Yuan, Gerhard Litscher, Si Hua Gao, Shu Feng Zhou, Zhi Ling Yu, Hou Qi Chen, Shuo Feng Zhang, Min Ke Tang, Jian Ning Sun, and Kam Ming Ko. 2014. "Historical Perspective of Traditional Indigenous Medical Practices: The Current Renaissance and Conservation of Herbal Resources." In: *Evidence-Based Complementary and Alternative Medicine*. Oxford University Press. https://doi.org/10.1155/2014/525340.

Pandian, Saravanakumar, Vinoth Jeevanesan, Chandrasekar Ponnusamy, and Subramanian Natesan. 2017. "RES-Loaded Pegylated CS NPs: For Efficient Ocular Delivery." *IET Nanobiotechnology* 11 (1): 32–39. https://doi.org/10.1049/iet-nbt.2016.0069.

Pannu, Naveet, and Archana Bhatnagar. 2019. "Resveratrol: From Enhanced Biosynthesis and Bioavailability to Multitargeting Chronic Diseases." *Biomedicine and Pharmacotherapy* 109: 2237–51. https://doi.org/10.1016/j.biopha.2018.11.075.

Pannu, Naveet, and Archana Bhatnagar. 2020. "Combinatorial Therapeutic Effect of Resveratrol and Piperine on Murine Model of Systemic Lupus Erythematosus." *Inflammopharmacology* 28 (2): 401–24. https://doi.org/10.1007/s10787-019-00662-w.

Peng, Yingqian, Luosheng Tang, and Yedi Zhou. 2017. "E-Mail Subretinal Injection: A Review on the Novel Route of Therapeutic Delivery for Vitreoretinal Diseases." *Ophthalmic Research*. https://doi.org/10.1159/000479157.

Pérez, Oscar E., Tatyana David-Birman, Ellina Kesselman, Sharon Levi-Tal, and Uri Lesmes. 2014. "Milk Protein-Vitamin Interactions: Formation of Beta-Lactoglobulin/Folic Acid Nano-Complexes and Their Impactoninvitro Gastro-Duodenal Proteolysis." *Food Hydrocolloids* 38 (July): 40–47. https://doi.org/10.1016/j.foodhyd.2013.11.010.

Pirhan, Dilara, NurŞen Yüksel, Esra Emre, Abdulkadir Cengiz, and Demir KürŞat Yildiz. 2016. "Riluzole- and Resveratrol-Induced Delay of Retinal Ganglion Cell Death in an Experimental Model of Glaucoma." *Current Eye Research* 41 (1): 59–69. https://doi.org/10.3109/02713683.2015.1004719.

Rabin, David M., Richard L. Rabin, Timothy A. Blenkinsop, Sally Temple, and Jeffrey H. Stern. 2013. "Chronic Oxidative Stress Upregulates Drusen-Related Protein Expression in Adult Human RPE Stem Cell-Derived RPE Cells: A Novel Culture Model for Dry AMD." *Aging* 5 (1): 51–66. https://doi.org/10.18632/aging.100516.

Rasouli, Hassan, Mohammad Hosein Farzaei, and Reza Khodarahmi. 2017. "Polyphenols and Their Benefits: A Review." *International Journal of Food Properties* 20 (2): 1700–1741. https://doi.org/10.1080/10942912.2017.1354017.

Razali, Norhafiza, Renu Agarwal, Puneet Agarwal, Sunil Kumar, Minaketan Tripathy, Sushil Vasudevan, Jonathan G. Crowston, and Nafeeza M. Ismail. 2015. "Role of Adenosine Receptors in Resveratrol-Induced Intraocular Pressure Lowering in Rats with Steroid-Induced Ocular Hypertension." *Clinical & Experimental Ophthalmology* 43 (1): 54–66. https://doi.org/10.1111/CEO.12375.

Ruan, Yue, Subao Jiang, and Adrian Gericke. 2021. "Age-Related Macular Degeneration: Role of Oxidative Stress and Blood Vessels." *International Journal of Molecular Sciences* 22 (3): 1–22. https://doi.org/10.3390/ijms22031296.

Rugină, Dumitriţa, Raluca Ghiman, Monica Focşan, Flaviu Tăbăran, Florina Copaciu, Maria Suciu, Adela Pintea, and Simion Aştilean. 2019. "Resveratrol-Delivery Vehicle with Anti-VEGF Activity Carried to Human Retinal Pigmented Epithelial Cells Exposed to High-Glucose Induced Conditions." *Colloids and Surfaces B: Biointerfaces* 181 (September): 66–75. https://doi.org/10.1016/j.colsurfb.2019.04.022.

Russo, Constanza Andrea, María Florencia Torti, Agostina Belen Marquez, Claudia Soledad Sepúlveda, Agustina Alaimo, and Cyebel Carina García. 2021. "Antiviral Bioactivity of Resveratrol against Zika Virus Infection in Human Retinal Pigment Epithelial Cells." *Molecular Biology Reports* 48 (1): 14. https://doi.org/10.1007/s11033-021-06490-y.

Saha, Mithun, Diti Rani Saha, Tahamina Ulhosna, Shazid Md Sharker, Md Hasanuzzaman Shohag, Muhammad Saiful Islam, Swapan K. Ray, G. M. Sayedur Rahman, and Hasan Mahmud Reza. 2021. "QbD Based Development of Resveratrol-Loaded Mucoadhesive Lecithin/Chitosan Nanoparticles for Prolonged Ocular Drug Delivery." *Journal of Drug Delivery Science and Technology* 63: 102480. https://doi.org/10.1016/J.JDDST.2021.102480.

Salehi, Bahare, Abhay Prakash Mishra, Manisha Nigam, Bilge Sener, Mehtap Kilic, Mehdi Sharifi-Rad, Patrick Valere Tsouh Fokou, Natália Martins, and Javad Sharifi-Rad. 2018. "Resveratrol: A Double-Edged Sword in Health Benefits." *Biomedicines* 6 (3): 1–20. https://doi.org/10.3390/biomedicines6030091.

Sarubbo, Fiorella, David Moranta, Victor J. Asensio, Antonio Miralles, and Susana Esteban. 2017. "Effects of Resveratrol and Other Polyphenols on the Most Common Brain Age-Related Diseases." *Current Medicinal Chemistry* 24 (38). https://doi.org/10.2174/0929867324666170724102743.

Singh, Amrita, and Surendra H. Bodakhe. 2020. "Resveratrol Delay the Cataract Formation against Naphthalene-Induced Experimental Cataract in the Albino Rats." *Journal of Biochemical and Molecular Toxicology* 34 (1). https://doi.org/10.1002/JBT.22420.

Singh, Rishi P., Michael J. Elman, Simran K. Singh, Anne E. Fung, and Ivaylo Stoilov. 2019. "Advances in the Treatment of Diabetic Retinopathy." *Journal of Diabetes and Its Complications* 33 (12). https://doi.org/10.1016/J.JDIACOMP.2019.107417.

Smith, Andrew J.O., Julie A. Eldred, and I. Michael Wormstone. 2019. "Resveratrol Inhibits Wound Healing and Lens Fibrosis: A Putative Candidate for Posterior Capsule Opacification Prevention." *Investigative Ophthalmology & Visual Science* 60 (12): 3863–77. https://doi.org/10.1167/IOVS.18-26248.

Tekade, Rakesh K. 2019. "Biomaterials and Bionanotechnology." *Biomaterials and Bionanotechnology*, 1–724. https://doi.org/10.1016/C2017-0-01066-2.

Tilburt, Jon C., and Ted J. Kaptchuk. 2008. "Herbal Medicine Research and Global Health: An Ethical Analysis." *Bulletin of the World Health Organization* 86 (8): 594–99. https://doi.org/10.2471/BLT.07.042820.

Tufarelli, Vincenzo, Elisabetta Casalino, Angela G D'Alessandro, and Vito Laudadio. 2017. "Dietary Phenolic Compounds: Biochemistry, Metabolism and Significance in Animal and Human Health." *Current Drug Metabolism* 18 (10): 905–13. https://doi.org/10.2174/1389200218666170925124004.

Veeresham, Ciddi. 2012. "Natural Products Derived from Plants as a Source of Drugs." *Journal of Advanced Pharmaceutical Technology and Research*. https://doi.org/10.4103/2231-4040.104709.

Vora, Deepal, Segewkal Heruye, Dunesh Kumari, Catherine Opere, and Harsh Chauhan. 2019. "Preparation, Characterization and Antioxidant Evaluation of Poorly Soluble Polyphenol-Loaded Nanoparticles for Cataract Treatment." *AAPS PharmSciTech* 20 (5). https://doi.org/10.1208/S12249-019-1379-Y.

Wang, S, Z Wang, S Yang, T Yin, Y Zhang, Y Qin, RN Weinreb, and M X Sun. 2017. "Tissue Distribution of Trans-Resveratrol and Its Metabolites after Oral Administration in Human Eyes." *Journal of Ophthalmology*. https://doi.org/10.1155/2017/4052094.

Watkins, Rebekah, Ling Wu, Chenming Zhang, Richey M. Davis, and Bin Xu. 2015. "Natural Product-Based Nanomedicine: Recent Advances and Issues." *International Journal of Nanomedicine* 10: 6055–74. https://doi.org/10.2147/IJN.S92162.

Weng, Yuhua, Juan Liu, Shubin Jin, Weisheng Guo, Xingjie Liang, and Zhongbo Hu. 2017. "Nanotechnology-Based Strategies for Treatment of Ocular Disease." *Acta Pharmaceutica Sinica B. Chinese Academy of Medical Sciences*. https://doi.org/10.1016/j.apsb.2016.09.001.

Wu, Lin Ping, Danyang Wang, and Zibiao Li. 2020. "Grand Challenges in Nanomedicine." *Materials Science and Engineering C*. https://doi.org/10.1016/j.msec.2019.110302.

Wu, Xianbo, Xinwei Yang, Qi Liang, Xiali Xue, Jianli Huang, Jie Wang, Yihua Xu, et al. 2021. "Drugs for the Treatment of Glaucoma: Targets, Structure-Activity Relationships and Clinical Research." *European Journal of Medicinal Chemistry* 226 (December). https://doi.org/10.1016/J.EJMECH.2021.113842.
Yasin, M. Naveed, Sachin S. Thakur, and Ilva D. Rupenthal. 2018. "Penetration Routes to Retina and Posterior Segment." In: Patel, J., Sutariya, V., Kanwar, J., Pathak, Y. (eds) *Drug Delivery for the Retina and Posterior Segment Disease*, 69–81. Springer. https://doi.org/10.1007/978-3-319-95807-1_4.
Yuan, Haidan, Qianqian Ma, Li Ye, and Guangchun Piao. 2016. "The Traditional Medicine and Modern Medicine from Natural Products." *Molecules* 21 (5). https://doi.org/10.3390/molecules21050559.

8 Phenolic Compounds Delivery Systems and Antiaging Application

Fábia Rafaella Silva Alves,
Bárbara Cristina Silva Holanda Queiroz,
Ana Beatriz da Silva Araújo Sena,
Gilson Cassiano de Góes Neto,
Simone Braga Carneiro, and
Ádley Antonini Neves de Lima

8.1 INTRODUCTION

Plants are an abundant source of several classes of bioactive components. Phenolic compounds such as phenolic acids, flavonoids, stilbenes, curcuminoids, and tannins are great representatives of the biological potential of plant species, acting as potent antioxidants, anti-inflammatory, antimicrobial, antiwrinkle, antiaging agents, lighteners, or even adjuvants in sun protection, characteristics required by the cosmetic industry that make them valuable ingredients for various applications in dermatological conditions. Substances of natural origin in antiaging products have reached greater proportions since 2011. In this context, bioproducts extracted from the plant matrix have been demonstrated as potential antiaging active ingredients, acting significantly through different mechanisms (Cherubim et al., 2019; Garavand et al., 2021; Kyriakoudi et al., 2021; Oliveira et al., 2021; Silva et al., 2020).

The largest interests for plant-derived products are those ingredients with DNA protective mechanisms, followed by actives that are capable of activating or inactivating enzymes or reducing inflammation. Several plants have been investigated for their potential and gained prominence for their high content of phenolics, such as *Vitis vinifera*, *Butyrospermum parkii*, *Glycine soja*, *Theobroma cacao*, *Glycyrrhiza glabra*, and *Passiflora edulis f. edulis*. Among these components, it is worth mentioning the flavonoids, phenolic acids, and stilbenes, which are part of the composition of these plants and have been targets of interest due to their efficacy demonstrated in *in vitro* and *in vivo* assays (Ferreira et al., 2021; Yepes et al., 2021).

Many molecules belonging to the large class of phenolic compounds have been widely researched and even used for antiaging purposes. Among them, we can highlight ferulic acid, resveratrol, quercetin, and anthocyanins, incorporated in products already on the market (Figueroa-Robles, Antunes-Ricardo, Guajardo-Flores, 2021), in addition to others widely spread such as caffeic, ellagic, gallic, *p*-coumaric, chlorogenic acid, and aglycone. Many of these even have potential inhibition of enzymes related to skin aging, such as tyrosinase, elastase, collagenase, and hyaluronidase, and also increased collagen synthesis, a representative set of properties that generate a high impact on the cosmetic industry (Carvalho et al., 2021; Yepes et al., 2021).

Given the great importance of the creation of antiaging products, the industrial interest in phenolic bioactive is a current and growing reality, mainly due to the constant need for new ingredients with multifunctional properties. However, despite providing interesting cosmeceutical characteristics,

 DOI: 10.1201/9781003292692-8

some aspects such as instability, low solubility, and bioavailability limit the maximum use of its benefits, making it necessary to use strategies that allow their sustained release, as well as their preservation. Therefore, a wide range of delivery systems (DS) has been explored in order to favor biological activities in target tissues, in addition to other benefits such as increased solubility and stability of these compounds (Taofiq et al., 2019; Vaiserman et al., 2020; Garavand et al., 2021; Kyriakoudi et al., 2021).

In view of the critical attributes that determine the best use of these substances, technological approaches such as encapsulation can be used in order to achieve the required purposes (Borges et al., 2020). Many antioxidants have already been encapsulated in solid entrainment systems (nano- and microparticles). In these systems, the assets have optimized properties, such as solubility in a hydrophilic medium, as well as greater preservation of their potential, contributing to greater stability and control of the release of the asset (Santos et al., 2016; Silva, Michniak-Kohn, Leonardi, 2017; Kanha et al., 2021).

In addition to the encapsulation strategy, which involves the use of natural or synthetic polymers to formulate micro- and nanocapsules, other technological approaches can be used, including liposomes, which have a multilayer structure, and cyclodextrins that contain a hydrophobic cavity, allowing both systems to encapsulate insoluble compounds, expanding the possibilities of incorporating the actives into aqueous systems. Additionally, nano- and microemulsions have also been shown to be efficient vehicles for the incorporation of phenolic compounds. All of these delivery systems, in turn, can influence, in addition to improvements in solubility, stability, and bioavailability, the increase in the permeability of phenolics in the skin and enhancement of the bioactive properties of incorporated substances, expanding their applications in different fields (Carneiro et al., 2019; Vaiserman et al., 2020; Figueroa-Robles, Antunes-Ricardo, Guajardo-Flores, 2021; Garavand et al., 2021; Li et al., 2021; Muñoz-Shugulí et al., 2021). Figure 8.1 shows the different delivery systems that can be applied to phenolic compounds, with the respective improvements provided by these technologies.

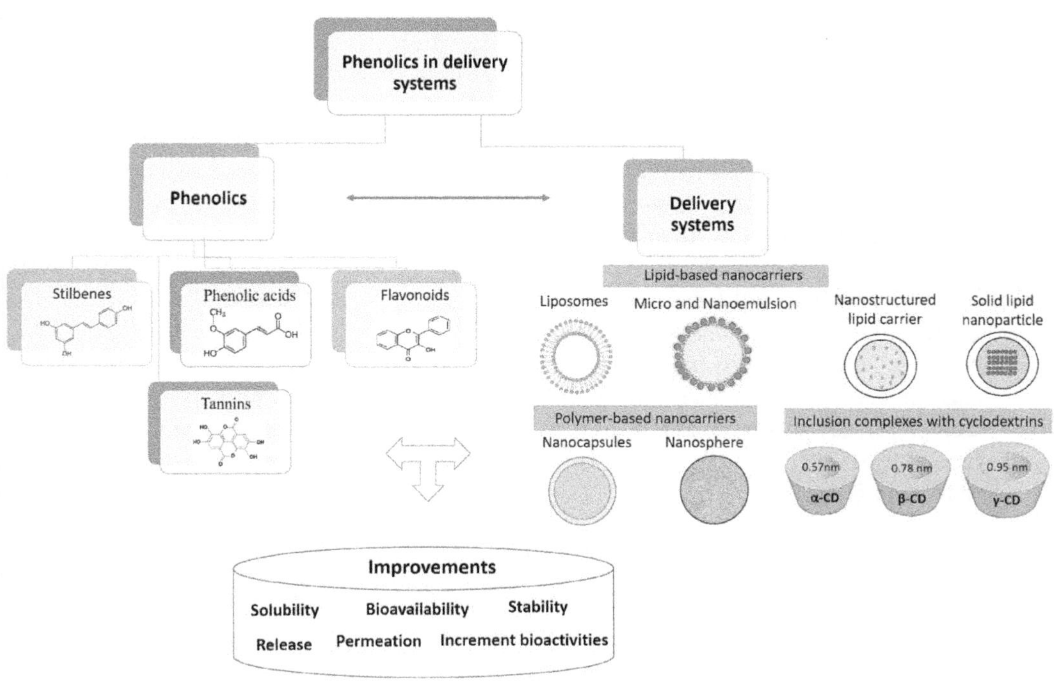

FIGURE 8.1 Improvements of phenolic bioactives inserted in DS.

Given the significance of polyphenols and their inherent bioactive potential, coupled with the cosmetic industry's keen interest in maximizing their properties and developing products with technological diversity (Kashyap et al., 2021; Tran et al., 2019), this chapter delineates the primary categories of phenolic compounds (flavonoids, phenolic acids, stilbenes, tannins, and curcuminoids) organized within delivery systems (DS) exhibiting antiaging properties and their clinical implications. This presentation is based on a compilation of literature spanning from 2011 to 2021, sourced from the ScienceDirect and PubMed databases.

8.2 PHENOLIC COMPOUNDS

An infinity of plants comprises thousands of polyphenolic molecules, which correspond to secondary metabolites, with a structure characterized by several hydroxyl groups in aromatic rings. These molecules are released by plant species in response to exogenous factors such as ultraviolet radiation, aggression by pathogens, or others that represent a threat, thus functioning as a defense mechanism. The classification of polyphenols is based on the number of phenolic rings and the substitution pattern (Kaur and Kaur, 2014). Thus, they can be divided into phenolic acids, stilbenes, flavonoids, lignans, and others (López-Fernández et al., 2020). Borges et al. (2020) add that phenolic compounds can be categorized as non-flavonoids, which include phenolic acids, stilbenes, and lignans, or flavonoids, which share a common structure, the flavanic nucleus. The great structural diversity of polyphenols allows us to understand their broad biological activities, among which the antioxidant action deserves a special focus (Silva et al., 2020). According to the data available in full, the classes of phenolic compounds for which there was a greater number of scientific publications associated with DS and antiaging activity will be discussed below.

8.2.1 Flavonoids

Flavonoids are compounds present in foods and beverages of plant origin, such as fruits, vegetables, cereals, pulses, cocoa, olive oil, tea, coffee, and wine, thus constituting an important component of human nutrition. This group demonstrates relevant activities such as antioxidants, antiaging, and free radicals reduction. The wide variety of substances in the flavonoid subgroup explains its wide use in the pharmaceutical industry and the growing interest in its use in the cosmetics industry (Singla et al., 2019; Lianza et al., 2020). Figure 8.2 presents the common structures of flavonoids.

Flavonoids have a chemical structure of diphenylpropanes, which consist of two benzene rings joined by a linear chain of three carbons (C6-C3-C6). Usually formed by a structure of 15 carbon atoms arranged in three rings, A, B, and C, the three central carbons form a pyran ring with one of the benzene rings (Singla et al., 2019).

The flavonoids differ in the oxidation degree and the pattern of substitution of the central ring. According to the variations in the ring structure, flavonoids can be classified into: flavanols (taxifolin), anthocyanidins, anthocyanins (delphinidinidins, pelargonidinidins), isoflavones (genistein, daidzein, glycitein, puerinidin), flavones (apigenin, luteolin, chrysin), flavonols (quercetin, kaempferol, myricetin, galangin, fisetin), flavanones (naringenin, hesperitin, eriodictyol), and flavanonols (Singla et al., 2019; Lianza et al., 2020). Figure 8.3 presents the structures of the biggest flavonoid groups.

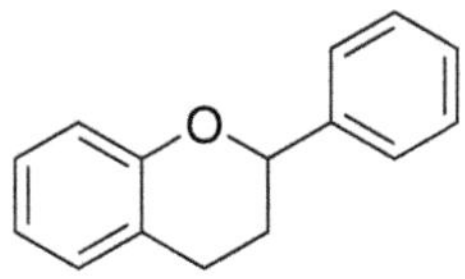

FIGURE 8.2 Structure of flavonoids.

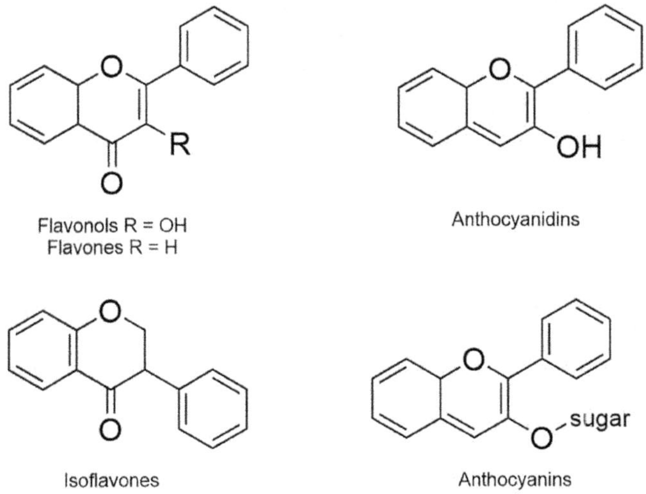

FIGURE 8.3 Structure of biggest flavonoid groups, including flavonols, flavones, and isoflavones, anthocyanidins, and anthocyanins.

Many biochemical properties were found for these compounds, explaining their physiological roles in plants and their biological roles in humans and animals. The antioxidant activity is the most notable among its functions, which comprises the capacity to modulate important cell signaling pathways and enzymatic activities involved in pathophysiological events such as cell proliferation, inflammation, immune response, platelet aggregation, and cytotoxicity. Much scientific research indicates that most of these effects of flavonoids are associated with their ability to bind to DNA and RNA (Blesso, 2019).

Several antioxidant mechanisms have been reported for flavonoids, demonstrating the action of these compounds in inhibiting or reducing melanogenesis, acting on epithelial cells, preventing DNA damage, and eliminating reactive oxygen species (ROS) such as peroxyl, alkyl, hydroxyl, and superoxide radicals. Active redox metal ion chelators such as copper and iron, given the presence of vicinal hydroxyl groups, prevent the formation of free radicals, in addition to the ability to inhibit enzymes that play important roles in the skin, demonstrating a promising role as an antiaging agent (Casarini et al., 2020; Davinelli et al., 2018; Zhao et al., 2020).

Bioavailability is a limiting point for flavonoids; some classes have better absorption than others, such as the set of isoflavones (genistein, daidzein) and catechins, which are better assimilated than the others. Despite the entire context to be considered, the bioavailability of flavonoids is low, a subject agreed upon by most authors, due to their chemical structure and molecular weight, factors that provide little water solubility, as well as low oral absorption or for topical application (Zhao et al., 2020).

The development of suitable molecular carriers, such as nanomolecules, to enhance the action of flavonoids, is already underway in scientific research, but more studies are still needed so that efficient and definitive solutions are developed and applied in humans (Costanzo and Angelico, 2019; Casarini et al., 2020; Zhao et al., 2020).

A study involving propolis extract, a compound that contains many flavonoids, with antioxidant and antiaging activity in the skin, was carried out with the intention of improving the bioavailability and transdermal permeation of these substances. For this purpose, a phytosome hydrogel was developed and tested. Phytosomes increased the dissolution of bioactive compounds by up to 4 times, improving skin permeability and retention by up to 6 times when compared to the non-encapsulated formulation with DS (Permana et al., 2020). Thus, revealing the importance and capacity of DS as carriers of phenolic compounds.

In this context, pharmaceutical technology has improved its tools using an extensive set of solubilization methodologies and techniques such as nanotechnology, with the purpose of imprisoning poorly water-soluble-active plant ingredients in complexes and facilitating their solubilization (Zhao et al., 2020). Currently, researchers have been pointing out the main and most efficient ways to increase the *in vivo* absorption strength of phenolic compounds. A large part of the successful formulation strategies refer to complexation with cyclodextrins, solid dispersions stabilized with polymers, micro, and nanoemulsions (Casarini et al., 2020).

Nanotechnology has been highlighted as a good alternative to improve the cutaneous availability of many flavonoids. Likewise, its association with other DS, with cyclodextrins and solid dispersions, has also shown promising results in enhancing the characteristics of the substances of interest (Mandrone et al., 2018). In a study by Mandrone et al. (2018), phloretin was associated with hydroxypropyl-β-cyclodextrin. This strategy increased the aqueous solubility of phloretin, and its stability was notably improved, where the cyclodextrin preserved the active principle, ensuring greater action of this flavonoid (Mandrone et al., 2018; Davinelli et al., 2018; Costanzo and Angelico, 2019).

8.2.2 Phenolic Acids

Phenolic acids are compounds naturally present in plant species, exhibiting functions that include plant growth, development, and defense. They are derived from the shikimic acid route, via the phenylpropanoid pathway, with a unique C6-C3 chemical structure. Some of the molecules in this category can also be obtained from microorganisms, containing C6-C1 bonds. This group of molecules can be divided into two categories: cinnamic acid derivatives, called hydroxycinnamic acids, and benzoic acid derivatives, which are hydroxybenzoic acids (Kaur and Kaur, 2014; Kumar and Goel, 2019). The subgroup of hydroxybenzoic acids includes to gallic acid, vanillic acid, protocatecic acid, ellagic acid, gentisic acid, and singic acid (Garavand et al., 2021). Likewise, hydroxycinnamic acid derivatives are *p*-coumaric, ferulic, caffeic, sinapic, and their conjugates/etherified acids, such as chlorogenic acid (Contardi et al., 2021; Zhao and Moghadasian, 2010).

Among the phenolic acids already reported in the literature with antioxidant activity, the following stand out: ferulic acid, *p*-coumaric acid (Yuan et al., 2021), chlorogenic acid, neochlorogenic acid, strongly associated with the ability to scavenge radicals (Agatonovic-Kustrin, Kustrin, Morton, 2018; Ding et al., 2020), caffeic acid, which is also highly efficient in capturing reactive species (Spagnol et al., 2019), and gallic acid (Agatonovic-Kustrin, Kustrin, Morton, 2018), which were verified in different plant sources. Additionally, some phenolic acids conjugated with chitosan have shown increased antioxidant potential, such as gallic and caffeic acids (Wang et al., 2019). Hence, hydroxycinnamic acids are more predominant compared to previously documented hydroxybenzoic acids.

Ferulic acid represents the subgroup of hydroxycinnamic acids and is well known for its valuable properties, acting as an antioxidant, antiwrinkle, and whitening agent. Such properties open space for exploration in topical formulations for application in skin rejuvenation (Cavalcanti et al., 2021). As a representative of hydroxybenzoic acids, gallic acid also demonstrated antiaging properties by different mechanisms, including suppression of melanin formation, and inhibition of MMP-2, exhibiting more significant effects when incorporated into DS. Figure 8.4 presents the structures of these two valuable molecules, which constitute this class with potential antiaging applications.

Several studies have evaluated the antiaging potential of different plant extracts containing phenolic acids in their composition. Bazzicalupo et al. (2021) suggested that *Carpobrotus edulis* (L.) extract may be suitable for incorporation in topical formulations with antiaging and healing properties, also acting in the prevention of damage to the molecular structure of the dermal matrix, as an anti-inflammatory. The extract showed hydroxycinnamic acids as the most abundant compounds, represented by dicaffeoylquinic acid, dihydrocaffeic acid 3-sulfate, and *p*-coumaric acid 4-*o*-glucoside.

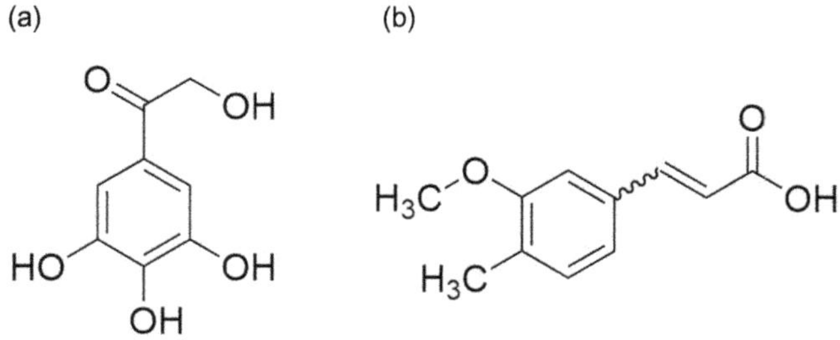

Gallic acid

Ferulic acid

FIGURE 8.4 Chemical structure of gallic acid (a) and ferulic acid (b), two important phenolic acids with numerous applications.

Like the other components belonging to the large group of phenolics, phenolic acids have limitations in terms of solubility, stability, and bioavailability, which are factors that prevent the maximum use of the inherent bioactive potential of this class, especially at an industrial level. Thus, the use of encapsulation systems, for example, makes it possible to overcome instabilities, provide protection against degradation, and enable controlled release (Khoshnoudi-Nia, Sharif, Jafari, 2020).

8.2.3 Tannins

Tannins are a class of polyphenols that can be divided into: condensed, comprising catechins and proanthocyanidins; hydrolyzable, composed of gallic or ellagic derivatives; and complex or mixed tannins, derived from condensed and hydrolyzable tannins. Condensed tannins are responsible for the astringent character of fruits, beverages, and teas; they are present in several plants in their bark, leaves, and fruits, especially when they are green. For the plant, they protect against predation, but when it comes to medicinal use, they have different properties such as anticancer and cardioprotective activities, importance in the elimination of free radicals, vasodilator, anti-inflammatory, antithrombotic, and protective action against UV rays (Ky et al., 2016). Among these functions, two draw attention to its antiaging potential: its antioxidant and UV-protector potential. Due to the intense action of the sun's rays on the skin, it begins to undergo changes that culminate in the appearance of spots and wrinkles.

During the research, no studies using isolated tannins with DS were observed. Generally, the studies performed used plant extracts that predominantly contained phenolic compounds, among them tannins, and mainly proanthocyanidins, which are from the class of condensed tannins derived from the biosynthetic pathway of flavonoids, with the basic formula (C6–C3–C6)n, where "*n*" represents the number of carbons (Zuiter, 2014). Figure 8.5 shows the chemical structure of proanthocyanidins.

In the study by Haddada et al. (2020), an extract of *Hubertia ambavilla* species, mainly composed of flavonoids and proanthocyanidin-type tannins, was incorporated into gold nanoparticles to enhance antioxidant and dermoprotective action for potential incorporation of this technology into cosmetic preparations. On the other hand, Garcia et al. (2021) developed microemulsions incorporating extracts of *Theobroma cacao* L., containing phenolic compounds and alkaloids, predominantly proanthocyanidins (58%), with the aim of using the system in cosmetics for skin application.

In the study by Haddada et al. (2020), an extract of *Hubertia ambavilla* species, mainly composed of flavonoids and proanthocyanidin-type tannins, was incorporated into gold nanoparticles to enhance antioxidant and dermoprotective action for potential incorporation of this technology into cosmetic preparations. On the other hand, Garcia et al. (2021) developed microemulsions incorporating extracts of *Theobroma cacao* L., containing phenolic compounds and alkaloids, predominantly proanthocyanidins (58%), with the aim of using the system in cosmetics for skin application.

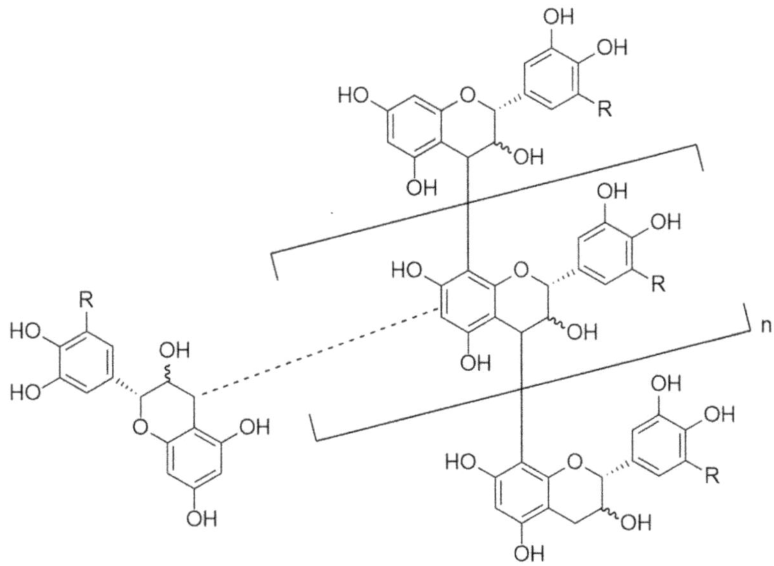

FIGURE 8.5 Molecular structure of proanthocyanidins, the most abundant tannin with antiaging activity, where "n" represents the repeating unit in the molecule.

Studies demonstrate the potential of microemulsions as topical vehicles to incorporate several bioactive compounds, including extracts containing phenolics. The compounds incorporated in microemulsions partition between the aqueous and hydrophobic phases, depending on their relative solubilities, making them attractive formulations for the incorporation of herbal medicines that bring extracts, considering the complex composition of these bioactives (Callender et al., 2017). Despite this evidence, the study still needs to be completed, considering that *in vitro* tests and clinical tests were not carried out to observe the absorption, antioxidant, protective potential, and differences obtained in the skin with topical use.

Another significant study describes antioxidant activity in four tests. Govindappa et al. (2021) aimed to incorporate the extract of the fleshy Polycarp of pomegranate (*Punica granatum* L. Lythraceae), a species rich in polyphenols, mainly tannins, into silver nanoparticles. The DPPH radical scavenging assays showed a DPPH• scavenging activity value of 81%. The hydrogen peroxide scavenging activity was obtained as 81.5%. For nitric oxide, the highest scavenging activity was recorded at 77.5% with a higher concentration of 100 μg/mL silver nanoparticles. Additionally, the reducing power testing demonstrated a reducing power of 74.51%, confirming their antioxidant potential.

8.2.4 Stilbenes

Stilbenes are phenolic compounds derived from phenylpropanoids, which have structural characteristics such as 1,2-diphenylethylene and 3,5,4′-trihydroxystilbene, which constitute resveratrol, being the main polyphenol in this class, and also the chemical composition of plant species from red grapes (Alonso et al., 2017; Vaiserman et al., 2020; Garavand et al., 2021). Resveratrol in plants acts as a phytoalexin, defending plants from phytopathogenic microorganisms. The stilbenes class has different functions, such as antioxidant, neuroprotection, and sirtuin activation, which are proteins in our body that regulate the rate of aging (Davies et al., 2020; Vaiserman et al., 2020; Das et al., 2020; Garavand et al., 2021).

Within the stilbenes class, resveratrol is a potent natural antioxidant constituent. This ability strongly depends on the redox of hydroxyl groups and the potential for electron delocalization in its chemical structure (Sedlak et al., 2018; Vaiserman et al., 2020).

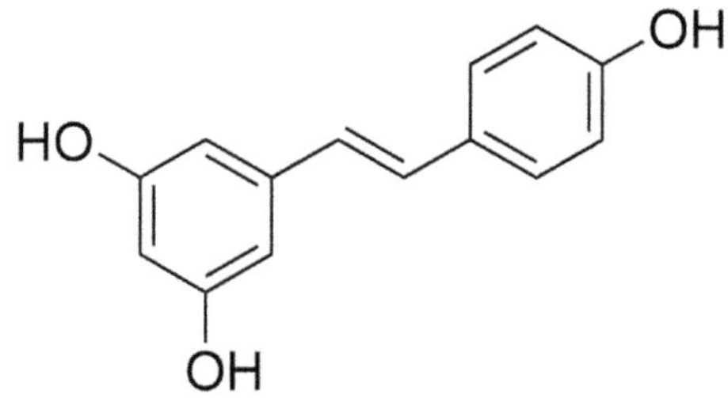

FIGURE 8.6 Chemical structure of resveratrol showing the three hydroxyl groups known to potentially eliminate ROS.

Resveratrol has three hydroxyl groups (Figure 8.6) known to eliminate ROS and act in the chelation of metal ions, being essential in preventing the production of ROS. Furthermore, resveratrol triggers cellular defense because, in addition to acting directly as an antioxidant, it also acts indirectly by inducing the endogenous antioxidant system (Salehi et al., 2018).

This phenolic acts on the transcription factor Nrf2, which regulates a range of antioxidant enzymes and also has roles in gene regulation effects, acting on the expression of NADPH oxidase, contributing to the inhibition of ROS (Smith et al., 2016; Xia et al., 2017).

Resveratrol is one of the most promising natural antiaging compounds. This can be attributed to its ability to activate sirtuin 1 and modulate regulatory proteins such as protein kinase B, peroxisome proliferator-activated receptor activator-1a, FoxO, and NF-kb, which are major players in the aging process (Camins et al., 2009; Wahl et al., 2018; Pannu and Bhatnagar, 2019).

However, the elaboration of products with this phenolic has some limitations, such as the water solubility of this phytochemical, which is very low, and also has low bioavailability (Chauhan, 2015; Davies et al., 2020).

Therefore, DS with resveratrol, have been developed and investigated so that they can minimize these problems and thus the active can be used more effectively. Some examples of resveratrol in DS are seen in literature such as Das et al. (2020), which incorporated resveratrol in microemulsions, thus obtaining better stability, and solubility of the active and also a formulation with better skin permeation. Another study with resveratrol together with lipoic acid was carried out with the incorporation of these actives in lipid core nanocapsules for cutaneous application. Nanoencapsulation increased the substance's stability in addition to controlling the release of actives. Phenolics were able to permeate in the skin matrix when nanoencapsulated, with faster permeation. The antioxidant activity was enhanced by the co-nanoencapsulation of resveratrol and lipoic acid, being promising for application in topical formulations aiming at antioxidant effects with the potential for antiaging product elaboration (Davies et al., 2020).

However, even with all this bioactive power that phenolics have for the production of antioxidant and antiaging products, the use of this class of natural substances is limited, due to their low stability, solubility, and bioavailability. It is necessary to include this class of actives in DS that are being increasingly investigated so that bioactives can be used effectively in cosmetic and pharmaceutical formulations.

8.2.5 Curcuminoids

Curcumin (1,7-bis(4-hydroxy-3-methoxyphenyl)-1,6-heptadiene 3,5-dione (Figure 8.7) is a hydrophobic polyphenol derived from the rhizome of the herb *Curcuma longa* (turmeric), Zingiberaceae family, that is widely used in traditional Indian and Chinese medicines. Previous research shows that curcumin has a wide range of beneficial biological and pharmacological activities including antioxidant, anti-inflammatory, antimicrobial, anticancer, and antiaging. Furthermore, the hepato- and nephroprotective, thrombosis suppressing, myocardial infarction protective, hypoglycemic, and antirheumatic effects of curcumin are also well established (Shah et al., 2016).

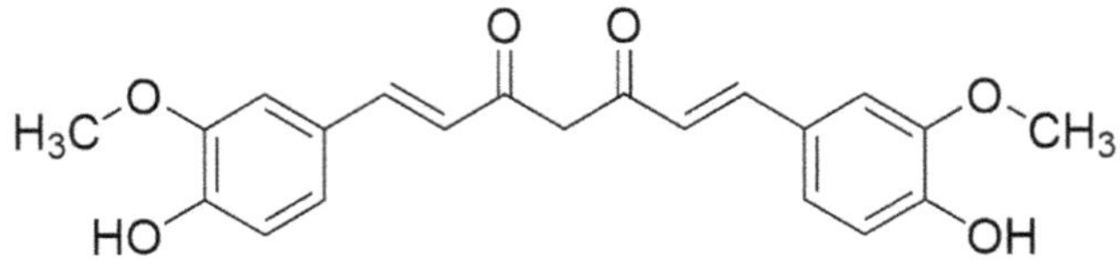

FIGURE 8.7 Chemical structure of curcumin [(E,E)-1,7-bis(4-hydroxy-3-methoxyphenyl)-1,6-heptadiene-3, 5-dione], a natural diketone compound extracted from the rhizomes of some plants in the *Zingiberaceae*. Curcumin contains a phenolic structure, which gives it unique antioxidant activity, making it a reducing antioxidant.

Although it is a promising compound, the clinical advancement of curcumin is hampered by its low water solubility and short biological half-life, resulting in low bioavailability in both plasma and tissues. The main reasons that are supposed to justify this low bioavailability are the poor solubility of curcumin in aqueous media, rapid hydrolysis followed by molecular fragmentation at physiological pH, and inactivity of its metabolic products. To solve the physicochemical instability of curcumin and improve its water solubility and bioavailability, a number of attempts have been made using methods like encapsulation of curcumin in surfactant micelles, phospholipids, cyclodextrin, hydrogel, liposomes, polymeric micelles, nanoparticles, among others (Shah et al., 2016; Liu et al., 2021).

In recent years, the investigations about curcumin have been focused on its application to aging and age-associated diseases, such as cancer, neuroinflammation, and infections due to low-grade chronic systemic inflammation. Curcumin, like other polyphenols, has pleiotropic activity. Among its mechanisms of action, stand out the interaction with many proteins, strong free radical scavenging activity, and the capacity to induce cellular response to external stimuli. Moreover, curcumin up- and downregulates different miRNAs and takes part in epigenetic changes in cells (Gupta et al., 2013).

With regard to studies demonstrating the successful application of curcumin into DS, it can be highlighted the one accomplished by Liu and Huang (2012) studied curcumin-encapsulated myristic acid microemulsion. The physicochemical properties of the curcumin microemulsion were explained, and the *in vitro* inhibitory effect on *Streptococcus epidermidis* was investigated. They proved that the curcumin-encapsulated myristic acid microemulsion is a good carrier to inhibit *Staphylococcus epidermidis*. Many studies have shown that the use of microemulsions to encapsulate curcumin can significantly increase its solubility and bioavailability. With this example, it can be seen that microemulsion systems not only can be used as a carrier for encapsulating curcumin to increase its solubility but also can improve the antioxidant activity of curcumin, which is of great significance for further expanding its application range (Acharya and Hartley, 2012).

Moghaddasi et al. (2018) formulated curcumin nanoemulsions to improve their oral bioavailability and solubility in water, using black pepper oil (BP oil) as the oil phase and Tween 80 as the surfactant to prepare curcumin nanoemulsions and used DPPH assay to study the antioxidant capacity of curcumin nanoemulsions. Their results showed that curcumin has good antioxidant activity in nanoemulsions, and its clearance rate is significantly higher than that of other solvents.

Such studies corroborate each other about the increase in solubility and bioavailability of curcuminoids, thus enhancing their antioxidant activity and other activities inherent to the latter.

Moreover, curcumin increased the ability of human epidermal keratinocytes to differentiate during replicative senescence, strengthening the idea of an antiaging mechanism. Therefore, it is possible that the positive effects of curcumin supplementation, observed on the organismal level, can be attributed to sirtuin and AMPK induction rather than the inhibition of cellular senescence. Although curcumin has been shown to exert antiaging effects in different experimental approaches, there are only a few data displaying its capacity to modulate cell senescence (Grabowska et al., 2015).

Other studies have shown that the antioxidant activity of curcumin incorporated into the silk fibroin film can be well retained. Curcumin was encapsulated in silk fibroin film (silk/cur film) and nanoparticles (silk/cur NPs), and its antiaging effect was compared to free curcumin in solution. The aim was to clarify the antiaging mechanism of silk-related curcumin and better serve future biomedical applications. The strong binding of curcumin molecules to the beta-sheet domains of silk fibroin resulted in the slow release of curcumin with better activity. For cell aging studies, rat bone marrow mesenchymal stem cells (rBMSCs) were cultured in the presence of free curcumin (FC), silk/cur film, and silk/cur NP, markers of cell proliferation and aging (P53 gene expression, P16, HSP70, and β-galactosidase activity).

The results indicated that cell aging was delayed in all FC, silk/Cur NP, and silk/cur film samples, with silk-associated curcumin activity superior to FC. The inclusion of silk curcumin nanoparticles helps clarify the antiaging mechanism of silk-related curcumin and provides data to support the future application of curcumin for oral administration (Yang et al., 2017).

The research made it possible to observe that curcumin activity depends on the concentration, formulation (pure natural or modified), cell type (differences in vulnerability), and context (organism, disease, manner of administration). Besides, data collected by curcumin researchers showed a large number of beneficial activities for this compound (Bielak-Zmijewska et al., 2019). They mostly concern its anticancer activity, but for several years, the amount of data showing curcumin's role in the modulation of aging has been intensively growing, and the issue has been widely discussed. More clinical trials are necessary to lead to the development of curcumin as a safe and effective chemopreventive agent.

8.3 DRUG DELIVERY SYSTEMS

In recent decades, there has been a growing interest in the development of effective DS in order to overcome the limitations of phenolic compounds and improve their incorporation into pharmaceutical products. The focus on the incorporation of DS by the cosmetic industry has been an innovation strategy, given the efficient performance in the creation of new active ingredients with improved efficacy through the use of these technologies, in addition to protection against active degradation, greater penetration into the skin, ensuring safety and greater effectiveness of the products developed.

Organic and inorganic, synthetic and natural, simple and complex, and nano- and micro-dimensional materials have been widely investigated as potential carriers for a wide range of plant actives, with different physicochemical, biological, and functional properties (Chu, Chew, Nyam, 2021; Kyriakoudi et al., 2021). The use of different DS to attribute better characteristics to phenolic compounds and antiaging applications was observed in the literature. Data on the most used technologies, as well as the potentials attributed to the incorporated assets, are shown in Table 8.1. For the plant extracts applied to carriers, the components presented were considered in the majority.

Considering the studies listed in Table 8.1, which involve the incorporation of phenolic compounds in DS and the antiaging potential with product prospecting for the market, the results were quite promising. The studies' authors used different assays to assess the antioxidant and/or antiaging capacity that will be discussed below.

In the study carried out by Karatoprak et al. (2020), an approach to the in vitro antioxidant activity of *Salvia aramiensis* extract (with rosmarinic acid as the major compound) was carried out, in addition to the study of the inhibition of two important enzymes in the context of aging, collagenase and elastase, which are activated by free radicals and cause the degradation of collagen and elastin, respectively. The authors demonstrated that the antioxidant potential presented by the extract in in vitro assays, the enzymatic inhibition, and its capacity for incorporation into nanocarriers, provide support for obtaining a pharmaceutical product.

It is highlighted that, in addition to the importance of studying the antioxidant activity of plant extracts, greater investment in research is needed for the application of these components in formulated systems. And, in order to obtain a pharmaceutical product in the end, it is necessary to carry out *in vivo* studies, which prove successful activity, opening space for this new product to become available and have certified safety.

The improvement of the isolated rosmarinic acid properties was also studied *in vitro* by Huerta-Madroñal et al. (2021) when formulating conjugates of this phenolic acid with chitosan.

TABLE 8.1
Phenolic Compounds Used in DS for Antiaging Application and Potential Attributes

Phenolic Class	Phenolic Name	DS	Antioxidant Assays	Antiaging Assays	Improved Characteristics	References
Phenolic acids	Rosmarinic acid	Nanoliposomes	DPPH scavenging activity, ABTS+scavenging activity (2,2-azino-bis acid (3-ethylbenzthiazoline-6-sulfonic acid)), Determination of inhibition of β-carotene/linoleic acid co-oxidation	*In vitro* assay: collagenase and elastase enzyme inhibition	From the *in vitro* release experiment, there was a release of 60%±0.85% of extract. The authors indicated that this amount released is above the EC50 value for scavenging the DPPH radical and that it would be able to inhibit collagenase and elastase by more than 66.64% and 76.68%, respectively	Karatoprak et al. (2020)
Flavonoids are the most presente substances	Not reported	Solid lipid nanoparticles (SLN)	*In vitro* tests: DPPH•, ABTS•+ and β-carotene	*In vitro* assays: estimation of antielastase, anticollagenase, and antityrosinase activities *In vivo* assays: antiwrinkle study in a mouse model with UV-induced photoaging	The nanoformulation containing extract of *Prunus persica* (L.) in SLN showed greater penetration power, reaching the dermal layer of the skin, where there was release of antioxidant constituents	Mostafa et al. (2021)
Phenolic acids	Gallic acid	Bacterial Cellulose Membrane (BC)	In vitro assays: DPPH• and ABTS•+	–	The membranes were used as carriers for the extract of *Epilobium angustifolium*, increasing the antioxidant potential exhibited by the extract and the content of phenolic acids. Furthermore, it demonstrated release and absorption of the actives through the skin model used	Nowak et al. (2021)
Phenolic acids	Chlorogenic acid	Liposomes	*In vitro* assays: DPPH• and evaluation of the protection of fibroblasts with induced cytotoxicity	–	Liposomes containing *Lycium barbarum* provided a slower release of polyphenols when compared to the free extract	Păvăloiu et al. (2021)

(Continued)

TABLE 8.1 (*Continued*)
Phenolic Compounds Used in DS for Antiaging Application and Potential Attributes

Phenolic Class	Phenolic Name	DS	Antioxidant Assays	Antiaging Assays	Improved Characteristics	References
Phenolic acids	Gallic acid	Niosomes	–	*In vitro* tests: (1) Melanogenesis assay, measuring the melanin content, measurement of tyrosinase activity, and tyrosinase-related protein (TRP-2) (2) Assay of antioxidant activity through the evaluation of cell viability (3) Evaluation of metalloproteinase inhibition (MMP-2)	The study demonstrated superior stability and skin antiaging activity of niosomes loaded with gallic acid compared to free gallic acid	Chaikul et al. (2019)
Flavonoids, Tannins and Phenols	Not reported	Water in oil emulsion (W/O)	*In vitro* DPPH• assay	–	The emulsion showed stability, antioxidant activity, and sun protection with 5% of *Beta vulgaris* leaf extract	Huma et al. (2020)
Phenolic acids	Cinnamic acid	Sodium alginate microparticles	*In vitro* DPPH• assay and reducing power	*In vitro* assays: Antityrosinase activity	The incorporation of microencapsulated extracts of *Agaricus bisporus* and *Pleurotus ostreatus* in creams did not increase the antioxidant capacity displayed by the extracts in free form, demonstrating that the release profile of the system was not sufficient over time to exert the effect	Taofiq et al. (2019)

(*Continued*)

TABLE 8.1 (*Continued*)
Phenolic Compounds Used in DS for Antiaging Application and Potential Attributes

Phenolic Class	Phenolic Name	DS	Antioxidant Assays	Antiaging Assays	Improved Characteristics	References
Tannins, flavonoids and phenolic acids	18 types of phenolic substances	Gold nanoparticles	*In vitro* DPPH• assay	*In vitro* anticollagenase activity assay, *ex vivo* evaluation of antioxidant activity in human skin explants, phototoxicity, Keratinosens test, SENS-IS test, *in vitro* skin irritation test with reconstructed human epidermis model, *in vitro* with epithelial model cornea for eye irritation test	The gold nanoparticles obtained using plant extract had a significantly higher antioxidant activity than vitamin E. The study also observed a significant dose-dependent decrease in collagenase production when the cells were treated with the gold nanoparticles containing extract. In the explants there was no change in cell viability, but there was a reduction in the formation of oxidized proteins and with the safety tests it was observed that they are non-toxic, non-sensitizing, non-phototoxic and non-irritating	Haddada et al. (2020)
Tannins, flavonoids and phenols	Not reported	Silver nanoparticles	DPPH• radical test, hydrogen peroxide scavenging test, nitric oxide radical scavenging activity and reducing power test	–	All tests showed and confirmed the antioxidant potential of silver nanoparticles containing extract from the pomegranate fruit pericarp	Govindappa et al. (2021)
Phenolic acids	Gallic acid	Niosomes	DPPH• test for free radical scavenging activity of gallic acid and the semi-purified fraction loaded in elastic and in elastic niosomes	Clinical antiaging skin test in human volunteers measuring skin elasticity and roughness	The DS tested was effective in terms of antioxidant activity, showed no significant irritation on rabbit skin and human skin, improved elasticity, and decreased skin roughness	Manosroi et al. (2011)
Flavonoids	Quercetin	Nanostructured lipid carriers (NLC)	The antioxidant activity of free and encapsulated quercetin was evaluated *in vitro* by analyzing the auto-oxidation of pyrogallol (1,2,3-benzenetriol)	Skin permeation study with swine ear skin and modified Franz diffusion cells were used (Hanson Corporation).	Encapsulated quercetin showed improved antioxidant activity compared to free. The greatest penetration of quercetin from the dermis and epidermis occurred in the encapsulated model	Pivetta et al. (2019)

(*Continued*)

TABLE 8.1 (*Continued*)
Phenolic Compounds Used in DS for Antiaging Application and Potential Attributes

Phenolic Class	Phenolic Name	DS	Antioxidant Assays	Antiaging Assays	Improved Characteristics	References
Phenolic acids	Gallic, caffeic and ellagic acid	Nanostructured cellulose membranes	DPPH• radical scavenging assay and *in vitro* assay to combat the oxidative stress induced by Lipopolysaccharide in macrophages	*In vitro* skin permeation assay using human abdominal skin samples were obtained from women undergoing abdominoplasty at Centro Hospitalar São João, Portugal, using Franz cells	The tests showed excellent antioxidant activity and a skin permeation that ensured that with DS the phenolics were released more slowly and sustainably	Morais et al. (2019)
Stilbene and Phenolic acids	Resvertrol and lipoic acid	Lipid core nanocapsules	Lipid peroxidation test	Skin permeation done in Franz diffusion cells using snake skin	The actives were able to permeate the skin when nanoencapsulated, with a faster permeation. Antioxidant activity was enhanced by co-encapsulation of resveratrol and lipoic acid	Davies et al. (2020)
Stilbene	Resveratrol	Nanostructuredlipid carrier gel loaded (NLC-Gel)	Anti-UV activity *in vitro*	Anti-UV irradiation activity and *in vivo* antioxidant activity in rat skin	The technology exhibited anti-UV irradiation and antioxidant activity *in vivo*. Resveratrol in NLC-Gel exhibited great potential for application in protecting the skin against UV radiation	Liu et al. (2021)
Stilbene	Resveratrol	Microemulsion	*In vitro* skin permeation study using Strat-M membrane	–	Permeation of the *in vitro* formulation in a Strat-M synthetic membrane indicated that microemulsion gel can aid in the penetration of resveratrol into the deeper layers of the skin	Das et al. (2020)
Curcuminoids	Curcumina	Silk fibroin films (silk/cur films) and nanoparticles (silk/cur NPs)	–	Rat bone marrow mesenchymal stem cells (rBMSCs) culture and cell proliferation, and markers of aging (P53, P16, HSP70 gene expression, and β-Galactosidase activity) were examined	The results indicated that cell aging was retarded in all FC, silk/cur NP, and silk/cur film samples, with the silk-associated curcumin superior to the FC	Yang et al., (2017)

It was observed that the resulting conjugates were water-soluble and demonstrated a three-fold increase in free radical scavenging activity, in addition to anti-inflammatory activity and reduction of UVB-induced damage and ROS production in fibroblasts, keratinocytes, and macrophages.

During the research, it was possible to perceive and confirm the potential of incorporating phenolic compounds in DS to increase the antiaging activity, showing itself to be expressive, especially for plant extracts, as seen in the study by Mostafa et al. (2021), in which the formulated system also exhibited greater penetration power of the active into the skin layers. Davinelli et al. (2018) emphasize that free phenolic compounds may even permeate through the skin layers, but it is known that this phenomenon is not optimized. Additionally, the enhancement of antioxidant and antiaging attributes in polyphenolic extracts linked with these methodologies can be substantiated through various evaluations, forecasting market potential grounded on *in vitro* assessments, predominantly employed for antioxidant appraisal. Subsequently, enzymatic assays imply endorsement for antiaging efficacy, alongside *in vivo* examinations, and, to a lesser extent, clinical trials. Enzymatic tests to assess antiaging activity generally assess the inhibition of enzymes such as elastase, responsible for the degradation of elastic fibers, inducing the aging process through ultraviolet radiation, with the consequent formation of wrinkles and reduced skin elasticity. And polyphenols are effective in slowing down these processes.

In the context of extract trapping, some recent articles have demonstrated bacterial cellulose as an important carrier agent, including the study by Nowak et al. (2021) that bring contributions to the increase in the antioxidant capacity of the *E. angustifolium* extract loaded in this medium, as well as positive release and absorption, which were studied using *in vitro* methods.

Accordingly, Perugini et al. (2020) reinforce the promising idea of the topical application of these antioxidant DS on the skin, bringing satisfactory results from clinical tests. Considering the exponential market for skincare products, this study encourages the development of innovation in products such as facial masks, based on the use of an alternative natural polymer with water retention capacity, tolerability, and biocompatibility. In addition, it demonstrates, in an elementary way, the versatility of incorporating different extracts and actives, allowing the customization of products, which can be an interesting point for the industry and even for compounding pharmacies.

However, despite the importance of the study carried out by Perugini et al. (2020) as a clinical trial, in the context of this study, the formulated masks present an extract composition that acts synergistically to produce the evaluated effects. In addition, other active ingredients, such as hyaluronic acid and peptides, are incorporated, so that it is not possible to assign which phenolic component is responsible for the greater expressiveness of the antioxidant and antiaging activity of the formulation.

Emulsified systems, especially micro and nanoemulsions, have advantages that attract attention, especially for their superior interface and availability. In addition, the droplet size of the emulsified system is an important factor to be considered for best results. As these dimensions are reduced, it is possible to achieve better solubility and bioavailability, which makes microemulsions efficient alternatives for encapsulating phenolic compounds as they offer better conditions than classic emulsions (Garavand et al., 2021). Garcia et al. (2021) discuss the advantages of using microemulsion to incorporate extracts, as it is a non-greasy system, which can be pleasant to the consumer's sensory perception, in addition to the intention of enhancing the extract penetration into the skin. However, there is no study of permeation and antioxidant activity after the incorporation of *Theobroma cacao* L. extract in the microemulsified system. The authors also emphasize the need to carry out clinical trials to validate this type of topical formulation and ensure its effectiveness and safety.

Considering the antioxidant activity tests, especially antiaging, some studies present a set of *in vitro* assays for evaluating and attributing this potential to phenolic substances disposed of in DS. Chaikul et al. (2019) include as tests for evaluating this parameter the suppression of melanin formation by tyrosinase and TRP-2 inhibitory activities, antioxidant activity by evaluating cell viability, and MMP-2 inhibitory activities. As discussed in the study, the increase in MMP-2-type enzymes contributes to the degradation of the matrix composed of proteins that are significant in the aging process, such as collagen and elastin. In addition, pigmentation is a process present in aging skin, which explains the performance of tests related to melanin. Thus, once the positive results for the tests performed are demonstrated, the authors conclude the antiaging potential of niosomes loaded with gallic acid, denoting a promising agent for incorporation in pharmaceutical and cosmetic products.

One of the most used encapsulation strategies for phenolic compounds, being demonstrated expressively, were the nanotechnology systems (Davies et al., 2020; Govindappa et al., 2021; Haddada et al., 2020; Karatoprak et al., 2020; Liu et al., 2021; Mostafa et al., 2021; Pivetta et al., 2019). According to Karatoprak et al. (2020), drug delivery nanosystems are means that enable the efficient transport of active substances with low stability to the lower layers of the skin, with exemption from the need for large amounts of active and enabling a longer or longer-lasting action. Inserted in this context, lipid nanoparticles have a differential prominence in the cosmetic area, providing varied applications. Given the similarity between the lipid matrix that makes up the material and the lipid composition of the skin, it is possible to achieve improvements in some aspects of the skin surface, such as the hydration factor, which results in improved elasticity.

Complementarily, the possibility of increasing the stability of active components defines why these systems are so promising for topical application (Borges et al., 2020). Furthermore, solid nanoparticles (SLNs) showed potential for the application of natural antioxidant extracts, functioning as effective carriers for cutaneous delivery, with high stability and without causing irritating effects on the skin, thus improving the extract's performance as a cosmetic ingredient. Its biocompatible properties, internalization capacity, stability, and deposition in target organs open doors for the development of new drugs based on natural substances (Castellani et al., 2018). As reviewed by Borges et al. (2020), among the phenolic acids, ferulic acid was a widely reported active, proving to be a natural compound of interest for encapsulation in SLNs.

Some research carried out with nanocarriers has even evaluated the permeability of actives incorporated in DS through the skin layers, especially the epidermis and dermis, among which the one by Rapalli et al. (2021) was successful. The authors investigated this parameter for carriers of nanostructured lipids (NLCs) loaded with Apremilast, incorporated into a gel for topical application. The authors observed that, in addition to the systems showing good absorption by cells *in vitro*, there was an improvement in permeation, given the display of smaller particle size, and high retention in the skin, also demonstrating greater efficacy for the treatment of a dermatological condition.

Figueroa-Robles, Antunes-Ricardo, and Guajardo-Flores (2021) also contextualize the increase in the permeability of phenolic compounds that can be favored by liposome-type systems, which are shown to be a great advantage over other cosmetic products that cannot be deeply absorbed. Chaturvedi and Garg (2021) add that the stratum corneum of the skin acts as a barrier for the absorption of drug molecules and that this permeation depends both on the physiological characteristics of the skin, as well as on the physicochemical properties of the permeating agent. The physicochemical properties of carriers, such as size, surface shape, porosity, and elasticity, as well as their interactions with biological barriers, can determine the therapeutic success of the active principles that are carried out (Zhao et al., 2020).

Considering the particle size, different release profiles in the skin layers can be obtained. In Figure 8.8 it is possible to observe that the active not incorporated in DS has a larger dimension, which, in turn, makes it difficult to permeate through the skin layers, being more retained in the first layer of the epidermis (a). On the other hand, when this active is incorporated in a DS, such as nanoparticles, a better distribution is obtained through the layers, reaching up to the dermis (b), which can favor the results depending on the intended objective.

Furthermore, Flieger et al. (2021) highlight the promising trend of developing nanoparticles functionalized with antioxidants, which offer potential benefits such as lower environmental impacts and reduced costs. These systems can be synthesized through green methods that avoid the use of high pressure, energy, temperature, or toxic reagents. This approach ensures the formation of nanoparticles free from chemical contaminants, surpassing traditional methods and offering

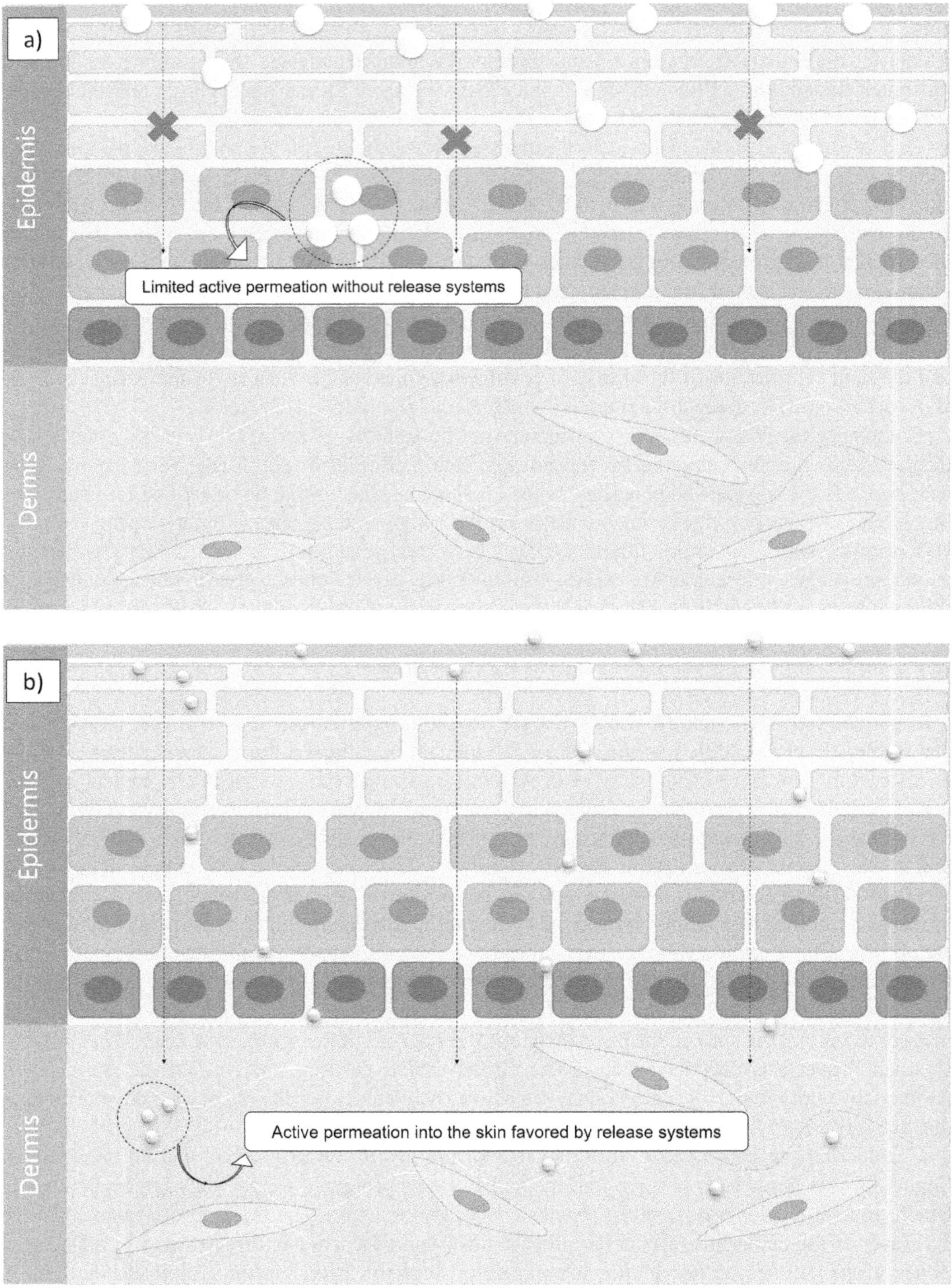

FIGURE 8.8 Topical application of the asset without a release system (a) and incorporated in a release system (b). Based on Rapalli et al. (2021) and Chaturvedi and Garg (2021).

potential cost-saving opportunities for industry production. In the context of more ecological DS, biodegradable bacterial cellulose polymers have also been shown to be promising agents for the release of actives into the skin, potentially being applied in dermatology and cosmetology (Haddada et al., 2020; Morais et al., 2019; Nowak et al., 2021).

8.4 ANTIAGING ACTIVITY

The aging process is widespread due to its association with increased oxidative stress and is also related to the process of systemic inflammation (Vaiserman et al., 2020). ROS are expressively formed in the body. In particular, large amounts of hydroxyl radicals, superoxide, peroxyl, singlet oxygen, and hydrogen peroxide are formed in the skin. Much has been studied about the influence that free radicals exert on the aging process, which can be a result of the products of oxidative reaction accumulation. In this process, many enzymes are activated when the skin is exposed to different factors, with UV radiation being a significant condition that indirectly leads to the production of ROS that generate oxidative stress. The body itself uses strategies to fight free radicals, either through enzymatic or non-enzymatic mechanisms of antioxidant activity. The superoxide dismutase, catalase, and glutathione peroxidase systems stand out as enzymatic mechanisms, with the addition of non-enzymatic ones, which can be endogenous components or obtained exogenously (Silva, Michniak-Kohn, Leonardi, 2017).

The use of phenolics acting on the enzymatic systems that regulate aging is very widespread. In the study by Xiao et al. (2018), procyanidin B2 was administered to mice of mimetic age induced by injection of d-galactose (500 mg/kg, sc daily). Very relevant results were obtained for antiaging activity, characterized by the improvement of impaired memory and, in particular, increased activities of the enzymes superoxide dismutase, glutathione peroxidase, and TAC in the tissues of mice, after 7 weeks.

As for the triggering factors for ROS production, considering that UV radiation is a factor closely linked to the skin aging process, the development of efficient photoprotective formulations is extremely necessary. In this context, plant extracts composed of phenolics have been seen as adjuvant ingredients to increase the sun protection factor (SPF) and reduce the need for the use of chemical filters in this type of formulation. With this focus, Mota et al. (2019) investigated the effect of the association of the extract of guava fruit (*Psidium guajava* L.) with the sunscreen 2-ethylhexyl methoxycinnamate (EHM), by carrying out SPF tests of creams with the incorporation of different percentages of extract. The results were positive, showing a notable enhancement in SPF compared to the formulation containing only the synthetic filter. This indicates a synergistic effect of the extract components, particularly flavonoids and tannins, with the EHM. Consequently, there was a reduction in the concentration of chemical filters, leading to decreased toxicity and production cost of the final product.

Many other factors contribute to the progress of skin aging, creating an imbalance between free radicals and the body's antioxidants. Boosting antioxidant levels represents a feasible strategy to mitigate the adverse effects of oxidative processes. This approach operates through various mechanisms, including shielding cells from oxidative stress, triggering genes responsible for collagen synthesis, and inhibiting metalloproteinases (MMPs) involved in the aging process. Additionally, it may elevate the concentration of human Telomerase Reverse Transcriptase (TERT), an enzymatic complex crucial for maintaining telomere length. This shortening can be accelerated due to the generation of ROS that damage DNA. Such effects were generated by niosomes loaded with punicalagin extracted from pomegranate (Mohamad et al., 2021).

Among the phenolic compounds, many flavonoids have shown antiaging properties and also depigmentation effects, such as phloretin. It is denoted that much of this ability is due to its antioxidant competence. Some of these compounds are available on the market, arranged in formulations generally classified as cosmetics or cosmeceuticals (Samanta et al., 2016; Singla et al., 2019).

Floretin is found mainly in fruits and is evaluated in dermatological treatments such as melasma, photoaging, acne, and melanoma (Wang et al., 2019). The mechanism of action is by inhibiting the activity of elastase and matrix metalloproteinase-1, reducing cell tyrosinase activity, and inducing apoptosis in melanoma cells (Casarini et al., 2020; Cherubim et al., 2019). Its action extends from the epidermis to the dermis, with anti-inflammatory effects evaluated in human keratinocytes (HaCaT) stimulated by tumor necrosis factor-alpha. Thus, in addition to preventing skin diseases, this action can also reduce the effects of aging (Casarini et al., 2020).

Likewise, ellagic acid, a phenolic acid, also showed antimelanogenesis activity by different mechanisms, observed in *in vitro* and *in vivo* assays with a zebrafish model, which evidenced its ability to act as a bleaching agent and a potential candidate for cosmetic application (Yang et al., 2021).

Among the studies found with phenolic compounds incorporated in DS and which demonstrated antiaging potential, nanoliposomes incorporated with *S. aramiensis* extract (Karatoprak et al., 2020), gallic acid niosomes (Chaikul et al., 2019) can be highlighted, nanoparticles with extract of *Hubertia ambavilla*, all with demonstration of the enzymatic inhibition potential. In addition, solid lipid nanoparticles incorporated with an extract of *P. persica* (L.) demonstrated more than the enzymatic inhibition potential, acting as a wrinkle-reducing system in mouse models (Mostafa et al., 2021). Finally, niosomes carrying a fraction of *T. chebula* extract containing gallic acid demonstrated improved skin elasticity and reduced roughness in clinical trials (Manosroi et al., 2011), confirming the market potential of these improved active ingredients.

Additionally, the promising potential of obtaining inclusion complexes with oils rich in phenolic compounds has also been demonstrated. *Euterpe oleracea* Martius oil, popularly known as "açaí", is rich in flavonoids and phenolic acids (Marques et al., 2016). In a study developed by Magalhães et al. (2021), the significant increase in the antioxidant activity of this oil complexed with β-cyclodextrin and hydroxypropyl-β-cyclodextrin is pointed out. The comparison of antioxidant test results between the pure oil and the complexes revealed significant improvements in the capacity to eliminate free radicals, total antioxidant capacity (TAC), and reducing power (RP). This indicates a promising study for formulating new ingredients with higher antioxidant activity, potentially applicable in the cosmetic industry.

Furthermore, Magalhães et al. (2020) present in their review the growing interest in *E. oleracea* Martius, however, they describe the lack of clinical studies, reporting the need for more efforts to clarify the efficacy and safety of crude extracts or isolated compounds to obtain pharmaceutical products.

Garcia et al. (2021) reinforce the need to carry out clinical trials to better prove the efficacy and safety of antiaging formulations, however, despite the significance of these very promising systems, as well as the importance and assurance generated by the results of clinical trials, the studies showed themselves quite meager.

According to Costa and Santos (2017), the innovation of cosmetic products in the market, with the incorporation of DS in the formulations, presents some critical points. As explored by the authors, the development of products with these carriers requires a lot of experimentation, as they are not compatible with all dosage forms. Furthermore, after incorporation, some parameters such as product uniformity need to be monitored and other aspects such as sensory and long-term stability need to be guaranteed. Despite this, as reviewed by the authors, several DS with antiaging purposes can already be found in different products on the market, namely: nanosomes, nanocapsules, liposomes, and nanoparticles, incorporating actives such as coenzyme Q10, vitamin C, and retinol. However, no phenolic compound was cited by the authors among the incorporated assets.

Additionally, Figueroa-Robles, Antunes-Ricardo, and Guajardo-Flores (2021) describe the limitation regarding the variety of products using specifically liposomes as carriers of phenolic compounds, opening perspectives for innovation in the cosmetics market. Despite this, they present five commercial products, from different brands, formulated with liposomes loaded with different phenolic compounds, such as ferulic acid, resveratrol, quercetin, and, being present in three of the five products presented, anthocyanins (class of flavonoids), extracted from different fonts.

Given the above, it was possible to observe that despite the majority of studies with phenolic compounds, incorporated or not in DS, showing potential application in cosmetic products, a minimum number of clinical trials are carried out. The need for these studies is also reported by the authors themselves, to guarantee the efficacy and safety of these products. Despite this, given the importance of proving the antioxidant and antiaging activity, different tests are widely explored to characterize the potential provided by these components, to demonstrate the effective delivery and benefits to the skin (Silva, Michniak-Kohn, Leonardi, 2017). Thus, some of the most discussed tests currently are described in the next topic.

8.5 ANTIOXIDANT TESTS

Distinctions regarding antioxidant activity and antioxidant capacity are described. While "antioxidant activity" refers to assays based on measuring the rate constant of a reaction between an antioxidant and a specific oxidant, "antioxidant capacity" is given by the efficiency of inhibiting the oxidative degradation of bio-compounds by the antioxidants. These measurements are based on the reaction between the antioxidants to be studied and free radicals (observing whether there is inactivation, extinction, or elimination of reactive species) or on the reaction of the sample with transition metals (Bunaciu et al., 2016; Flieger et al., 2021).

Differences between *in vitro* and *in vivo* test results for the evaluation of phenolic compounds are presented by Dong et al. (2021) and Fierascu et al. (2018). According to the authors, different types of assays can result in divergent data on antioxidant capacity, since for this determination by *in vitro* tests, for example, aqueous or alcoholic solutions are used, differently from what occurs in *in vivo* assays, which depending on the test performed, if you have a complex biological fluid medium, consisting of an emulsion of lipids, proteins, water, minerals, and other substances. *In vitro* assays represent an important tool in the primary evaluation of the studied compounds, however, it is emphasized that more research must be carried out with a focus on cell lines or in vivo assays.

Thus, it is possible to infer the idea that the tests carried out together add complementarity to the results obtained, attributing greater confidence when compared to tests carried out in isolation. Considering, then, the granting of antiaging activity to phenolic compounds or extracts contained with these substances, especially those incorporated in DS, we sought to evaluate the antioxidant assays, as well as those that direct the antiaging potential that has been more currently employed, described as follows.

8.5.1 *In Vitro* Assays

The *in vitro* antioxidant activity evaluation methods are widely used for several compounds. Alam, Bristi, and Rafiquzzaman (2013) described 19 types of tests used for this purpose, among which DPPH stands out as the most used method of scavenging free radicals. Other tests cited include: hydrogen peroxide (H_2O_2) scavenging tests, nitric oxide scavenging activity, peroxynitrite radical scavenging, Trolox equivalent antioxidant capacity (TEAC) method/ABTS+radical cation decolorization test, method of the total radical scavenging antioxidant parameter (TRAP), FRAP, superoxide radical scavenging (SOD), hydroxyl radical scavenging, hydroxyl radical scavenging capability (HORAC), oxygen radical scavenging capability (ORAC), potency reduction (RP), phosphomolybdenum method, ferric thiocyanate method (FTC), thiobarbituric method (TBA), DMPD method (*N,N*-dimethyl-*p*-phenylene diamine dihydrochloride), β-carotene linoleic acid method, xanthine oxidase, the antioxidant capacity of cupric ion reduction (CUPRAC) and metal chelating activity.

The *in vitro* antioxidant activity tests performed for phenolic compounds and the analysis, mainly, of plant extracts observed in the articles, corroborate the findings of Alam, Bristi, and

Rafiquzzaman (2013), revealing that the DPPH elimination test is still the *in vitro* method most used in current studies, followed by the ABTS•+ method, in a much smaller proportion.

Other authors confirm the frequent analysis of the antioxidant capacity of phenolic compounds, such as flavonoids, based on their DPPH scavenging activity. Studies demonstrate that the ability to scavenge free radicals, such as DPPH•, hydroxyl, SOD, and others, has supported the conception of the antioxidant action of these components (Zhao et al., 2020; Casarini et al., 2020). In addition, the DPPH• method is described as simple and easy, in addition to being fast and economical (Oliveira, 2015), which may explain the frequency with which it is used, as well as its predominance in the analyzed studies.

Another notable observation is that the majority of studies employing the DPPH antioxidant assay have supplemented it with other in vitro methods to validate antioxidant activity. These include ABTS•+ assay, inhibition of β-carotene/linoleic acid co-oxidation, hydrogen peroxide elimination, nitric oxide radical scavenging, reducing power determination, and assessment of oxidative stress induced by lipopolysaccharide in macrophages (Govindappa et al., 2021; Karatoprak et al., 2020; Morais et al., 2019; Mostafa et al., 2021; Nowak et al., 2021).

Only Huma et al. (2020) used the DPPH method in isolation to assess the antioxidant activity. In this study, the authors suggest an emulsion formulation with *Beta vulgaris* extract with antioxidant properties, in addition to protecting against UV radiation, however, they point out the need for *in vivo* tests to prove its use on the skin. According to Baschieri and Amorati (2021), assays such as DPPH• and ABTS•+ are simplified methods, which have many limitations but are widely used to assess antioxidant activity.

Thus, although some articles have presented different SD as a polyphenol delivery strategy, many are limiting in terms of cosmeceutical application. The study by Păvăloiu et al. (2021), presents liposomes as an excellent strategy to obtain a better action of the extract of *L. barbarum*, however, the authors use the DPPH assay as an antioxidant assay, in addition to demonstrating the protective effect of fibroblasts with cytotoxicity induced by H_2O_2, which, given what has been exposed, does not provide the necessary robustness for attributing the concrete antiaging activity of the formulated system, as presented by other authors, such as Damasceno et al. (2020), who carry out a series of tests, including *in vitro* methods and clinical studies, to guarantee this proof.

For better composition of the antiaging activity, it was found that most of the studies proposed the associated evaluation of enzyme inhibition, skin permeation with Franz diffusion cells, evaluation of cell viability, in vivo antioxidant activity, antiradiation activity UV, antiaging activity in vivo and/or clinical test in humans of DS developed with plant extracts or isolated phenolic compounds.

Another type of *in vitro* test is exposed by Haddada et al. (2020) who used cultivated human skin to form explants and evaluate the antioxidant activity when subjected to UV radiation. It was seen that the topical application of gold nanoparticles with *H. ambavilla* extract did not change cell viability compared to control explants, but the formation of oxidized proteins decreased. It was also observed that in the absence of UV-A irradiation, there was a decrease in the rate of endogenous oxidized proteins in the dermis and that this decrease also occurs with radiation exposure, which characterizes the antioxidant protection of the dermis by nanoparticles.

8.5.1.1 Enzymatic Assays

The determination of antioxidant activity *in vitro* using enzymes related to skin aging has also been a widely considered method for researching antiaging activity. Given the potential for degradation of the skin matrix, the inhibition of enzymes such as elastase, collagenase, and hyaluronidase, has been investigated for different inputs, considering extracts, isolated phenolics, as well as the DS generated from the incorporation of these actives (Bazzicalupo et al., 2021; Yepes et al., 2021).

A study developed by Karatoprak et al. (2020), evaluated the inhibitory effect of 70% methanolic extract of *S. aramiensis* on the collagenase and elastase enzymes, which are induced by the formation of free radicals and are responsible for the degradation of collagen and elastin. The results showed that collagenase was inhibited at different concentrations of the *S. aramiensis* extract, but there was no significant difference in the concentrations of 100 and 200 µg/mL when compared to the standard rosmarinic acid, with percentage inhibition between 66.64% and 72.66%. On the other hand, the anti-elastase activity of the extract at different concentrations was statistically significant when compared to the standard, with the concentration of 200 µg/mL of the extract being more active than rosmarinic acid, with an inhibition percentage of 86.9%. In this same study, it was seen that rosmarinic acid, a phenolic acid, was the main compound found, with a concentration of 122.32±0.21 mg/g.

Yepes et al. (2021) sought to observe the antiaging potential of purple passion fruit (*Passiflora edulis* f. Edulis Sims) seed extract, which is characterized by the presence of flavonoids and phenolic acids. The enzymatic assay performed showed inhibition for collagenase, elastase, and tyrosinase enzymes, with optimal IC50 values (inhibitory capacity). Complementarily, docking studies proved this evident binding capacity of extract constituents to collagenase and tyrosinase. Molecular dynamics with MM/PBSA calculations (Poisson - Boltzmann surface area molecular mechanics method) provide strong evidence for its specific strong binding and the stabilization of the enzyme-ligand complex binding. Therefore, by evaluating different perspectives, *in vitro* and *in silico*, the authors provide support for the incorporation of *P. edulis* extract, as a promising antiaging cosmetic ingredient, by proving its enzymatic specificity.

The study by Pinto et al. (2021) also sought to establish the cosmetic potential of extracts obtained from the nut husk, which is even considered a by-product. For the enzymatic test, elastase inhibition results were obtained from 33.56% to 82.70%, depending on the extractive method used, at a concentration of 0.4 mg/mL. These values were higher than the positive controls used. This study also evaluated the inhibition of hyaluronidase, for which the IC50 determined for the extracts varied from 0.76 mg/mL to 54.36 mg/mL, depending on the extraction method, proving *in vitro* the potential that these extracts possess antiaging and antioxidant activity.

The medicinal plant *Salvia officinalis*, a species rich in phenolic compounds and other secondary metabolites, popular in Asia and some parts of Latin America, was also analyzed with the intention of cosmetic application. After successive extraction processes, the extract was used for antioxidant, antiaging, and antiwrinkle tests. The antiaging potential of the extract was studied through several inhibitory effects estimated enzymatically, namely the inhibitory effects of collagenase enzyme, elastase enzyme, and hyaluronidase enzyme on prematurely aging human skin fibroblasts. An ultraviolet-induced photoaging model was used to evaluate the antiwrinkle potential of the *Salvia miltiorrhiza* plant. And the antioxidant potential through the DPPH assay (Khare et al., 2021).

Raspberries, *Rubus idaeus*, rich in phenolic compounds, are known for their varied therapeutic activities, with an emphasis on their antioxidant action on human cells. The phenolics present in the plant are oil soluble and are also primary components of the skin's hydrolipid film barrier. Aiming at cosmetic applications, some researchers have already developed liquid suspension cultures of *R. idaeus* leaves, to obtain an active ingredient to improve the skin's hydration capacity. The plant extract was tested on cultures of keratinocytes and fibroblasts and later on the skin of volunteer women, for analysis of its application in cosmetics (Tito et al., 2015).

One work carried out tests on skin cell cultures, the extract induced the genes, aquaporin 3, filaggrin, involucrin, and hyaluronic acid synthase, responsible for skin hydration. It was also efficient in stimulating the expression of the glucocerebrosidase enzyme, involved in the production of ceramide. The increased synthesis of extracellular matrix components in cultured fibroblasts also demonstrated the skin's ability to moisturize, thus revealing its aptitude as a cosmetic product with efficacy as a moisturizing and antiaging ingredient (Tito et al., 2015).

8.5.2 Skin Permeation Tests

8.5.2.1 Skin Permeation Studies Using Biological Skin Matrices

The beneficial effects of antioxidants as functional ingredients are of great importance to the pharmaceutical and cosmetic industry. To carry out their biological activities, topically applied substances need to be released from the formulation, to reach the skin and overcome the stratum corneum barrier. The release of active substances and further penetration into the skin depend on the properties of the molecules, such as molecular weight and lipophilicity, but also on the vehicle of the formulation (Zillich et al., 2015; Davies et al., 2020).

Biological matrices such as skin and synthetic membranes are used in a Franz diffusion cell apparatus to investigate the release, permeation, and retention of active substances such as, for example, in skin layers and thus be potential for the elaboration of cosmetic products such as antiaging. Permeations of substances in various skin matrices with an antiaging focus are investigated (Alonso et al., 2017; Davies et al., 2020).

A study developed by Nowak et al. (2021) evaluated the viability of using bacterial cellulose membranes as a matrix for trapping the extract of *E. angustifolium* L. (for which gallic acid was identified in greater quantity), through the penetration of bioactive compounds, using abdominal skin porcine, given its permeability similar to human skin. After application to pig skin, it was found that many bioactive compounds released from the membranes were absorbed through the skin.

Morais et al. (2019) performed studies with Franz cells (Hanson vertical diffusion cells) using abdominal skin samples to observe the skin permeation of nanostructured cellulose membranes with phenolic-based ionic liquids for cutaneous application, to observe this factor as important for this system application in products with diverse functionalities, among them the antiaging activity. It was concluded with the test that the flow and the permeated amount showed that, with the use of this system, a slower and more sustained release of the phenolic compounds was achieved.

Pivetta et al. (2019) also carried out studies with Franz cells, using porcine ear skin as a model to assess the permeation of drugs in the skin. In this particular study, quercetin - a flavonoid known for its broad antioxidant activity - encapsulated or not in NLC was tested to evaluate the difference in permeability obtained when incorporation into DS was carried out, given the difficulty that quercetin has to maintain dermis and epidermis for greater topical action. This study resulted in greater penetration of quercetin from the dermis and epidermis, which occurred in the encapsulated model, showing that encapsulation in lipid nanoparticles enhances the penetration of the active into the skin. It is noteworthy that the study is related to the antiaging function, as it seeks with the tests to formulate a semi-solid pharmaceutical form for topical use and, in addition to the presented test, it carried out other experiments such as *in vitro* tests to observe antioxidant and antiallergic action.

8.5.2.2 Skin Permeation Studies in Artificial Skin Models and Synthetic Membranes

In vitro skin reconstruction has become an alternative to tests carried out on animals, gaining visibility in recent years. These are controlled and characterized skin models in which human skin cells grow within a three-dimensional (3D) structure organized and surrounded by other cells, such as fibroblasts, and substances that support and influence the differentiation of keratinocytes. Thus, these equivalents of artificial human skin provide an object of study for the assessment of cell interactions, melanogenesis regulation, proliferation and differentiation of keratinocytes, and post-trauma reepithelialization, in addition to being a relevant strategy in the evaluation of cosmetic products, being very well applied to photoaging models. Given the similarity of skin reconstructed in 3D with human skin, mainly due to its thickness and differentiation of layers, it is possible to study the SPF of photo protectors, as well as the changes caused in cells and extracellular matrix resulting from radiation (Brohem et al., 2011).

The use of this model was observed in the study carried out by Hwang et al. (2017), which sought to observe the effect of coumestrol, a compound derived from isoflavones known for its antiaging effect, observing its action against skin darkening. An African human artificial skin model (MelanoDerm) was used for which treatments were performed with different concentrations of coumestrol, diluted with saline solution buffered with Dulbecco's phosphate. The results showed the skin-lightening effect provided by coumestrol.

The 3D model was also used by Pinto et al. (2021), this time we used the EpiSkin™ and SkinEthic™ Human Corneal Epithelium models no longer to assess the whitening effect, as

Hwang et al. (2017), but to assess the irritation that the SWE chestnut husk extract offered to the skin and eyes. In this test, positive results were obtained, this extract being non-irritating and safe for further tests.

Another test observed with the same EpiSkin™ model was performed by Haddada et al. (2020) testing gold nanoparticles, for which skin and eye irritation has not been demonstrated. Haddada et al. (2020) and Pinto et al. (2021) emphasize that the skin irritation test is important for cosmetic products for topical use to avoid problems during treatment, while the eye irritation test is performed to avoid adverse effects and possible problems such as cell death in this region, which is very fragile, having considering that it is possible that during the use of the product, the user takes their hands with the product to their eyes.

Another study of skin permeation with Franz cells was carried out by Duarte et al. (2021), this time testing the isolated compound ferulic acid in gel and gel-cream formulations with a synthetic or optimized membrane impregnated on it. This test was performed to observe the kinetic parameters of acid release, vehicle interference, and the influence of the synthetic membrane. Thus, it was shown that the release of ferulic acid increased over time, and using the optimized membrane obtained better results, with a faster release of ferulic acid.

Skin permeation studies with synthetic membranes are shown in the work by Uchida et al. (2015) where they compared the permeation profiles of substances using different matrices, both animal and synthetic. Thirteen chemical compounds with different lipophilicity and molecular weights were tested, the synthetic matrix evaluated was the synthetic membrane Strat-M™, and as a comparison of their efficiency, the same compounds were studied before their permeation in excised human skin and hairless rat skin, the assay was done in Franz diffusion cells. The study demonstrated a correlation in permeability coefficient values between human skin, rat skin, and the Strat-M™. Furthermore, the permeation and diffusion parameters of the chemicals tested in Strat-M™ were not significantly different from excised human and rat skin.

Guided by this study, Das et al. (2020) to verify the permeation of the phenolic resveratrol incorporated in a microemulsion, used the synthetic membrane Strat-M™ which is an artificial membrane already proven in studies by Uchida et al. (2015) as an alternative to human skin and animals for the study of skin permeation of chemical substances. The results of release and *in vitro* permeation of resveratrol in microemulsion suggest a more gradual release of the active and constant penetration in the epidermis and dermis, that is, in deeper layers of the skin.

Based on the study models highlighted in this topic, using synthetic matrices and artificial skin models, to investigate the release and permeation of substances in the skin, it is possible to observe the efficiency of using these models as an alternative to the use of biological membranes as the matrix of animal skins. This shows the effectiveness of the development of these matrices and their potential in studies involving the release and permeation of actives in the skin. It also highlights and strengthens the importance of further research in the development of new synthetic membranes and models of artificial skin and their use in research involving studies with the permeation of actives in the skin layers, as well as in research in the and cosmetic areas that also focus on antioxidant and antiaging, taking care to carry out research and develop new products with the cruelty-free segments.

8.5.3 *In Vivo* Assays

The investigation of the cosmeceutical potential of phenolic compounds, either in extracts or in isolation, incorporated or not in DS, has also been investigated using animal models. In the study by Mostafa et al. (2021), an animal model was used to evaluate the antiwrinkle effect of solid lipid nanoparticles incorporated with *P. persica* (L.) extract. The systems were incorporated into creams for topical application, at concentrations of 2% and 5%, and further evaluated against ultraviolet-induced photoaging in a mouse model, using a wrinkle-scoring method, tissue biomarkers, and histopathology. The authors describe that at high or low doses it was possible to verify an improvement in the appearance of wrinkles, a decrease in the thickness of the dermis and epidermis, collagen content and fibroblast content increase, and consequent preservation of elastic fibers, also offering a protective effect significant against UV rays, expressing very satisfactory and promising results for the use of these systems in cosmetic products.

Researchers performed *in vivo* tests with the medicinal plant *Salvia officinalis*, rich in phenolic compounds (coumarins, flavonoids, tannins). Rutin is a polyphenol flavonoid that has been widely studied for its promising inflammation-fighting effects, which may preserve the skin. Rutin also can neutralize vitamin C and regenerate free radicals, thus helping to restore its antioxidant potential. Since vitamin C plays a key role in collagen production, collagen is an important part of the capillary wall and the supporting structure of the epidermis, it can have a major impact on the skin's health (Khare et al., 2021).

Albino *Swiss* rats were used, where 4 cm^2 of hair was removed from the back and the drug solution obtained from *S. officinalis* was applied to the skin of these animals, using micropipettes. It was found that the pure methanol extract of *S. officinalis* was able to inhibit 50% of the activity of enzymes related to aging (collagenase, elastase, and hyaluronidase). The extract of *S. officinalis* confirmed the effect of polyphenols against the enzymes evaluated, demonstrating high antioxidant potential and inhibitory potential *in vitro* and *in vivo*, therefore, they can be used in the development of cosmetics (Khare et al., 2021).

Another study was carried out with *Hubertia ambavilla* extract incorporated in gold nanoparticles, the purpose is the same, to favor cosmetic treatments with the implementation of natural extracts that contain phenolic compounds to provide antioxidant and antiaging activities. Haddada et al. (2020) unlike Khare et al. (2021) do not perform tests directly on animals, considering that currently, animal tests are banned for cosmetics, but use *in vitro* tests to find safety in the research product, thus the gold nanoparticles containing extract have proven to be non-toxic to fibroblast cells and dermal cells; are also able to efficiently scavenge free radicals, being also effective in protecting against UV-A damage to fibroblast cells and dermal cells. Regulatory tests to ensure the safety of gold nanoparticles as ingredients in cosmetics have shown that nanoparticles are non-toxic, phototoxic, genotoxic, irritant, or sensitizing and are therefore ideal for cosmetic skin treatments.

An *in vivo* model that has been widely used for antioxidant evaluation is the zebrafish. This model has already been used to evaluate phenolic compounds with antiaging potential, demonstrating the suppressive effect of oxidative stress and the inflammatory response, contributing to the targeting of possible candidates for the pharmaceutical and cosmetic industries (Wang et al., 2020). In addition, zebrafish has also been used as a depigmenting model, which has attracted special attention from the cosmetic industry. The antimelanogenic properties of ellagic acid were investigated by Yang et al. (2021) using this *in vivo* model. When evaluating the effects of this phenolic on tyrosinase activity and the percentage of melanin produced, it was demonstrated that there is a reduction in endogenous pigmentation in zebrafish embryos. These results were confirmed by assays of tyrosinase activity, which was shown to be decreased, and by the percentage of melanin data.

8.5.4 Clinical Studies

Several studies point to the importance and necessity of conducting clinical tests to obtain efficient and safe antiaging products (Garcia et al., 2021; Magalhães et al., 2020; Yepes et al., 2021). In general, it was observed that most articles contain a composition of *in vitro* assays and, to a lesser extent, their association with *in vivo* tests or clinical trials. These clinical trials were carried out to assess the safety of the tested ingredients and/or to verify the improvements of some parameters adopted for the skin, such as hydration and wrinkle reduction, to support the antiaging activity.

The study by Perugini et al. (2020) evaluated the effectiveness of biocellulose face masks as active DS for the skin. The effect on skin parameters of three biocellulose masks, formulated to have different cosmetic effects (antiaging, lifting, and cell renewal) was tested on 69 healthy white volunteers, between 25 and 64 years old, divided into three separate studies. Different actives were used in the masks, such as hyaluronic acid and peptides, with combinations of different plant extracts. The skin parameters evaluated were hydration, color, viscoelastic properties, surface smoothness, wrinkle reduction, dermal homogeneity, and stratum corneum renewal. Face masks were applied using the split face method three times a week for 4–8 weeks, depending on the study. The incorporation of different extracts associated with other actives in facial masks promoted results of improvement in homogeneity, dermal firmness, elasticity, and cell renewal, depending on the actives used.

Another very robust study developed by Damasceno et al. (2020), demonstrates *Prosopis juliflora*, popularly known as "algaroba" (mesquite), as a new cosmetic ingredient. The clinical trials developed sought to evaluate both the safety, by the Patch method, as well as the antiaging efficacy presented by the formulation of a gel extract formed from a solid core. For this, 14 volunteers, aged between 20 and 67 years, were selected to carry out the study characterized as a randomized unilateral blind placebo-controlled study. The application took place on both forearms of the volunteers. The results showed that the extracts contained a high content of phenolics (flavonoids and phenolic acids) and the clinical evaluation reinforced their potential for skin hydration, in addition to skin micro-relief improvements, presenting antiaging activity and proving to be a prototype for an innovative product.

Among the studies that evaluated the safety of plant products, such as extracts, for antiaging application, is the one by Pinto et al. (2021). The authors evaluated *Castanea sativa* extracts obtained by three different extraction methods for antiaging application. After characterization, it was seen that the ideal extract proved to be rich in ellagic acid (40.4 μg/mg dw), caffeic acid derivative (15.4 μg/mg dw), and epigallocatechin (15.3 μg/mg dw). For safety evaluation, the Patch test was used, in which skin irritation was observed in human volunteers using adhesives containing or not extracts from the chestnut husk. Only mild and transient irritant reactions for both the extract and controls were identified, showing no significant adverse effects. According to the results, the authors describe that the chestnut husk extract can be applied in antiaging cosmetics and indicate as a next step the incorporation of this extract in creams for topical use (Pinto et al., 2021).

The incorporation of plant extract with abundant phenolic content in topical formulations, such as emulsion and gel, for antiaging purposes was also seen by Marques et al. (2017). *Cynara scolymus* L. extracts were investigated for safety through the human repeat insult contact test (HRIPT) and, unlike the studies described so far, the protective effect against oxidative stress after UV radiation was also investigated by chromometer evaluation in 50 healthy human volunteers. The results were positive, since, for the HRIPT test, no allergenic potential was observed and there was no skin irritation, proving the safety of the extract. The protective effect test showed a decrease in the oxidizing activity of UV-A radiation on the skin with the formulation's use.

The study by Tito et al. (2015) also included tests with 20 female volunteers with dry or very dry skin to observe the hydration of the fat-soluble extract of *R. idaeus*. The extract was obtained from the culture of plant cells from the leaves of the species. The results for the observed parameter proved to be quite satisfactory for the product tested, attributing hydration with prolonged action and a moisturizing effect observed after 7 h of application, which can be implemented in the market, available in facial creams with antiaging objective.

Considering the evaluation of phenolic components isolated in DS, niosomes loaded with a semi-purified fraction containing gallic acid from *T. chebula* galls were formulated and tested using two tests to verify irritation: one *in vivo* with rabbits, for observation of erythema and edema and another with 31 healthy human volunteers, in which the areas of the forearm were divided for application of seven gel formulations containing or not niosomes loaded with gallic acid or semi-purified fraction, with application twice a day for 8 weeks. From the test on rabbits, the gels to be tested on humans were selected. The human test showed that most of the gels applied showed improvement in the skin elasticity of the volunteers. It was also observed that all gels containing gallic acid, regardless of the DS being present, showed an improvement in skin roughness. The last assessment was about skin pigmentation and hydration, but all gel formulations developed did not show significant differences in this regard (Manosroi et al., 2011).

8.6 CONCLUSION

Plant extracts have been used more significantly in distribution systems when compared to phenolic compounds alone. A great diversity of species is used to obtain these extracts, using different parts of plants such as leaves, bark, and even residues, which demonstrates the richness of phytochemical components that can be obtained from different sources. Among the phenolic compounds, gallic acid, quercetin, lipoic acid, and resveratrol were incorporated and isolated in DS, showing satisfactory results for antiaging application.

Different DS has been used to incorporate these assets, among which nanoencapsulation systems have shown greater applicability today. However, all the DS studies contributed positively to improving the characteristics of both the isolated actives and enhanced the action of extracts, improving aspects such as skin permeation, solubility, bioavailability, and/or activity.

Furthermore, the antiaging potential of extracts containing the phenolic compounds discussed in this chapter (flavonoids, phenolic acids, tannins, stilbenes, and curcuminoids), whether or not incorporated into DS, was also studied. From the analysis performed, it was found that some species or phenolic compounds have the potential for antiaging applications, but there is little development of current and complete research that prospects these products for the market.

It is noteworthy that antioxidant activity is an important parameter and one of the first steps to see if a compound or extract might offer any advantage for antiaging application. This fact can be observed in most articles, in which the *in vitro* antioxidant tests were carried out a priori. However, the antioxidant activity by itself often does not support the attribution of the antiaging activity, given the importance of the complementarity of different assays, which provide subsidies to guarantee the activity. Additionally, in most articles, a set of tests is carried out, attributing robustness and greater security to the antiaging potential, especially for those who carry out the clinical tests.

It was found that most studies aimed at antiaging applications incorporate the evaluation of enzyme inhibition assays, involving enzymes that act in the aging process, as well as assays with fibroblasts and permeation tests with Franz cells. Few studies used *in vivo* models with mice and some clinical tests have been performed, for which hydration, antiwrinkle effect, roughness, and pigmentation skin were generally evaluated as parameters for attributing antiaging activity. In addition, research strongly points to the need for investments in clinical trials, which provide greater support and ensure greater safety and effectiveness of the developed products.

Thus, this chapter brings a recent analysis of the most studied DS, incorporating flavonoids, phenolic acids, tannins, stilbenes, and curcuminoids, as well as the respective extracts that contain them, emphasizing their antiaging potential, especially for the cosmetic market, bringing, also, an approach to the systems already commercialized. With this analysis, it is possible to identify that these systems and their association with the aforementioned compounds are still little explored, which opens doors for innovation in the industrial field. Furthermore, the clinical implications of these systems describe the importance of deepening this theme, given the lack of clinical studies in the area. Therefore, with the advancement of research, the use of DS with phenolic compounds becomes a strategy that, in addition to delivering stable and improved activity inputs, can contribute to the formulation of innovative products for the cosmetic industries.

REFERENCES

Acharya DP, Hartley PG. Progress in microemulsion characterization. *Current Opinion in Colloid & Interface Science*. 2012, 17:274–280.

Agatonovic-Kustrin S, Kustrin E, Morton DW. Phenolic acids contribution to antioxidant activities and comparative assessment of phenolic content in mango pulp and peel. *South African Journal of Botany*. 2018, 116:158–163.

Alam MN, Bristi NJ, Rafiquzzaman M. Review on *in vitro* methods evaluation of antioxidant activity. *Saudi Pharmaceutical Journal*. 2013, 21:143–152.

Alonso C, Marti M, Barba C, Carrer V, Rubio L, Coderch L. Skin permeation and antioxidant efficacy of topically applied resveratrol. *Archives of Dermatological Research*. 2017, 309:423–431.

Baschieri A, Amorati R. Methods to determine chain-breaking antioxidant activity of nanomaterials beyond DPPH. A Review. *Antioxidants*. 2021, 10:1551.

Bazzicalupo M, Cornara L, Burlando B, Cascini A, Denaro M, Smeriglio A, et al. *Carpobrotus edulis (L.)* N.E.Br. extract as a skin preserving agent: From traditional medicine to scientific validation. *Journal of Integrative Medicine*. 2021, 19:526–536.

Bielak-Zmijewska A, Grabowska W, Ciolko A, Bojko A, Mosieniak G, Bijoch Ł, Sikora E. The role of curcumin in the modulation of ageing. *International Journal of Molecular Sciences*. 2019, 20:1239.

Borges A, Freitas V, Mateus N, Fernandes I, Oliveira J. Solid lipid nanoparticles as carriers of natural phenolic compounds. *Antioxidants*. 2020, 9:998.

Blesso CN. Dietary anthocyanins and human health. *Nutrients*. 2019, 11:2107.

Brohem CA, Cardeal LBA, Tiago M, Soengas MS, Barros SBM, Maria-Engler SS. Artificial skin in perspective: Concepts and applications. *Pigment Cell & Melanoma Research* (Print). 2011, 24:35–50.

Bunaciu AA, Danet AF, Fleschin S, Enein HYA. Recent applications foi *in vitro* antioxidant activity assay. *Critical Reviews in Analytical Chemistry*. 2016, 46:389–399.

Callender PS, Mathews AJ, Kobernyk K, Wettig DS. Microemulsion utility in pharmaceuticals: Implications for multi-drug delivery. *International Journal of Pharmaceutics*. 2017, 526:425–442.

Camins A, Junyent F, Verdaguer E, Beas-Zarate C, Rojas-Mayorquín AE, Ortuño-Sahagún D. Resveratrol: An antiaging drug with potential therapeutic applications in treating diseases. *Pharmaceuticals*. 2009, 2:194–205.

Carneiro SB, FIC Duarte, L Heimfarth, S JS Quintans, JL Quintans-Júnior, VF Veiga Júnior, Lima ÁAN. Cyclodextrin-drug inclusion complexes: *In vivo* and *in vitro* approaches. *International Journal of Molecular Sciences*. 2019, 20:642.

Carvalho MJ, Oliveira AL, Pedrosa SS, Pintado M, Madureira AR. Potential of sugarcane extracts as cosmetic and skincare ingredients. *Industrial Crops and Products*. 2021, 169:113625.

Casarini TPA, Frank LA, Pohlmann AR, Guterres SS. Dermatological applications of the flavonoid phloretin. *European Journal of Pharmacology*. 2020, 889:173593.

Castellani S, Trapani A, Spagnoletta A, di Toma L, Magrone T, Di Gioia S, Mandracchia D, Trapani G, Jirillo E, Conese M. Nanoparticle delivery of grape seed-derived proanthocyanidins to airway epithelial cells dampens oxidative stress and inflammation. *Journal of Translational Medicine*. 2018, 16:140.

Cavalcanti GR, Duarte FIC, Converti A, Lima ÁAN. Ferulic acid activity in topical formulations: Technological and scientific prospecting. *Current Pharmaceutical Design*. 2021, 27:2289–2298.

Chaikul P, Khat-udomkiri N, Iangthanarat K, Manosroi J, Manosroi A. Characteristics and *in vitro* anti-skin aging activity of gallic acid loaded in cationic CTAB niosome. *European Journal Pharmaceutical Sciences*. 2019, 131:39–49.

Chauhan AS. Dendrimer nanotechnology for enhanced formulation and controlled delivery of resveratrol. *Annals of the New York Academy of Sciences*. 2015, 1348:134–140.

Chaturvedi S, Garg A. An insight of techniques for the assessment of permeation flux across the skin for optimization of topical and transdermal drug delivery systems. *Journal of Drug Delivery Science and Technology*. 2021, 62:102355.

Cherubim DJ, Martins CV, Fariña L, Lucca RA. Polyphenols as natural antioxidants in cosmetics applications. *Journal of Cosmetic Dermatology*. 2019, 19:33–37.

Chu CC, Chew SC, Nyam KL. Recent advances in encapsulation technologies of kenaf (*Hibiscus cannabinus)* leaves and seeds for cosmeceutical application. *Food and Bioproducts Processing*. 2021, 127:99–113.

Contardi M, Kossyvaki D, Picone P, Summa M, Guo X, Heredia-Guerrero JA, et al. Electrospun polyvinylpyrrolidone (PVP) hydrogels containing hydroxycinnamic acid derivatives as potential wound dressings. *Chemical Engineering Journal*. 2021, 409:128144.

Costa R, Santos L. Delivery systems for cosmetics - From manufacturing to the skin of natural antioxidants. *Powder Technology*. 2017, 322:402–416.

Costanzo A, Angelico R. Formulation strategies for enhancing the bioavailability of silymarin: The state of the art. *Molecules*. 2019, 24:2155.

Damasceno GAB, Barreto SMAG, Reginaldo FPS, Souto AL, Negreiros MMF, Viana RLS, et al. *Prosopis juliflora* as a new cosmetic ingredient: Development and clinical evaluation of a bioactive moisturizing and anti-aging innovative solid core. *Carbohydrate Polymers*. 2020, 233:115854.

Das S, Lee SH, Chow PS, Macbeath C. Microemulsion composed of combination of skin beneficial oils as vehicle: Development of resveratrol-loaded microemulsion based formulations for skin care applications. *Colloids and Surfaces B: Biointerfaces*. 2020, 194:111161.

Davies S, Contri CV, Guterres SS, Pohlmann RA, Guerreiro KCI. Simultaneous nanoencapsulation of lipoic acid and resveratrol with improved antioxidant properties for the skin, *Colloids and Surfaces B: Biointerfaces*. 2020, 192:111023.

Davinelli S, Bertogliob JC, Polimenic P, Scapagninia G. Cytoprotective polyphenols against chronological skin aging and cutaneous photodamage. *Current Pharmaceutical Design*. 2018, 24:99–105.

Ding T, Cao K, Fang W, Zhu G, Chen C, Wang X, et al. Evaluation of phenolic components (anthocyanins, flavanols, phenolic acids, and flavonols) and their antioxidant properties of peach fruits. *Scientia* Horticulturae *(Amsterdam)*. 2020, 268:109365.

Dong W, Chen D, Chen Z, Sun H, Xu Z. Antioxidant capacity differences between the major flavonoids in cherry (*Prunus pseudocerasus*) *in vitro* and *in vivo* models. *LWT*. 2021, 141:110938.

Duarte FIC, Costa ABSM, Vieira Filho JF, Freite VLP, Freire JVA, Converti A, Ferrari M, Gomes APB, Ostrosky E, Lima, ÁAN. *In vitro* release studies of ferulic acid in semi-solid formulations with optimized synthetic membrane. *Journal of Drug Delivery Science and Technology*. 2021, 61:102106.

Ferreira MS, Magalhães MC, Oliveira R, Sousa-Lobo JM, Almeida IF. Trends in the use of botanicals in anti-aging cosmetics. *Molecules*. 2021, 26:3584.

Fierascu RC, Ortan A, Fierascu IC, Fierascu I. *In vitro* and *in vivo* evaluation of antioxidant properties of wild-growing plants. A short review. *Current Opinion in Food Science*. 2018, 24:1–8.

Figueroa-Robles A, Antunes-Ricardo M, Guajardo-Flores D. Encapsulation of phenolic compounds with liposomal improvement in the cosmetic industry. *International Journal of Pharmaceutics*. 2021, 593:120125.

Flieger J, Flieger W, Baj J, Maciejewski R. Antioxidants: Classification, natural sources, activity/capacity measurements, and usefulness for the synthesis of nanoparticles. *Materials* (*Basel*). 2021, 14:4135.

Garavand F, Jalai-Jivan M, Assadpour E, Jafari SM. Encapsulation of phenolic compounds within nano/microemulsion systems: A review. *Food Chemistry*. 2021, 364:130376.

Garcia L, Pires G, Oliveira D, Silva L, Gomes A, Amaral J, Pereira G, Ruela A. Incorporation of glycolic extract of cocoa beans (*Theobroma cacao L.*) into microemulsions and emulgels for skincare. *Industrial Crops and Products*, 2021, 161:113181.

Govindappa M, Tejashree S, Thanuja V, Hemashekhar B, Srinivas C, Nasif O, Pugazhendhi A, Raghavendra VB. Pomegranate fruit fleshy pericarp mediated silver nanoparticles possessing antimicrobial, antibiofilm formation, antioxidant, biocompatibility and anticancer activity. *Journal of Drug Delivery Science and Technology*. 2021, 61:102289.

Grabowska W, Kucharewicz K, Wnuk M, Lewinska A, Suszek M, Przybylska D, Mosieniak G, Sikora E, Bielak-Zmijewska A. Curcumin induces senescence of primary human cells building the vasculature in a DNA damage and ATM-independent manner. *Age*. 2015, 37:1–17.

Gupta SC, Kismali G, Aggarwal BB. Curcumin, a component of turmeric: From farm to pharmacy. *Biofactors*. 2013, 39:2–13.

Haddada BM, Gerometta E, Chawech R, Sorres J, Bialecki A, Pesnel S, Spadavecchia J, Morel, A-L. Assessment of antioxidant and dermoprotective activities of gold nanoparticles as safe cosmetic ingredient. *Colloids and Surfaces B: Biointerfaces*. 2020, 189:110855.

Hwang JA, Park NH, Na YJ, Lee HK, Lee JH, Kim YJ, Lee CS. Coumestrol down-regulates melanin production in melan-a murine melanocytes through degradation of tyrosinase. *Biological and Pharmaceutical Bulletin*. 2017, 40:535–539.

Huerta-Madroñal M, Caro-León J, Espinosa-Cano E, Aguilar MR, Vázquez-Lasa B. Chitosan - Rosmarinic acid conjugates with antioxidant, anti-inflammatory and photoprotective properties. *Carbohydrate Polymers*. 2021, 273:118619.

Huma S, Khan HMS, Sohail M, Akhtar N, Rasool F, Majeed F, et al. Development, in-vitro characterization and assessment of cosmetic potential of *Beta vulgaris* extract emulsion. *Journal of Herbal Medicine*. 2020, 23:100372.

Kanha N, Regenstein JM, Surawang S, Pitchakarn P, Laokuldilok T. Properties and kinetics of the *in vitro* release of anthocyanin-rich microcapsules produced through spray and freeze-drying complex coacervated double emulsions. *Food Chemistry*. 2021, 340:127950.

Karatoprak GŞ, Yücel Ç, Göger F, Sobarzo-Sánchez E, Akkol EK. Potential antioxidant and enzyme inhibitory effects of nanoliposomal formulation prepared from *Salvia aramiensis* rech. f. extract. *Antioxidants*. 2020, 9:293.

Kashyap N, Kumari A, Raina N, Zakir F, Gupta M. Prospects of essential oil loaded nanosystems for skincare. *Phytomedicine Plus*. 2021, 2:100198.

Kaur H, Kaur G. A critical appraisal of solubility enhancement techniques of polyphenols. *Journal of Pharmaceutics*. 2014, 2014:1–14.

Khare R, Upmanyu N, Jha M. Exploring the potential effect of methanolic extract of *Salvia officinalis* against UV exposed skin aging: *In vivo* and *in vitro* model. *Current Aging Science*. 2021, 14:46–55.

Khoshnoudi-Nia S, Sharif N, Jafari SM. Loading of phenolic compounds into electrospun nanofibers and electrosprayed nanoparticles. *Trends in Food Science & Technology*. 2020, 95:59–74.

Kumar N, Goel N. Phenolic acids: Natural versatile molecules with promising therapeutic applications. *Biotechnology Reports*. 2019, 24:e00370.

Ky I, Le Floch A, Zeng L, Pechamat L, Jourdes M, Teissedre P-L. Tannins. In: edited by Benjamin Caballero, Paul M. Finglas, and F. Toldra *Encyclopedia of Food and Health*. Elsevier, 2016. p. 247–255.

Kyriakoudi A, Spanidi E, Mourtzinos I, Gardikis K. Innovative delivery systems loaded with plant bioactive ingredients: Formulation approaches and applications. Plants *(Basel)*. 2021, 10:1238.

Li H, Zhao Q-S, Chang S-L, Chang T-R, Tan M-H, Zhao B. Development of cannabidiol full-spectrum oil/2,6 -di-O-methyl-β-cyclodextrin inclusion complex with enhanced water solubility, bioactivity, and thermal stability. *Journal of Molecular Liquids*. 2021, 347: 118318.

Lianza M, Mandrone M, Chiocchio I, Tomasi P, Marincich L, Poli F. Screening of ninety herbal products of commercial interest as potential ingredients for phytocosmetics. *Journal of Enzyme Inhibition and Medicinal Chemistry*. 2020, 35:1287–1291.

Liu CH, Huang HY. Antimicrobial activity of curcumin-loaded myristic acid microemulsions against *Staphylococcus epidermidis*. *Chemical & Pharmaceutical Bulletin*. 2012, 60:1118–1124.

Liu W, Pan N, Han Y, Li D, Chai J. Solubilization, stability and antioxidant activity of curcumin in a novel surfactant-free microemulsion system. *LWT*. 2021, 147:111583.

López-Fernández O, Domínguez R, Pateiro M, Munekata PES, Rocchetti G, Lorenzo JM. Determination of polyphenols using liquid chromatography-tandem mass spectrometry technique (LC-MS/MS): A review. *Antioxidants*. 2020, 9:479.

Magalhães TSSA, Macedo PCO, Costa ÉCP, Tavares EA, Silva VC, Guerra GCB, Pereira JR, Lemos TMAM, Negreiros MMF, Rocha HAO. Increase in the antioxidant and anti-inflammatory activity of *Euterpe oleracea* martius oil complexed in β-cyclodextrin and hydroxypropyl-β-cyclodextrin. *International Journal of Molecular Sciences*. 2021, 22:11524.

Magalhães TSSA, Macedo PCO, Converti A, Lima ÁAN. The use of *Euterpe oleracea* mart. As a new perspective for disease treatment and prevention. *Biomolecules*. 2020, 10:813.

Mandrone M, Coqueiro A, Poli F, Antognoni F, Choi YH, Identification of a collagenase-inhibiting flavonoid from *Alchemilla vulgaris* using NMR-based metabolomics. *Planta Medica*. 2018, 84:941–946.

Manosroi A, Jantrawut P, Akihisa T, Manosroi W, Manosroi, J. *In vitro* and *in vivo* skin anti-aging evaluation of gel containing niosomes loaded with a semi-purified fraction containing gallic acid from *Terminalia chebula* galls. *Pharmaceutical Biology*. 2011, 49:11.

Marques ES, Froder JG, Carvalho JCT, Rosa PCP, Perazzo FF, Maistro, EL. Evaluation of the genotoxicity of *Euterpe oleraceae* Mart. (*Arecaceae*) fruit oil (açaí), in mammalian cells *in vivo*. *Food and Chemical Toxicology*, 2016, 93:13–19.

Marques P, Marto J, Gonçalves LM, Pacheco R, Fitas M, Pinto P, Serralheiro MLM, Ribeiro H. *Cynara scolymus* L.: A promising Mediterranean extract for topical anti-aging prevention. *Industrial Crops and Products*. 2017, 109:699–706.

Moghaddasi F, Housaindokht MR, Darroudi M, Bozorgmehr MR, Sadeghi A. Synthesis of nano curcumin using black pepper oil by O/W nanoemulsion technique and investigation of their biological activities. *LWT - Food Science and Technology*. 2018, 92:92–100.

Mohamad EA, Aly AA, Khalaf AA, Ahmed MI, Kamel RM, Abdelnaby SM, et al. Evaluation of natural bioactive-derived punicalagin niosomes in skin-aging processes accelerated by oxidant and ultraviolet radiation. *Drug Design, Development and Therapy*. 2021, 15:3151–3162.

Morais ES, Silva NHCS, Sintra TE, Santos SAO, Neves BM, Almeida IF, et al. Anti-inflammatory and antioxidant nanostructured cellulose membranes loaded with phenolic-based ionic liquids for cutaneous application. *Carbohydrate Polymers*. 2019, 206:187–197.

Mostafa ES, Maher A, Mostafa DA, Gad SS, Nawwar MAM, Swilam N. A unique acylated flavonol glycoside from *Prunus persica* (L.) var. florida prince: A new solid lipid nanoparticle cosmeceutical formulation for skincare. *Antioxidants*. 2021, 10:1–20.

Mota MD, Costa RYS; Guedes Aaa S, Silva LCRC, Chinalia FA. Guava-fruit extract can improve the UV-protection efficiency of synthetic filters in sun cream formulations. *Journal of Photochemistry and Photobiology B: Biology*. 2019, 201:111639.

Muñoz-Shuguli C, Vidal CP, Cantero-López P, Lopez-Polo J. Encapsulation of plant extract compounds using cyclodextrin inclusion complexes, liposomes, electrospinning and their combinations for food purposes. *Trends in Food Science & Technology*. 2021, 108:177–186.

Nowak A, Ossowicz-rupniewska P, Rakoczy R, Konopacki M, Perużyńska M, Droździk M, et al. Bacterial cellulose membrane containing *Epilobium angustifolium* L. Extract as a promising material for the topical delivery of antioxidants to the skin. *International Journal of Molecular Sciences*. 2021, 22:1–21.

Oliveira ALS, Gondim S, Gómez-García R, Ribeiro T, Pintado M. Olive leaf phenolic extract from two Portuguese cultivars - bioactivities for potential food and cosmetic application. *Journal of Environmetal Chemical Engineering.* 2021, 9:106175.

Oliveira G. Determinação da Capacidade Antioxidante de Produtos Naturais *in vitro* pelo Método de DPPH: Estudo de Revisão. *Revista Brasileira de Plantas Medicinais.* 2015, 17:36–44.

Pannu N, Bhatnagar A. Resveratrol: From enhanced biosynthesis and bioavailability to multitargeting chronic diseases. *Biomedicine & Pharmacotherapy.* 2019, 109:2237–2251.

Păvăloiu RD, Sha'at F, Neagu G, Deaconu M, Bubueanu C, Albulescu A, et al. Encapsulation of polyphenols from *Lycium barbarum* leaves into liposomes as a strategy to improve their delivery. *Nanomaterials.* 2021, 11:1–13.

Permana AD, Utamic RN, Courtenaya AJ, Manggaud MA, Donnellya RF, Rahmanb L. Phytosomal nanocarriers as platforms for improved delivery of natural antioxidant and photoprotective compounds in propolis: An approach for enhanced both dissolution behaviour in biorelevant media and skin retention profiles. *Journal of Photochemistry & Photobiology*, B: Biology. 2020, 205:111846.

Perugini P, Bleve M, Redondi R, Cortinovis F, Colpani A. *In vivo* evaluation of the effectiveness of biocellulose facial masks as active delivery systems to skin. *Journal of Cosmetic Dermatology.* 2020, 19:725–735.

Pinto D, Cádiz-Gurrea M, Garcia J, Saavedra M, Freitas V, Costa P, Sarmento B, Delerue-Matos C, Rodrigues F. From soil to cosmetic industry: Validation of a new cosmetic ingredient extracted from chestnut shells. *Sustainable Materials and Technologies*, 2021, 29:e00309.

Pivetta T P, Silva L B, Kawakami CM, Araújo MM, Del Lama M PFM, Naal RMZG, Maria-Engler SS, Gaspar LR, Marcato PD. Topical formulation of quercetin encapsulated in natural lipid nanocarriers: Evaluation of biological properties and phototoxic effect. *Journal of Drug Delivery Science and Technology*, 2019, 53:101148.

Rapalli VK, Sharma S, Roy A, Singhvi G. Design and dermatokinetic evaluation of Apremilast loaded nanostructured lipid carriers embedded gel for topical delivery: A potential approach for improved permeation and prolong skin deposition. *Colloids Surfaces B Biointerfaces.* 2021, 206:111945.

Samanta A, Chanda S, Bandyopadhyay B, Das N. Establishment of drug delivery system nanocapsulated with an antioxidant (+)-catechin hydrate and sodium meta borate chelator against sodium fluoride induced oxidative stress in rats, *Journal of Trace Elements in Medicine and Biology.* 2016, 33:54–67.

Santos PP, Flôres SH, de Oliveira Rios A, Chisté RC. Biodegradable polymers as wall materials to the synthesis of bioactive compound nanocapsules. *Trends in Food Science & Technology.* 2016, 53:23–33.

Salehi B, Mishra AP, Nigam M, Sener B, Kilic M, Sharifi-Rad M. Resveratrol: A double-edged sword in health benefits. *Biomedicines.* 2018, 6:E91.

Sedlak L, Wojnar W, Zych M, Wygledowska-Promienska D, Mrukwa-Kominek E, Kaczmarczyk-Sedlak I. Effect of resveratrol, a dietary-derived polyphenol, on the oxidative stress and polyol pathway in the lens of rats with streptozotocin-induced diabetes. *Nutrients.* 2018, 10:E1423.

Shah BR, Zhang C, Li Y, Li B. Bioaccessibility and antioxidant activity of curcumin after encapsulated by nano and Pickering emulsion based on chitosan-tripolyphosphate nanoparticles. *Food Research International.* 2016, 89:399–407.

Singla RK, Dubey AK, Garg A, Sharma RK, Fiorino M, Ameen SM, Haddad MA, Al-Hiary M. Natural polyphenols: Chemical classification, definition of classes, subcategories, and structures. *Journal of AOAC International.* 2019, 102:1397–1400.

Silva AS, Reboredo-Rodríguez P, Süntar I, Sureda A, Belwal T, Loizzo MR, et al. Evaluation of the status quo of polyphenols analysis: Part I-phytochemistry, bioactivity, interactions, and industrial uses. *Comprehensive Reviews in Food Science and Food Safety.* 2020, 19:3191–3218.

Silva SAM e, Michniak-Kohn B, Leonardi GR. An overview about oxidation in clinical practice of skin aging. *Anais Brasileiros de Dermatologia.* 2017, 92:367–374.

Smith R E, Tran K, Smith C C, McDonald M, Shejwalkar P, Hara K. The role of the Nrf2/ARE antioxidant system in preventing cardiovascular diseases. *Diseases.* 2016, 4:E34.

Spagnol CM, Assis RP, Brunetti IL, Isaac VLB, Salgado HRN, Corrêa MA. *In vitro* methods to determine the antioxidant activity of caffeic acid. *Spectrochimica Acta Part A: Molecular and Biomolecular Spectroscopy.* 2019, 219:358–366.

Taofiq O, Heleno SA, Calhelha RC, Fernandes IP, Alves MJ, Barros L, et al. Phenolic acids, cinnamic acid, and ergosterol as cosmeceutical ingredients: Stabilization by microencapsulation to ensure sustained bioactivity. *Microchemical Journal.* 2019, 147:469–477.

Tito MB, Carola A, Lucia A, Barbulova AA, Tortora GC, Apone A. An oil-soluble extract of *Rubus idaeus* cells enhances hydration and water homeostasis in skin cells. *International Journal of Cosmetic Science.* 2015, 37:588–594.

Tran VV, Nguyen TL, Moon J-Y, Lee Y-C. Core-shell materials, lipid particles and nanoemulsions, for delivery of active anti-oxidants in cosmetics applications: Challenges and development strategies. *Chemical Engineering Journal*. 2019, 368:88–114.

Uchida T, Kadhum WR, Kanai S, Todo H, Oshizaka T, Sugibayashi K, Prediction of skin permeation by chemical compounds using the artificial membrane, Strat.-M. *European Journal of Pharmaceutical Sciences*. 2015. 67:113–118.

Vaiserman A, Koliada A, Zayachkivska A, Lushchak O. Nanodelivery of natural antioxidants: An anti-aging perspective. *Frontiers in Bioengineering and Biotechnology*. 2020, 7:447.

Wang L, Lee W, Jayawardena TU, Cha S-H, Jeon Y-J. Dieckol, an Algae-derived phenolic compound, suppresses airborne particulate matter-induced skin aging by inhibiting the expressions of pro-inflammatory cytokines and matrix metalloproteinases through regulating NF-κB, AP-1, and MAPKs signaling pathways. *Food and Chemical Toxicology*. 2020, 146:111823

Wang Y, Xie M, Ma G, Fang Y, Yang W, Ma N. The antioxidant and antimicrobial activities of different phenolic acids grafted onto chitosan. *Carbohydrate Polymers*. 2019, 225:115238.

Wahl D, Bernier M, Simpson SJ, Cabo R, Le Couteur, DG. Future directions of resveratrol research. *Nutrition and Healthy Aging*. 2018, 4:287–290.

Yang H-L, Lin C-P, Gowrisankar YV, Huang P-J, Chang W-L, Shrestha S, Hseu Y-C. The anti-melanogenic effects of ellagic acid through induction of autophagy in melanocytes and suppression of UVA-activated α-MSH pathways via Nrf2 activation in keratinocytes. *Biochemical Pharmacology*. 2021, 185:114454.

Yang L, Zheng Z, Qian C, Wu J, Liu Y, Guo S, Li G, Liu M, Wang X, Kaplan DL. Curcumin-functionalized silk biomaterials for anti-aging utility. *Journal of Colloid and Interface Science*. 2017, 496:66–77.

Yepes A, Ochoa-Bautista D, Murillo-Arango W, Quintero-Saumeth J, Bravo K, Osorio E. Purple passion fruit seeds (*Passiflora edulis* f. *edulis Sims*) as a promising source of skin anti-aging agents: Enzymatic, antioxidant and multi-level computational studies. *Arabian Journal of Chemistry*. 2021, 14:102905.

Yuan Y, Xiang J, Zheng B, Sun J, Luo D, Li P, et al. Diversity of phenolics including hydroxycinnamic acid amide derivatives, phenolic acids contribute to antioxidant properties of proso millet. *LWT*. 2021, 154:112611.

Xia N, Daiber A, Förstermann U, Li H. Antioxidant effects of resveratrol in the cardiovascular system. *British Journal of Pharmacology*. 2017, 174:1633–1646.

Xiao Y, Dong, J. Yin, Z. Wu, Q. Zhou, Y. Zhou, X. Procyanidin B2 protects against D-galactose-induced mimetic aging in mice: Metabolites and microbiome analysis. *Food and Chemical Toxicology*. 2018, 119, 141–149.

Zhao Z, Cui X, Ma X, and Wang Z, Preparation, characterization, and evaluation of antioxidant activity and bioavailability of a self-nanoemulsifying drug delivery system (SNEDDS) for buckwheat flavonoids. *Acta Biochimica et Biophysica Sinica*. 2020, 52:1265–1273.

Zhao Z, Moghadasian MH. Bioavailability of hydroxycinnamates: A brief review of *in vivo* and *in vitro* studies. *Phytochemistry Reviews*. 2010, 9:133–145.

Zillich OV, Schweiggert-Weisz U, Eisner P, Kerscher M. Polyphenols as active ingredients for cosmetic products. *International Journal of Cosmetic Science*. 2015, 37:455–464.

Zuiter, AS. Proanthocyanidin: Chemistry and biology: From phenolic compounds to proanthocyanidins. In: *Reference Module in Chemistry, Molecular Sciences and Chemical Engineering*, 2014,1–29.

9 Delivery of Cannabidiol to and through the Skin
Cannabidiol as Phytopharmaceutical Ingredient and the Challenging Task of Developing CBD-Loaded Topical and Transdermal Formulations

Thaís Nogueira Barradas, Fabrício Felippe dos Santos, and Stephani Araujo Cardoso

9.1 INTRODUCTION: BACKGROUND AND DRIVING FORCES

Cannabis sativa has been used for centuries due to its therapeutic properties, which are essentially a consequence of special bioactive molecules, known as cannabinoids. The two best-known cannabinoids are Δ9-tetrahydrocannabinol (Δ9-THC or simply THC), which represents the main psychoactive compound, and cannabidiol (CBD), which is known for its high therapeutic value (Crippa et al., 2018). Over time and with advances in research, cannabinoid receptors, known as cannabinoid receptor type 1 (CB1) and cannabinoid receptor type 2 (CB2), were discovered. Both of them belong to the G-protein-coupled receptors superfamily and, thus, contain seven transmembrane typical spanning domains. Additionally, two endogenous cannabinoids are widely recognized as the main natural ligands for cannabinoid receptors, namely, *N*-arachidonoylethanolamine (AEA) and 2-arachidonoylglycerol (2-AG). Both AEA and 2-AG, together with endogenous enzymes responsible for the synthesis and degradation of these components, comprise the endocannabinoid system (ECS). The ECS is an evolutionarily conserved system that is being increasingly related be part of a multifaceted homeostatic regulation system, impacting in a wide range of physiological processes through all human organism, including the skin (Baswan et al., 2020; Tóth et al., 2019).

The cutaneous ECS is important for regulating various skin homeostasis functions, for example, inflammatory processes, immune responses, and nociception., in addition to modulating cell differentiation (Baswan et al., 2020; Nikan et al., 2016; Scheau et al., 2021; Tóth et al., 2019). In addition to endogenous ligands, the ECS features exogenous ligands, often considered phytocannabinoids, such as cannabis-derived bioactive molecules and synthetic molecules, as well (Lee, 2010).

Since ECS plays a critical modulatory function in the skin and its appendages, the research on the application of CBD as a potential treatment for certain skin disorders has increased. In this context, CBD can provide several potential benefits when used alone or in combination with other cannabinoids such as THC (Itin et al., 2020; Baswan et al., 2020). Evidence shows that topical delivery of CBD can offer benefits to patients with different dermatological conditions. However, most of the clinical evidence is based on systemic administration of CBD, with limited evidence to the therapeutic potential of CBD topical or transdermal applications (Baswan et al., 2020).

DOI: 10.1201/9781003292692-9

According to studies present in the literature, CBD is effective for the treatment of dermatological diseases, a clinical study carried out with 20 volunteers suffering from different skin conditions showed that the use of an ointment with CBD for 3 months improved their clinical results, providing anti-inflammatory action. In view of the commercial interest in the topical application of CBD, several formulations were produced and patented (Casiraghi et al., 2020).

In addition to anti-inflammatory, antiproliferative activity, CBD is reported to show antioxidant activity, with potential protective activity against UVA radiation, preserving cell viability of both keratinocytes and melanocytes (Vacek et al., 2021), being also beneficial in tumor regression. The activation of the cutaneous ECS is reported to block the evolution and progression of several types of cancer, including skin cancer (Llanos Casanova et al., 2003). In this context, CBD can be used as an auxiliary therapy in patients who use chemotherapy, alleviating common symptoms related to side effects related to the chemotherapy treatments (Simmerman et al., 2019). In addition to these applications, CBD can promote or suppress hair shaft elongation, in a dose-dependent pattern, with mechanistic activity on valinoid-type transient potential receptors 4 (TRPV4).

Due to its lipophilic profile, CBD can show great affinity for the *stratum corneum*, making it a promising agent for cutaneous administration for skin care and dermatological treatments (Casiraghi et al., 2020; Itin et al., 2020), showing good penetration into the skin most external layers. Therefore, the topical and transdermal route represent an important alternative for CBD topical delivery for the treatment of skin diseases or transdermal delivery, as is the case for systemic applications (Bíró et al., 2009).

Although topical and transdermal routes seem to be promising approaches for exploring CBD's potential benefits to skin disorders and CBD features an adequate molecular weight (314.46 Da), its high lipophilicity (log P ~ 6.3) (ChemSpider) poses a significant challenge to the development of aqueous and transdermal formulations. Limitations such as low solubility and drug targeting problems can also be overcome by employing nanotechnological approaches to drug delivery. There are different types of nanocarriers with different release profiles, strategies are currently being developed to improve the specificity of nanostructures, to ensure that they reach different areas in the body, and studies to reduce immunogenicity through coating or functionalization of nanocarriers (Patra et al., 2018).

In recent years, studies with cannabis have been highlighted in nanomedicine, that cannabinoids can be loaded in nanoparticles, facilitating biodistribution and targeted delivery to the target site. The use of nanoparticles allows for increased drug action and aids in theranostics, a strategy that combines diagnosis and therapy. Technological development has allowed the development of synthetic cannabinoids in a cheaper way facilitating their use as nanoconjugated cannabinoids. Studies in this area can help in the production of multifunctional nanocarriers with the ability to treat various diseases (Onaivi et al, 2020).

Nanoencapsulation constitutes an important and well-studied strategy to overcome pharmacotechnical challenges such as high lipophilicity, limited bioavailability at target site, instability, unacceptable texture, viscosity, or odor, and systemic or local side effects. Proper formulation designing and nanoencapsulation are effective strategies to enable dermal and transdermal delivery of CBD, successful reports have previously shown (Hammell et al., 2016; Lodzki et al., 2003; Stella et al., 2021).

Currently, there is a range of topical and transdermal CBD products commercially available in the USA, Europe, Canada, and Australia, as OTC products. An avalanche of ever-increasing launches of CBD-based skin care cosmetics with a wide range of claims can be observed year after year. However, despite not being properly regulated by regulating agencies, CBD-based cosmetic products and ingredients are not subject to approval for commercialization, not being considered prohibited. Both natural and synthetic versions of CBD are being formulated in topical and transdermal products for the treatment of a wide range of dermatological conditions, such as acne, psoriasis, rosacea, atopic dermatitis, and other proinflammatory or proliferative disorders. For instance,

Botanix Pharmaceuticals have developed both gel and spray formulations containing from 5% to 25% of CBD for treating several skin disorders (Eugene & Matthew, 2018).

In this chapter, we conducted an extensive literature review on the proven and therapeutic uses of CBD for the treatment of skin conditions and other cosmetic applications. In addition, we provided on the CBD cosmetic market and a bibliometric analysis of the topical and transdermal application of CBD.

9.2 CANNABIS AND THE ECS

C. sativa has been used for thousands of years for recreational purposes and also due to its therapeutic properties reported since antiquity. Plants of the *Cannabis* genus contain a wide range of bioactive components called cannabinoids, with more than 500 biologically active molecules being identified so far (Tóth et al., 2019). Of these, Δ9-THC, or simply THC, is the main component with psychoactive properties acting in the central nervous system (CNS), which was isolated and identified in 1964 (Crippa et al., 2018). Additionally, CBD is a compound that has therapeutic applications, both when administered alone and with THC or other cannabinoids (Itin et al., 2020).

Given the beneficial effects arising from the administration of *Cannabis* plants and seeking to understand how cannabinoids exercise their biological and pharmacological functions, the main cannabinoid receptors CB1 and CB2 were subsequently cloned and identified (Matsuda et al., 1990; Mnekin & Ripoll, 2021). Both CB1 and CB2 receptors can recognize and respond to three classes of ligands: endogenous ligands (endocannabinoids), naturally found in plants, such as *Cannabis* spp. and often referred as phytocannabinoids, and synthetic cannabinoids (Lee, 2010; Nikan et al., 2016).

Cannabinoids are able to activate, antagonize, or inhibit a wide range of cellular targets, including transmembrane receptors, ionotropic and nuclear receptors (Tóth et al., 2019). Cannabinoid receptors, endocannabinoids, and the enzymes involved in their biosynthesis and degradation are known as the ECS (Di Marzo, 2008). ECS comprises a molecular signaling system involved in regulating a range of biological functions, contributing to cutaneous homeostasis. (Pisanti et al., 2017; Tóth et al., 2019).

All cannabinoid receptors belong to the G-protein-coupled receptors family and are closely related to the biological effects of phytocannabinoids, endocannabinoids, and synthetic cannabimimetic substances. Both CB1 and CB2 are GPCRs linked, via Gi/0, to the adenylyl cyclase enzyme, promoting its inhibition or activity reduction. Consequently, the activation of such receptors promotes the reduction of intracellular cAMP, causing the suppression of protein kinase A (PKA) activity and alteration of the gene expression of several interleukins, chemokines and growth factors (Kupczyk et al., 2009).

CB1 receptors can be found, mainly, in the CNS, and are related to pain regulation. CB1 receptors are found in greater concentration in the brain and have a well-distributed pattern, except in the encephalic throne, where their distribution has become less homogeneous. CB2 receptors, instead, relate to the immune system and their activation factors, as well as, anti-inflammatory responses (Shao et al., 2021). The presence of CB2 receptors in peripheral tissues is consistent with the fact that no severe depression of respiratory function with the use of cannabinoids has ever been described, for example. CB2 receptors are located in lymphoid tissue with greater expression and differ in their responsiveness to cannabinoid ligands (Ritter et al., 2016; Zheng et al., 2012).

CBD binding is also possible to take place with non-cannabinoid receptors, such as the transient potential receptors (TRPs), more specifically, of the potential transient receptors of the valinoid-type (TRPV). Those are ion channels directly related to physiological processes, such as pain, heat, and itching, and their dysfunction is channels associated with pathological conditions, being expressed in the skin. Any activity dysfunction of such receptors is commonly associated with pathological conditions (Caterina, 2014; Szabó et al., 2019). TRPs are channels of transient potential receptors present in sensory neurons that respond to thermal and chemical stimuli (Caterina, 2014).

Regarding the perception of pain, the transient receptor channel of potential vanilloid type 1 (TRPV1), is one of the most important channels, being a non-CB1/CB2 receptor that can be activated by cannabinoids. Such receptors are related to thermal and pain sensations and are expressed in immune system cells, keratinocytes, hair follicles, skin sensory nerves, and smooth muscle cells (Szabó et al., 2019). TRPV1 is regulated by phosphorylation and G protein-coupled substances that cause pain, such as bradykinin, can activate it (Ritter et al., 2016).

CBD can reduce the release of inflammatory mediators and decrease immune system flares by desensitizing TRPV1 receptors (Stinchcomb et al., 2004), behaving like a TRPV1 agonist (Pisanti et al., 2017). Furthermore, TRPVs can be involved in other physiological functions such as the formation and maintenance of the skin barrier, growth, cell differentiation, and immunological, and inflammatory processes (Baswan et al., 2020).

The discovery of cannabinoid receptors aroused the interest of researchers in the search for endogenous ligands related to both CB1 and CB2 receptors. Thus, as the first groundbreaking advance in this scientific field, the discovery of the first endogenous ligand, known as N-arachidonylethanolamide or simply anandamide (AEA), was reported and a few years later, the second ligand, 2-arachidonoyl glycerol (2-AG) was identified (Figure 9.1). Such endogenous ligands are known as endocannabinoids and are produced on demand from arachidonic acid and a complex biochemical cascade (Nikan et al., 2016; Scheau et al., 2021; Shao et al., 2021).

AEA is produced after an enzymatic reaction by a selective enzyme, known as N-acylphosphatidylethanolamine—selective phospholipase (NAPE-PLD). Additionally, 2-AG is produced by two selective enzymes (DAGLs). All enzymes responsible for cannabinoid biosynthesis are sensitive to calcium ions. Moreover, those endocannabinoids are fat-soluble molecules and are hydrolyzed by an enzyme known as fatty acid amide hydrolase (FAAH). Such biochemical reactions produce arachidonic acid, ethanolamine, and glycerol, as products (Baswan et al., 2020; Ritter et al., 2016).

Exogenous cannabinoids comprise both phytocannabinoids and synthetic ones. The most known and studied phytocannabinoids include THC, CBD, and cannabigerol (CBG). Specifically, CBD (Figure 9.2) is not related to any psychoactive activity. Besides, CBD has been pointed out as a potential active pharmaceutical ingredient (API), due to its potential application in medicinal and therapeutic use and its cosmetic properties. In this context, the prescription of CBD-based products for the treatment of some neurological disorders, such as anxiety, pain, and seizures is already a reality (Nahar et al., 2020).

One of the reasons that may explain the lack of psychoactive effects associated with CBD may be due to its low binding affinity to CB1 receptors (mostly present in the brain). Such absence of psychoactivity was observed in a study carried out by Hammell et al. (2016), in which the use of transdermal CBD for the treatment of monoarthritis was evaluated and the exploratory behavioral activity was similar to that of the study's control group. This result indicates that psychoactive alterations were avoided with the use of CBD. Interestingly, THC has been pointed out as a partial

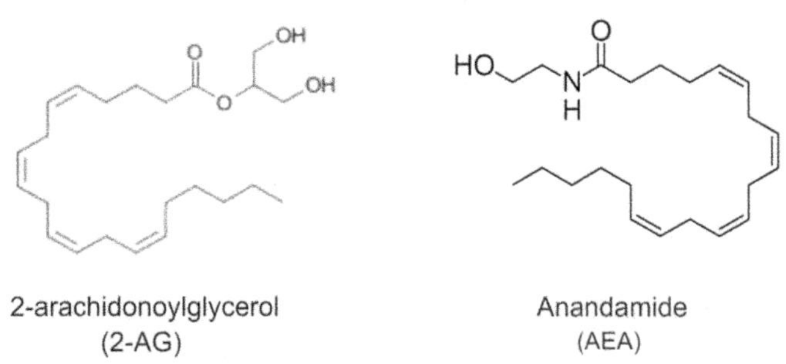

FIGURE 9.1 Chemical structure of the main endocannabinoids.

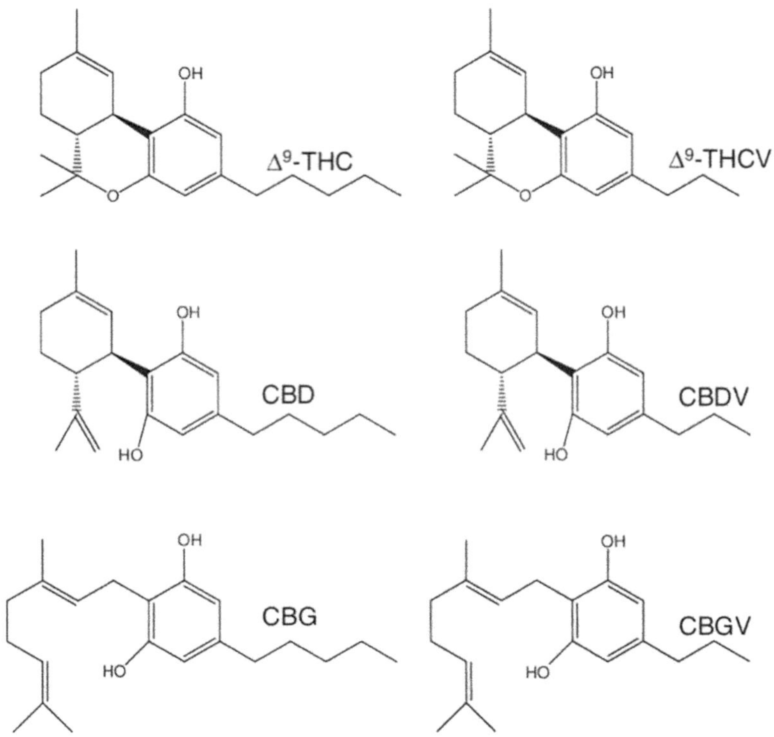

FIGURE 9.2 Chemical structures of the main known phytocannabinoids: cannabidiol (CBD), cannabidivarine (CBDV), Δ9-tetrahydrocannabivarin (THCV), and cannabigerol (CBG).

CB1 agonist, whereas CBD is a CB1 antagonist, being able to block any psychotropic side effects from the administration of cannabis-based medicines (Tóth et al., 2019).

However, CBD shows an intriguing behavior, as it is considered a functional CB1 activator, by indirectly stimulating CB1 receptors, increasing AEA levels by inhibiting the action of the FAAH enzyme. Such capacity leads to a local increase in endogenous cannabinoids, providing a local ECS stimulation. Therefore, it can be considered an important AEA reuptake inhibitor (Petrosino et al., 2018; Tóth et al., 2019).

9.2.1 Cutaneous ECS

The skin constitutes the largest human organ, comprising three major layers: epidermis, dermis, and hypodermis. The skin provides an important role as a physical barrier against pathogens and a variety of environmental challenges, in addition to regulating temperature and water loss, and dehydration (Tóth et al., 2019). The epidermis shows different layers, also known as *strata*. The basal layer is the deepest portion of the epidermis, where keratinocytes and melanocytes originate. Keratinocytes are keratin producers and play an important role in the formation of the epidermal barrier, as they produce and secrete lipids, in addition to being involved in the process of vitamin D obtention. Melanocytes, on the other hand, are responsible for the production of melanin, which gives color to the skin. The epidermis is also composed of the spinous layer, made up of irregularly shaped cells that come into contact with neighboring cells by desmosomes. In this layer, dendritic cells can be found that play an essential role in the defense against pathogens, presenting antigens to lymphocytes (Raiker et al., 2020).

The granular layer is composed of cells that are loaded with keratin precursors. Finally, the *stratum corneum*, the upper portion of the epidermis, is composed of keratin and dead keratinocytes.

These, in turn, secrete defensins that are part of the first immune defense system. The *stratum corneum* is composed of keratin and a lipid matrix, representing the main limitation for the absorption of API into and through the skin. The other layers are more hydrophilic in nature and so, in this case, IPAs are required to be adequately lipophilic, allowing their solubility within the lipid matrix of the *stratum corneum*, although they must be hydrophilic enough to allow passage through the other layers (Tijani et al., 2021).

The skin and together with its appendages have been emerging as an active neuro-immuno-endocrine organ as they produce several hormones and cytokines, which comprise an important neuronal complex network with a wide range of functions, an immunological organized system involved in detecting external signals and in the production and release of important growth factors, hormones, and bioactive substances (Bíró et al., 2009; Tóth et al., 2019). The skin barrier is a complex, cooperative, and sophisticated system that includes physical, immunological, and microbiological barriers finely tuned. Such precise regulation requires a well-defined, controlled, and sequential communication cascade connecting different cell types, comprising immune cells, e.g., immune cells Langerhans cells, dendritic cells, macrophages, mast cells, various T cell populations and other specialized skin cells, as keratinocytes, fibroblasts, melanocytes, sebocytes, adipocytes (Tóth et al., 2019).

Recent discoveries have reported that the skin shows a functional ECS called the cutaneous ECS, being involved in several biological and homeostatic cellular processes, such as proliferation, growth, differentiation, and apoptosis, among others (Bíró et al., 2009). Cutaneous ECS comprises endocannabinoids, receptors, and degrading enzymes are expressed. The cutaneous ECS system is responsible for modulating the proliferation, differentiation, growth, and apoptosis of different cell types, as well as tumorigenesis (Bíró et al., 2009; Tijani et al., 2021) and local cytokine production (Tóth et al., 2019).

In general, the main physiological role of the cutaneous ECS appears to provide the balance among cell proliferation, differentiation, and survival. Additionally, cutaneous ECS have proved to be involved in skin immunological and allergenic processes (Bíró et al., 2009). In this context, cutaneous ECS has been repeatedly investigated regarding its role in undesired cell proliferation and growth, sebum production, and inflammatory conditions. Maccarone et al. (2003) reported evidence that human keratinocytes are part of a functional ECS. Furthermore, it was established that differentiating keratinocytes have reduced levels of AEA, due to a decrease in its enzymatic synthesis and an increase in its degradation (Caterina, 2014; Sheriff et al., 2020). Additionally, cannabinoid receptors proved to be widely distributed in nerve fibers in the skin and mast cells, reinforcing their potential role in treating inflammatory conditions and other skin disorders (Ständer et al., 2005).

The distribution and expression of CB2 receptors suggest that they are involved in the wound-healing process in mice, as demonstrated by Zheng et al. (2012). The CB2 receptor is a component of the so-called immunocannabinoid system, as they are expressed in lymphocytes, NK cells, and monocytes, in addition to lymphoid organs (Kupczyk et al., 2009; Cabral & Griffin-Thomas, 2009). Li et al. (2016) showed the wound re-epithelialization ability of cannabinoid ligands, Moreover, they investigated the role of CB2 receptors in the processes of inflammation and fibrogenesis using CB2 agonists and antagonists. Their results evidenced the anti-inflammatory and anti-fibrogenic properties of CB2 receptor agonists during wound healing.

Sebaceous glands produce lipid-rich sebum, which is part of the skin barrier system. Besides, sebum is also an important factor for the microbiological barrier function of the skin, due to its antimicrobial lipid content, which controls the skin microbiota growth (Tóth et al., 2019). Sebaceous glands are the main targets for the treatment of common dermatoses such as acne and seborrhea. Additionally, their dysfunction leads to skin dryness-related skin disorders, which can greatly impact life quality and be the leading cause or symptom of several skin diseases, such as psoriasis, itching, and atopic dermatitis (Bíró et al., 2009; Tóth et al., 2019). The expression of CB1 in differentiated sebocytes and CB2 in on-differentiated basal sebocytes was demonstrated in 2005 by Ständer et al (Ständer et al., 2005). CB2 is related to the maintenance of homeostatic sebaceous lipid production. Endocannabinoids can promote the suppression of cellular proliferation, induce apoptosis, and modify the transcription of cytoskeleton proteins, which affect the cell cycle. An intriguing capacity of CB2 agonists to promote sebum/lipid production, and increase endocannabinoid production, besides inhibiting their degradation has been reported in a dose-dependent manner. Indeed, CB2 agonists, such as AEA and 2-AG can be FAAH or monoacylglycerol lipase (MAGL) inhibitors, being promising therapeutic agents for dry skin-related conditions and potential cosmetic ingredients (Czifra et al., 2012; Zákány et al., 2018).

The expression of CB1 is well-proven in both hair follicles and scalp (Mercati et al., 2012; Ständer et al., 2005; Zheng et al., 2012). Besides, endocannabinoids and their derivatives were reported to be present in hair fiber (Telek et al., 2007). Nevertheless, the presence of CB2 in hair follicles is still open to discussion (Telek et al., 2007; Yu et al., 2010). It is still necessary to clarify the actual role of each receptor in hair growth. CB2 agonists can lead to lipogenesis in the scalp, being contraindicated in cases of androgenic alopecia, which is often marked by abnormal lipogenesis and chronic agonism of the peroxisome proliferator-activated receptor gamma (PPAR-γ) (Sadgrove & Simmonds, 2021, Ramot et al., 2020; Ho et al., 2019). CB2 receptors are not expressed in hair follicle epithelium, unlike CB1, which is widely distributed in this region. In this sense, CB1 agonists are negative growth regulators. Therefore, it is important to select CB2 agonists for hair growth potential treatments (Sadgrove & Simmonds, 2021).

Hair follicles feature prostamide receptors. Bimatoprost is an AEA metabolite and mimics the effects of prostamide. Thus, a cover concept arises, as hair follicles can be considered both targets and sources of endocannabinoids (Bíró et al., 2009). In this sense, several studies have observed the positive effect of treatment with bimatoprost resulting in follicle growth, favoring hair growth. However, there are reports that endocannabinoids negatively regulate hair growth via CB1 receptors, which justifies careful investigation to clarify the effect of CB1 agonists on hair growth and the possible application of CB1 antagonists in treating alopecia areata (Bíró et al., 2009).

Besides CB1, other cannabinoid-sensitive receptors are expressed in human follicles (Bíró et al., 2006; Borbíró et al., 2011; Szabó et al., 2019). The activation of TRPV1 receptors is shown to induce the catagen phase, leading to inhibition of keratinocyte proliferation in the cell matrix. Consequently, this sequence of events delays hair shaft elongation (Bíró et al., 2006; Telek et al., 2007). Similar to TRPV1, the TRPV4 receptor is expressed in hair follicles. Such a finding was published by Szabó et al. (2019) when evaluating the effects of TRPV4 agonists. In fact, the activation of TRPV4 receptors caused the reduction in hair shaft reduction *in vitro*, inducing premature catagen regression. Moreover, the effects resulting from the activation of TRPV4 receptors can be reversed with a specific antagonistic agent, as was shown in this work.

Based on the investigations of Telek et al. (2007), it was demonstrated that AEA inhibits, in a dose-dependent manner, the elongation of the hair shaft, in addition to inhibiting the keratinocytes proliferation, without affecting the pigmentation of hair follicles. This finding is of great relevance, since CB1 and not CB2 receptors are expressed in the follicle. Moreover, their expression increases at the catagen phase. Interestingly, hair follicles not only respond but express endocannabinoids AEA and 2-AG. As part of the Telek et al. work, the effect of THC on the follicles was investigated. It has been shown that THC acts similarly to AEA, having the capacity to inhibit hair growth, the difference, however, is that, unlike AEA, THC demonstrated inhibitory effects on follicular melanogenesis. Additionally, the reversion of the inhibitory effects on hair growth was achieved with AM-251, a CB1 antagonist. Thus, it can be assumed that in cases of alopecia, the expression of endocannabinoids might be increased.

Marquart et al. (2010) evaluated the role of CB1 in systemic sclerosis, since despite the similarities found between the receptors, the expression patterns of CB1 and CB2 are distinct, causing them to have different effects. Their work used bleomycin as an inducing agent of fibrosis in rats and it was shown that the inactivation of CB1 receptors resulted in protection against fibrosis. Their data suggest that CB1 is related to profibrotic processes and its inactivation led to a reduction in leukocyte infiltration, preventing the release of profibrotic mediators and, thus, leading to protection against fibrosis.

The activation of the ECS can block the evolution and progression of several types of cancer, including skin cancer. Furthermore, AEA seems to be related to induced apoptosis in non-melanoma tumor cells of murine cell lines (Shao et al., 2021). Llanos Casanova et al. (2003) conducted several studies, to verify the capacity of cannabinoids to inhibit both skin tumor growth and angiogenesis. Their results showed the presence of CB1 and CB2 receptors in human skin, in keratinocytes, and even in skin carcinomas.

In vitro and *in vivo* studies claim that the use of cannabinoids promotes beneficial effects in the treatment of different cancers that can affect the skin, becoming a promising therapeutic agent for anticancer chemotherapy (Shao et al., 2021). Accumulated evidence shows that the cutaneous ECS is closely involved in the regulation of skin cell proliferation and apoptosis *in vivo*. For instance, different types of skin tumors, such as basal cell carcinoma, squamous cell carcinoma, and melanomas express both CB1 and CB2 receptors (Bíró et al., 2009).

CB1 and CB2 receptors are involved in cannabinoid-induced apoptosis. This was one of the important conclusions that Llanos Casanova et al. (2003) found in their study when analyzing the effect of selective antagonists on these receptors. Antagonists of cannabinoid receptors were able to prevent the apoptotic effect induced by their agonists. Tumors need an adequate supply of oxygen and nutrients for their growth and act in the production of pro-angiogenic factors such as the Endothelial Growth Factor (VEGF). In their experiments, tumors treated with cannabinoids showed a reduction in the expression of pro-angiogenic factors such as VEGF and narrowing of blood vessels. Furthermore, in tumors treated with cannabinoids, the expression levels of Epidermal Growth Factor Receptors (EGF-R) were found to be low (Llanos Casanova et al., 2003). EGF-R inhibitors can block tumor growth, as the signaling pathway of this receptor appears to be one of the essential pathways in the malignant progression of tumors, in addition to being important in the development of skin epithelial diseases (Ali et al., 2012; Khan et al., 2020).

Moreover, topical administration of CB1 and CB2 agonists can inhibit the growth of malignant skin tumors in mice, which has been closely related to induced tumor cell apoptosis and impaired local vascularization (Llanos Casanova et al., 2003). Corroborating such evidence, CB1 and CB2 agonists also inhibited melanoma growth *in vivo*, which was also accompanied by reduced signaling for proliferation and angiogenesis, besides the observation of unusual blood vessel morphology (Blázquez et al., 2006). In healthy skin, there is expression of CB1 and CB2 receptors and in cancer cells, there is an overexpression of cannabinoid receptors. Blázquez et al. (2006) evaluated the antitumor potential of cannabinoids. Their results found the presence of CB1 and CB2 receptors in melanoma cell lines of murine and human models. In addition, *in vivo* tests demonstrated a reduction in tumor progression.

Other published results demonstrate that activation of such receptors resulted in apoptosis induction in tumorigenic cells, inhibiting tumor growth (Eagelston et al., 2018), without affecting healthy epidermal cells and preventing metastases through the activation of CB2 receptor agonists, which can reduce the expression of VEGF (Kupczyk et al., 2009; Nikan et al., 2016).

Thus, given the physiological and regulatory role of cutaneous ECS in a wide range of cell functions, such as proliferation, differentiation, survival, and cytokine production or skin immune competence, for example, and also its relation to different skin disorders, ECS is configured as a key determinant of cutaneous homeostasis and, therefore, a novel target to address different skin disorders and pathological conditions (Baswan et al., 2020, Bíró et al., 2009).

9.3 THERAPEUTIC POTENTIAL OF DERMAL ADMINISTRATION OF CBD

CBD by means of cutaneous ECS receptors can contribute to the treatment of some skin disorders and it is becoming considered a trend in cosmetic products, which bring enthusiastic claims to promote skin healthy appearance and nutrition (Nahar et al., 2020).

Due to its lipophilic constitution, CBD has a great affinity for the *stratum corneum*, making it a promising agent for cutaneous administration (Casiraghi et al., 2020; Itin et al., 2020). Therefore, dermal application of CBD appears to constitute a good delivery route for the treatment of skin diseases (Bíró et al., 2009). Topical formulations for dermal application, allow the delivery of API at a local level to precisely treat dermatological conditions. The solubility of the API in the formulation vehicle must be adequate to promote the availability of the bioactive molecule in the ideal concentration on the skin.

In order to study the CBD solubility and the influence of the type of vehicle used in topical preparations, Casiraghi et al. (2020) evaluated the behavior of two preparations. The first formulation presented a lipophilic composition, as in an ointment-like formulation. The second one was a hydrophilic propylene glycol-based gel, which provided greater CBD skin permeation and retention compared to the ointment. However, the ointment-like formulation provides interesting results, proving that the skin permeation and retention profile of CBD can be modulated by changing the composition and the hydrophilic character of the formulation.

Due to the crosstalk between adenosine and cannabinoid receptors, CB agonists are increasingly being considered in dermal treatments for a wide range of applications from lipogenesis, hair growth, and anti-inflammatory to antioxidant activity, which often leads to aesthetically pleasing effects in the skin. In this sense, cannabinoids used to target dermal conditions are often called "cutannbinoids" (Sadgrove & Simmonds, 2021).

Given the relevance of the subject, some of the most common skin disorders and the possibility of treatment with the use of phytocannabinoids, specifically CBD, will be presented next, with a focus on CBD.

9.3.1 Anti-Inflammatory and Antioxidant Properties

The skin constitutes the ultimate barrier against environmental conditions that can generate reactive oxygen species (ROS). Oxidative stress is known to induce cell damage and lead to inflammatory conditions that can become chronic and are involved in several skin disorders, including skin cancer and aging (Bickers & Athar, 2006). The topical application of CBD-loaded formulations can be an important tool to treat inflammatory, hyperproliferative, and pro-oxidative skin disorders. Over the past few years, research has been carried out to investigate the use of cannabinoids in pro-oxidative conditions such as itching, inflammation, and even skin cancer (Mounessa et al., 2017).

CBD features both antioxidant and anti-inflammatory effects, being considered as a cytoprotective agent able to hinder oxidative metabolic changes in skin cells (Gęgotek et al., 2019). CBD has shown an anti-inflammatory character, being related to the release of cytokines and lymphocyte proliferation (Aizpurua-Olaizola et al., 2017). One of the reasons why CBD can produce an anti-inflammatory effect is the fact that it is considered an inverse agonist of the CB2 receptor, blocking the migration of macrophages, and neutrophils, and reducing inflammation (Pisanti et al., 2017). Monocyte apoptosis may interfere with macrophage development, thus the pro-apoptotic effect of CBD may be an important mechanism contributing to anti-inflammatory properties (WU et al., 2018).

The antioxidant effect of CBD can be understood through different mechanisms. One of which lies in the fact that CBD, like other antioxidants, blocks free radical chain reactions, transforming them into less reactive molecules and also reducing the activity of enzymes that produce ROS. CBD is reported to show chelating properties, which reduces ROS production due to the inactivation of transition metal ions involved in the Fenton reaction that can generate highly reactive hydroxyl radicals (Hamelink et al., 2005). Such direct antioxidant properties are similar to those found in butylated hydroxytoluene (BHT), a classic antioxidant agent (Atalay et al., 2020). Still, CBD can significantly increase the enzymatic activity of superoxide dismutase and thioredoxin reductase, both important antioxidant enzymes in keratinocytes (Jastrząb et al., 2019).

Additionally, the antioxidant activity of CBD can occur through binding to cannabinoid receptors, of which it is a weak agonist, through other receptors such as TRPVs or even increasing the level of endocannabinoid 2-AG indirectly, which has a reduced level due to lipid imbalance occasioned (Biernacki et al., 2021; Scheau et al., 2021). CBD has been demonstrated to induce the expression of the stress-induced enzyme Hemeoxygenasel (HMOX1), one of the key NRF2 target genes, with antioxidant and anti-inflammatory properties (Juknat et al. 2012). Additionally, in normal human keratinocytes, CBD is related to an increased level of NRF2 target genes, besides upregulating the levels of HMOX1 (Casares et al., 2020).

The expressive antioxidant profile attributed to CBD, makes it to be considered the phytocannabinoid with the most effective potential protective activity against UVA radiation, increasing the viability of keratinocytes and melanocytes (Vacek et al., 2021), while it can penetrate and permeate the upper skin layers and, due to its lipophilic profile (Biernacki et al., 2021). Excessive exposure to UVA and UVB radiation, comprise a major risk factor for the development of skin cancer, given its cumulative effect, which can promote the activation of oncogenes and block tumor suppressor genes (Chudnovsky et al., 2005). It is well-known that UVB rays penetrate the epidermis, affecting the metabolism of keratinocytes, while UVA radiation reaches the dermis, mainly affecting cells of the immune system and microvessels of the circulatory system, reaching the organism as a whole. Together, UV radiation can increase the generation of ROS causing harmful changes, such as the appearance of skin lesions, photoaging, and the development of skin cancer (Biernacki et al., 2021; Gegotek et al., 2016).

The antioxidant activity of CBD was studied by analyzing the proteomic profile of 2D and 3D fibroblast cultured cells exposed to UVA and UVB. The cytoprotective effect of CBD was found to follow different patterns in cultured cells according to the type of UV radiation and the number of cell layers. The extension of cell damage and reaction was significantly different in multilayered cell cultures, which should be taken into consideration when designing UV-protective topical formulations (Gęgotek et al., 2019). CBD can affect the expression of inflammatory proteins, as well as those involved in redox balance and apoptosis originating from UV exposure. Besides, CBD treatment seemed to counteract and even partially eliminate UV cell damage and inflammatory response by modifying anti-apoptotic and pro-apoptotic cytokines (Atalay et al., 2020). Thus, CBD has proved to maintain keratinocyte proteomic profile, being pointed as a cytoprotective agent against UV-induced metabolic and biochemical cell damage in epidermal cells.

Łuczaj et al. (2020) studied the changes in keratinocyte phospholipid profile in rat skin following UV exposition and the influence of topical application of CBD. Topical application of CBD could reverse some of the effects of UV radiation over sphingomyelin and sphingomyelinase and ceramides. Thus, CBD proved to be able to prevent the changes that affect phospholipids *in vitro*, reducing the levels of ROS and the action of phospholipase A2 (PLA2) on keratinocytes, preventing lipid peroxidation and providing an improvement in the state of the transepidermal barrier (Jastrząb et al., 2019; Łuczaj et al., 2020).

Moreover, CBD can induce the expression of erythroid-derived nuclear factor 2, a transcription factor that influences the transcription of antioxidant proteins which, in turn, feature a protective effect against oxidative stress caused by UV radiation (Ryšavá et al., 2020; Vacek et al., 2021).

9.3.2 Psoriasis

Psoriasis is a hyperproliferative skin disease that affects specialized epithelial cells, keratinocytes, and immune cells. It is a chronic and non-contagious inflammatory disorder, often referred to as an autoimmune disease, that is considered a multifactorial disease since it does not result from a specific cause for its development. Psoriasis manifests as inflammation accompanied by hyperproliferation of keratinocytes and skin inflammation, with an accumulation of dead skin on the affected areas, appearing primarily on the knees, and elbows (Stinchcomb et al., 2004). The most common form of occurrence is the so-called plaque psoriasis, presenting as silvery/whitish areas in scales surrounded by reddish margins (Ritter et al., 2016). Itching is also common, being present in more than 70% of patients (Castilho et al., 2021).

In psoriasis, external causes, such as trauma or infection, can sensitize keratinocytes, inducing the activation of dendritic cells and macrophages which, consequently, by producing cytokines and activating T lymphocytes, contribute to triggering an inflammatory cascade (Caterina, 2014). The proliferation of keratinocytes in the epidermis is increased and persistent in psoriatic skin (Donetti et al., 2012). Besides, patients with psoriasis can develop psoriatic arthritis, in addition to dermatological manifestations. Moreover, there are frequent reports of inflammation in peripheral joints, such as synovitis (Castilho et al., 2021). Additionally, studies have shown that patients suffering from psoriasis have a higher incidence of cardiovascular disease and prevalence of metabolic syndrome (Lima & Lima, 2011).

In general, hyperproliferative disorders impact and reduce cell cycle time. For instance, in psoriatic skin, the keratinocyte cell cycle takes 4 days. Nevertheless, the time required for both keratinocyte maturation and elimination to occur, under normal conditions, is around 26 days (Sugawara et al., 2012). Keratinocytes can produce neurotransmitters and, therefore, can regulate a skin barrier function (Baroni et al., 2012). In addition to playing a role in the maintenance of the skin's physical barrier, keratinocytes can constitute an immuno-microbiological barrier to pathogens, producing different cytokines involved in the inflammation process, such as interleukins 6 and 13 (IL-6, IL-33) (Balato et al., 2012) and IFN-γ that stimulates the activity of antigen-presenting cells (Abbas et al., 2019).

It is known that certain types of keratin, produced by keratinocytes, are associated with psoriasis, such as k6 and k16 keratins (Ramot et al., 2013). Key cytokines, interleukins 17 and 22 (IL-17 and IL-22), which trigger the proliferation of keratinocytes and the expression of keratins k6 and k16 are produced in inflammatory reactions resulting from psoriasis (Eagelston et al., 2018). It is noteworthy that k6 keratin, in addition to regulating the migration of keratinocytes in the wound repair process, is related to hyperproliferative processes (Rotty & Coulombe, 2012). In order to investigate the behavior and interaction of k6 and k16 keratins with cannabinoid receptors, Ramot et al. (2013) analyzed the effects of agonists and antagonists on CB1 receptors against the proliferation of keratinocytes. In their work, it was shown that the CB1 agonists were able to inhibit the production of both k6 and k16, reducing the proliferation of keratinocytes. Thus, the results demonstrated that CB1 receptors can act as important regulators of keratinocytes. Previously, Maccarrone et al. (2003) had already shown that the activation of receptors by the endocannabinoid ligand, AEA, decreased the production of keratin k6.

Mast cells are inflammatory cells that participate in innate and adaptive immunities. They are involved in autoimmune diseases and wound-healing processes (Stone et al., 2009). The number of mast cells is known to be markedly increased in psoriatic patients (Suttle et al., 2012). In this context, the activation of CB1 receptors can limit the hyperproliferation of mast cells. Therefore, CB1 agonists constitute an attractive class of bioactive agents in the future treatment of psoriasis (Sugawara et al., 2012). Thus, cannabinoid agents that activate CB1 can act as promising pharmacological agents in the treatment of psoriasis (Ramot et al., 2013).

A considerable study carried out by Jonhathan D. Wilkinson and Elizabeth M. Williamson has been conducted in order to evaluate the use of cannabinoids in the treatment of psoriasis. It was demonstrated that phytocannabinoids inhibited the proliferation of keratinocytes in a dose-dependent manner. A wide range of cannabinoids was evaluated: THC, CBN (Canabinol), CBD, and CBG, with CBD and cannabigerol (CBG). In this study, CBD and CBG stood out for their antiproliferative activity. An important conclusion to consider is that phytocannabinoids do act to inhibit keratinocytes' proliferative activity and may be potential agents for the treatment of psoriasis. However, this predominant action does not take place through the CB1 and CB2 receptors (Wilkinson & Williamson, 2007).

9.3.3 Acne and Seborrhea

Acne is an inflammatory disease that affects the pilosebaceous unit, an important skin appendage. It is a multifactorial disorder, and it most frequently affects adolescents predominantly, being both the face and back, the most affected regions (Picardo et al., 2017). The face features the highest distribution of sebaceous glands (Ali et al., 2015). *Propionibacterium acnes* is commonly involved in acne-borne infection and inflammation. CBD can inhibit *P. acnes* proliferation, emphasizing its antimicrobial function. Furthermore, such phytocannabinoids, being highly absorbable by the skin, can contribute to improving the appearance of acne (Ferreira et al., 2023).

The initial step related to the pathogenesis of acne is stimulation of the sebaceous gland, increasing the sebum production (Oláh et al., 2017), followed by qualitative changes in the sebum constitution, such as an increased proportion of saturated and monounsaturated fatty acids, and the establishment of an inflammatory condition (Picardo et al., 2017). Sebum plays an important role in promoting and maintaining the skin barrier. Squalene represents one of the major components of sebaceous secretion and acts as a protective factor against sunburn and UV rays. In addition, sebum contains lipids with notable antimicrobial activity besides contributing to thermoregulation. Sebocytes, modified keratinocytes, produce cytokines and inflammatory mediators (Rotsztejn et al., 2011).

Given the current treatments available for acne, the search for innovative solutions that are at the same time effective and with reduced side effects emerges. In this context, a remarkable study carried out by Oláh et al. (2014) showed that CBD promoted a complex anti-acne effect in human sebocytes. Such property is related to lipostatic, antiproliferative, and anti-inflammatory activities mediated by the activation of non-cannabinoid receptors, and TRPV4 receptors. CBD normalized the excessive lipogenesis in sebocytes, without affecting homeostatic sebum lipogenesis (Oláh et al., 2016).

Several *in vitro* evidence showed that CBD could be an innovative therapeutic alternative to the management of acne due to its action on sebum production, sebocyte proliferation, and inflammatory conditions. Oláh et al (2014) showed the sebostatic effects of CBD on sebaceous gland function, by inhibiting the lipogenic actions of various compounds and suppressing sebocyte proliferation via the activation of TRPV4 receptors. They showed that in 24 h CBD alone was able to prevent lipogenesis when sebocytes were stimulated with AEA, a lipogenic endocannabinoid (Oláh et al., 2014).

In addition, CBD features anti-inflammatory activities that are independent of the activation of the TRPV4 receptor and cannabinoid receptors. Briefly, the anti-inflammatory effect can occur through the reduction in transcription activity of nuclear factor (NF-κB) (Kozela et al., 2010; Oláh et al., 2014), a transcription factor, that enables genetic inducing and production of cytokines. In this way, CBD effectively controls inflammation (Li et al., 2016). The anti-inflammatory properties of CBD have been previously reported. The influence of CBD on inflammatory signaling has been demonstrated by Sangiovanni et al, who studied the effects of *C. sativa* extract (with 5% CBD) on human keratinocytes and dermal fibroblasts. Their results showed that *Cannabis* extract downregulated 26 inflammatory genes, while CBD alone downregulated 15 genes involved in wound healing and inflammatory conditions (Sangiovanni et al., 2019).

In 2015, Ali et al. published a single-blind, split-face study of the topical application of a 3% cannabinoid-loaded creams and their capacity to reduce erythema and sebum content on the skin of healthy male volunteers with acne lesions and erythema for 12 weeks. It was found that the topical formulations containing CBD promoted the reduction of erythema and showed efficacy in reducing sebum production, compared to the control (i.e., the vehicle-treated side. Such results showed that CBD proved to be a safe and effective treatment for the treatment of common conditions that affect the skin, such as acne and psoriasis. Such findings were later supported by the work conducted by Spleman et al. (2018) with a synthetic CBD-loaded formulation, known as Permetrex™, in 20 healthy volunteers for 28 days. On Day 14 and Day 28, safety, including adverse events and tolerability (erythema, scaling, dryness, burning/stinging, and contact dermatitis) were measured. The phase I clinical results showed that a 5% formulation of CBD applied twice daily was well tolerated, suggesting that its use in moderate to severe acne is safe. Additionally, the impact on acne progression has been observed.

The effects of non-psychotropic phytocannabinoids, notably CBD, to treat acne include the impact on sebocyte viability, which, together with its antibacterial, antiproliferative, and anti-differentiation effect on human keratinocytes, make CBD an efficient anti-acne agent (Tóth et al., 2019).

9.3.4 Eczema and Dry Skin Conditions

Eczema is a universal term to describe a set of symptoms that affect the skin including xerosis, itching, scaling, and inflammation. Also known as atopic dermatitis, it is a common, non-contagious skin disease and is related to genetic susceptibility associated with environmental factors, such as allergens and infections (Nutten, 2015). It commonly develops in childhood, although some initial symptoms might appear in adulthood. Eczema also can occur due to changes in the skin barrier components, specifically with ceramides endogenous production (Addor & Aoki, 2010). Atopic dermatitis is usually accompanied by other immune-borne disorders, such as allergic rhinitis, asthma, and food allergy (Palmieri et al., 2019).

The pathophysiology of eczema involves changes in skin barrier function leading to an increased transepidermal water loss, which is accompanied by itching and inflammation (Ritter et al., 2016). Di Nardo et al. (1998) demonstrated that the levels of ceramides 1 and 3 are reduced, while the levels of free cholesterol are increased. Ceramides, together with fatty acids and cholesterol, structurally compose the epidermal lipid matrix, filling spaces between the keratinocytes, and preventing water loss (Meckfessel & Brandt, 2014; Moore & Rawlings, 2018). It is also known that filaggrin, a protein that contributes to the integrity of the skin barrier, is found in reduced levels in patients with atopic dermatitis (Nutten, 2015).

To access barrier integrity, transdermal water loss (TEWL) can be measured. TEWL is an important parameter for evaluating eczema progression. In atopic dermatitis, TEWL is increased, resulting in dry skin, leading to a predisposition to aggression, that leads to inflammation, desquamation, and itching (Addor & Aoki, 2010). TEWL is considered one of the most important skin integrity measurement parameters that can be evaluated by measuring instruments (Akdeniz et al., 2018; Montero-Vilchez et al., 2021).

By definition, itching is characterized by an "unpleasant sensation that elicits the desire or reflex to scratch". Pruritus can become chronic when the itching is persistent, usually over 6 weeks, significantly affecting the quality of life (Ständer et al., 2005). The development of therapeutic alternatives for both topical and systemic treatments for chronic itch has been emerging recently (Yosipovitch et al., 2018; Ständer et al., 2006). Topical cannabinoid agonists can alleviate itching in an antipruritic effect from the activating cannabinoid receptors or by enhancing the activity of endocannabinoids (Ständer et al., 2005). In this way, N-palmitoyl ethanolamine (PEA), an endocannabinoid, has been standing out as the most promising cannabinoid agonist in clinical trials (Ständer et al., 2006; Eberlein et al., 2008; Visse et al., 2017).

CBD is reported to show a similar neuromodulator effect as capsaicin, a well-known option to treat neuropathic pain and itching through the activation of TRPV1 receptors, widely known as capsaicin receptors. Both CBD and PEA are believed to have indirect effects on cannabinoid receptors, as they proved to increase AEA levels (Avila et al., 2020).

In a pre-post observational study, the effects of a topical gel formulation containing 1% CBD in patients with atopic dermatitis were evaluated. Patients were evaluated regarding disease severity and pruritus using the Eczema Area and Severity Index (EASI), Visual Analogue Scale-Pruritus (VAS), and 5-D Pruritus Scales, respectively, during 14 days. Besides, subjects were asked to take a questionnaire, which included satisfaction with their current eczema medication, frequency of application, difficulty in topical use, and the occurrence of side effects. After 2 weeks of treatment, a significant improvement was observed in the improvement of pruritus, excoriations, erythema, lichenification, and edema (Maghfour et al., 2021).

In order to identify the possible role of cannabinoid receptors in the treatment of atopic dermatitis, Nam et al. (2016) evaluated the use of CB1 receptor agonists, using oxazolone for eczema induction in nude mice. Oxazolone causes epithelial hyperplasia and influx of inflammatory cells into the skin (such as mast cells) and higher histamine levels, known to increase capillary permeability and regulate immune reactions. CB1 receptor agonists reduced skin thickness, suppressed the inflammatory effects of oxazolone, and inhibited serum histamine concentrations. Taken together, the results of this study suggested that CB1 receptors play an important anti-inflammatory role by modulating mast cell degranulation, preventing the sequence of events and symptoms of eczema.

Additionally, Palmieri et al. (2019), investigated the effects of a treatment based on the topical application of a CBD-based ointment for the treatment of skin diseases and scars. They evaluated the evolution of volunteers with psoriasis (*n*: five patients), atopic dermatitis (*n*: five patients), and outcome scars (*n*: ten patients), for 3 months with a twice-daily application. The topical use of CBD-based ointment resulted in positive effects on skin parameters, such as increased hydration in the treated area and reduced TEWL, reduction of blemishes, scarring, papules, and pustules, as well as enhanced elasticity. Supporting previous findings, this study showed that the topical application of CBD-based formulations is a safe, effective, and patient-friendly approach to improving the quality of life of patients with skin disorders.

It is believed that, due to the high lipophilicity of CBD, it could be able to achieve good penetration by accumulating within sebaceous glands and hair follicles, which is considered the preferential penetration route for steroids and other components with similar chemical properties. Furthermore, cannabinoid receptor inhibitors can increase AEA activity by inhibiting the FAAH enzyme, being a promising target for the treatment of atopic dermatitis (Eagelston et al., 2018).

Contact dermatitis, on the other hand, is a sensitivity reaction that occurs due to exposure to an allergen. In a first contact, there is activation of the inflammatory process with activation of T lymphocytes, which enables the amplification of the inflammatory response (Ritter et al., 2016). Petrosino et al. (2018) sought to assess the anti-inflammatory capacity of CBD against allergic contact dermatitis. *In vitro* experiments showed that CBD was able to inhibit MCP-2, a chemokine able to recruit macrophages and mast cells to the infection site. CBD promoted a reduction in the level of other cytokines, such as IL-6 and IL-8. Additionally, their results found that CBD can act indirectly by activating CB2 receptors, stimulating an increase in AEA levels, in addition to desensitizing TRPV1 receptors (Palomino & Marti, 2015; Petrosino et al., 2018).

9.3.5 Alopecia and Hirsutism

Both androgenetic alopecia and alopecia areata comprise the most common disorders that lead to hair loss. Androgenetic alopecia leads to miniaturization of hair follicles making the hair shaft thinner and shorter. Androgenic alopecia is known to be related to an increase in the level of androgen hormones and has a strong genetic influence (Mulinari-Brenner et al., 2011). Alopecia areata is considered an inflammatory disease and may also be associated with a genetic factor. In this case, hair loss is visible through the formation of plaques on the scalp. On the other hand, such a disorder does not destroy the follicles and the hair can grow back (Hordinsky, 2013).

A considerable list of treatments for alopecia is available however, they can cause local or systemic unwanted toxic effects, such as dermatitis, depression, liver alteration, or reduced libido, among others. The efficiency is also limited due to several drawbacks including limited drug penetration into hair follicles, the target for any therapy for alopecia (Cardoso & Barradas, 2020).

Hair follicles are unique structures with well-defined life-long growth (anagen), regressive (catagen), and resting (telogen) phases that are controlled by vanilloid receptor-1 (Oh et al., 2016). Misregulation of any of the phases is directly related to hair growth disorders such as hair loss, in cases of alopecia, or undesired hair growth, known as hirsutism, a cause related to increased androgen levels (Park et al., 2021; Tóth et al., 2019).

The abuse of certain synthetic cannabinoids has already been related to hair loss and discoloration, leading to the conclusion that hair follicles are influenced by cannabinoids (Inci et al., 2017). The well-established anti-inflammatory and immunosuppressive effects of cannabinoids can justify the observation of cannabinoid-related development of alopecia areata (Tóth et al., 2019). Catenins are important proteins during the anagen phase and activate essential hair growth factors present in follicular cells. Thus, the beta-catenin signaling pathway can be related to the induction of the anagen phase, which is essential for hair growth, in addition to the proliferation and differentiation of keratinocytes in hair follicles. The high expression of beta-catenin in the hair shaft of hair tissues without alopecia has been identified, while the expression was negative in the hair shaft in cases of alopecia. Besides, testosterone-treated dermal papilla cells showed reduced beta-catenin expression, proving that such protein is, in fact, regulated by testosterone. Recently, it was demonstrated that the low expression of beta-catenin induced by testosterone can be reversed with treatment based on CBD, having a modulating function in alopecia (Park et al., 2021). Besides, additional data show that CBD has a wide therapeutic window, with few adverse events and quite a high tolerance for topical application (Bíró et al. 2009; Tóth et al., 2019).

Preliminary studies show that CBD seems to exert a dose-dependent effect on hair growth. Szabò et al., published a hair shaft elongation effect of CBD in low concentrations (0.1 μM), whereas, higher concentrations (10 μM) were able to suppress hair growth. Such dual activity can possibly occur due to the CBD effect on adenosine receptors for hair elongation and on TRPV-4 for inhibition of hair elongation (Szabó et al., 2017). However, Baswan et al. showed that CBD can activate the prostamide receptor, positively affecting hair shaft growth. On the other hand, small doses of CBD can reduce the intrafollicular production of catagen-promoting cytokines, which can be useful in the treatment of capillary loss related to inflammatory processes, where the levels of these cytokines are found to be increased. In addition, CBD can promote or suppress hair shaft elongation, in a dose-dependent manner, acting on adenosine and TRPV4 receptors (Baswan et al., 2020).

It was shown that the application of controlled concentrations of CBD can help to strengthen hair growth, differentially influencing hair elongation, as well as the cycle of human hair follicles (Szabó et al., 2019). Szabó et al observed that CBD can produce dose-dependent opposing effects on hair growth dynamics. in ex vivo studies with human hair follicles. 0.1 and 1.0 μM doses of CBD induced hair shaft growth similar to controls. Nevertheless, 10 μM of CBD, causes significant suppression, besides inducing the follicle catagen phase. Their results showed that an increase in CBD dosage caused a reduction in keratinocyte proliferation. Thus, the authors hypothesized that CBD concentration may lead to differential receptor activation, with low doses favorably affecting hair growth pathways and higher doses activating suppression targets such as TRPV4 (Szabó et al., 2019).

Topical formulations can increase CBD concentration in hair follicles, as it is a fat-soluble molecule and poorly absorbed through the epidermis, which favors penetration into the pores (Szabó et al., 2017). For example, a recent study conducted by Smith and Satino, showing the topical use of hemp oil, CBD rich, in hair growth of patients with androgenetic alopecia was published. Hemp is a variety of cannabis, which contains no more than 0.3% THC and is considered legal in the USA. The participants were instructed to apply the oily extract every day on bald areas for 6 months. As a result, all individuals showed some hair growth resulting from the treatment, with men experiencing slightly more expressive results than women. Overall, there was a nearly 93.5% increase in nonvellus hair after 6 months of therapy. Moreover, no side effects associated with the topical use of CBD were reported during the investigation. Such case reports support the benefits of CBD in producing hair regrowth in joint therapies for alopecia since it acts through different mechanisms in comparison with FDA-approved treatment, minoxidil, and finasteride (Smith & Satino, 2021).

Topically applied, CBD can offer benefits to patients with hair growth disorders through targeting the cutaneous ECS. Thus, the application of CB1 agonists constitutes a promising approach to prevent and reverse hair loss. Importantly, a cautious selection of CBD doses can also be beneficial to activate hair growth. Further investigation on this dose-dependent effect regarding CBD, as well as controlled clinical trials are necessary to clarify the actual therapeutic potential of CBD in alopecias.

9.3.6 Skin Cancer and Skin Tumorigenesis

The recreational use of medicinal marijuana has been intensely observed as a palliative treatment in numerous cases of tumors and cancer treatment (Birdsall et al., 2016; Davis, 2016). A large body of evidence demonstrates that there is a complex cannabinoid signaling pathway that, when properly modulated, exerts significant antitumor activity in a dose-dependent pattern. (Mckallip et al., 2005; Oláh et al., 2017).

Both skin cancer and skin tumorigenesis are proliferative conditions that affect skin cells and can be of great morbidity. In this context, evidence shows that cannabinoids can provide pharmacological modulation and, thus, show an antitumor effect, besides palliative care (Tóth et al., 2019). This is supported by the pro-apoptotic and anti-proliferative activity often associated with cannabinoids and CBD (Kis et al., 2019). However, *in vitro* studies and their results should be carefully analyzed, as they are usually based on essays with extremely high cannabinoid doses and do not show an *in vivo* correlation.

It is also noteworthy that there are different types of skin cancers, such as melanoma and non-melanoma, which promote the malignant transformation of melanocytes, and non-melanoma (basal and squamous cell carcinoma) (Apalla et al., 2017). In fact, non-melanoma skin cancer is one of the most common malignancies in humans (Llanos Casanova et al., 2003). Its emergence can occur through the influence of several factors, such as human papillomavirus infection, exposure to chemical carcinogens, in addition to UV radiation (Scheau et al., 2021).

Topical skin cancer treatment is beneficial, as it allows more accentuated levels of anticancer drugs to be present at the tumor site, which may result in a reduced occurrence of toxic events compared to the systemic route (Micali et al., 2010). In this sense, CBD can be used in cancer therapies by providing a reduction in the production of ROS, as previously observed (Singer et al., 2015). Excessive exposure to UVA radiation can reach the dermis, affecting blood and lymphatic vessels, while UVB rays can be a stimulatory factor for the release of cytokines. As already mentioned, UV radiation leads to the production of ROS, resulting in oxidative stress and resulting in modifications of DNA, proteins, and lipids.

It has been demonstrated that the treatment of mice with melanoma using CBD was able to increase survival time and quality of life compared to the control group treated with cisplatin, an anticancer agent widely used in chemotherapy of different types of cancer. The quality of life factor was analyzed based on the observation of movement patterns and less aggressive behavior among other rats. Furthermore, CBD was able to reduce tumor growth and increase the survival time of tumor-bearing mice significantly compared to non-treated animals. Treatment with cisplatin showed the longest survival time. However, the quality of life of mice treated with cisplatin was significantly lower than that of the other two groups. These data raise the possibility of cannabinoids, such as CBD, being used as auxiliary therapy in patients under chemotherapy, alleviating common symptoms related to treatment-related side effects (Simmerman et al., 2019).

According to data obtained from cell cultures and animal models, CBD can induce apoptosis and favor the prevention of metastasis by inhibiting biochemical factors such as VEGF and EGF-R in tumors, confirming the role of potential adjuvants in cancer therapy, with beneficial effects for pain treatment and inflammatory conditions, which are often related to that (Nikan et al., 2016).

9.3.7 Wound Healing and Other Keratinocyte-Proliferative Disorders

Given the cutaneous ECS role in keratinocyte proliferation and in controlling cellular functions of fibroblasts, topical application of CBD oil is a beneficial approach for the treatment and alleviating different keratinocyte-proliferation-related disorders, such as wound healing and epidermolysis bullosa, among others (Tóth et al., 2019). The beneficial aspects of CBD in wound healing have been reviewed by Weigelt et al. (2021).

Wound healing comprises a complex sequence of sequential phases. During skin injury, keratinocytes and fibroblasts are significantly affected by a wide range of biological mediators, cytokines, and growth factors. During the wound process, several proinflammatory mediators are released, including TNF-a, TGF-a, and VEGF, among others. Preclinical research has shown that dysregulation in the ECS can be related to chronic wounds. Cannabinoids can interact with the ECS and promote wound healing and analgesia (Maida et al., 2020). A prospective study in elderly female patients with non-uremic calciphylaxis and venous leg ulcers showed that topical cannabis promoted rapid complete wound closures when applied daily to both wound beds and peri-wound tissue (Maida et al., 2020, 2021).

In vitro and *in vivo* studies with HaCaT cells have been recently shown that CBD is able to inhibit inflammatory mediators involved in wound inflammatory process, especially those whose expression is strongly induced by TNF-a, such as NF-kB (Esposito et al., 2006; Sangiovanni et al., 2019). Interestingly, the pronounced anti-inflammatory effect of CBD to accelerate wound healing seems to be more important during the first 3 days of post wounding, when CBD-treated injuries showed a significant reduction on inflammatory markers, which was not observed at 7 days of post wounding (Klein et al., 2018).

Chelliah et al (2018) published an observational study where three pediatric patients with epidermolysis bullosa used topical CBD oil on a daily basis. As a result, All patients reported faster wound healing, a significant reduction of skin blistering since day 1, and amelioration of pain related to keratodermas with CBD use. Additionally, two of them were able to stop oral analgesic medication after topical use of CBD with no side effects reported. These results were preliminary and self-reported with possible bias, although they shed a light in elucidating and opening a novel use for CBD-loaded products.

Later, Schräder et al (2019) investigated the potential of CBD-based medicine in pain related to epidermolysis bullosa in three adult patients. Pain is the most debilitating symptom in adult epidermolysis bullosa. CBD has been previously reported to alleviate pain in children with epidermolysis bullosa, through its interaction with ECS in CNS. The anecdotal reports of three adults prescribed with CBD include improved pain scores, reduced pruritus and reduction in analgesic intake. Once again, those results can be biased, as they are based on self-reported outcomes and observations. In this context, it is necessary to further investigate and establish the actual contribution that CBD treatment can bring to epidermolysis bullosa patients.

In order to produce more reliable scientific data of such benefits for the point of care treatment of epidermolysis bullosa, Schräder et al. (2019) recently published an international cross-sectional survey, still based on anecdotal reports from 71 patients with epidermolysis bullosa from five different continents using CBD to treat their condition. CBD improved the perception of pain by 3 points in pain score for 94% of the patients. Moreover, 95% reported the improvement in overall symptoms, 91% observed a reduction in pruritus and 81% in wound healing. The reduction in analgesic medication was noted in 79% of individuals, with the most common side-effect reported being dry mouth.

Casares et al. (2020) reported that topical formulations loaded with CBD increased the levels of HMOX1 and to induce the expression of wound-healing-related keratins 16 and 17 in mice. Their results enabled them to characterize the CBD effects and the mechanisms on human keratinocytes at molecular level. Indeed, CBD was shown to modulate keratinocyte differentiation, skin repair and epidermal cell differentiation, all processes involved in wound-healing pathways. Altogether, these findings led to the conclusion that the *in vivo* application of topical CBD exert a significant anti-inflammatory activity, especially in the early phase of wound-healing process. Therefore, it can be an alternative effective treatment for wound-healing conditions and other skin disorders as atopic dermatitis and keratin-related dysfunctions (Casares et al., 2020; Klein et al., 2018).

9.4 TRANSDERMAL APPLICATION OF CBD

CBD has limited bioavailability in preparations for both oral and systemic routes. Its oral bioavailability is limited due to an extensive first-pass metabolism in the liver, in addition to instability under the influence of pH found through the gastrointestinal tract and high lipophilicity (Mazzetti et al., 2020). The routes of administration have an influence on the speed and efficiency of drug action. The access of the active compound to the target and the drug treatment also depend on the route of administration used. Currently, an increase in the number of studies and developments in the area can be observed, aiming at increasing the bioavailability of drugs and reducing their toxic effects (Mignani et al., 2013).

Because it is a lipophilic compound CBD, has its absorption hampered by the digestive system, Epidyolex® is a CBD-based medicine in the pharmaceutical form of an oral solution approved by the European Medicines Agency (EMA) for the therapy of seizures associated with the syndrome. Lennox-Gastaut or Dravet syndrome). Due to the low solubility of CBD in water and the first-pass effect, the drug showed low bioavailability. In this context, administration by other routes, such as topical and transdermal, is very interesting from a pharmacokinetic point of view (Casiraghi et al., 2020).

These administration routes have the advantages of avoiding the first-pass effect and the degradation of bioactive molecules by intestinal enzymes, as well as being generally more accepted by patients. The actives administered by the transdermal route need to meet certain requirements, such as having low molecular weight and moderate lipophilicity, such characteristics greatly limit the number of molecules capable of passively diffusing through the skin to the systemic circulation (Tijani et al., 2021).

The transdermal route has advantages over the oral and nose-to-brain routes, as it provides the API reaches the bloodstream in greater bioavailability. This noninvasive route allows controlled delivery with stable plasma concentration, reducing the occurrence of side effects, providing a sustained-release pattern, in cases where long-term treatments are needed. Additionally, transdermal route can be a noninvasive approach for a nonaddictive nonopioid-based therapy for both localized and systemic conditions (Paudel et al., 2010). The evidence supporting transdermal application has been increasing due to the potential to improve CBD efficacy. In this way, several companies are making considerable efforts to develop transdermal CBD-loaded products. Although most studies for transdermal CBD delivery are intended to overcome the GIT challenges, several works and patents reports the transdermal application of CBD for CNS disorders, such as epilepsy, developmental and epileptic encephalopathy, fragile-X syndrome, autism spectrum disorder and others (Stella et al., 2021).

Zynerba Pharmaceuticals has patented a permeation-enhanced 4% CBD gel (Zygel™) using propylene glycol and ethanol as CBD solvents and Transcutol-HP as permeation enhancer for focal epilepsy treatment in adults (Gutterman et al., 2019). Additionally, transdermal CBD significantly reduced anxiety and other behavioral symptoms in children and adolescents with fragile-X syndrome in 12 weeks, creating a meaningful effect in quality of life of the participants (Heussler et al., 2019; Bonn-Miller et al., 2019).

A 5% CBD transdermal gel produced a 48.8% reduction in neurodegeneration in a rodent model of alcohol use disorder. Besides, transdermal CBD absorption maintained drug plasma concentration of about 100.0 ng/mL for 3 days, comparable with intraperitoneal injection. Such results demonstrate the feasibility of transdermal drug delivery systems for the treatment of neurological disorders (Liput et al., 2013).

9.4.1 Pain Relief Purposes

Pain is a common condition, which, when not properly addressed, can evolve to chronification. Both acute and chronic pain are associated with a wide range of limitations of everyday life, with significant economic impacts (Stockbridge et al., 2015; Langley et al., 2013; Axon & Kamel, 2021). Particularly, chronic pain is a common and yet complex disabling problem with countless impacts on both individuals and society (Stella et al., 2021). Total health care expenditures associated with pain treatment are estimated to increase around $2,500 for chronic pain patients and by $1,008 for non-chronic pain in the USA (Stockbridge et al., 2015).

Conventional treatment for chronic and acute rheumatic, neuropathic and musculoskeletal pain comprises systemic administration of NSAIDs drugs, opioids, antidepressants and anticonvulsants, among others. Such drugs are often related to undesirable and possibly toxic side effects, being opioid overdose and self-medication with pain killers the major concern. Besides, pain is often the main symptom of a wide range of disorders such as fibromyalgia, endometriosis, rheumatoid arthritis and cancer. Nociceptive pain occurs at the peripheral level as a result of nociceptive nerves subjected to external stimuli that overcome the threshold sensibility. Inflammatory conditions lead to an increased release of inflammatory mediators that can cause an over-sensitization of peripheral nociceptors. Such sensitization causes a nociceptive response to low intensity stimuli or even in the absence of stimuli due to the reduction in sensibility threshold (Stella et al., 2021).

Several studies have suggested that topical and transdermal therapy to address peripheral inflammation, neuropathic pain and muscular pain can be an effective approach to reduce localized pain avoiding side effects, especially in the CNS and gastrointestinal tract (Burness et al., 2016; D'Andre et al., 2021; Eskander et al., 2020; Jones et al., 2011).

Recently, an increasing number of clinical data has indicated cannabinoids are promising therapies for pain management (Donvito et al., 2018; Whiting et al., 2015). Given that ECS is related to nociceptive response, it can be a promising target for novel therapeutic strategies with cannabinoid agonists. Indeed, both cannabinoid receptor agonists and inhibitors of endocannabinoid-regulating enzymes FAAH and MAGL are known to produce an antinociceptive effect in preclinical pain models (Kinsey et al., 2009; Schlosburg et al., 2009; Wilkerson et al., 2017). Several anecdotal studies and randomized controlled trials report the medicinal use of cannabis for pain relief in a wide range of painful conditions, such as cancer, HIV, multiple sclerosis and fibromyalgia (Allan et al., 2018; Aly & Masocha, 2021; Donvito et al., 2018; Rog et al., 2005; Wang et al., 2021).

There are several CBD-based transdermal formulations being developed, clinically tested and commercially available for pain relief (Stella et al., 2021). Most of them rely on the development of CBD-loaded patches for local delivery of CBD to ameliorate musculoskeletal pain, which is a convenient approach to ensure patient adherence and enables the utilization of several permeation-enhancing strategies. Greene Street Pharma LLC has informed the development of a 24 h sustained-release CBD transdermal patch, which seems to be able to maintain CBD plasma concentration. Such results are a consequence of their novel CBD-Cubed™ technology that has been developed in partnership with VerdePharmHealth and provides precise, steady, and sustained-release transdermal delivery of CBD by combining the use of adhesives and enhancers. Their product has been tested in clinical trials, for other therapeutic indications such as Alzheimer's, autism, rheumatoid arthritis and muscle pain, as well.

The application of CBD in painful conditions of chemotherapy-induced peripheral neuropathy, a common burden in cancer therapy, has been recently reported in a series of oncologic patients in a randomized placebo-controlled clinical trial (D'Andre et al., 2021; Xu et al., 2019). Commercially available CBD-based cosmetic products containing 120–600 mg of CBD were used. In general, the relief in pain/burning of tingling sensations has been reported. However, such an effect lasted for a limited time period, demonstrating a transient effect. The presence of penetration enhancers seemed to have a beneficial effect, as the analgesic effect seemed to last all night time with no relevant side effects reported, suggesting the potential of transdermal absorption of CBD even from conventional formulations (D'Andre et al., 2021).

Additionally, plain CBD oil skin application is reported to statistically reduce intense pain, sharp pain, cold and itchy sensations in CBD-treated groups of patients with symptomatic peripheral neuropathy after 4 weeks, with no adverse effects reported (Xu et al., 2019). Collectively, these data demonstrate that both crude oil and formulated CBD-loaded pharmaceutical products can effectively deliver CBD through the skin layers. However, it is worth noting that highly lipophilic molecules tend to partition in the lipid matrix of *stratum corneum* instead of diffuse toward the deeper layers of the skin. In general, drug skin absorption is facilitated by small molecular weight substances with moderate lipophilicity (logP 1–5) (Stella et al., 2021). As a strategy, permeation enhancers are often used to favor CBD transdermal absorption, avoiding its retention on upper skin layers, such as the *stratum corneum*. Enhancers are known to improve the transdermal permeation of lipophilic compounds, mainly due to the fluidizing capacity of the lipid matrix in the *stratum corneum*.

Paudel et al. studied the bioavailability of CBD-loaded patches after transdermal absorption with and without permeation enhancers. Their results showed that Transcutol-HP provided a steady state of CBD plasma concentrations after 15.5 ± 11.7 h, representing an increase of 3.7-fold in plasma concentrations in comparison to the formulations with CBD in a propylene glycol-hydroxyethyl cellulose gel (Paudel et al., 2010). Curcio et al. studied the influence of different vehicles in the permeability of CBD in *ex vivo* experiments. The most successful formulation regarding permeation rate and skin retention was a hydrogel constituted by water, caprylyl glycol, decylene glycol, carbomer and glycerin in comparison with hydrophilic ointments (Curcio et al., 2021). Such evidence leads to the conclusion that proper formulation design and development is a crucial factor to enable transdermal drug absorption.

The well-known anti-inflammatory and few side effects of CBD transdermal application, makes it a potential and effective treatment of arthritis and rheumatic conditions. Transdermal application of CBD showed good results in cases of arthritis models in rats. Hammell et al. (2016) evaluated the progression of arthritis induced in rats after transdermal application of an ethanol-based gel containing different concentrations of CBD for 4 consecutive days after arthritis induction. Their results showed that CBD plasma concentration increased after the application of the formulations, indicating that transdermal absorption took place. Besides, transdermal CBD gel significantly reduced joint swelling, pain, immune cell infiltration and thickening of synovial membrane in a dose-dependent pattern. After treatment with transdermal formulations containing CBD, there were no noticeable changes in behavior, which suggest no CNS-related side effects. Such results showed that transdermal administration of CBD might provide a long-lasting therapeutic effect and that the transdermal application of CBD can be an effective approach in reducing pain and the biomarkers related to arthritis-borne inflammation.

9.5 NANOTECHNOLOGY-BASED FORMULATIONS

From the increased lipophilicity of phytocannabinoids, CBD included, arise several drawbacks, such as low permeation through the deeper layers of the skin, reducing the possibility of pharmacological effects at the systemic level (Biernacki et al., 2021). Moreover, highly lipophilic molecules tend to be retained in the lipid matrix in the *stratum corneum*. For instance, Bartner et al. reported the PK profile of transdermal formulations containing CBD, aiming to achieve high bioavailability. Their results suggested that CBD-loaded transdermal creams did not promote CBD skin penetration and systemic absorption, which was a consequence of the accumulation of CBD within the stratum corneum (Bartner et al., 2018).

It is known that the nanostructured systems and nanoencapsulation techniques can be used as strategies that could be applied to facilitate transdermal and topical treatment with phytocannabinoids, such as CBD. For example, flexible vesicles, as ethosomes of transferosomes, or nanoemulsions can interact with *stratum corneum* lipids, and enhance phytocannabinoids skin permeation (Tijani et al., 2021). Given the feasibility of administering CBD through the skin, nanostructured formulations are novel approaches that can increase the penetration and permeation of phytocannabinoids into and through the skin (Osborne & Musakhanian, 2018). Besides, cannabinoids show low aqueous solubility, are sensitive to light and temperature, and may undergo oxidation when in solution, which could be prevented with nanoencapsulation techniques.

The characteristics of drug carriers and the formulation vehicle is known to influence both drug-release pattern and its skin permeation. The development of nanostructured drug carriers using penetration enhancers such as propylene glycol, ethanol, and oleic acid can favor the drug diffusion through the epidermis due to their interaction cell membrane and lipid matrix by altering their complex structure and changing the partition of drug into stratum corneum and other skin strata. Moreover, different useful excipients such as solvents, co-solvents, surfactants, humectants, thickening agents, as they can modulate the thermodynamic activity and drug partition coefficient, which significantly affect the extent of drug absorption. For instance, CBD-loaded O/A microemulsions with size of about 35 nm were produced by Vanti et al. (2021). The developed nanostructured drug carriers were produced with 20% of Solutol HS 15, as a surfactant, 9% of Transcutol P, as a cosolvent, 5% of isopropyl myristate, as oil phase, 66% of water and 1% of CBD. Microemulsions were dispersed in a hydrogel matrix, which allowed sustained *in vitro* release for at least 24 h, following the Higuchi model.

CBD distillate is a product with a high purity, and it is most commonly used in modern formulations. It has a viscous characteristic and is soluble in vegetable oils such as coconut, olive and palm oil. It can be used in lipophilic formulations such as body butters and also emulsions. Those kinds of formulations are highly hydrophobic and require lipophilic compatible excipients, such as polyethylene glycol and stearyl alcohol, which, instead, are compatible with most cannabinoids. Additionally, the addition of essential oils to CBD distillate-based formulations can be beneficial and provide novel or enhanced CBD basic properties. The addition of linalool to CBD distillate, for example, is reported to be an effective approach in the treatment of acne. Moreover, oleic acid combined with CBD can improve skin elasticity and hydration. Additionally, Stinchcomb et al. showed the permeation-enhancing effect of ethanol on CBD skin permeation. Ethanol concentrations of about 30% significantly increased CBD transdermal flux, with a small time lag (Stinchcomb et al., 2004), demonstrating the potential of ethanol in producing improved transdermal formulations.

Lodzki et al. reported the production of CBD-loaded ethosomes for transdermal therapy of rheumatic diseases. Ethosomes are flexible nanosized vesicles mainly composed by phospholipids and ethanol, which is a well-known permeation enhancer. Their characterization results showed that CBD and phosphatidylcholine formed an eutectic mixture, which can further enhance CBD skin permeation. After CBD-loaded ethosomal carriers transdermal absorption, CBD was found in the skin, in the underlying muscle tissue, liver, and pancreas, confirming the effective transdermal absorption. The steady state of plasma concentrations was reached after 24 h and was maintained for 72 h. As a consequence, CBD-loaded nanocarriers prevented the inflammation and edema in animal models and provided a skin accumulation depot, being promising vehicles for CBD transdermal and systemic treatment (Lodzki et al., 2003). Such benefit from ethosomes relies in the fact that ethanol is a well-tolerated organic solvent, capable of extracting the lipid fraction of epidermal barrier, reducing the effectiveness of skin barrier and favor drug diffusion through the skin strata.

CBD-loaded lipid nanocapsules, with 20 and 50 nm, were produced in order to overcome pharmacotechnical drawbacks usually associated with cannabinoids and provide extended drug release. The *in vitro* antitumor effect and *in vitro* CBD release showed to be dependent on particle size. In this way, 20 nm particles reduced the IC_{50} on tumor cells by threefold. Besides, CBD-decorated lipid nanoparticles proved to increase tumor targeting by 3.4-fold, compared to undeclared lipid nanoparticles. Such targeting effect is justified by the overexpression of cannabinoid receptors in tumor cells. Moreover, the combination of CBD-decorated surface and nanoencapsulated CBD contributed to a synergic reducing effect on IC_{50} values, which highlights the efficiency of combining strategies in nanoformulations (Aparicio-Blanco et al., 2019).

CBD-loaded oil-in-water (O/W) nanoemulsions have been reported in a patent described by Docherty and Bunka (2020). The nanoemulsions were composed of highly purified CBD crystals dissolved in sunflower oil in a 1:1 ratio, as the oil phase. Tween 80 was used as a surfactant and *Quillaja saponaria* extract, rich in natural surfactant bioactive saponins, composed the aqueous phase. Size reduction was achieved by high shear homogenization and led to nanoscale sizes that increased CBD solubility and bioavailability.

Lewinska produced O/W nanoemulsions for the CBD transport for topical application of the formulation. Nanoemulsions can be applied for diagnostic and therapeutic purposes, they are characterized by being colloidal systems with kinetic stability used for the protection, encapsulation, and delivery of actives. In this study, O/W nanoemulsions were produced, where the oil phase is composed of CBD oil containing the active CBD, and the aqueous phase is composed of a mixture of non-ionic surfactants that help in the emulsification of the system, the formulation components need to be compatible in order to remain stable during storage. Toxicity tests in cell culture were performed in order to assess the toxicity of the formulations. The nanoemulsions remained stable for at least 30 days, and the production method directly influenced the size and polydispersity index of the samples, *in vitro* tests with cell lines demonstrated the compatibility of the formulations with the skin, therefore, they are useful for use in cosmetics (Lewińska, 2021).

Matarazzo et al. have reported the development of nanostructured lipid carriers containing CBD to increase its bioavailability. Both stearic acid and oleic acid were used as lipid excipients and absorption enhancers. Span 20 was used as a surfactant, which helps in obtaining nanosized particles smaller than 200 nm with a positive superficial charge. Mucoadhesion was ensured by electrostatic interactions with negatively charged mucin molecules. *In vitro* drug-release tests indicated a sustained CBD release, which was reflected in an *in vivo* long-lasting antinociceptive effect in neuropathic pain animal models (Matarazzo et al., 2021).

Cyclodextrins (CD) are nanosized cyclic oligosaccharides that can provide molecular inclusion complexes with lipophilic molecules that are entrapped into their hydrophobic cavity. They are widely studied and explored in pharmaceutical preparations, especially to improve a wide range of physicochemical properties of hydrophobic molecules, including the solubility of poorly water-soluble drugs (Ramalho et al., 2021).

A self-assembled cyclodextrin system was produced for targeted delivery of CBD by Zhu et al. (2021). A graft of biotin on β-cyclodextrin (β-CD) provided a functionalized surface for enhanced anticancer activity against three human cancer cell strains. Besides, the inclusion complex was shown to promote a pH-triggered drug release in weakly acidic conditions, which is beneficial to targeted anticancer therapy.

α-CDs, β-CDs, and γ-CDs were used to produce inclusion complexes with CBD. Such a strategy significantly increased water solubility to about 3.7, 2.1, and 5.3 mg/mL, respectively, which is a promising approach to the development of aqueous formulations. *In vitro* cytotoxicity against human cancer cell lines Hep G2, and A549 showed that CBD-loaded inclusion complexes decreased the viability of cancer cells and increased cell viability in normal human cells, in comparison with free CBD. Such results suggest a more effective and safer anticancer formulation (Lv et al., 2019). The increased solubility of CBD in an aqueous media can optimize the interactions among the nanostructured drug carriers and the cells and cell membrane, enabling an increased permeation across the cell membrane (Ramalho et al., 2021).

Similar results have been reported in the work published by Li et al. (2021). Inclusion complexes between CBD and β-CD were obtained through an endothermic spontaneous process. Moreover, the aqueous solubility of CBD was increased by 614-fold after the complexation of CBD with cyclodextrins, which was reflected in the increased dissolution rate. The antioxidant activity of CBD was also favored by complexation with CD, as scavenging ability was higher than vitamin C, the golden pattern for antioxidant activity.

Momekova et al. (2021) produced super-macroporous cryogels containing β-CD inclusion complexes with CBD for topical application to treat neoplastic diseases. Such cryogel-based drug delivery systems produced sustained local delivery of CBD in a bimodal delivery profile, which was a combination of burst and sustained release. That can be beneficial since CBD produces a dose-dependent effect on tumor cells. At lower drug concentrations, the cytotoxic effect of the free CBD can be more pronounced. In this context, bi-phasic release profiles might offer new perspectives for efficient skin cancer therapy. Such an approach provides an immediate exposure of cancer cells to the initially released CBD and sustained inhibitory activity, provided by the second drug-release phase (Momekova et al., 2021). Similar results have been reported to other nanocomposite hydrogels (Barradas et al., 2017). The inclusion of nanocarriers into a hydrogel matrix originates a biphasic system producing a primary drug vehicle, that is, a drug nanocarrier and a secondary vehicle, which is the hydrogel matrix (Barradas et al., 2017).

A transdermal formulation for the treatment of chronic pain, employing colloidal nanosilica was produced to promote CBD delivery, aiming at better targeting the compound at the target site, as this system facilitates skin penetration. These systems allowed the interactions between the ligand and the receptor to be designed aiming at reducing the toxicity of the CNS and improving the therapeutic effect (Wine & Wine, 2010).

Although a significant amount of evidence and interesting results in animal models have been reported regarding nanostructured formulations, a comparative efficacy study among nanosized CBD carriers still lacking. In this way, both patents and clinical studies with CBD-loaded nanoformulations seem to be still preliminary. Several factors can justify this, with a highlight on the difficulty of scaling up the manufacturing processes to produce nanotechnology-based products and the need for a deep and wide physicochemical material characterization combined with toxicity and efficacy tests to standardize the CBD therapy with nanostructured formulations.

9.6 OPPORTUNITIES AND TRENDS ANALYSIS

9.6.1 Temporal Evolution and Scientific Scenario Analysis with VOSviewer® Software

The research was conducted in the Web of Science® database using the terms: "Cannabidiol" and "Transdermal". Of the 18 articles retrieved, the first was published in 2003. Even with a small amount of articles, it is possible to observe that, over the years, there has been an increase in the number of publications (Figure 9.3). The data support the idea that it's an underexplored area with a growth trend, justified by all the physiological applications explored above.

All those 18 articles and their bibliographical data such as keywords, abstracts, and titles were analyzed by the software VOSviewer® (Figure 9.4). From Figure 9.4, it can be observed that one cluster has been created by this analysis. The cluster describes a relation between the words "CBD", "Cannabidiol," and "Study". This result is expected, given the few articles found by this search. Justifying, one more time, that's an unexplored area with a big innovation potential.

The same research in the Web of Science® database was conducted with the terms "Topical" and "Cannabidiol." This time, a total of 45 articles have been published relating those words since 2003. In this analysis, the growth trend is clearer than the first one. One more thing that could be noticed is a higher increase from 2020 (Figure 9.5):

Analyzing all those 45 articles with VOSviewer® software, a more complex result could be noticed. Not one, but two clusters could be found. One cluster with the words related to the

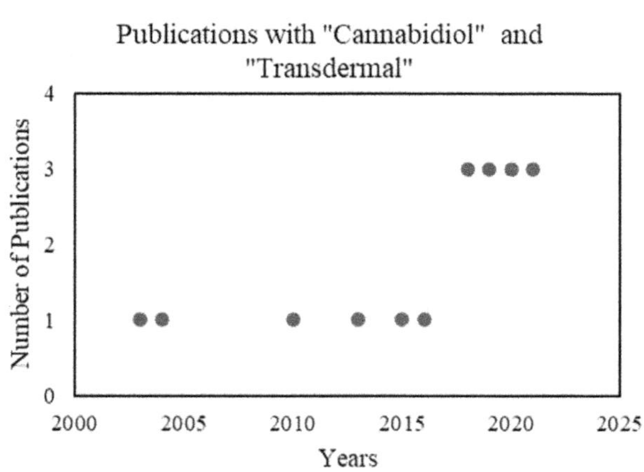

FIGURE 9.3 Number of Publications relating "Cannabidiol" and "Transdermal."

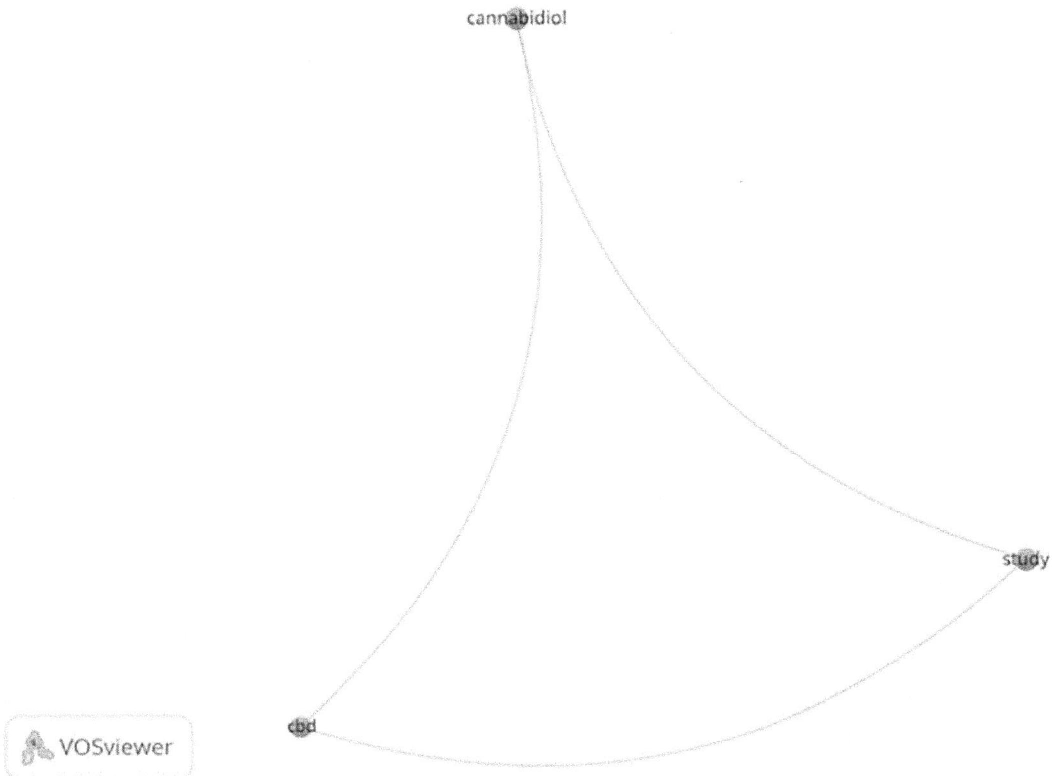

FIGURE 9.4 Cluster of VOSviewer® from "Cannabidiol" and "Transdermal" analysis.

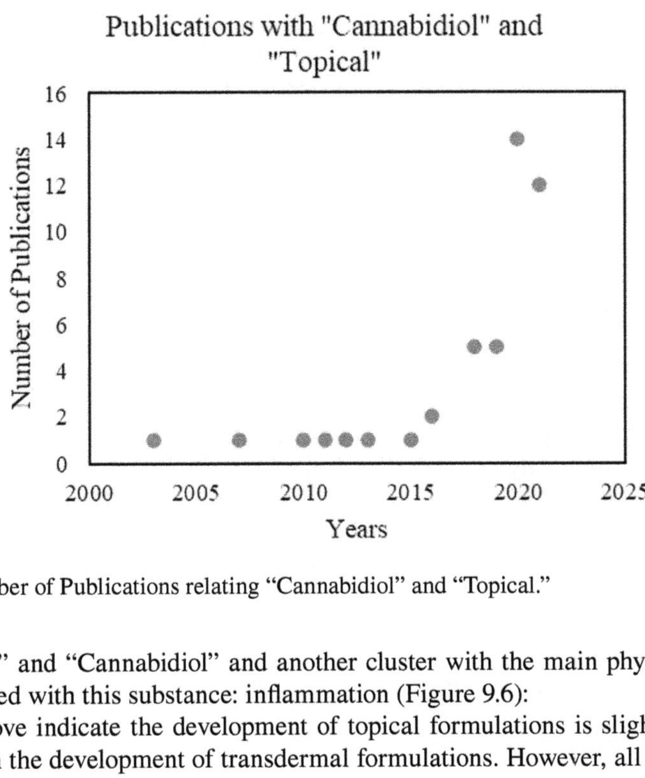

FIGURE 9.5 Number of Publications relating "Cannabidiol" and "Topical."

compound: "CBD" and "Cannabidiol" and another cluster with the main physiological condition that could be treated with this substance: inflammation (Figure 9.6):

The results above indicate the development of topical formulations is slightly advanced when it's compared with the development of transdermal formulations. However, all those areas are still

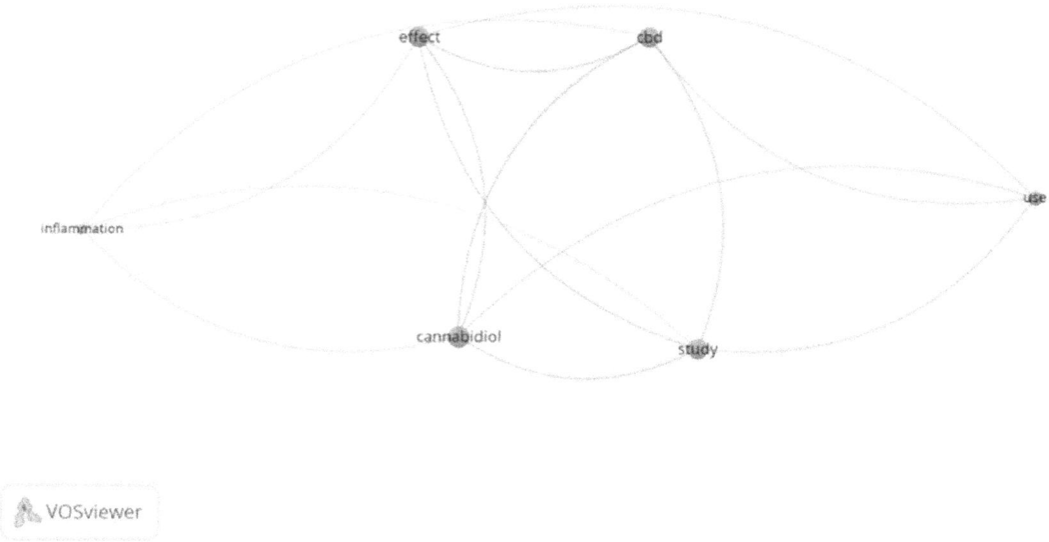

FIGURE 9.6 Clusters of VOSviewer® from "Cannabidiol" and "Topical" analysis.

underexplored, even with the significant increase in 2020 observed in topical formulations. A relevant result observed in VOSviewer® analysis from topical formulations is inflammation as the main physiological condition related to this kind of administration route. As the transdermal cluster couldn't present a physiological condition related to this kind of formulation, a title and abstract overview were conducted. From 18 articles, 4 articles (22%) study the effect of CBD on inflammation (Palmieri *et* al., 2019; Hammell et al., 2016; Giacoppo et al., 2015; Lodzki et al., 2003). And the major part of the other articles study the pharmacotechnical methods to develop a stable formulation. So, the analysis of the most common terms in the title, keywords, and abstract indicates the study of proinflammatory conditions comprises the main focus of the study of transdermal formulations too. Not only this but also indicates the efforts are concentrated on searching for different approaches to make a formulation with enhanced drug permeation and bioavailability of CBD.

9.7 CONCLUDING REMARKS

Chemical compounds extracted from *C. sativa* have drawn attention due to their unique therapeutic potential. In particular, CBD is a non-psychoactive cannabinoid that has been raising attention for its empirical therapeutic potential for a wide range of medical applications, including both dermatological and cosmetic purposes.

Skin disorders can be deeply uncomfortable and impact the quality of life of millions of people in the world. The cutaneous ECS has undoubtedly been proven to modulate several homeostatic and biological processes in the skin. Cannabinoid receptors and non-cannabinoid receptors have been identified in the skin and its appendages, which justify dermatological manifestations observed from cannabinoid abuse. Additionally, any disbalance in such cannabinoid signaling cascades can profoundly affect skin homeostasis and contribute to the pathogenesis of a wide range of skin diseases.

In this context, *in vitro* and preclinical and a few clinical studies suggest that the beneficial effects of topical dermal CBD comprise acne and seborrhea, pigmentation disorders, epidermolysis bullosa, allergic and inflammatory conditions, hair growth disorders and epidermal barrier dysfunction. Clinical efficacy and the elucidation of molecular mechanisms involved in CBD therapy are yet to be confirmed. Moreover, *in vitro* and *ex vivo* studies have demonstrated CBD potential for skin permeation. Nevertheless, few clinical data regarding the topical absorptive ability in human subjects are available.

Additionally, CBD poses several pharmacotechnical challenges namely, high lipophilicity, instability in solution and certain pH, high susceptibility to first-pass metabolism, poor permeability across the biological barriers, and lack of specificity to target the pharmacological sites. Such characteristics hinder formulation development and require novel strategies to design proper drug carriers that enable to achievement of CBD's full therapeutic potential. Nanoencapsulation techniques can bring important benefits such as increased CBD penetration into the cells, reduced administered dose, as well as decreased toxic events and side effects.

However, to date, no clinical trials investigating nanostructured CBD-loaded formulations have been identified. In this context, future work is necessary to enable advances in CBD research in topical and transdermal formulations involving nanoencapsulation strategies, comprising novel potential areas for future research exploration. Although countless clinical data have proved CBD has therapeutic potential for dermatology, besides being safe, well-tolerated, and non-psychotropic, its practical use still finds significant resistance.

Through text mining analysis, we showed that systematic research together with further large-scale clinical double-blind experiments are welcomed to unveil the actual clinical potential of CBD in managing skin disorders and also to enable the development of novel therapeutic approaches and pharmaceutical formulations to topical and transdermal applications.

A significant amount of CBD-loaded cosmetic formulations is commercially available with a wide range of claims that vary from antioxidant effects to UV protection. In 2018, 55% of dermatologists reported a patient-initiated conversation regarding the use of cannabinoids in the USA (Robinson et al., 2018). Accordingly, the medicinal cannabis market is expected to reach US $ 12.5 billion by 2025 (New Frontier Data, 2018). Nevertheless, a conservative part of the average population still rejects CBD-based products due to their preconceived notion against cannabis derivatives and their psychoactive effects arising from cannabis recreational use. In this context, scientific communication and sharing of the benefits of CBD is fundamental to prompt advances in more topical and transdermal formulations, among others.

REFERENCES

Abbas, A. B. et al. Design and synthesis of A PD-1 binding peptide and evaluation of its anti-tumor activity. *International Journal of Molecular Science*, v. 20, n. 3, p. 1–19, 2019.

Addor, F. A.; Aoki, V. Barreira cutânea na dermatite atópica Skin barrier in atopic dermatites. *Anais Brasileiros de Dermatologia*, v. 85, p. 184–194, 2010.

Aizpurua-Olaizola, O. et al. Targeting the endocannabinoid system: Future therapeutic strategies. *Drug Discovery*, v. 22, p. 105–110, 2017.

Akdeniz, M. et al. Transepidermal water loss in healthy adults: A systematic review and meta-analysis update. *British Journal of Dermatology*, v. 179, p. 1049–1055, 2018.

Allan, G. M. et al. Systematic review of systematic reviews for medical cannabinoids pain, nausea and vomiting, spasticity, and harms. *Canadian Family Physician*, v. 64, p. 78–94, 2018.

Ali, A.; Akhtar, N.; Khan, M. S. In vivo evaluation: The effects of a cream containing Acacia bark extract on skin melanin and erythema content. *Advances in Dermatology and Allergology*, v. 29, n. 5, p. 369–372, 2012.

Aly, E.; Masocha, W. Targeting the endocannabinoid system for management of HIV-associated neuropathic pain: A systematic review. *IBRO Neuroscience Reports*, v. 10, p. 109–118, 2021.

Apalla, Z. et al. Skin cancer: Epidemiology, disease burden, pathophysiology, diagnosis, and therapeutic approaches. *Dermatology and Therapy*, v. 7, p. 5–19, 2017.

Aparicio-Blanco, J. et al. Lipid nanocapsules decorated and loaded with cannabidiol as targeted prolonged release carriers for glioma therapy: In vitro screening of critical parameters. *European Journal of Pharmaceutics and Biopharmaceutics*, v. 134, p. 126–137, 2019.

Atalay, S. et al. Antioxidative and anti-inflammatory properties of cannabidiol. *Antioxidants*, v. 9, 2020.

Axon, D. R.; Kamel, A. Patterns of healthcare expenditures among older united states adults with pain and different perceived health status. *Healthcare (Switzerland)*, v. 9, n. 10, p. 1327, 2021.

Avila, C. et al. Cannabinoids for the treatment of chronic pruritus: A review. *Journal of the American Academy of Dermatology*, v. 82, p. 1205–1212, 2020.

Balato, A. et al. IL-33 is secreted by psoriatic keratinocytes and induces pro-inflammatory cytokines via keratinocyte and mast cell activation. *Experimental Dermatology*, v. 21, p. 892–894, 2012.

Baroni, A. et al. Structure and function of the epidermis related to barrier properties. *Clinics in Dermatology*, v. 30, p. 257–262, 2012.

Barradas, T. et al. Hydrogel-thickened nanoemulsions based on essential oils for topical delivery of psoralen: Permeation and stability studies. *European Journal of Pharmaceutics and Biopharmaceutics*, v. 116, p. 38–50, 2017.

Bartner, R. L. et al. Pharmacokinetics of cannabidiol administered by 3 delivery methods at 2 different dosages to healthy dogs. *Canadian Journal of Veterinary Research*, v. 82, n. 3, p. 178–183, 2018.

Baswan, S. M. et al. Therapeutic potential of cannabidiol (CBD) for skin health and disorders. *Clinical, Cosmetic and Investigational Dermatology*, v. 13, p. 927–942, 2020.

Bickers, D. R.; Athar, M. Oxidative stress in the pathogenesis of skin disease. *Journal of Investigative Dermatology*, v. 126, n. 12, p. 2565–2575, 2006.

Biernacki, M. et al. Oxidative stress and its consequences in the blood of rats irradiated with UV: Protective effect of cannabidiol. *Antioxidants*, v. 10, n. 6, p. 821, 2021.

Birdsall, S. M. et al. The use of medical marijuana in cancer. *Current Oncology Reports*, v. 18, n. 7, p. 40, 2016.

Bíró, T. et al. The endocannabinoid system of the skin in health and disease: Novel perspectives and therapeutic opportunities. *Trends in Pharmacological Sciences*, v. 30, p 411–420, 2009.

Blázquez, C. et al. Cannabinoid receptors as novel targets for the treatment of melanoma. *The FASEB Journal*, v. 20, n. 14, p. 2633–2635, 2006.

Bonn-Miller et al. inventors. Treatment of fragile X syndrome with cannabidiol. Zynerba Pharmaceuticals, Inc, assignee. Patent US10213390. 2019.

Borbíró, I. et al. Activation of transient receptor potential vanilloid-3 inhibits human hair growth. *Journal of Investigational Dermatology*, v. 131, n. 8, p. 1605–1614, 2011.

Burness, C. B.; Mccormack, P. L. Capsaicin 8% patch: A review in peripheral neuropathic pain. *Drugs*, v. 76, p. 123–134, 2016.

Cabral, G. A.; Griffin-Thomas, L. Emerging role of the cannabinoid receptor CB 2 in immune regulation: Therapeutic prospects for neuroinflammation. *Expert Reviews in Molecular Medicine*, v. 11, p. e3, 2009.

Cardoso, S. A.; Barradas, T. N. Developing formulations for drug follicular targeting: Nanoemulsions loaded with minoxidil and clove oil. *Journal of Drug Delivery Science and Technology*, v. 59, p. 101908, 2020.

Casares, L. et al. Cannabidiol induces antioxidant pathways in keratinocytes by targeting Bach1. *Redox Biology*, v. 28, p. 101321, 2020.

Casiraghi, A. et al. Topical administration of cannabidiol: Influence of vehicle-related aspects on skin permeation process. *Pharmaceuticals*, v. 13, n. 11, p. 1–12, 1 Nov. 2020.

Castilho, A. C. DA S.; Lopes, C. DE O. P.; Salles, B. C. C. Fisiopatologia da psoríase e seus aspectos imunológicos: Uma revisão sistemática. *Research, Society and Development*, v. 10, n. 11, p. e256101119346, 2021.

Caterina, M. J. TRP channel cannabinoid receptors in skin sensation, homeostasis, and inflammation. *ACS Chemical Neuroscience. American Chemical Society*, v. 5, p. 1107–1116, 2014.

Chelliah, M. P. et al. Self-initiated use of topical cannabidiol oil for epidermolysis bullosa. *Pediatric Dermatology*, v. 35, n. 4, p. e224–e227, 2018.

Chudnovsky, Y.; Khavari, P. A.; Adams, A. E. Melanoma genetics and the development of rational therapeutics. *Journal of Clinical Investigation the American Society for Clinical Investigation*, v. 115, p. 813–824, 2005.

Crippa, J. A. et al. Translational investigation of the therapeutic potential of cannabidiol (CBD): Toward a new age. *Frontiers in Immunology*, v. 9, p. 2009, 2018.

Curcio, A. et al. In vitro study evaluating the influence of vehicle in the permeability process of a topical composition containing cannabis sativa oil, escin, bromelain, glucosamine sulphate, methylsulfonylmethane, methylsalicylate and boswellia extract, designed for local treatment of musculoskeletal painful and inflammatory conditions. *Journal of Clinical Chemistry and Laboratory Medicine*, v. 4, p. 1–5, 2021.

Czifra, G. et al. Endocannabinoids regulate growth and survival of human eccrine sweat gland-derived epithelial cells. *Journal of Investigative Dermatology*, v. 132, n. 8, p. 1967–1976, 2012.

D'Andre, S. et al. Topical cannabinoids for treating chemotherapy-induced neuropathy: A case series. *Integrative Cancer Therapies*, v. 20, p. 153473542110617, 2021.

Davis, M. P. Cannabinoids for symptom management and cancer therapy: The evidence. *Journal of the National Comprehensive Cancer Network*, v. 14, n. 7, p. 915–922, 2016.

Di Marzo, V. New approaches and challenges to targeting the endocannabinoid system. *Nature Reviews Drug Discovery*, v. 17, p. 623–639, 2018.

Di Nardo, A. et al. Ceramide and cholesterol composition of the skin patients with atopic dermatitis. *Acta Dermatology Venereology*, v. 78, p. 27–30, 1998.
Docherty, J.; Bunka, C. A., inventors. Nanoemulsion compositions comprising biologically active ingredientes. Poviva Corp., Assignee. Patent WO2020236798, 2020.
Donetti, E. et al. Etanercept restores a differentiated keratinocyte phenotype in psoriatic human skin: A morphological study. *Experimental Dermatology*, v. 21, p. 549–551, 2012.
Donvito, G. et al. The endogenous cannabinoid system: A budding source of targets for treating inflammatory and neuropathic pain. *Neuropsychopharmacology*, v. 43, n. 1, p. 52–79, 2018.
Eagelston, L. R. M. et al. Cannabinoids in dermatology: A scoping review. *Dermatology Online Journal*. v. 24, p. 15–24, 2018.
Eberlein, B. et al. Adjuvant treatment of atopic eczema: Assessment of an emollient containing N-palmitoylethanolamine (ATOPA study). *Journal of the European Academy of Dermatology and Venereology*, v. 22, n. 1, p. 73–82, 2008.
Eugene, C.; Matthew, C., inventors. Formulations of cannabinoids comprising siloxane for the treatment of dermatitis and inflammatory skin diseases. Botanix Pharmaceuticals Ltd. Botanix Pharmaceuticals, Inc, assignee. Patent WO2018148785. 2018.
Eskander, J. P. et al. Cannabidiol (CBD) as a treatment of acute and chronic back pain: A case series and literature review. *Journal of Opioid Management*, v. 16, n. 3, p. 215–218, 2020.
Esposito, G. et al. Cannabidiol inhibits inducible nitric oxide synthase protein expression and nitric oxide production in β-amyloid stimulated PC12 neurons through p38 MAP kinase and NF-κB involvement. *Neuroscience Letters*, v. 399, n. 1–2, p. 91–95, 2006.
Ferreira, B. P. et al. Skin applications of cannabidiol: Sources, effects, delivery systems, marketed formulations and safety. *Phytochemical Reviews*, v. 22, p. 781–828, 2023.
Gegotek, A. et al. The cross-talk between electrophiles, antioxidant defence and the endocannabinoid system in fibroblasts and keratinocytes after UVA and UVB irradiation. *Journal of Dermatological Science*, v. 81, n. 2, p. 107–117, 2016.
Gęgotek, A. et al. The differences in the proteome profile of cannabidiol-treated skin fibroblasts following UVA or UVB irradiation in 2D and 3D cell cultures. *Cells*, v. 8, n. 9, p. 1–17, 2019.
Giacoppo, S. et al. A new formulation of cannabidiol in cream shows therapeutic effects in a mouse model of experimental autoimmune encephalomyelitis. *Daru-Journal of Pharmaceutical Sciences*, v. 23, p. 23–48, 2015.
Gutterman, D. et al. inventors; Zynerba Pharmaceuticals, Inc., assignee. Synthetic transdermal cannabidiol for the treatment of focal epilepsy in adults. patent US20190083388. 2019.
Hamelink, C. et al. Comparison of cannabidiol, antioxidants, and diuretics in reversing binge ethanol-induced neurotoxicity. *Journal of Pharmacology and Experimental Therapeutics*, v. 314, n. 2, p. 780–788, 2005.
Hammell, D. C. et al. Transdermal cannabidiol reduces inflammation and pain-related behaviours in a rat model of arthritis. *European Journal of Pain (United Kingdom)*, v. 20, n. 6, p. 936–948, 2016.
Heussler, H. et al. A phase 1/2, open-label assessment of the safety, tolerability, and efficacy of transdermal cannabidiol (ZYN002) for the treatment of pediatric fragile X syndrome. *Journal of Neurodevelopmental Disorders*, v. 11, n. 1, p. 16, 2019.
Ho, B. S. Y. et al. Progressive expression of PPARGC1α is associated with hair miniaturization in androgenetic alopecia. *Scientific Reports*, v. 9, n. 1, p. 8771, 2019.
Hordinsky, M. K. Overview of alopecia areata. *Journal of Investigative Dermatology Symposium Proceedings*, v. 16, n. 1, p. 13–15, 2013.
Inci, R. et al. Dermatological aspects of synthetic cannabinoid addiction. *Cutaneous and Ocular Toxicology*, v. 36, n. 2, p. 125–131, 2017.
Itin, C. et al. Prolonged oral transmucosal delivery of highly lipophilic drug cannabidiol. *International Journal of Pharmaceutics*, v. 581, p. 119276, 2020.
Jastrząb, A.; Gęgotek, A.; Skrzydlewska, E. Cannabidiol regulates the expression of keratinocyte proteins involved in the inflammation process through transcriptional regulation. *Cells*, v. 8, n. 8, p. 827, 2019.
Jones, V. M.; Moore, K. A.; Peterson, D. M. Capsaicin 8% topical patch (qutenza)-a review of the evidence. *Journal of Pain and Palliative Care Pharmacotherapy*, v. 25, n. 1, p. 32–41, 2011.
Juknat, A. et al. Differential transcriptional profiles mediated by exposure to the cannabinoids cannabidiol and Δ9-tetrahydrocannabinol in BV-2 microglial cells. *British Journal of Pharmacology*, v. 165, n. 8, p. 2512–2528, 2012.
Khan, R. et al. The therapeutic role of Cannabidiol in mental health: A systematic review. *Journal of Cannabis Research*, v. 2, n. 1, p. 2, 2020.

Klein, M. et al. Effects of cannabidiol, a Cannabis sativa constituent, on oral wound healing process in rats: Clinical and histological evaluation. *Phytotherapy Research*, v. 32, n. 11, p. 2275–2281, 2018.

Kozela, E. et al. Cannabinoids Δ9-tetrahydrocannabinol and cannabidiol differentially inhibit the lipopolysaccharide-activated NF-κB and interferon-β/STAT proinflammatory pathways in BV-2 microglial cells. *Journal of Biological Chemistry*, v. 285, n. 3, p. 1616–1626, 2010.

Kinsey, S. G. et al. Blockade of endocannabinoid-degrading enzymes attenuates neuropathic pain. *Journal of Pharmacology and Experimental Therapeutics*, v. 330, n. 3, p. 902–910, 2009.

Kis, B. et al. Cannabidiol-from plant to human body: A promising bioactive molecule with multi-target effects in cancer. *International Journal of Molecular Sciences*, v. 20, n. 23, p. 5905, 2019.

Kupczyk, P.; Reich, A.; Szepietowski, J. C. Cannabinoid system in the skin - A possible target for future therapies in dermatology. *Experimental Dermatology*, v. 18, n. 8, p. 669–679, 2009.

Langley, P. C. et al. The burden associated with neuropathic pain in Western Europe. *Journal of Medical Economics*, v. 16, n. 1, p. 85–95, 2013.

Lee, M. A. The discovery of the endocannabinoid system. *Medical Cannabis Handbook*, v. 1, p. 215–2017, 2019.

Lewińska, A. Optimizing the process design of oil-in-water nanoemulsion for delivering poorly soluble cannabidiol oil. *Processes*, v. 9, n. 7, p. 1180, 2021.

Li, H. et al. Inclusion complexes of cannabidiol with β-cyclodextrin and its derivative: Physicochemical properties, water solubility, and antioxidant activity. *Journal of Molecular Liquids*, v. 334, p. 116070, 2021.

Li, S. S. et al. Cannabinoid CB2 receptors are involved in the regulation of fibrogenesis during skin wound repair in mice. *Molecular Medicine Reports*, v. 13, n. 4, p. 3441–3450, 2016.

Lichtman, A. H. Reversible inhibitors of fatty acid amide hydrolase that promote analgesia: Evidence for an unprecedented combination of potency and selectivity. *Journal of Pharmacology and Experimental Therapeutics*, v. 311, n. 2, p. 441–448, 2004.

Lima, E. D.; Lima, M. D. Reviewing concepts in the immunopathogenesis of psoriasis. *Anais Brasileiros de Dermatologia*, v. 86, p. 1151–1159, 2011.

Liput, D. J. et al. Transdermal delivery of cannabidiol attenuates binge alcohol-induced neurodegeneration in a rodent model of an alcohol use disorder. *Pharmacology Biochemistry and Behavior*, v. 111, p. 120–127, 2013.

Llanos Casanova, M. et al. Inhibition of skin tumor growth and angiogenesis in vivo by activation of cannabinoid receptors. *Journal of Clinical Investigation*, v. 111, n. 1, p. 43–50, 2003.

Lodzki, M. et al. Cannabidiol - Transdermal delivery and anti-inflammatory effect in a murine model. *Journal of Controlled Release*, v. 93, n. 3, p. 377–387, 2003.

Łuczaj, W. et al. Changes in lipid profile of keratinocytes from rat skin exposed to chronic uva or uvb radiation and topical application of cannabidiol. *Antioxidants*, v. 9, n. 12, p. 1–15, 2020.

Lv, P. et al. Structural analysis and cytotoxicity of host-guest inclusion complexes of cannabidiol with three native cyclodextrins. *Journal of Drug Delivery Science and Technology*, v. 51, p. 337–344, 2019.

Maccarrone, M. et al. The endocannabinoid system in human keratinocytes: Evidence that anandamide inhibits epidermal differentiation through CB1 receptor-dependent inhibition of protein kinase C, activating protein-1, and transglutaminase. *Journal of Biological Chemistry*, v. 278, n. 36, p. 33896–33903, 2003.

Maghfour, J. et al. Assessing the effects of topical cannabidiol in patients with atopic dermatites. *Dermatology Online Journal*, v. 27, n. 2, p. 1–4, 2021.

Maida, V. et al. Topical cannabis-based medicines - A novel paradigm and treatment for non-uremic calciphylaxis leg ulcers: An open label trial. *International Wound Journal*, v. 17, n. 5, p. 1508–1516, 2020.

Maida, V. et al. Topical cannabis-based medicines - A novel adjuvant treatment for venous leg ulcers: An open-label trial. *Experimental Dermatology*, v. 30, n. 9, p. 1258–1267, 2021.

Marquart, S. et al. Inactivation of the cannabinoid receptor CB1 prevents leukocyte infiltration and experimental fibrosis. *Arthritis and Rheumatism*, v. 62, n. 11, p. 3467–3476, 2010.

Matarazzo, A. P. et al. Mucoadhesive nanostructured lipid carriers as a cannabidiol nasal delivery system for the treatment of neuropathic pain. *European Journal of Pharmaceutical Sciences*, v. 159, p. 105698, 2021.

Matsuda, L. A. et al. Structure of a cannabinoid receptor and functional expression of the cloned cDNA. *Nature*, v. 346, p. 561–564, 1990.

Mazzetti, C. et al. Quantification of the content of cannabidiol in commercially available e-liquids and studies on their thermal and photo-stability. *Scientific Reports*, v. 10, n. 1, p. 1–6. 2020.

Mckallip, R. J.; Nagarkatti, M.; Nagarkatti, P. S. Δ-9-Tetrahydrocannabinol enhances breast cancer growth and metastasis by suppression of the antitumor immune response. *The Journal of Immunology*, v. 174, n. 6, p. 3281–3289, 2005.

Meckfessel, M. H.; Brandt, S. The structure, function, and importance of ceramides in skin and their use as therapeutic agents in skin-care products. *Journal of the American Academy of Dermatology*, v. 71, n. 1, p. 177–184, 2014.
Mercati, F. et al. Identification of cannabinoid type 1 receptor in dog hair follicles. *Acta Histochemica*, v. 114, n. 1, p. 68–71, 2012.
Micali, G. et al. Treating skin cancer with topical cream. *Expert Opinion on Pharmacotherapy*, v. 11, n. 9, p. 1515–1527, 2010.
Mignani, S. et al. Expand classical drug administration ways by emerging routes using dendrimer drug delivery systems: A concise overview. *Advanced Drug Delivery Reviews*, v. 65, n. 10, p. 1316–1330, 2013.
Mnekin, L.; Ripoll, L. Topical use of cannabis sativa L. biochemicals. *Cosmetics*, v. 8, p. 85, 2021.
Momekova, D. et al. Polysaccharide cryogels containing β-cyclodextrin for the delivery of cannabidiol. *Pharmaceutics*, v. 13, n. 11, p. 1774, 2021.
Montero-Vilchez, T. et al. Skin barrier function in psoriasis and atopic dermatitis: Transepidermal water loss and temperature as useful tools to assess disease severity. *Journal of Clinical Medicine*, v. 10, n. 2, p. 1–12, 2021.
Moore, D. J.; Rawlings, A. V. The chemistry, function and (patho)physiology of stratum corneum barrier ceramides. *International Journal of Cosmetic Science*, v. 39, n. 4, p. 366–372.
Mounessa, J. S. et al. The role of cannabinoids in dermatology. *Journal of the American Academy of Dermatology*, v. 77, p. 188–190, 2017.
Mulinari-Brenner et al. Entendendo a alopecia androgenética. *Surgical & Cosmetic Dermatology*, v. 3, n. 4, p. 329–337, 2011.
Nahar, L.; Onder, A.; Sarker, S. D. A review on the recent advances in HPLC, UHPLC and UPLC analyses of naturally occurring cannabinoids (2010-2019). *Phytochemical Analysis*, v. 31, p. 413–457, 2020.
Nam, G. et al. Selective cannabinoid receptor-1 agonists regulate mast cell activation in an oxazolone-induced atopic dermatitis model. *Annals of Dermatology*, v. 28, n. 1, p. 22–29, 2016.
Nikan, M.; Nabavi, S. M.; Manayi, A. Ligands for cannabinoid receptors, promising anticancer agents. *Life Sciences*, v. 146, p. 124–130, 2016.
Nutten, S. Atopic dermatitis: Global epidemiology and risk factors. *Annals of Nutrition and Metabolism*, v. 66, p. 8–16, 2015.
Oh, J. W. et al. A guide to studying human hair follicle cycling in vivo. *Journal of Investigative Dermatology*, v. 136, n. 1, p. 34–44, 2016.
Oláh, A. et al. Cannabidiol exerts sebostatic and antiinflammatory effects on human sebocytes. *Journal of Clinical Investigation*, v. 124, n. 9, p. 3713–3724, 2014.
Oláh, A.et al. Targeting cannabinoid signaling in the immune system: "High"-ly exciting questions, possibilities, and challenges. *Frontiers in Immunology*, v. 8, p. 1–4, 2017.
Onaivi, E. S.; Singh Chauhan, B. P.; Sharma, V. Challenges of cannabinoid delivery: How can nanomedicine help? *Nanomedicine Future Medicine Ltd.*, v. 15, p. 2023–2028, 2020.
Osborne, D. W.; Musakhanian, J. Skin penetration and permeation properties of transcutol(r)-neat or diluted mixtures. *AAPS PharmSciTech*, v. 19, n. 8, p. 3512–3533, 2018.
Palmieri, B. et al. A therapeutic effect of cbd-enriched ointment in inflammatory skin diseases and cutaneous scars. *Clinica Terapeutica*, v. 170, n. 2, p. 93–99, 2019.
Palomino, D. C. Arolina T.; Marti, L. C. Avalheiro. Chemokines and immunity. *Einstein*, v. 13, n. 3, p. 469–473, 2015.
Park, Y. J. et al. Regulatory effect of cannabidiol (CBD) on decreased β-catenin expression in alopecia models by testosterone and PMA treatment in dermal papilla cells. *Journal of Pharmacopuncture*, v. 24, n. 2, p. 68–75, 2021.
Patra, J. K. et al. Nano based drug delivery systems: Recent developments and future prospects. *Journal of Nanobiotechnology*, v. 16, 19 set. 2018.
Paudel, K. S. et al. Cannabidiol bioavailability after nasal and transdermal application: Effect of permeation enhancers. *Drug Development and Industrial Pharmacy*, v. 36, n. 9, p. 1088–1097, 2010.
Petrosino, S. et al. Anti-inflammatory properties of cannabidiol, a nonpsychotropic cannabinoid, in experimental allergic contact dermatitis. *Journal of Pharmacology and Experimental Therapeutics*, v. 365, n. 3, p. 652–663, 2018.
Picardo, M. et al. Acne and Rosacea. *Dermatology and Therapy*, v. 7, p. 43–52, 2017.
Pisanti, S. et al. Cannabidiol: State of the art and new challenges for therapeutic applications. *Pharmacology and Therapeutics*, v. 175, p. 133–150, 2017.
Raiker, R. et al. 18853 Associations between cannabis use and cutaneous disorders. *Journal of the American Academy of Dermatology*, v. 83, n. 6, p. AB223, 2020.

Ramalho, Í. M. DE M. et al. Current trends on cannabidiol delivery systems: Where are we and where are we going? *Expert Opinion on Drug Delivery*, v. 18, n. 11, p. 1577–1587, 2021.

Ramot, Y. et al. A novel control of human keratin expression: cannabinoid receptor 1-mediated signaling down-regulates the expression of keratins K6 and K16 in human keratinocytes in vitro and in situ. *Peer J*, v. 19, n. 3, p. 1–12, 2013.

Ramot, Y. et al. PPAR-γ signalling as a key mediator of human hair follicle physiology and pathology. *Experimental Dermatology*, v. 29, p. 312–321, 2020.

Ritter, J. et al. *Rang & Dale's Pharmacology*. 8th edition. Edinburgh: Churchill Livingstone. Elsevier, 2016.

Robinson, E. et al. Knowledge, attitudes, and perceptions of cannabinoids in the dermatology community. *Journal of Drugs in Dermatology*, v. 17, p. 1273–1278, 2018.

Rog, D. J. et al. Randomized, controlled trial of cannabis-based medicine in central pain in multiple sclerosis. *Neurology*, v. 65, n. 6, p. 812–819, 2005.

Rotsztejn, H. et al. Skin surface lipids and their measurements. *Postępy Dermatologii i Alergologii*, v. 28, n. 6, p. 498–505, 2011.

Rotty, J. D.; Coulombe, P. A. A wound-induced keratin inhibits Src activity during keratinocyte migration and tissue repair. *Journal of Cell Biology*, v. 197, n. 3, p. 381–389, 2012.

Ryšavá, A. et al. Effect of UVA radiation on the Nrf2 signalling pathway in human skin cells. *Journal of Photochemistry and Photobiology B: Biology*, v. 209, p. 1–12, 2020.

Sadgrove, N. J.; Simmonds, M. S. J. Topical and nutricosmetic products for healthy hair and dermal antiaging using "dual-acting" (2 for 1) plant-based peptides, hormones, and cannabinoids. *FASEB BioAdvances*, v. 3, p. 601–610, 2021.

Sangiovanni, E. et al. Cannabis sativa L. extract and cannabidiol inhibit in vitro mediators of skin inflammation and wound injury. *Phytotherapy Research*, v. 33, n. 8, p. 2083–2093, 2019.

Scheau, C. et al. Personalized medicine cannabinoids and inflammations of the gut-lung-skin barrier. *Journal of Personalized Medicine*, v. 11, p. 494, 2021.

Schräder, N. H. B. et al. Combined tetrahydrocannabinol and cannabidiol to treat pain in epidermolysis bullosa: A report of three cases. *British Journal of Dermatology*, v. 180, n. 4, p. 922–924, 2019.

Schlosburg, J. E. et al. Targeting fatty acid amide hydrolase (FAAH) to treat pain and inflammation. *AAPS Journal*, v. 11, n. 1, p. 39–44, 2009.

Shao, K.; Stewart, C.; Grant-Kels, J. M. Cannabis and the skin: The effect of Cannabis on skin. *Clinics in Dermatology*, v. 39, p. 784–795, 2021.

Sheriff, T. et al. The potential role of cannabinoids in dermatology. *Journal of Dermatological Treatment*, v. 31, p. 839–845, 2020.

Simmerman, E. et al. Cannabinoids as a potential new and novel treatment for melanoma: A pilot study in a murine model. *Journal of Surgical Research*, v. 235, p. 210–215, 2019.

Singer, E. et al. Reactive oxygen species-mediated therapeutic response and resistance in glioblastoma. *Cell Death & Disease*, v. 6, n. 1, p. e1601, 2015.

Smith, G.; Satino, J. Hair regrowth with Cannabidiol (CBD)-rich hemp extract - A case series. *Cannabis*, v. 4, n. 1, p. 53–59, 2021.

Spleman, L. et al. 1061 The safety of topical cannabidiol (CBD) for the treatment of acne. *Journal of Investigative Dermatology*, v. 138, n. 5, p. S180, 2018.

Song. B. New Frontier Data Projects U.S. Legal Cannabis Market to Grow to $25 Billion by 2025. Frontier Financial Group. Cannabis Insights, New Frontier Data Updates Web site, 2018. Available from: https://newfrontierdata.com/cannabis-insights/new-frontier-data-projects-u-s-legal-cannabis-market-grow-25-billion-2025. Accessed April 23rd 2020.

Ständer, S. et al. Distribution of cannabinoid receptor 1 (CB1) and 2 (CB2) on sensory nerve fibers and adnexal structures in human skin. *Journal of Dermatological Science*, v. 38, n. 3, p. 177–188, 2005.

Ständer, S.; Reinhardt, H. W.; Luger, T. A. Topische cannabinoidagonisten. Eine effektive, neue möglichkeit zur behandlung von chronischem pruritus. *Hautarzt*, v. 57, n. 9, p. 801–807, 2006.

Stella, B. et al. Cannabinoid formulations and delivery systems: Current and future options to treat pain. *Drugs*, v. 81, p. 1513–1557, 2021.

Stinchcomb, A. L. et al. Human skin permeation of Δ8-tetrahydrocannabinol, cannabidiol and cannabinol. *Journal of Pharmacy and Pharmacology*, v. 56, n. 3, p. 291–297, 2004.

Stockbridge, E. L.; Suzuki, S.; Pagán, J. A. Chronic pain and health care spending: An analysis of longitudinal data from the medical expenditure panel survey. *Health Services Research*, v. 50, n. 3, p. 847–870, 2015.

Stone, K. D.; Prussin, C.; Metcalfe, D. D. Ig E, Mast cells, basophils, and eosinophils. *Journal of Allergy and Clinical Immunology*. v. 125, p. 73–80, 2009.

Sugawara, K. et al. Endocannabinoids limit excessive mast cell maturation and activation in human skin. *Journal of Allergy and Clinical Immunology*, v. 129, n. 3, 2012.

Suttle, M. et al. Experimentally induced psoriatic lesion associates with interleukin (IL)-6 in mast cells and appearance of dermal cells expressing IL-33 and IL-6 receptor. *Clinical & Experimental Immunology*, v. 169, p. 311–319, 2012.

Szabó, I. L. et al. 263 (-)-cannabidiol differentially influences hair growth. *Journal of Investigative Dermatology*, v. 137, n. 10, p. S238, 2017.

Szabó, I. L. et al. TRPV4 is expressed in human hair follicles and inhibits hair growth in vitro. *Journal of Investigative Dermatology*, v. 139, n. 6, p. 1385–1388, 2019.

Telek, A. et al. Inhibition of human hair follicle growth by endo-and exocannabinoids. *The FASEB Journal*, v. 21, n. 13, p. 3534–3541, 2007.

Tijani, A. O. et al. Delivering therapeutic cannabinoids via skin: Current state and future perspectives. *Journal of Controlled Release*, v. 334, p. 427–451, 2021.

Tóth, K. F. et al. Cannabinoid signaling in the skin: Therapeutic potential of the "c(ut)annabinoid" system. *Molecules*, v. 24, p. 1–56, 2019.

Vacek, J. et al. Antioxidant function of phytocannabinoids: Molecular basis of their stability and cytoprotective properties under UV-irradiation. *Free Radical Biology and Medicine*, v. 164, p. 258–270, 2021.

Vanti, G. et al. Development and optimisation of biopharmaceutical properties of a new microemulgel of cannabidiol for locally-acting dermatological delivery. *International Journal of Pharmaceutics*, v. 607, p. 1–13, 2021.

Visse, K. et al. Efficacy of body lotion containing N-palmitoylethanolamine in subjects with chronic pruritus due to dry skin: A dermatocosmetic study. *Acta Dermato-Venereologica Medical Journals*, v. 97, n. 5, p. 639–641, 2017.

Wang, L. et al. Medical cannabis or cannabinoids for chronic non-cancer and cancer related pain: A systematic review and meta-analysis of randomized clinical trials. *The BMJ*, v. 374, p. 1–15, 2021.

Weigelt, M. A.; Sivamani, R.; Lev-Tov, H. The therapeutic potential of cannabinoids for integumentary wound management. *Experimental Dermatology*, v. 30, p. 201–211, 2021.

Whiting, P. F. et al. Cannabinoids for medical use: A systematic review and meta-analysis. *JAMA - Journal of the American Medical Association*, v. 313, n. 24, p. 2456–2473, 2015.

Wilkerson, J. L. et al. The endocannabinoid hydrolysis inhibitor SA-57: Intrinsic antinociceptive effects, augmented morphine-induced antinociception, and attenuated heroin seeking behavior in mice. *Neuropharmacology*, v. 114, p. 156–167, 2017.

Wilkinson, J. D.; Williamson, E. M. Cannabinoids inhibit human keratinocyte proliferation through a non-CB1/CB2 mechanism and have a potential therapeutic value in the treatment of psoriasis. *Journal of Dermatological Science*, v. 45, n. 2, p. 87–92, 2007.

Wine, W.A.; Wine, H. inventors, Transdermal formulations of synthetic cannabinoids and nano colloidal silica. US Patent 12/373,632, 2010.

Wu, H. Y. et al. Cannabidiol induced apoptosis in human monocytes through mitochondrial permeability transition pore-mediated ROS production. *Free Radical Biology and Medicine*, v. 124, p. 311–318, 2018.

Xu, D. H. et al. The effectiveness of topical cannabidiol oil in symptomatic relief of peripheral neuropathy of the lower extremities. *Current Pharmaceutical Biotechnology*, v. 21, n. 5, p. 390–402, 2019.

Yosipovitch, G. et al. Itch: From mechanism to (novel) therapeutic approaches. *Journal of Allergy and Clinical Immunology*, v. 142, n. 5, p. 1375–1390, 2018.

Yu, T. S. et al. The cannabinoid receptor type 2 is time-dependently expressed during skeletal muscle wound healing in rats. *International Journal of Legal Medicine*, v. 124, n. 5, p. 397–404, 2010.

Zákány, N. et al. Endocannabinoid tone regulates human sebocyte biology. *Journal of Investigative Dermatology*, v. 138, n. 8, p. 1699–1706, 2018.

Zheng, J. L. et al. Cannabinoid receptor type 2 is time-dependently expressed during skin wound healing in mice. *International Journal of Legal Medicine*, v. 126, n. 5, p. 807–814, 2012.

Zhu, P. et al. Self-assembly system based on cyclodextrin for targeted delivery of cannabidiol. *Frontiers in Chemistry*, v. 9, p. 754832, 2021.

10 Novel Phytopharmaceuticals Used for Wound Healing

Meghna A. Singh, Abhishek Wahi, Anil Kumar Teotia, Prasad Thota, Raunak Varshney, and Madhu Gupta

10.1 INTRODUCTION

The relationship between humans and plants is as ancient as the first human civilization on earth. Besides several uses of plants in human survival, the most crucial role of plants is that they are the great source of several medicinal principles known today for the prevention and mitigation of ailments (Chah et al., 2006). Plants such as *Azadirachta indica, Allium sativum, Aloe vera, curcumin, Withania somnifera, Carica papaya,* etc. are widely used throughout the world for their medicinal properties (Badukale et al., 2021; Wadekar et al., 2021). Even though advances in herbal drug delivery have opened up research opportunities, it has several advantages over conventional formulations. Basically, the problems in the conventional formulation were the drug's solubility, sustainability, stability, and toxicity. The novel drug delivery system introduces nanoparticles, nanoemulsions, nanogels, hydrogels, and pytosomes, etc., which overcome the problems regarding earlier practices of drug development. These advanced techniques improve drug sustainability, enhance solubility, and defend against physical and chemical degradation. Additionally, this promotes the tissue macrophages' distribution, which is vital for higher pharmacological and therapeutic efficacy (Saraf, 2010).

Globally, herbal medicines are considered a complementary alternative to modern allopathic medicines (Zhang et al., 2016). During the last 20 years, herbal medicines have seen enormous recognition and market value. This has also resulted in some malpractices involving adulteration and poor quality control, necessitating regulatory interventions all over the world by authorities such as the United States Food and Drug Administration (USFDA) and the European Medicines Agency (EMA). The word "Phytopharmaceuticals" is defined in Chapter 1 of the New Drugs & Clinical Trials Rules, 2018 (GSR 227(E), dated March 19, 2019, Central Drugs Standards Control Organization, India) as a purified and standard fraction with a defined minimum of four bioactive or phytochemical compounds (qualitatively and quantitatively assessed) of an extract of a medicinal plant or its part, for internal or external use of human beings or animals for diagnosis, treatment, mitigation, or prevention of any disease or disorder, but does not include administration by parenteral route (FAQ on Phytopharmaceutical drugs—CDSCO, assessed on 03-02-2022). Several phytopharmaceuticals have proven beneficial pharmacological effects in managing systemic diseases and topical ailments, including the healing of skin wounds.

Healing of wounds is a complex biophysical phenomenon comprising exchangeable mechanisms at the biochemical and cellular levels. Initiating at the very time of tissue injury, the complete process stages through hemostasis, inflammation, and proliferation followed by remodeling (Gosain and DiPietro, 2004; Mathieu, 2006). The process of wound healing can be further complicated by factors such as; inappropriate blood and oxygen supply to the wound area and the formation of biofilm by microbial growth (Raina et al., 2008). The attack of microorganisms to cause infiltration and hamper tissue growth is the major challenge for complete and faster wound healing (Edwards and Harding, 2004; MacNeil, 2008). Hence, the protection of healing wounds from microbial growth is one of the major concerns where modern synthetic medicinal products have failed due to significant microbial resistance. However, phytopharmaceuticals are immune to this action because microbes cannot develop resistance to them (Moffatt, 2005). Therefore, phytopharmaceuticals have

DOI: 10.1201/9781003292692-10

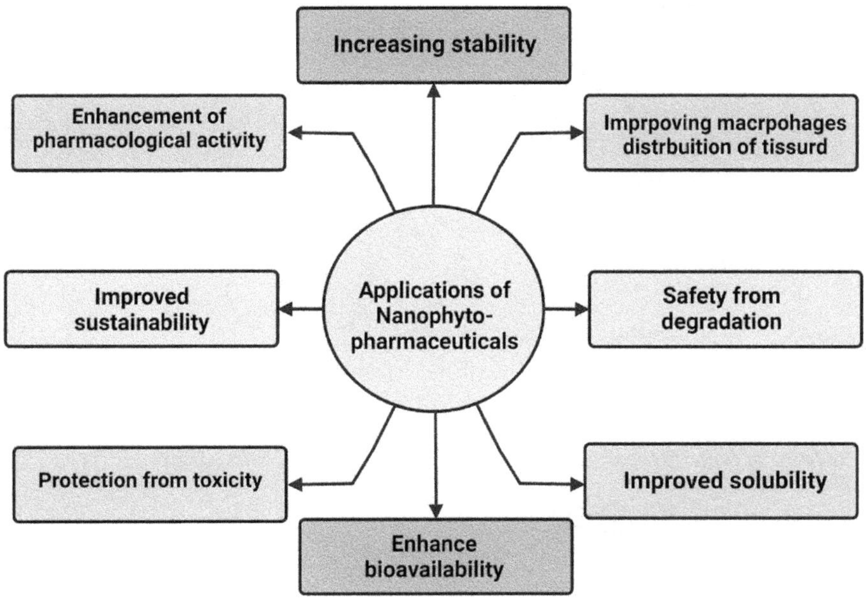

FIGURE 10.1 Applications of nanophyto-pharmaceuticals in wound healing.

greater potential in the field of wound care and management, and the market for these products is exponentially growing around the world. The schematic diagram (Figure 10.1) shows the potential advantages of advanced drug delivery of phytopharmaceuticals.

10.2 WOUND PATHOPHYSIOLOGY AND HEALING

Wound healing is a multifaceted phenomenon that involves cellular and biological interactions within the tissue and organs of the body (Gonzalez et al., 2016). There are four phases of wound healing: hemostasis, inflammation, proliferative phase, and remodeling which are diagrammatically represented in Figure 10.2. The brief explanations regarding the wound healing stages are given below (Velnar et al., 2009; Singer and Clark, 1999):

10.2.1 Hemostasis

This phase of healing takes place immediately after an injury or wound to stop bleeding and prevent excess blood loss. To begin with, platelets are activated through integrin receptors found in extravascular collagen, which results in the release of soluble mediators (fibrinogen, fibronectin, and thrombospondins) that make them clump and stick together (Velnar et al., 2009). Further, a fibrin plug forms at the side of the wound trapping platelets within it and thus maintaining hemostasis while also protecting against bacteria (Falanga, 2004). A number of growth factors (including platelet-derived growth factor (PDGF), transforming growth factor (TGF), and cytokines) are produced from the cytoplasm of platelets and contribute to the wound cleaning and healing process (inflammation) (Raina et al., 2008).

10.2.2 Inflammation

When growth factors are released into the body, inflammatory cells (leukocytes, neutrophils, monocytes, and macrophages) are activated. During inflammation, these cells are activated and play an important role. Moreover, it increases through the endothelial gap and stimulates pain receptors

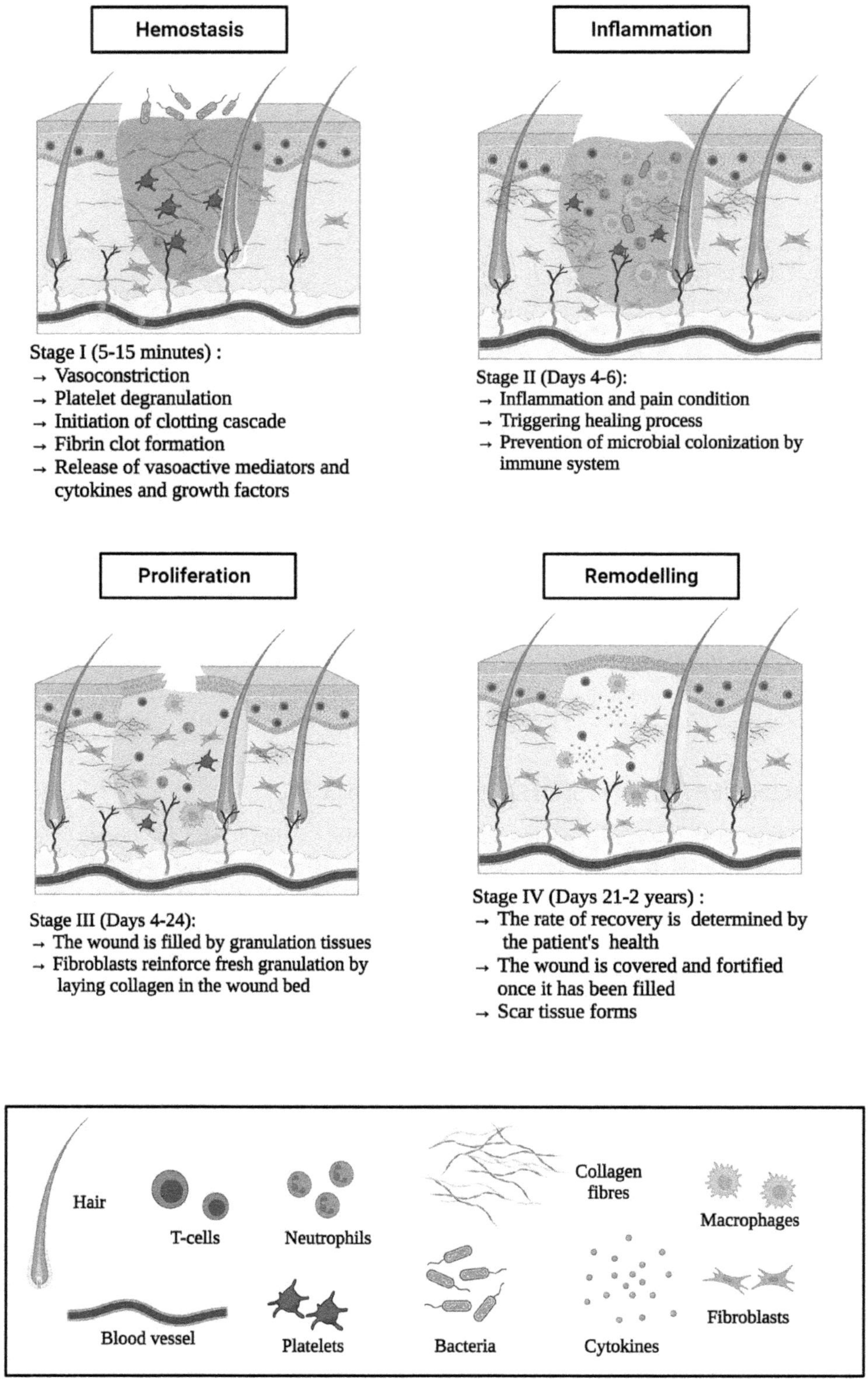

FIGURE 10.2 Phases of wound healing.

in the extracellular space, causing swelling and inflammation at the site of the wound (Barchitta et al., 2019). Cells often respond within 24 h, but it can take up to 2 days. These activated cells act as phagocytes and produce pro-inflammatory cytokines, antimicrobial substances such as reactive oxygen species (ROS), and protease to clean and heal a wound (Pazyar et al., 2014).

10.2.3 Proliferation

During this phase, the cellular structure begins to synthesize proteins such as collagen, fibroblasts move to the damaged site, and small blood vessels develop to aid in the granulation of wounded tissues. Blood cells migrate fibroblasts to form new tissues in response to the release of growth factors and oxygen (Kumar et al., 2018). Collagen production aids in the creation of elastic tissues, which are necessary for wound closure. Several cytokines generated during hemostasis can induce angiogenesis in freshly created tissues. Indeed, angiogenic factors such as fibroblast growth factors (FGFs), vascular endothelial growth factors (VEGFs), PDGFs, angiogenin, and TGFs are essential for wound scarring and closure (Falanga, 2004).

10.2.4 Remodeling

The remodeling or maturation phase of wound healing, which includes the creation of new epithelial layers and scar growth, takes a few weeks to 2 years (Falanga, 2004). During the production of the extracellular matrix, a highly disordered form of collagen III arises. For faster and more efficient wound healing, new collagen will be structured and tensile strength will be generated.

10.3 HERBS USEFUL IN VARIOUS STAGES OF WOUND HEALING

Various intrinsic biochemical mechanisms and extrinsic physical barriers such as remodeling of tissues, antibacterial, anti-inflammatory, and analgesic functions, and bacterial infiltration at the site of injury are deemed critical to the complete healing of chronic wounds. For decades, medicinal plants have been used as first-line therapy for inflammation, burns, ulcers, and surgical wounds (Sharma et al., 2021). Wound-healing activities of medicinal plants are exhibited by several bioactive molecules that are responsible for bringing about the desired molecular actions that hasten the healing process. The medicinal properties of some important herbs for wound healing are discussed here:

10.3.1 Clove (Syzygium aromaticum)

Syzygium aromaticum (clove) is a traditionally used spice proven for various pharmacological actions due to the presence of "clove oil" which is an essential oil (Drugs.com, 2020). Several studies around the world have demonstrated the beneficial effects of clove oil in wound healing. Hameed et al. (2021), developed and tested nanofiber (NF) mats consisting of essential oils of clove with enhanced antibacterial activity, regenerative properties, and wound-healing potential. Using chitosan and polyethylene oxide (PEO) polymers, clove essential oil (CEO) was encapsulated to form nanofibers and scanning electron microscopy (SEM) confirmed the diameters of the finest nanofiber 154 ± 35. Moreover, the encapsulation efficacy was $87.6 \pm 13.1\%$ and loading capacity of CEO was observed $8.9 \pm 0.98\%$. The drug release profile of CEO at acidic pH 5.5 was found to be 79% with a degree of swelling of 117%. *Staphylococcus aureus* and *Escherichia coli* nanofibers exhibited significant antibacterial action against *S. aureus* and expressed no cytotoxic signs on the human fibroblast cell lines. The wound-healing effect of CEO nanofibers was statistically significant ($p > 0.05$) compared to the control group in both the incision and excision wound models of rats (Hameed et al., 2021).

Alam et al. (2017) prepared nanoemulsions of clove oil and tested these nanoemulsions for their wound-healing potential. Optimized nanoemulsion (droplet size of 29.10 nm) was used to examine wound healing, collagen proportions, and histopathology in rats. Considerable healing effects were observed with the optimized nanoemulsion of clove oil compared to pure clove oil. In addition, the nanoemulsion had significantly higher leucine content (0.61 mg/g) as compared to pure clove oil (0.50 mg/g) and the negative control (0.31 mg/g). In histopathology, rats treated with nanoemulsion showed no signs of inflammation. Results also indicated that the nanoemulsion of clove oil was safe and nontoxic (Alam et al., 2017).

10.3.2 Licorice (*Glycyrrhiza glabra* (L.))

The liquorice plant is an herbaceous perennial legume belonging to the family Leguminosae, it has inhabitant to Western Asia, North Africa, and Southern Europe (Kaur et al., 2021) The Studies have used glycyrrhizin and glycyrrhetinic acid to reduce skin inflammation. Other ingredients, such as licochalcone A. and isoliquiritigenin (Yu et al., 2017), also have anti-inflammatory effects. Additionally, the roots of licorice contain a number of different bioactive compounds (Yang et al., 2015). The inflammatory properties of licorice are best understood through the saponin, glycyrrhetinic acid, and the steroid saponin of licorice.

Al-Radadi. (2021) produced a licorice root extract gold nanoparticles technology that was both cost-effective and unique. After two and a half hours at 25°C, the formulation was adjusted by using 6 mL of salt concentration and 4 mL of extract with pH 5. The zeta potential was detected at –26.5, which indicated the stability of green synthesized gold nanoparticles. Furthermore, the TEM characterization showed a 43.17-diameter with a size range between 2.647 and 16.25 nm, and SEM showed a spherical shape. Further, antibacterial studies were performed on five strains, including Gram-positive and Gram-negative bacteria, and the results showed good activity in *E. coli* (Gram-negative bacteria). In conclusion, licorice root extract gold nanoparticles showed noteworthy antibacterial and antifungal activity toward bacterial and fungal strains used, and this formulation was also effective in anticancer activity (Al-Radadi, 2021).

10.3.3 Birch Bark (*Betula* spp.)

Birch Bark is well known for its medicinal and chemical activities. It also showed wound-healing effects at the cellular level, particularly at the inflammatory phase as well as during keratinocyte migration and differentiation phases (Ebeling et al., 2014). It was demonstrated that betulin was effective in the healing of split-thickness wounds when applied topically to a water-free betulin oleogel (Metelmann et al., 2015). The effect of oleogel on superficial wounds and secondary burns was studied in a controlled and randomized study (Frew et al., 2019) (LOE-A). In 2016, the EMA authorized betulin oleogel for the treatment of superficial wounds and burns (Scheffler, 2019).

10.3.4 Turmeric (*Curcuma longa*)

Turmeric (*C. longa*) is a perennial, rhizomatous, and herbaceous category of plant. For many years, it has been used as a remedy and in food. It belongs to the ginger family. Curcumin was used by Ayurvedic practitioners to treat skin injuries, asthma, respiratory disorders, and diabetes. Anti-inflammatory, antiseptic, anticarcinogenic, antiangiogenic, antispasmodic, antimicrobial, and antiparasitic properties were discovered (Aggarwal et al., 2007).

According to Shariati et al. (2019), nanocurcumins have exhibited better effectiveness at 256 g/mL in DMSO as compared to suppression of multidrug-resistant *P. aeruginosa* growth at 128 g/mL in water. At non-inhibitory doses, nanocurcumin decreased biofilm formation, and at 64 g/mL, 58% of formed biofilms were destroyed. Furthermore, curcumin nanoparticles suppressed transcription of virulence genes (except nfxB) and had no cytotoxic effects on human epithelial cell lines (A549) (Shariati et al. 2019).

10.3.5 Aloe vera (*Aloe barbadensis*)

Aloe vera *(A. barbadensis Miller)* is a perennial plant of the Asphodelaceae family. Glycosides, polysaccharides, saponins, pyrocatechol, anthraquinones, acemannan, oleic acid, and water-soluble polysaccharides are among the natural bioactives found in them (Sánchez et al. 2020). Acemannan was replaced via the cyclin D1 and AKT/mTOR signaling pathways, resulting in a substantial reduction in wound closure time (Ahmad et al., 2010).

Rajbar R. and his team formulated chitosan nanoparticles and aloe vera for full-thickness infected wounds that had developed Gram-positive bacteria resistance. They selected the rat as an animal model and were randomized into five different groups consisting of six rats in each group. Animals were applied with the dressing of chitosan nanoparticles with topically applied aloe vera. Further, wound size at 6, 9, 12, 15, 18, and 21 days after surgery was measured. Overall, the results indicated a significant reduction in wound size (Ranjbar and Yousefi, 2018).

10.3.6 Propolis (Natural Resins)

Propolis is a resin, also known as bee glue. Due to its antibacterial and antifungal properties, it has been used for many years in traditional medicines (Ghavidel et al., 2021). The researchers developed a nanoemulsion loaded with propolis, tea tree oil, and clindamycin hydrochloride for effective healing of wounds. The formulation (F-3) composed of tea tree oil (9%) and propolis extracts (0.4%) was found effective in wound closures. The zeta potential and viscosity were found to be 24.5 0.2 mV and 69.4 1.8 mP. The characterization of TEM confirmed the spherical shape. The study concluded that the nanoemulsions composed of propolis and tea tree oil, loaded or unloaded with clindamycin hydrochloride, showed effective local therapy for wound healing (Abdellatif et al., 2021).

10.3.7 Chinaberry (*Melia azedarach*)

The Chinaberry tree *(Melia azedarach)* is a popular name for *M. azedarach* (Linnaeus, 1737), **which** belongs to the family Meliaceae. In the study, it was found that the silver nanoparticles of *M. azedarach* (MA-AgNPs) were effective in wound healing. The formulation was prepared at room temperature using a stable brown colloidal solution with a maximum absorbance of 420 nm (UV-Vis Spectroscopy). The 1,000 g/mL dose of MA-AgNPs showed a significant inhibitory effect on the growth of *Bacillus cereus* and *Escherichia coli,* and the minimum inhibitory concentration (MIC) was found to be 34 and 37 mm, respectively. Additionally, human dermal fibroblast cells were scratched to determine the potential for wound healing. The result concluded that MA-AgNPs were effective in wound-healing management (Chinnasamy et al., 2019).

10.3.8 Ginkgo Tree (*Ginkgo biloba*)

G. biloba, also known as ginkgo or ginkgo, contains unique bioactive like biflavones, alkylphenols, and polyprenols (Van Beek and Montoro, 2009). The leaf aqueous extract of *G. biloba has been* proven to be effective in the treatment of diabetic wound healing. The study developed an emulsion-based cream containing 1% and 5% bioactive aqueous extract of *G. biloba.* There were three groups of alloxan-induced diabetic rats that had completed excisional skin wounds. First, the control group was treated with normal saline. Second, the reference drug, that is, Cytol Centella cream®," and the third group was cream-based *G. biloba* extract.

Hence, among the three groups, *G. biloba* extract cream showed a non-newtonian pseudoplastic pattern and uniform size distribution during the stability testing protocol as well as a higher wound closure rate. Research suggested that the formulation of *G. biloba* extract cream was effective for the management of acute diabetic wounds (Rani et al., 2017).

10.3.9 Chamomile (*Matricaria chamomilla*)

M. chamomilla, which is commonly known as chamomilla, is a member of the Asteraceae family. Chamomile is a Greek word that means "earth apple." It contains the uncommon bioactive component known as Apigenin, which has shown significant results in wound healing (Singh et al., 2011).

Behrooz M and his team developed a nanofiber wound dressing containing chamomile. The combination of caprolactone and polystyrene (PCL/PS) was used as the polymer, and the nanofibers were prepared by the electrospun method. The formulation used 15% chamomile extract in combination with PCL/PS and showed an approximately 99% recovery rate of wound closure within the 14th day post-treatment period. Overall, the study found rapid and effective improvement in dermis tissue supported by re-epithelization and collagen formation, leading to wound closure (Motealleh et al., 2014).

10.3.10 Neem (*Azadirachta indica*)

A. indica, commonly known as neem, belongs to the family Meliacea. The ideal properties of neem are antifungal, antimicrobial, antimalarial, anti-infertility, antiulcerogenic activity, antihypertensive, and antihyperglycaemic effects (Subapriya and Nagini, 2005). Chinnasamy et al. (2021) developed silver nanoparticles from Azadirachta indica (AI-NPs). The electron microscopy determined the particle diameter, that is, 33 nm, and spherical shape of AI-AgNPs. The biocompatible polymer PF127 is involved in the AI-AgNPs formulation to form a viscous fluid. The AI-AgNPs-PF127 hydrogel showed low-toxic and eco-friendly delivery with effective wound management behavior (Chinnasamy *et al.* 2021). Chandran and colleagues performed in-vitro studies on hybrid leaf extracts of *A. indica* (Neem) and developed the film as a novel wound dressing. They reported that a wound dressing made from *A. indica* (Neem) leaf extract has the potential to close wounds (Chandran *et al.* 2015; Table 10.1).

10.4 CHALLENGES AND FUTURE ADVANCEMENTS

10.4.1 Challenges

The following are the challenges with herbal drugs that need to be overcome (Teja et al., 2021):

- Herbal substances must be handled carefully and protected during the entire extraction process. Adequate drying conditions must be observed as improper drying can lead to natural adulteration. For example, the enzymatic hydrolysis of digitalis glycosides above 65°C leads to their decomposition.
- The lack of methods for analyzing absorption sites, finding animal models for toxicity and safety testing, and legal and regulatory problems associated with herbal medicines.
- Clinical trials are challenging with herbal drugs.
- Pharmacological and toxicological evaluations of herbal formulations can be challenging.

10.5 PATENTED HERBAL FORMULATIONS FOR MANAGEMENT OF WOUNDS

With the advancement in research and analytical evaluation of medicinal herbs, researchers around the world are aggressively exploring the potential of phytopharmaceutical-based formulations in the healing of various types of wounds. This has also paved the way for researchers to register their intellectual properties related to newer combinations of plant-based bioactive compounds and novel formulation designs that accelerate complex and chronic wound healing. Some of the important patents registered in this domain during the last decade are shown in Table 10.2.

TABLE 10.1
Phyto-Pharmacological Profiles of Medicinal Herbs Explored for Wound Healing

S. No	Herbs Common Name	Botanical Name and Family	Chemical Constituent	Animal Model Used in Wound Healing	Pharmacological Action in Wound Healing	References
1	Curcumin	*Curcuma longa*, family, Zingiberaceae	Curcunin, Demethoxycurcumin, BisdeMethoxycurcumin	Excision, Incision	Anti-inflammatory-Cyclooxygenase-2↓, inhibit the generation of ROS like superoxide anions, H_2O_2, and nitrite radical generation by activated macrophages, which play an important role in inflammation.	Chattopadhyay et al. (2004)
2	Neem	Azardirachta Indica Family: Meliaceae	Azadirachtin, azadirone, nimbin, nimbidin, nimbinin	Excision, Incision, Burn	Anti-inflammatory, antimalarial	Munir et al. (2021)
3	Liquorice	Glycyrrhiza glabra Family: Fabaceae	Glycyrrhizin, glycyrrhetinic	Acute/chronic wound healing	Antibacterial, anti-inflammatory, antiviral, antioxidant, and antidiabetic activities	Ameri et al. (2013)
4	Marigold	Calendula officinalis Family: Asteraceae.	Flavonol, glycoside, triterpene	Excision, Cutaneous, Acute, chronic, Burn	Anti-inflammatory	Nicolaus et al. (2017)
5	Damask rose	Rosa damascene Family: Rosaceae	α-Pinene β-Pinene Linalool	Excision, Incision	Anti-inflammatory, wound healing	Fahimi et al. (2015)
6	Papaya	Caricapapaya Family: Caricaceae	Papain	Diabetic, burn, soft tissue wounds	Anti-inflammatory, antiviral, antifungal	Mahmood et al. (2005)
7	Indian olive	Elaeocarpus serratus Family:Santalaceae	Biophenolics, Oleuropein secoiridoid, Luteolin	Incision wound healing	Anti- inflammatory	Koca et al. (2011)
8	Indian mulberry	Morinda citrifolia Family: Rubiaceae	Anthraquinones, steroid, phenol, tannin, and terpenoids	Closed wound healing	Anti- inflammatory	Taheri and Amiri-Farahani (2021); Nayak et al. (2007)
9	Indian penny Wort	Centella asiatica family Apiaceae	Asiatic acid, asiaticoside madecassoside, madecassic acid	Incision wound healing	Anti-inflammatory	Priya et al. (2004)
10	Indian sandalwood	*Santalum album* Family: Santalaceae	Fusariumin, terpenoids, asperterpenoid	Incision wound healing	*showed activity against Staphylococcus aureus* and *Pseudomonas aeruginosa*	Ahmed et al. (2013); Yan et al. (2018)

TABLE 10.2
List of Recent Patents Registered for Herbs-Based Formulations for Wound Healing

S. No	Application No	Publication Date	Inventors	Title	Herbs Used in Formulation	Key Points	References
01	202011046441	17.12.2021	Rohit Raina, Shweta Raina	Herbal wound healing composition	Babchi, Chaulmoogra, Cedrus Deodara, Marigold	Developed herbal formulation for skin infection, it showed antibacterial and antifungal activities	Raina and Raina (2021)
02	WO/2022/015374	20.01.2022	Faxon Meisner, Lorraine	Topical composition for the treatment of psoriasis and related skin disorders	Cannabidiol, Oleuropein and Berberine	The herbal formulation effectively resulted into Antioxidant and anti-inflammatory properties	Meisner (2022)
03	202011050143	01.01.2021	Khanna, Vibhu Singh, Saurabh Singh, Dileep Singh, Prabhjot Vij, Rohit	Novel herbomineral nanoemulsion for the treatment of burns		Easy and less expensive herbal nanoformualtion for burn treatment	
04	202111054212	03.12.2021	Anuj Kumar Dubey Ahsas Goyal Anuj Garg	Formulation and evaluation of nano-emulgel containing moringa oleifera seed oil for wound healing	Moringa seed oil	Nanoemulsion-based hydrogel embedded with Moringa seed oil used for topical treatment of wounds infection, it showed activity against Gram-positive and Gram-negative bacteria	Dubey et al. (2021)
05	202111017969	23.04.2021	Dr. Mohit	Synthesis and formulation of silver nanoparticles from extracted and characterized woodfodia fructiosa flowers for wound healing	Woodfodia fruticose	The results showed the synergistic activity of the prepared silver nanoparticles in combination with extract of Woodfodia fruticose in the management of wound	Mohit (2021)
06	201741023245	11.08.2017	Dr. Giridharan, Bupesh Dr. K. Pandian, Dr. Chinnaiah Amutha, Dr. P. Sasikumar, Dr. N.S. Muthiah, Dr. P. Subramanian	A topical wound healing formulation for faster and beneficial antimicrobial and wound healing effect	Aloe vera gel and curcumin	Novel, safe and effective topical formulation for faster wound healing formulation	Bupesh et al. (2017)

10.6 CONCLUSION

The unique importance of herbal medicines has been widely known for decades. Indeed, herbal remedies play a vital role in the management of wound closures. The phytopharmaceutical leads a scientific approach to avoid frequent administration of drugs and maintain patient compliance in a novel manner. However, the conventional use of herbal formulations has various limitations such as solubility, bioavailability, sustainability, therapeutic efficacy, drug release pattern, etc. As a result, advanced drug delivery systems introduce better therapeutic indexes with decreasing toxicity and rising bioavailability values to overcome these limitations. pharmaceutical researchers explore novel methods for herbal drug delivery such as silver and gold nanoparticles, nanoemulgels, ointments, nanoemulsions, nanocreams, nanofilms, and nanomembranes. However, there are various challenges with herbal formulations that need to be conquered by introducing various animal models for toxicity and safety parameters.

REFERENCES

Abdellatif MM, Elakkad YE, Elwakeel AA, et al. "Formulation and Characterization of Propolis and Tea Tree Oil Nanoemulsion Loaded with Clindamycin Hydrochloride for Wound Healing: In-Vitro and In-Vivo Wound Healing Assessment." *Saudi Pharmaceutical Journal.* 29 no.11 (2021):1238–1249.

Aggarwal BB, Surh YJ, Shishodia S, editors. *The Molecular Targets and Therapeutic Uses of Curcumin in Health and Disease.* Springer Science & Business Media, LLC, 233 Spring street, New York NY 10013, USA (2007):1–76.

Ahmad O, Aboutorab TN, Naeini T, Behrooz NE. "Effect of Aqueous Extract of Aloe Vera on Experimental Cutaneous Wound Healing in Rat. *Veterinarski Arhiv.* 80 no.4 (2010):509–522.

Ahmed N, Khan MSA., Jais AMM, Mohtarrudin N, Ranjbar M, Amjad MS, Nagaraju B, Faraz M, Pathan F, Chincholi A. "Anti-Ulcer Activity of Sandalwood (Santalum album L.) Stem Hydro-Alcoholic Extract in Three Gastric-Ulceration Models of Wistar Rats." *Boletín Latinoamericano y del Caribe de Plantas Medicinales y Aromáticas.* 12 no. 1 (2013):81–91.

Alam P, Ansari MJ, Anwer MK, et al. "Wound Healing Effects of Nanoemulsion Containing Clove Essential Oil." *Artificial Cells, Nanomedicine, and Biotechnology.* 45 no.3 (2017):591–597.

Al-Radadi NS. "Facile One-Step Green Synthesis of Gold Nanoparticles (AuNp) Using Licorice Root Extract: Antimicrobial and Anticancer Study Against Hepg2 Cell Line." *Arabian Journal of Chemistry.* 14 no.2 (2021):102956:1–25.

Ameri A, Rajive BB, Vaidya JG, Apte K, Deokule SS. "Anti-Staphylococcal and Wound Healing Activities of Ganoderma Praelongum and Glycyrrhiza Glabra Formulation in Mice." *International Journal of Applied Research in Natural Products.* 6 (2013):27–31.

Badukale NA, Panchale WA, Manwar JV, Gudalwar BR, Bakal RL. "Phytochemistry, Pharmacology and Botanical Aspects of Madhuca Indica: A Review." *Journal of Pharmacognosy and Phytochemistry* 10 no.2 (2021):1280–1286.

Barchitta M, Maugeri A, Favara G, et al. "Nutrition and Wound Healing: An Overview Focusing on the Beneficial Effects of Curcumin." *International Journal of Molecular Sciences* 20 no.5 (2019):1119. https://doi.org/10.3390/ijms20051119.

Bupesh G, Pandian K, Amutha C, Sasikumar P, Muthiah NS, Subramanian P. "A Topical Wound Healing Formulation for Faster and Beneficial Antimicrobial and Wound Healing Effect", 201741023245, 2017.

Chah KF, Eze CA, Emuelosi CE, Esimone CO. "Antibacterial and Wound Healing Properties of Methanolic Extracts of Some Nigerian Medicinal Plants." *Journal of Ethnopharmacology* 104 no.1–2 (2006):164–167.

Chandran VS, Amritha TS, Sujatha RG, Pandimadevi M. "Collagen-Azadirachta indica (Neem) Leaves Extract Hybrid Film as a Novel Wound Dressing: In Vitro Studies." *International Journal of Pharmaceutical Sciences Review and Research* 32 (2015):193–199.

Chattopadhyay I, Biswas K, Bandyopadhyay U, Banerjee RK. "Turmeric and Curcumin: Biological Actions and Medicinal Applications." *Current Science* 10 (2004):44–53.

Chinnasamy G, Chandrasekharan S, Bhatnagar S. "Biosynthesis of Silver Nanoparticles from Melia Azedarach: Enhancement of Antibacterial, Wound Healing, Antidiabetic and Antioxidant Activities." *International Journal of Nanomedicine* 14 (2019):9823–9836.

Chinnasamy G, Chandrasekharan S, Koh TW, Bhatnagar S. "Synthesis, Characterization, Antibacterial and Wound Healing Efficacy of Silver Nanoparticles from Azadirachta Indica." *Frontiers in Microbiology.* 19 (2021):1–14.

Drugs.com. 3 July 2020. Retrieved 27 September 2020.

Drugs/FAQs/New_Drugs_FAQs.doc. Assessed on 03-02-2022. FAQ on Phytopharmaceutical drugs. CDSCO. Available from https://cdsco.gov.in/opencms/opencms/en/Drugs/New-Drugs/ref.

Dubey AK, Goyal A, Garg A, "Formulation and Evaluation of Nano-Emulgel Containing Moringa Oleifera Seed Oil for Wound Healing", 202111054212, 2021.

Ebeling S, Naumann K, Pollok S, Wardecki T, et al. "From A Traditional Medicinal Plant to A Rational Drug: Understanding the Clinically Proven Wound Healing Efficacy of Birch Bark Extract." *PLoS One*. 9 no.1 (2014):e86147:1–18

Edwards R, Harding KG. "Bacteria and Wound Healing." *Current Opinion in Infectious Diseases*. 17 no.2 (2004):91–96.

Fahimi S, Abdollahi M, Mortazavi SA, Hajimehdipoor H, Abdolghaffari AH, Rezvanfar MA. "Wound Healing Activity of A Traditionally Used Poly Herbal Product in a Burn Wound Model in Rats. *Iranian Red Crescent Medical Journal*. 17 no.9 (2015):1–8.

Falanga V. " The Chronic Wound: Impaired Healing and Solutions in the Context of Wound Bed Preparation." *Blood Cells, Molecules, and Diseases* 32 no.1 (2004):88–94.

Frew Q, Rennek ampff HO, Dziewulski P, et al., "BBW-11 Study Group. Betulin Wound Gel Accelerated Healing of Superficial Partial Thickness Burns: Results of A Randomized, Intra-Individually Controlled, Phase III Trial With 12-Months Follow-Up." *Burns* 45 no.4 (2019):876–890.

Ghavidel F, Javadi A, Anarjan N, Jafarizadeh-Malmiri H. "New Approach in Process Intensification Based on Subcritical Water, As Green Solvent, In Propolis Oil in Water Nanoemulsion Preparation." *Green Processing and Synthesis* 10 no.1 (2021):208–218.

Gonzalez AC, Costa TF, Andrade ZD, Medrado AR. "Wound Healing-A Literature Review." *Anais brasileiros de dermatologia* 91 no.5 (2016):614–620. https://doi.org/10.1590/abd1806-4841.20164741

Gosain A, DiPietro LA. "Aging and Wound Healing." *World Journal of Surgery*. 28 no.3 (2004):321–326.

Hameed M, Rasul A, Waqas MK, et al. "Formulation and Evaluation of a Clove Oil-Encapsulated Nano-fiber Formulation for Effective Wound-Healing." *Molecules* 26 no.9 (2021):2491.

Kaur L, Kaur R, Singh A, Kaur N. "A brief review on ethnobotanical, pharmaceutical and therapeutical uses of Glycyrrhiza glabra." *The Pharma Innovation Journal* 10 no. 3 (2021):997–1006.

Koca U, Süntar I, Akkol EK, Yılmazer D, Alper M. "Wound Repair Potential of Olea europaea L. Leaf Extracts Revealed by In Vivo Experimental Models and Comparative Evaluation of the Extracts' Antioxidant Activity. *Journal of Medicinal Food*. 14 no.1–2 (2011):140–146.

Kumar SS, Rajendran NK, Houreld NN, Abrahamse H. "Recent Advances on Silver Nanoparticle and Biopolymer-Based Biomaterials for Wound Healing Applications." *International Journal of Biological Macromolecules* 115 (2018):165–175.

Linnaeus C. (*Hortus cliffortianus*: 1737), 1753:493.

MacNeil S. "Biomaterials for Tissue Engineering of Skin." *Materials Today* 11 no.5 (2008):26–35.

Mahmood AA, Sidik K, Salmah I. Wound Healing Activity of Carica papaya L. Aqueous Leaf Extract in Rats. *International Journal of Molecular Medicine and Advance Sciences*. 1 no.4 (2005):398–401.

Mathieu D, editor. "Physiologic Effects of Hyperbaric Oxygen on Wound Healing Processes." *Handbook on Hyperbaric Medicine*, 135–145. New York: Springer, 2006.

Meisner LF, "Topical Composition for the Treatment of Psoriasis and Related Skin Disorders", WO/2022/015374, 2022.

Metelmann HR, Brandner JM, Schumann H, et al. "Accelerated Re-epithelialization by Triterpenes: Proof of Concept in the Healing of Surgical Skin Lesions." *Skin Pharmacology and Physiology* 28 no.1 (2015):1–1.

Moffatt PG. "Stochastic Choice and the Allocation of Cognitive Effort." *Experimental Economics*. 8 no.4 (2005):369–388.

Mohit, "Synthesis and Formulation of Silver Nanoparticles from Extracted and Characterised Woodfodia Fructiosa Flowers for Wound Healing", 202111017969, 2021.

Motealleh B, Zahedi, P, Rezaeian I, et al. "Morphology, Drug Release, Antibacterial, Cell Proliferation, and Histology Studies of Chamomile-Loaded Wound Dressing Mats Based on Electrospun Nanofibrous Poly (Ɛ-Caprolactone)/Polystyrene Blends." *Journal of Biomedical Materials Research Part B: Applied Biomaterials*, 102 no.5 (2014):977–987.

Munir M, Shah SN, Almas U, Khan FA, Zaidi A, Bukhari SM, Murtaza G. "An Assessment of the Wound Healing Potential of a Herbal Gel Containing an Azadirachta Indica Leaf Extract." *Veterinární Medicína*. 66 no.3 (2021):99–109.

Nayak BS, Isitor GN, Maxwell A, Bhogadi V, Ramdath DD. "Wound-Healing Activity of Morinda Citrifolia Fruit Juice on Diabetes-Induced Rats." *Journal of Wound Care*. 16 no.2 (2007):83–86.

Nicolaus C, Junghanns S, Hartmann A, Murillo R, Ganzera M, Merfort I. "In Vitro Studies to Evaluate the Wound Healing Properties of Calendula Officinalis Extracts. *Journal of Ethnopharmacology*. 196 (2017):94–103.

Pazyar N, Yaghoobi R, Rafiee E, Mehrabian A, Feily. "A Skin Wound Healing and Phytomedicine: A Review." *Skin Pharmacology and Physiology* 27 no.6 (2014):303–310. https://doi.org/10.1159/000357477

Priya KS, Arumugam G, Rathinam B, Wells A, Babu M. Celosia Argentea Linn. Leaf Extract Improves Wound Healing in a Rat Burn Wound Model. *Wound Repair and Regeneration* 12 no.6 (2004):618–625.

Raina R, Parwez S, Verma PK, Pankaj NK. "Medicinal Plants and Their Role in Wound Healing. *Online Veterinary Journal* 3 no.1 (2008):21. https://doi.org/10.3389/fphar.2017.00544.

Raina R, Raina S. "Herbal Wound Healing Composition". 202011046441, 2021.

Rani R, Dahiya S, Dhingra D., Dilbagh, N., Kim KH, Kumar S. "Evaluation of Anti-Diabetic Activity of Glycyrrhizin-Loaded Nanoparticles in Nicotinamide-Streptozotocin-Induced Diabetic Rats." *European Journal of Pharmaceutical Sciences* 106 (2017):220–230.

Ranjbar R, Yousefi A. "Effects of Aloe Vera and Chitosan Nanoparticle Thin-Film Membranes on Wound Healing in Full Thickness Infected Wounds with Methicillin Resistant Staphylococcus Aureus." *Bulletin of Emergency & Trauma*. 6 no.1 (2018):8–15.

Sánchez M, González-Burgos E, Iglesias I, Gómez-Serranillos MP. "Pharmacological Update Properties of Aloe Vera and its Major Active Constituents." *Molecules* 25 no.6 (2020):1324:1–37.

Saraf S. "Applications of Novel Drug Delivery System for Herbal Formulations." *Fitoterapia*. 81 no.7 (2010):680–689.

Scheffler A. "The Wound Healing Properties of Betulin from Birch Bark from Bench to Bedside." *Planta Medica*. 85 no.7 (2019):524–527.

Shariati A, Asadian E, Fallah F, et al. "Evaluation of Nano-Curcumin Effects on Expression Levels of Virulence Genes and Biofilm Production of Multidrug-Resistant Pseudomonas Aeruginosa Isolated from Burn Wound Infection in Tehran, Iran." *Infection and Drug Resistance* 12 (2019):2223–2235.

Sharma A, Khanna S, Kaur G, Singh I. "Medicinal Plants and their Components for Wound Healing Applications." *Future Journal of Pharmaceutical Sciences*. 7 no.1 (2021):1–3.

Singer AJ, Clark RA. "Cutaneous Wound Healing." *New England Journal of Medicine*. 341 no.10 (1999):738–746.

Singh O, Khanam Z, Misra N, Srivastava MK. "Chamomile (*Matricaria chamomilla* L.): An Overview." *Pharmacognosy Reviews* 5 no.9 (2011):82–95.

Subapriya R, Nagini, S. "Medicinal Properties of Neem Leaves: A Review". *Current Medicinal Chemistry-Anti-Cancer Agents,* 5 no.2 (2005):149156. https://doi.org/10.2174/1568011053174828

Taheri M, Amiri-Farahani L. "Anti-Inflammatory and Restorative Effects of Olives in Topical Application." *Dermatology Research and Practice*. n. 1 (2021):9927976.

Teja PK, Mithiya J, Kate AS, et al. "Herbal Nanomedicines: Recent Advancements, Challenges, Opportunities and Regulatory Overview." *Phytomedicine*. 96 (2021):153890.

Van Beek TA, Montoro P. "Chemical Analysis and Quality Control of Ginkgo Biloba Leaves, Extracts, and Phytopharmaceuticals." *Journal of Chromatography A*. 1216 no.11 (2009):2002–2032.

Velnar T, Bailey T, Smrkolj V. " The Wound Healing Process: An Overview of the Cellular and Molecular Mechanisms." *Journal of International Medical Research*. 37 no.5 (2009):1528–1542. https://doi.org/10.1177/147323000903700531

Wadekar AB, Nimbalwar MG, Panchale WA, Gudalwar BR, Manwar JV, Bakal RL. "Morphology, Phytochemistry and Pharmacological Aspects of Carica Papaya, an Review." *GSC Biological and Pharmaceutical Sciences*. 14 no.3 (2021):234–248.

Yan C, Liu W, Li J, Deng Y, Chen S, Liu H. Bioactive Terpenoids from Santalum Album Derived Endophytic Fungus *Fusarium* sp. YD-2. *RSC Advances*. 8 no.27 (2018):14823–14828.

Yang R, Wang LQ, Yuan BC, Liu Y. "The Pharmacological Activities of Licorice." *Planta Medica*. 81 no.18 (2015):1654–1669.

Yu H, Li H, Li Y, Li M, Chen G. "Effect of Isoliquiritigenin for the Treatment of Atopic Dermatitis-Like Skin Lesions in Mice." *Archives of Dermatological Research*. 309 no.10 (2017):805–813.

Zhang DK, Han X, Li RY, Niu M, Dong Q, Yang M, Wang JB, Xiao XH. "Investigation of the Chemical Markers for Experiential Quality Evaluation of Crude Aconite by UHPLC-Q-TOF-MS." *Journal of Separation Science*. 39 no.22 (2016):4281–4289.

11 Novel Phytopharmaceuticals Used as Antipsoriatic

Asma Praveen, Arya Lakshmi Marisetti, Saurabh Dahiya, Vijender Kumar, Jovita Kanoujia, Madhu Gupta, Saveena Chauhan, Saurabh Bhatia, and Ajay Sharma

11.1 INTRODUCTION

Psoriasis is a common chronic autoimmune or inflammatory skin disease. Psoriasis shows symptoms like redness, scaling, flaking, red and white patches, itching, and lesions. All the symptoms can depress the patient and negative effect on the physical and mental health of the patient (Takahashi & Yamasaki 2020). Based on geographical pattern disease prevalence differs across the region, it depends based on exposure to UV radiation and clinical amelioration of psoriasis (Di Meglio & Villanova 2014). Psoriasis worldwide prevalence according to the region. 1.5%–3% in Europeans, 0.05%–3% in Africans and 0.1%–0.5%in Asian (Takahashi &Yamasaki 2020), 2%–3% population in the UK (Higgins 2017), and 3%–4% adult population in the US approximately (Farahnik et al. 2017). A patient who suffers from psoriasis feels embarrassed, frustrated, helpless, lack of confidence, etc. (Kimball et al. 2010). Triggers factor- common trigger factors for psoriasis include stress, illness, injury to the skin, lesions, obesity, and genetic factors (Takahashi & Yamasaki 2020). Although triggers vary from person to person. Certain drugs, certain foods, alcohol, or environmental factors can trigger psoriasis. Certain drugs are β-blocker, lithium, and anti-malarial such as chloroquine, interferon, imiquimod, and terbinafine. Some other new associations are TNF-α antagonists and anti-programmed cell death protein 1 immune checkpoint inhibitors (Balak et al. 2017). Psoriasis can develop at any age; it has been reported at birth time and in old age but generally occurs before 40 years of age. The normal identified age range of psoriasis—is 15–20 years and the second peak at 55–60 years (Smith et al. 1993; Ferrándiz et al. 2002). Henseler and Christophers, present clinical data on the mean age of psoriasis patients—Type I & Type II. Type I starts before 40 years of age and Type II starts after 40 years of age (Langley et al. 2005). In type I psoriasis, 75% of cases and more severe than type II, but more cases are mild and can be treated with primary medication by medical therapy (Higgins 2017).

11.1.1 Pathophysiology

Psoriasis primarily considered as over-proliferation of keratinocytes and developed new cells in the outermost layer of skin (epidermis). Over proliferation develop hyperkeratosis (skin thickening) and parakeratosis (reduced or absent the granular layer) and form skin-covering plaques, silvery white scales, redness, and psoriatic lesions (Aghmiuni & Khiavi, 2017; Di Meglio & Villanova, 2014). It is an autoimmune disorder. T-cells are found to highly contribution to psoriasis pathology T-lymphocytes cells (CD4+ & CD8+ cells) stimulate inflammatory cytokine TNF-α, IFN-γ, IL-12, IL-17A, IL-22 & IL-23 (Figures 11.1 and Table 11.1) enhance the proliferation of keratinocytes (Ferenczi et al. 2000; Nestle et al. 2009). Dendritic cells involve from the beginning of psoriasis pathology. Cathelicidin (LL-37) is an anti-microbial peptide (AMP), expressed by keratinocytes after

 DOI: 10.1201/9781003292692-11

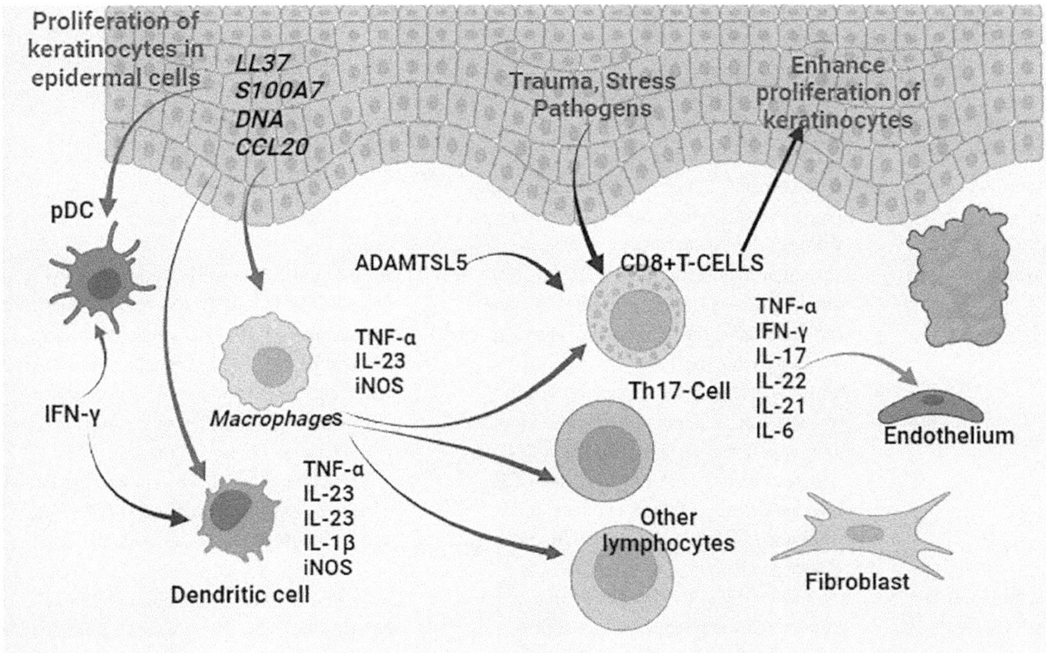

FIGURE 11.1 Pathology of psoriasis.

TABLE 11.1
Classification of Psoriasis with Variants Pathology

Type of Psoriasis	Description	Pathology of Variants
Psoriasis vulgaris (plaque-type psoriasis)	This is the most common form of psoriasis, approximately 85%–90% cases are reported. In these type of psoriasis dark red lesions is covered by white silvery patches. Plaques cover large skin surface, these types of lesions are generally located on the trunk, elbows, knees, scalp, lower back& limbs (Nestle et al. 2009; Ortonne et al. 2009)	TNF-α, IL-23, Th17 (IL-17, Il-21 & IL-22) cytokine enhance the keratinocytes proliferation in skin epidermis and important role play for disease progression (Matsuzaki et al. 2018)
Inverse psoriasis (Flexural Psoriasis)	This is the chronic plaque-type psoriasis, dark red, smoothy and shiny patches found on inverse like, axilla of arm, anogenital region, and skin folds of breast (Langley et al. 2005)	Chronic form of plaque psoriasis
Guttate psoriasis	In this type of psoriasis many small spots, scaly, pink lesions on the skin. This psoriasis can treat topically at initial stage otherwise it can convert in to chronic plaque psoriasis (Di Meglio & Villanova 2014; Griffiths & Barker 2007)	T-cell activation stimulate by the streptococcal superantigens (T-cell mitogens) secreted by streptococcus pyogenes, and form a MHC with class II molecules and T-cell receptor (TCR), PSORS1 (psoriasis susceptibility gene 1) located on the chromosomes of this complex, and PSORS1 locus allele is HLA-Cw6 that involve in the pathogenesis and activate the CD8 (+) T-cell & IFN-γ response (Rendon & Schäkel 2019; Ran et al. 2019)

(*Continued*)

TABLE 11.1 (*Continued*)
Classification of Psoriasis with Variants Pathology

Type of Psoriasis	Description	Pathology of Variants
Erythrodermic psoriasis	It is the rarest form of psoriasis (1–2.25) approximately. It can be associated with any type of psoriasis and requires emergency therapy (Rendon & Schäkel 2019).	
Pustular psoriasis	In pustular psoriasis observed are small, sterile, yellow pustules with erythema and scaling effects. Pustular psoriasis has two distinct phenotype Psoriasis pustulosa palmoplantaris (PPP) Acrodermatitis continua of Hallopeau (ACH) Both types affect the hands and feet, PPP was observed on the palm and soles and ACH affected the tip of fingers and toes. It is associated with high fever and fatigue (Navarini et al. 2017)	Pustular psoriasis is stimulated by the activation of IL-1β, IL-36α, and IL-36RN (Johnston et al. 2017). Cathelicidin and its residual peptide LL-37 stimulate the cytokine IL-36 and other peptide production and enhance the proliferation of keratinocytes (Takahashi & Yamasaki 2020). IL-36 is secreted by epithelial cells via the stimulation of inflammatory cytokines (IL-17, TNF-α, and IL-36 itself) (Johnston et al. 2017)
Psoriatic arthritis	It is observed in (20%–30%) of patients, It can affect the different types of tissue like synovium cartilage, bones, joints, and tendons. Psoriatic arthritis resembles rheumatoid arthritis because both affect joints (painful, warm, and swollen joints) (Gladman et al. 2005) Risk factors of PsA are Nail psoriasis and family history (Gullick et al. 2010)	It is stimulated by the activation of TNF-α NFκB, Il-6, and IL-8 cytokines (Goldminz et al. 2013). Other proinflammatory cytokines that are expressed by synovial tissue are IL-1, IFN-γ, and TNF-α are involved in PsA pathology (Boutet et al. 2018; Sakkas et al. 2017) Genetic factors that trigger the PsA are HLA-C, IL-12B, IL-23R, and TNIP1 (Hüffmeier et al. 2010; Ellinghaus et al. 2012; Liu et al. 2008), and PsA stimulated by TNF, IL-23, and IL-7 cytokines

injury and associated with pathology of psoriasis, It expressed by destroyed keratinocytes and form a complex with DNA receptor (TLR) 9of plasmacytoid dendritic cells (pDC) (Morizane et al. 2012). pDC start the development of plaque psoriasis and stimulate the cytokine, type 1 IFN (IFN-α & IFN-β). Type 1IFN encourage the mDC and stimulate the function of T-cells, and enhance the secretion of inflammatory cytokine IFN-γ, IL-17, Il-12, Il-23 & TNF-α (Nestle et al. 2005; Gregorio et al. 2010; Santini et al. 2011; Hänsel et al. 2011).

11.1.2 Comorbidities

Psoriasis has been associated with several diseases. Psoriasis is not limited only on skin or inflammation; it can affect different organ of our body. Comorbidities are reported generally in 15% of patients can occur in moderate to severe psoriasis (Kalkan 2017) (Table 11.2).

Psoriasis comorbidities are classified as:-

i. Classical comorbidities
ii. Emerging comorbidities

TABLE 11.2
Classification of Psoriasis-Associated Comorbidities

Classification of Comorbidities	
Classical comorbidities	• Psoriatic arthritis—Nail psoriasis associated with joints and increases the 80% chance of PsA (Langenbruch et al. 2014; Maejima et al. 2010) • Chron's disease (Inflammatory Bowel Disease) • Psychological and psychiatric disorders • Uveitis
Chronic systemic inflammation (Emerging comorbidities)	• Metabolic syndrome—including high blood pressure, hyperlipidemia, coronary artery disease, abnormal cholesterol level, Increased body mass index (BMI), and enhanced chances of cardiac attack and heart stroke (Sommer et al. 2007; Gerdes et al. 2016) • Atherosclerosis • Non-alcoholic fatty liver disease (NAFLD)- factors are obesity (contraindicated from MTX) (Vena et al. 2010), high cholesterol, and type 2 diabetes • Lymphoma—A survey reported that patients who have been affected by long-term psoriasis have a maximum chance of developing the cancer (colon, bladder, kidney cancer) due to the immunogenicity (Brauchli et al. 2009). And a patient who has taken UV (PUVA) therapy sessions for long and many times approximately (250 times), has the maximum chance of skin cancer (melanoma& non-melanoma) (Stern et al. 1994; Griffiths & Barker 2007). PUVA, MTX & cyclosporine therapy may show carcinogenic effects (Herrinton et al. 2012) • Chronic obstructive pulmonary disease—(inflammatory lung disease) • Obstructive sleep apnea/Hypopnea syndrome • Osteoporosis—Psoriasis arthritis reduces the bone mass and it depends on the severity of psoriasis and the number of affected joints (Attia et al. 2011) • Parkinsonism—It is associated with severe psoriasis, and life style factors • Erectile Dysfunction—Negative impact on sexual life due to the genital lesions (Goulding et al. 2011)
Lifestyle comorbidities	• Smoking, Drinking—Pustular form of psoriasis influences the smoking habit, and increases the chance of respiratory disease (Fortes et al. 2005), and alcohol consumption enhances the risk of depression, anxiety, hepatic steatosis, and cirrhosis (Poikolainen et al. 1990) • Anxiety
Treatment-related comorbidities	• Cyclosporine—cause dyslipidemia, Nephrotoxicity & Hypertension • Acitretin—cause dyslipidemia and hepatotoxicity (Vena et al. 2010) • Phototherapy causes skin cancer

iii. Life style and
iv. Treatment comorbidities (Oliveira et al. 2015; Menter et al. 2008).

11.1.3 Medical Treatments Available for Psoriasis

The decision of psoriasis treatment depend on the disease severity, mild to moderate psoriasis can treat topically, vitamin D analogue, and phototherapy. Moderate to severe psoriasis can be treated systemic therapy.

In this table, we have discussed topical therapy, phototherapy, systemic therapy, and biological therapy (Table 11.3).

TABLE 11.3
Medical Treatments Available for Psoriasis

Treatments with Mechanism of Action	
Topical therapy	• **Vitamin D analogue**—(calcipotriol, tacalcitol and calcitriol) block cell differentiation in skin epidermis • **Coal tar**—Suppress the DNA synthesis and inhibit keratinocyte proliferation • **Calcineneurin inhibitors** like Tacrolimus and Pimecrolimus both are prescribed for severe flexural or facial psoriasis • **Dithranol (anthralin derivative)**—It regulates the synthesis of cell DNA and inhibits cell proliferation (Aghmiuni & Khiavi, 2017; Smith & Barker, 2006) • **Corticosteroids**—combination therapy with calcipotriene and tazarotene are more effective (Regulate the cell proliferation) (Uva et al. 2012) • **Tazarotene**—(Retinoid analogue) reduces PASI effect and controls cell proliferation, showing more prominent activity in combination with corticosteroids and light therapy (Dando & Wellington 2005)
Phototherapy	• Treatment of severe psoriasis PUVA (Psoralen-UVA) systemic, PUVA-balneo are preferred. PUVA treatment, which is preferred for UV therapy, Psoralen has an active moiety derived from the natural product Psoralea corifolia. If UV treatment is combined with vitamin D analogue or with topical steroids is more effective rather than alone (Uva et al. 2012; Claes et al. 2006)
Systemic therapy	• **Methotrexate**—Approximately 20 years, Methotrexate is preferred in psoriasis treatment. (MTX analogue of folic acid block DNA synthesis via inhibiting the synthesis of thymidine and purine and proliferation of skin cells, and MTX can improve the HLA-Cw6 associated psoriasis (Saurat, et al. 2008) • **Acitretin**—In pustular psoriasis generally used Acitretin, which is known as retinoid these are Vitamin A analogue (blocks the proliferation of cells) • **Cyclosporine**—is a T-cell inhibiting immunosuppressant, It is a modern drug for psoriasis (It blocks cytokine stimulation and cell proliferation), It is potent but highly toxic many times, it is preferred when psoriasis is not controlled but it has shown more side-effects like kidney failure, hypertension & immunosuppressant (Ho et al. 2001)
Biological therapy	These agents block the molecular step in the pathogenesis of psoriasis. • **TNF-α inhibitors**—are Etanercept, infliximab, adalimumab, and certolizumab they can block TNF-α and treat psoriasis. They can also reduce the PASI effect (Lucka et al. 2012) • **IL-12, IL-23 inhibitors**—Ustekinumab, guselkumab, tildrakizumab and risankizumab.IL-12& IL-23chemokines stimulate the function T-cells. IL-23 is composed of two subunit p40 and p19. P40 subunit is common in IL-23 & IL-12. Ustekinumab bind with p40 subunit and block the function both receptors (Benson et al. 2011). Guselkumab, tildrakizumab and risankizumab these monoclonal antibodies bind with p19 subunit of IL-23 cytokine and inhibit the function and reduced PASI effect (Blauvelt et al. 2017) • **IL-17 inhibitors**—Secukinumab, ixekizumab and brodalumab. Secukinumab and ixekizumab block IL-17A. (Langley et al. 2014), and show better efficacy for scalp and nail psoriasis. Whereas brodalumab is inhibit IL-17 (A, F, &A/F). These inhibitors block the step of psoriasis pathogenesis and reduce PASI effect (Papp et al. 2016; Puig 2017)

(*Continued*)

TABLE 11.3 (*Continued*)
Medical Treatments Available for Psoriasis

Treatments with Mechanism of Action	
Biological therapy	• **JAK Inhibitors**—JAK inhibitors are the newer drug with good efficacy. Janus kinase proteins (JAK) have four proteins JAK1, JAK2, JAK3, & TYK2. These inhibitors block the JAK-STAT mechanism, block intracellular proteins and conjugate with STAT and inhibit the intracellular molecular step of psoriasis pathogenesis (Schwartz et al. 2017). Tofacitinib is JAK 1, 2, 3 inhibitors, this is already present in the market for RA, PsA and ulcerative colitis (He et al. 2019; Szilveszter et al. 2019). From the meta-analysis data reported that tofacitinib show well efficacy with psoriasis pathology (Papp et al. 2012). BMS-986165 selective TYK2 inhibitor, (phase 3 trials under pipeline) but phase 2 trial shows the highest efficacy for psoriasis (Papp et al. 2018). Other JAK inhibitors are peficitinib (ASP015K)-inhibit directed JAK. Solcitinib (GSK 2586184) & baricitinib influence the function of JAK1, baricitinib associated with JAK1 & 2 (these drugs currently under pipeline in phase 2 and phase 3 clinical trials) (Kvist-Hansen et al. 2020) • **NGF inhibitors**—NGF strongly associated with psoriasis pathogenesis TrkA kinase associated with NGF receptor and enhance the keratinocytes proliferation, stimulate T-cells. K252 Block the Trk association with NGF and inhibits keratinocyte proliferation (Raychaudhuri et al. 2004) • **Phosphodiesterase inhibitor**—Apremilast works intracellularly, PDE4 inhibitor blocks the stimulation of AMP (Adenosine monophosphate) and inhibits the secretion of proinflammatory cytokines, and reduces mDC and T-cells, count in psoriasis skin (Gottlieb et al. 2013; Papp et al. 2015)

11.2 CHALLENGES IN THE TREATMENT OF PSORIASIS

Current challenges associated in the drug delivery system of psoriasis

11.2.1 The Patient Perspective

11.2.1.1 Psychosocial Aspects (Satisfaction with Treatment)

This is the main and primary challenge, patient who has affected from psoriasis feel depressed 1.5 times more than normal person (Dowlatshahi et al. 2014). So, this is a very important task to make the patient aware of this issue, and what will be the clinical outcomes and feel easy to the patient. So the patient needs frequent counseling and psychoactive medication (Hoffman et al. 2016). Treatment type is also dependent on patient satisfaction. Patients is more satisfied with systemic therapy rather than topical and UV therapy (Poulin et al. 2010; Dubertret et al. 2006). Whereas maximum patients are more satisfied with biological therapy (Lebwohl et al. 2014).

11.2.1.2 Patient Adherence

Adherence is a link between treatment and results in healthcare system. Patient adherence is a big challenge because it is important for prescribed treatment effectiveness and positive results. Physicians can instruct and encourage to the patient for proper prescribed treatment regimen (Hodari et al. 2006).

11.2.2 Challenges in Treatment

11.2.2.1 Challenges Associated with Topical Therapy

Topical therapy is the primary (first-line) therapy for psoriasis. It targets only skin-localized processes; it cannot cover cardiovascular disease while other therapy regulates the immune response and are associated with cardiovascular disease (Prodanowich et al. 2005). Poor penetration into the psoriatic skin is also a big challenge because psoriatic skin is hard due to the low moisture level, cholesterol level is high, low ceramides level, and silvery patches on the skin. Even if the patient follows a proper medication regimen, the skin cannot absorb enough medication to clinically improve (Katare et al. 2010).

11.2.2.2 Challenges Associated with Phototherapy

Ultraviolet rays can enhance the lesion of psoriasis, which is light-sensitive. Phototherapy UVB & PUVA are preferred for moderate to severe psoriasis (Aghmiuni & Khiavi, 2017), but it has many limitations it's a time-consuming, costly, and long-term treatment process. It enhances drug interaction and is associated with harmful toxicities like skin cancer, acute sunburn, and erythema (Matsumura & Ananthaswamy 2004).

11.2.2.3 Challenges Associated with Immunogenicity

In immunogenicity, anti-idiotype-antibody decreases the therapeutic efficacy or neutralizes the drug antibody by inhibiting the binding capacity of the drug antibody with antigen binding site or formation of immunocomplex. Immunogenicity also depends on drug concentration, if the drug concentration is low then lowers the clinical efficacy, down retention rates, and increases the possibility of adverse effects such as perfusion and anaphylactic reaction (Carrascosa 2013; Carrascosa et al. 2014; Puig 2014; De Simone et al. 2013). Immunogenicity shows a big challenge with drug efficacy profile, ADAs for antigen IL-12/23, p40 reduce the therapeutic level of Ustekinumab (Zhu et al. 2010), and ADAs form an immunocomplex with infliximab antibody and neutralize (Van der Laken et al. 2007). ADAs of etanercept are not effective for drug antibodies and etanercept treatment shows the normal efficacy and safety profile (Tyring et al. 2007). Adalimumab anti-idiotypic antibody is associated with adalimumab antibody and reduces the efficacy of adalimumab therapy (Papp et al. 2011). The conclusion is that it's a big challenge for psoriasis therapy, and ADAs testing can helpful technique for psoriasis therapy although immunogenicity affects the efficacy of the drug, comorbidities, and safety profile.

11.2.2.4 Psoriasis Is Difficult to Treat Location

Psoriasis on the scalp, Pustular psoriasis (PPP and ACH), and nail psoriasis, these areas show low effects of topical treatment. PPP can emotionally affect the patient, it can affect life style comorbidities. These induce hyperkeratosis, fissures, and yellow pustules, causing pain and difficulty in walking. This is the most unsatisfactory location (Sánchez-Regaña et al. 2014).

11.2.3 Challenge Associated with Patient Profile

11.2.3.1 Age Factor

Psoriasis treatment in old age is challenging, these age populations more at risk for comorbidities, drug interaction, and dose adjustment (Balato et al. 2014).

11.2.3.2 Challenge during Pregnancy

Treatment during pregnancy and breastfeeding take more precautions. MTX and acitretin cannot be prescribed during pregnancy and breastfeeding, Cyclosporine can be used in this situation. Biological therapy can be prescribed during pregnancy but cannot be prescribed during lactation (Mervic 2014).

11.2.3.3 Psoriasis in Children

Cases of psoriasis in children are limited, so few reports are available for this type of treatment. But on the current management of pediatric psoriasis in the UK organize a multicenter national audit with (NICE) the National Institute for Health and Care Excellence, and indicate the pediatric psoriasis therapy, topical or phototherapy primarily prescribed to MTX, acitretin, and then cyclosporine (Cather 2014; Burden-Teh et al. 2015).

11.2.3.4 Obesity

Obesity is challenging for psoriasis treatment. Psoriasis is associated with obesity comorbidity, obesity can influence cardiovascular disease and metabolic syndrome (Duarte 2013). During psoriasis treatments obese patients increase the chance of adverse effects, eg: - hepatotoxicity to methotrexate (Puig 2011).

11.3 PHYTOPHARMACEUTICALS

11.3.1 Phytopharmaceuticals

(Phytomedicine), Phyto "Plant" and pharmaceutical means "Medicinal drug" According to WHO guidelines purified and standardized fraction of plant extract with at least four therapeutically active constituents of a medicinal plant or any part of the plant. Herbal drugs are prepared by extraction, fractionation, purification, characterization, or any other biological process. Herbal materials that are used in phytopharmaceutical preparation are gums, resins, fresh juice, fixed oils, and dry powder of herbs. Finished herbal products are prepared from one or more herbs if more than one herb can be a mixture of herbal products.

Worldwide phytopharmaceuticals are a strong alternative to synthetic medicine because mostly synthetic medicine is associated with mild to severe adverse effects. So synthetic drugs remain challengeable and replace by phytopharmaceutical medicine (Parmar et al. 2017; Bhatt 2016). According to WHO approximately 80% population in developing countries depend on phytopharmaceuticals, because of low cost affordable, and better accessibility (Asase et al. 2008). According to WHO guidelines to development of a new phytopharmaceutical formulation primary focus on the safety issue for health care and quality control (Narayana & Katiyar 2013). Problems with regulation of phytopharmaceutical because herbal drug have not consistently data and lack of documentation, sometime natural origin formulation may be harmful, this toxic effect associated with quality of herb like adulteration and selection mistakenly with wrong species, incorrect dosing, interaction with other medicine, toxic metals, microorganism and inorganic residue (Nooreen et al. 2018). Problems with regulation of herbal product are quality control of herbal product, safety and efficacy of herbal product and proper documentation and regulation. Systemetic preparation process of phytopharmaceuticals are given below (Latief & Ahmad 2017; Figure 11.2).

11.3.2 Phytopharmaceutical in Psoriasis

Many conventional treatments (prescribed medicine) for psoriasis are available in the market, these conventional drugs can lower the disease symptoms but cannot cure properly because psoriasis has no proper medication. Most of the conventional drug associated with mild to severe side-effect, so synthetic drugs remains challengeable (Rout et al. 2017), in that situation herbal medicine for psoriasis treatment with precaution to ignore harmful or dangerous interaction with psoriasis treatment.

11.3.3 Plants Reported to Possess Antipsoriatic Activity

1. ***Aloe vera* (L.)** (Ghritakumari), it is a perennial herb, belonging to family liliaceae. Anthraquinones, polysaccharides, vitamins, and salicylic acid are the active ingredients of aloe vera exhibiting anti-inflammatory and anti-pruritic activities. It has been used in dermatology since ancient time. So many investigations done on this plant and invented antipsoriatic products, and reported positive clinical efficacy against psoriasis (Syed et al. 1996; Paulsen et al. 2005; Choonhakarn et al. 2010). In a recent study, 95% ethanolic leaf gel extract of *aloe vera* tested on psoriasis positive mouse—tail group. Initiation of granular layer formation by the induction of orthokeratosis (inhibition of parakeratosis), and after 14 days in the histopathological section a well-developed granular layer was found in mouse tail skin after treated with *aloe vera* extract (Dhanabal et al. 2012a).
2. ***Astragalus sinicus (L)*** is a perennial legume, belonging to the family Fabaceae, found in Japan, Korea, and China. It contains flavonoids, triterpenoids, glycosides, and alkaloids (Cui et al. 1992; Nagasawa et al. 2001). Dichloromethane and methanol extract of *A. sinicus* further divided with water and butanol and then redistribution of butanol extract with methanol and n-Hexane then further prepare 8 methanolic fraction into rf1, rf2, rf3, rf4, rf5, rf6, rf7, & rf8., rf3& rf4 fraction show the strong anti-inflammatory and antioxidant

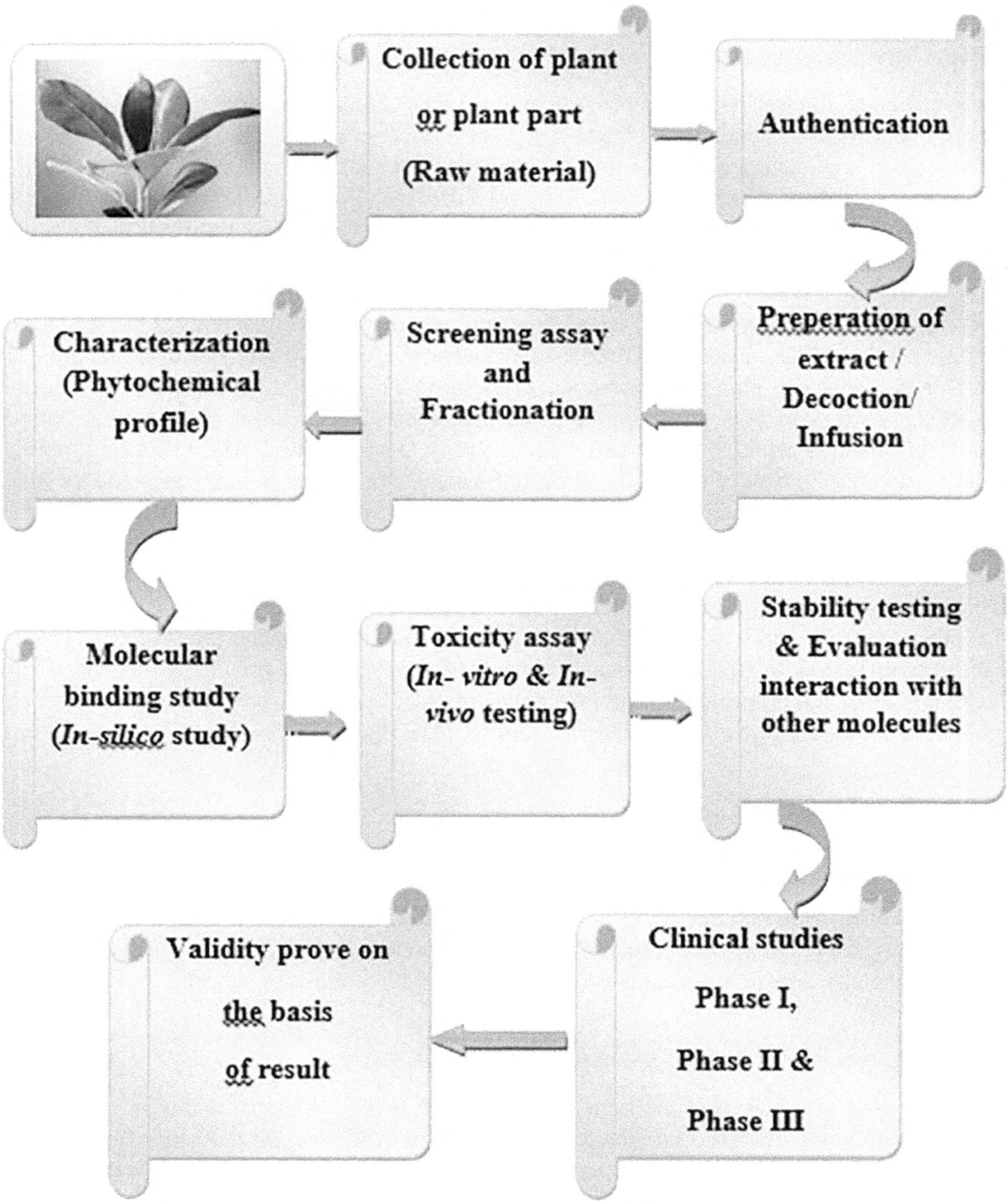

FIGURE 11.2 Procedure following for phytopharmaceutical preparation.

activity on using HaCaT cells. Interpret the inhibition of molecular pathogenesis of psoriasis, and interrupt the JAK/STAT mechanism associated with T-cell stimulation. Inhibit the T-cell stimulated cytokines such as IL-4, IL-17 & IL-22. And fraction rf3 & rf4, especially rf4 block the NFκB expression via blocking the phosphorylation and degradation of IκBα proteins, and also interrupt NFκB 65 transcriptional activity. So, blocking these steps of molecular pathogenesis inhibits the secretion of proinflammatory cytokines and chemokines (Kim et al. 2014).

3. ***Cesalpinia bonduc***: belongs to the family Cesalpiniaceae, found in hot temperate regions in India, Myanmar, and Sri Lanka. Prepare the extract fraction-CBD—Decoction of

C. bonduc with water, CBHA—hydroalcoholic extract of *C. bonduc* with 50% ethanol, CBDB—CBD extract fractionated with n-Butanol, CBHAB—CBHA extract fractionated with n-Butanol, CBHAW—CBHA extract fractionated with distilled water. These fractions experimented on the mouse tail method; CBHAB & CBHAW inhibit the parakeratosis in epidermal skin, and induction of orthokeratosis. CBHAW stimulates the thickness of the epidermal layer. Experimented on external human keratinocytes *(ex vivo)* with *C. bonduc* extract, the entire fraction shows antiproliferative activity, except the CBD fraction. CBHAW fraction of *C. bonduc* blocks the lipoxygenase activity (LOXs) and is associated with the development of psoriasis pathology. So CBHAW fraction shows more potent antipsoriatic activity (Muruganantham et al. 2011).

4. ***Hypericium perforatum*** **(L)** (St. John's wort), belongs to the family Hyperiaceae. It possesses many pharmacological activities and is used dermatologically. *H. perforatum* contains hypericin and hyperforin (Kang et al. 2001). *H. perforatum* extract and its constituent's hyperforin inhibit the secretion of inflammatory cytokines COX-1, 5-lipoxygenase, and PGE2 (Albert et al. 2002; Hammer et al. 2007; Koeberle et al. 2011). A recent study reported that an ointment of *H. perforatum* used for a feasibility study for plaque psoriasis afflicted ten patients, and reduced the PASI effect with low adverse effects (Najafizadeh et al. 2012). Hypericin can penetrate lesional skin easily. *H. perforatum* oil is used dermatologically (Schempp et al. 2000).
5. ***Illicium verum*** **Hook** (Star anise) is from the family Schisandraceae, distributed in China and Vietnam. It possesses anti-inflammatory activity. A study reported that ethanolic extract of *I. verum* (IVE) tested on HaCaT cell lines (Human keratinocytes) and block TNF-α and IFN-γ stimulated protein TARc, MDC, IL-6, IL-1β & ICAM-1 inhibition depends on the dose concentration. Effect on NFκB & STAT1, these mechanisms stimulate the TNF-α & IFN-γ production. So, when HaCaT cells are treated with IVE, inhibit NFκB P65 unit and phosphorylation and degradation of 1κBα and depend on the dose concentration. As well as it inhibits the NFκB luciferase and STAT1 mechanism.IVE interrupts the TNF-α/IFN-induced psoriasis mechanism, inhibits the phosphorylation of ERK & P38, and decreases the activation of MAPK and AKT (Sung et al. 2012a; Sung et al. 2012b). In another study of I.verum, ethanolic extract of I.verum test on human keratinocytes (HaCaT cell line), IVE inhibited the IFN-γ stimulated ICAM-1 and also interrupted the JAK/STAT signaling mechanism via inhibiting the IFN-γ (Sung & Kim 2013).
6. ***Indigo naturalis*** (Natural Indigo) is dry powder or paste from the leaves of *Baphic acanthuscusia (Nees)* family Acanthaceae, *Isatis indigotica* (Brassicaceae), *Polygonum tinctorium* (Polygonaceae), *Indigofera tinctoria* (Fabaceae). It is used as a dermatologically and anti-inflammatory (Reuter et al. 2010). Ointments of I. naturalis were used by 14 positive patients of plaque psoriasis and observed reduced PASI effect and upgraded the histopathology of skin (Lin et al. 2007). Claudin 1 is a protein that is present in the tight junction of keratinocytes and is associated with psoriasis pathology. Extract of *I. naturalis* enhances the stimulation of CLDN-1 and regulates the function of TJ and also enhances the activity of protein kinase C (PKC). Indirubin is a major component of *I. naturalis,* that inhibits EGF (Epidermal growth factor). EGF promotes the cell proliferation and secretion of CDC25B (cell division cycle 25 B) in epidermal keratinocytes (Blažević et al. 2015; Hsieh et al. 2012). *I. naturalis* affect IL-17 pathways, it inhibits IL-17 mRNA in the psoriasis pathology (Cheng et al. 2017).
7. ***Memecylon malabarum*** (Iron wood) belongs to the family Melastomataceae. Water fraction of hydroalcoholic extract of *M. malabarum* (MMHAW) induction of orthokeratosis in mouse tail model and another fraction reduced epidermal thickness and chloroform and butanol fraction of hydroalcoholic extract experimented with external human keratinocytes (HaCaT) block the lipoxygenase enzyme. This enzyme controls the psoriasis development. Quercetin was identified by phytochemical investigation of *M. malabarum* and quercetin is a potent inhibitor of phospholipase A2 (PA2). PA2 is associated with the arachidonic pathway (Dhanabal et al. 2012b).

8. ***Picea mariana*** is from (Black spruce), family Pinaceae. It possesses anti-inflammatory activity polyphenolic extract of *P. mariana* tested with psoriatic keratinocytes (HaCaT) inhibit TNF-α stimulating response. Extract of *P. mariana* interrupt the NFĸB protein regulation (NFĸB stimulates the TNF-α), decreases the production of nitric oxide, and reduces the stimulation of nitric oxide synthase, ICAM-1, IL-6, VEGF, IL-8, and trappin-2 elafingeneration (García-Pérez et al. 2014).
9. ***Viola tricolor*** family violaceae, and also possess many pharmacological activities. Aq. extract of *V. tricolor* tested on human keratinocytes (psoriatic positive cells) inhibit the activation of T-lymphocytes. It decreases the secretion of IL-12 and inhibits IFN-γ and TNF-α production and reduces the PASI effect in psoriatic skin (Hellinger et al. 2014; Gründemann et al. 2013).
10. ***Nigella sativa*** (Black cumin) is from family Ranunculaceae. It possesses many pharmacological activities anti-inflammatory, antibacterial, antifungal, and anthelmintic (Ghosheh et al. 1999). In a recent study ethanolic extract of *N. sativa* used on a mouse tail psoriatic model, reported the induction of orthokeratosis in the epidermal layer, and inhibited the hyperproliferation of keratinocytes (Dwarampudi et al. 2012).

Phytoconstituents as antipsoriatic activity

1. **Ginsenoside**: It is a steroidal saponin obtained from the plant *"Ginseng" (Panax ginseng radix or Panax notoginseng)* (Kim et al. 2017). A variety of ginsenoside are present in plant, Rg1, Rg5, G-Rb2, G-Rd etc.
   i. **Ginsenoside Rg1 (Rg1)**: In one study, the researcher did an experiment in a psoriasis mice model (Imiquimod-induced model). Positive control mice treated with GRg1 (50 mg/kg orally till 14 days), after histopathological analysis reduced PASI effect and thickness of psoriatic skin. It Inhibits inflammatory cytokines like IL-23, 22, 17A, 1β & TNF-α. It blocks the NFĸB expression and reduces inflammatory activity (Shi et al. 2019).

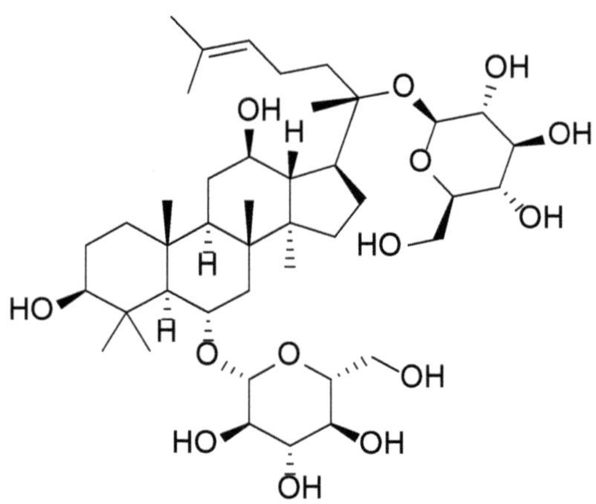

(a) GinsenosideRg1

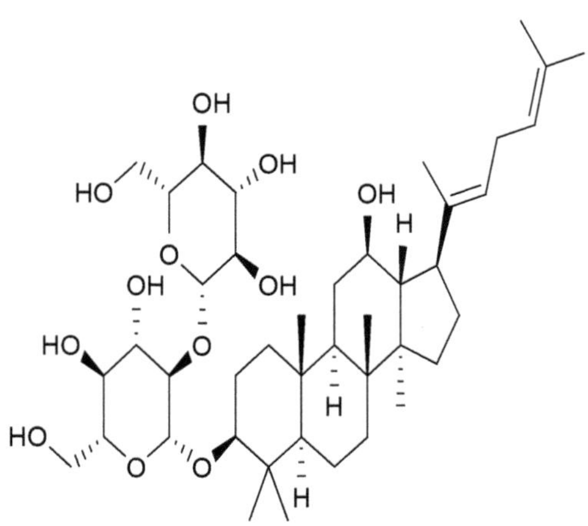

(b) GinsenosideRg5

ii. **Ginsenoside Rg5 &metabolite Rh3**: Experimented withoxazolone induced animal model. It reduces PASI, (erythema, scaling, and redness) and decreases the thickness of psoriatic lesional skin. Cycloxygenase regulates the secretion of inflammatory cytokines, Rg5 & Rh3 reduce the COX-2 expression and inhibit TNF-α & IL-1β.Rh3 inhibits the stimulation of IFN-γ, Rh3 is more potent than Rg5. Interrupt the mechanism of NFĸB and reduce the expression of inflammatory cytokine (Shin et al. 2006).

2. **Baicalein**: It is a trihydroxy flavone. It is obtained from different plants *Scutelleria baicalensis, Scutelleria lateriflora, & Rhus mysorensis.* Baicalein reduces the proliferation of epidermal cells and down regulated the inflammatory cytokines, In a dose-dependent manner (Chao et al. 2007). *Baicalein* treated with HaCaT-positive keratinocytes cells observed inhibition with keratinocyte proliferation, baicalein interrupts the phosphorylation of ERK and stimulation of TRPV4 & blocks the calcium influx mechanism (Huang et al. 2016).

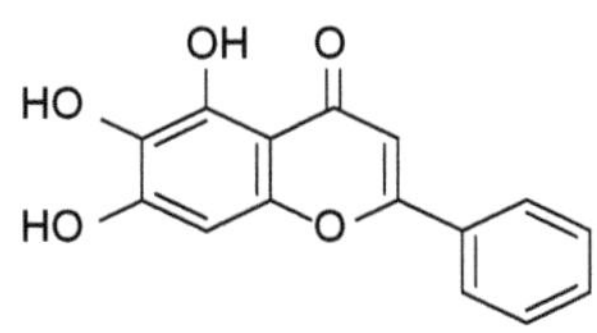

Baicalein

3. **Kaempferol**: Kaempferol, Quercetin and fisetin both are 3-hydroxy flavones. Kaempferol is found in most fruits and vegetables. Kaempferol mostly attaches to the sugar moiety as compared to the free form (Liu et al. 2019). Kaempferol was experimented with imiquimod-induced psoriatic mice model, kaempferol was applied topically on psoriatic skin and after histological analysis, observed regeneration of the epidermal layer. Inhibit T-cell mediated inflammatory cytokines such as IL-6, IL-17A & TNF-α. NFĸB mechanism enhances the stimulation of T-cells. Kaempferol interrupts the NFĸB mechanism and reduce the level of T-cell mediated cytokine and reduces inflammation in psoriasis lesional skin. Kaempferol reduces the level of IL-17A+ & CD4+T-cells in lymph nodes and spleen (Liu et al. 2019).

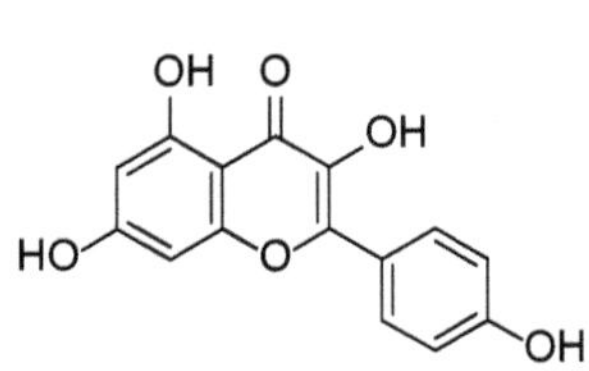

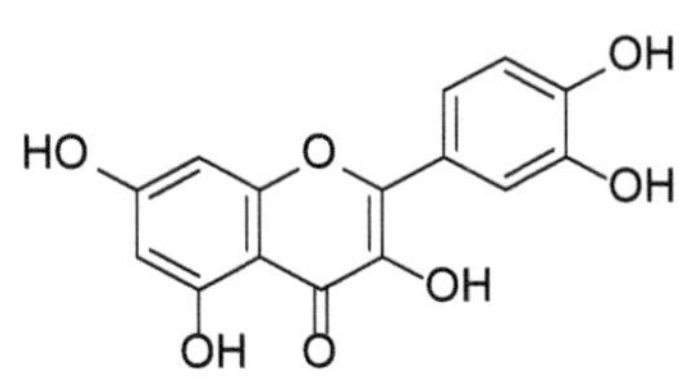

(a) Kaempferol **(b) Quercetin**

4. **Berberine**: It is obtained from *Berberis aristata (herb), Mahonia aquifolium* Rubisan Family (Berberidaceae) (Huang et al. 2019; Potdar et al. 2012). Prepare the topical formulation of (Liquid nano reservoir of berberine, berberine oleate) berberine and LCNP "Liquid crystalline nanoparticles" (Brb-ol) tested on imiquimod-induced mice model and show better efficacy for skin regeneration with high penetration, reduced PASI score. Berberine inhibits the production of proinflammatory cytokines i.e.- IL-17 & IL-23 and reduces inflammation in the psoriatic lesion, Brb-ol more easily penetrates the skin rather than berberine alone (Freag et al. 2019). Transferosome formulation prepared from *B. aristata* extracts treated with imiquimod-induced mice model. Histopathological analysis observed reduced thickness of the epidermal layer and reduced PASI effect (Nimisha et al. 2017).

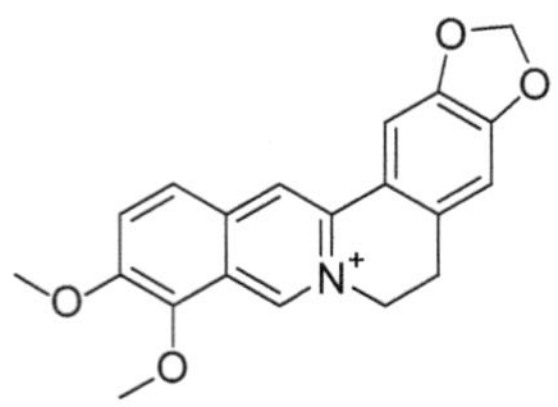

Berberine

5. **Curcumin**: It is obtained from rhizomes of *Curcuma longa* and *Curcuma xanthorrhiza* family Zingeberaceae. Clinical trial data of 20 patients, two groups of psoriasis patients—first group 31–40 years and second group 51–59 years treated with 1% *C. xanthorrhiza* ointment and observed improvement in PASI score (including redness, scaling, inflammation) and reduce the proliferation of the keratinocytes by inhibit the stimulation of K6 markers (Rahmayunita et al. 2018). Another study on 32 patients from a total of 63 patient, during the third phase trial, treated with curcumin orally and observed improvement in PASI and decrease the level of inflammation (inflammatory cytokines) IL-17, 22 (Antiga et al. 2015). In another study, a keratin 14-VEGF transgenic mice model treated with curcumin orally (40 mg/kg) was observed to reduce the stimulation of T-cells and inhibit T-cell stimulated cytokines IL-17, 22, 2, 8, IFN-γ and TNF-α and inhibit epidermal keratinocytes proliferation and reduce PASI effect (Kang et al. 2016). In this study, prepare the curcumin gel and topically applied on the imiquimod mice model was observed to reduce the level of inflammatory markers, IL-17A, F, 22, 1β, 6, and TNF-α. And block the T-cell subunit γδT and inhibit cell proliferation (Sun et al. 2013).

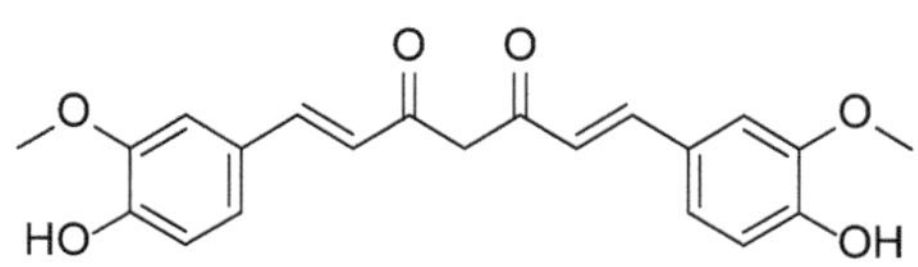

Curcumin

6. **Glycyrrhetinic acid (GL)**: Glycyrrhizin obtained from the root of *Glycyrrhiza glabra* (Liquorice) belonging to the family Fabaceae. In a study GL treated with HaCaT cell line observed interruption of the pathogenesis of psoriasis and it regulated the ICAM-1 and inhibited TNF-α stimulation via T-cell inhibition and down regulated the NFĸB mechanism and blocked phosphorylation of ERK/p38 MAPK mechanism (Xiong et al. 2015). Gao et al. in another study 18 β-Glycyrrhetinic acid treated with HaCaT cells and observed reduce the proliferation of keratinocytes, and induction of cell apoptosis. Down the ROS level (ROS associated with inflammation) via interruption PI3K-Akt mechanism (Gao et al. 2020).

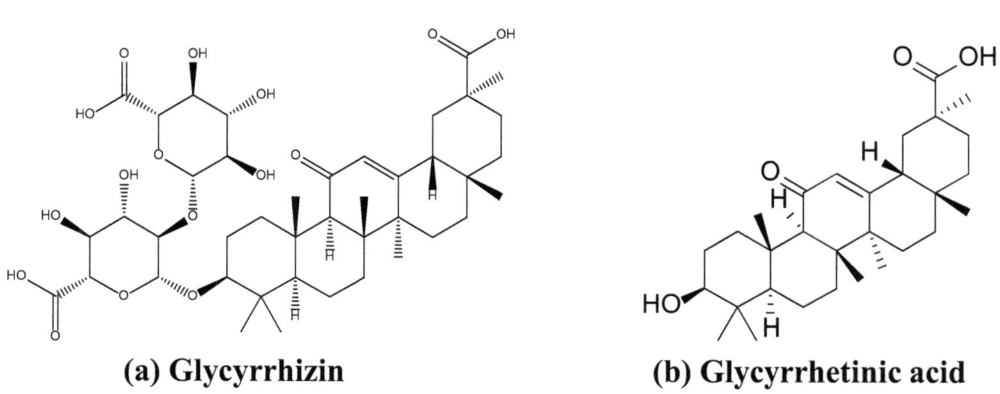

(a) Glycyrrhizin **(b) Glycyrrhetinic acid**

7. **Triptolide**: Obtained from an herbal medicinal plant, *Tripterygium wilfordii* Hook F. In this study, HaCaT keratinocytes treated with triptolide were observed to inhibit rhIFN-γ, and decrease IFN-γRα stimulation because rhIFN-γ enhances the stimulation of IFN-γRα and SOCS1. It interrupts the phosphorylation mechanism associated with the JAK/STAT mechanism and ICAM-1 production by inhibition of rhIFN-γ stimulated mechanism (Hongqin et al. 2011).

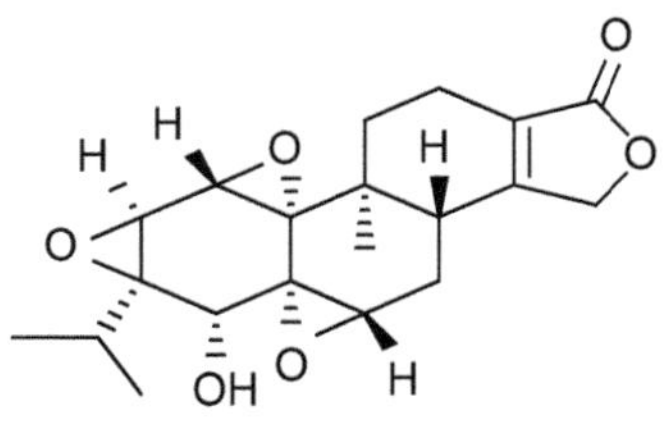

Triptolide

8. **Rottlerin (Mallotoxin)**: Natural polyphenolic compound obtained from *Mallotus philippinensis* belonging to the family euphorbiaceae (Gangwar et al. 2014). *In vitro* and *in vivo* study on psoriasis model- NHEKs cells (Primary normal human epidermal keratinocytes), Edu positive cells treated with rottelerin were observed to reduce the proliferation of keratinocytes cells. When observed via FACS analysis (fluorescence-activated cell sorting analysis) inhibition of cell growth proliferation, down-regulate the NFκB mechanism, this mechanism stimulates the proliferation of keratinocytes. It reduces the stimulation of miR-21 and miR-3 (mi-R21 and miR-31 enhancement of keratinocyte proliferation). Rottlerin decreases the stimulation of the ROS mechanism (Role of oxidative stress). Rottlerin treated with in vivo model (Imiquimod-induced psoriasis model) analyzes the reduction in PASI, reduced acanthosis, and initiation of orthokeratosis in the epidermal layer of skin and reduce the level of T-cell and decreased stimulation of inflammatory markers, that is, TNF-α, IL-6, 23 (Min et al. 2017).

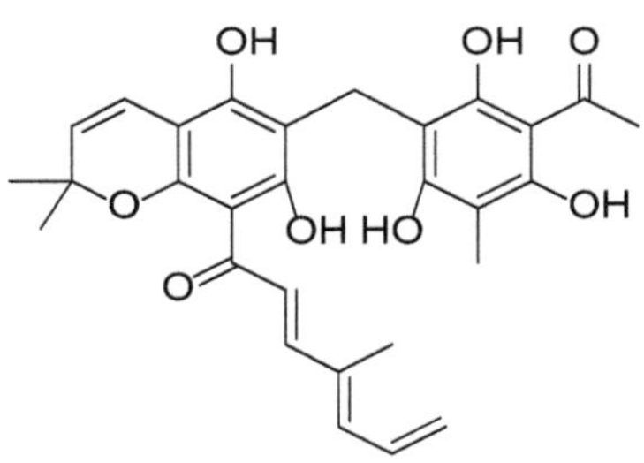

Rottlerin

11.4 COMBINATION OF PHYTOPHARMACEUTICAL WITH SYNTHETIC DRUG IN THE MANAGEMENT OF PSORIASIS

When a herbal formulation combine with synthetic formulation may increase or decrease the pharmacological action. Meta-analysis data explain the herb drug interaction. Nanoreservoir formulation of berberine with oleate (Brb-ol), penetrate the skin easily rather than berberine alone. Synthetic components enhance the skin penetration rate of herbal berberine compound (Freag et al. 2019).

A Chinese meta-analysis data report, Main plants of these combination (CHM decoction) *Salvia miltiorrhiza root, Lithospermum erythrorhizon root, Reh mania glutinosa root* (Zhang et al. 2014a).

A Chinese herbal medicine combination with acitretin (oral route). Acitretin show the more efficacy in combination with CHM rather than acitretin alone, and no severe adverse effect observed with this combination.CHM decoction with acitretin orally observed higher efficacy on psoriatic lesions and no adverse effect (Zhang 2012). An another study CHM decoction combined with 20 mg capsule acitretin/day orally show better therapeutic value, higher improvement in PASI with minor adverse effect (Zheng 2011). When CHM decoction combine with antihistaminic, vitamin-b, vitamin-B6, till 2 months, show better efficacy on psoriasis activity and improvement in PASI without any adverse effect (Yuan 2011). When CHM decoction combined with topical calcipotriol show better efficacy and reduced PASI effect, with few side-effects like redness because of calcipotriol and herbal medicine dissolved rashes easily. So this combination shows better effectivity (Jin et al. 2009).

In another clinical trial data observed, methotrexate combines with glycyrrhizin. MTX show the toxicity with glycyrrhizin, when MTX administered alone to the animal model then shows the normal concentration in blood. Whereas when it combines with glycyrrhizin show hepatic toxicity, induce the level of hepatic enzyme interrupt the clearance of MTX (Abe et al. 2021). A meta-analysis data observed combination with Chinese topical herbal medicinal plants with synthetic medicine (Deng et al. 2014).

Liubaibaibi is a herbal medicinal cream combination with clobetasol, vitamin E, and folic acid orally. Show better clinical efficacy reduced PASI effect. No adverse effect observed this combination is safe and effective (Gao et al. 2006; Yang et al. 2011). E-Bei-ointment is a (herbal formulation) combine with calcipotriol ointments and placebo. These same combinations used orally (tablet form), observed reduce PASI effect and show better efficacy level, with minor adverse effect like rashes, and rashes remove easily and improvement in itching because of herbal formulation (Song et al. 2007). Xiaoxuanling formula: combine with tretinoin cream (0.025% vitamin A acid cream) observed quickly improvement in PASI score, show better efficacy in low cost rather than conventional treatment and no adverse effect observed (Zhu et al. 2008). Qinbai ointment shows the better effect on psoriasis lesion and improvement in PASI rather than placebo group. But in both ointment group ‘fine’ qinbai ointment show better efficacy compare to conventional qinbai ointment (Xu et al. 2009). On the basis of these report topical herbal formulation show the better efficacy with synthetic drug combination or with placebo (approx 50% or more than) with minor adverse effect and same as oral formulation but few data available.

11.4.1 Clinical Studies of Treating Psoriasis Using Phytopharmaceuticals

Clinical trials are the final step in determining the effectiveness of new treatments. The development of new technologies, advancement in delivery system and start-up inventions able to foster more clinical trial in future. Despite substantial research on phytopharmaceuticals in the treatment of psoriasis, rigorous clinical trials are still required.

11.4.2 Phytopharmaceuticals-Based Novel Drug Delivery System for Psoriasis Management

The amalgamation of phytopharmaceuticals with novel drug delivery systems (NoDDS) delivers therapeutic benefits of plant-derived chemicals with the goal of increasing therapeutic efficiency and overcoming clinical application limitations. Several scientists are investigating NoDDS such as ethosomes, niosomes, transferosomes, liposomes, polymeric micelles and nanoparticles. Furthermore, these phytoconstituent-loaded particulate systems have been included in various gelling systems such as liposomal gel, nanogel, nano-emulgels, and hydrogels, which have become the latest trend in recent years due to their capacity to improve efficiency. Table 11.4 provides a summary of some of the most advanced psoriasis delivery technologies.

TABLE 11.4
Phytopharmaceuticals-Based Novel Drug Delivery System for Psoriasis Management

S. No	Formulation	Phytopharmaceuticals	Result	Reference
1	Liposomal gel	Curcumin and Ibrutinib	The combination was to show synergistic action in psoriasis management	Jain et al. (2022)
2	Liposphere gel	Curcumin and tacrolimus	Liposphere gel resulted in improved phenotypic and histological character of psoriatic skin	Jain et al. (2016)
3	Liposomal nanocarriers	Psoralen	Improved permeation, decrease in psoriasis system	Doppalapudi et al. (2017)
4	Liposomes and ethososmes	Psoralen	Ethosomes were found to improve high permeability	Zhang et al. (2014b)
5	Niosomal hydrogel	Celastrol	Increases celastrol permeation into the skin and improves the antipsoriatic activity.	Meng et al. (2019)
6	Niosomal hydrogel	Celastrol	Inhibit inflammation and hyperproliferation of keratinocytes in the skin	Qiu et al. (2021)
7	Nano-sponge based gel	Curcumin and caffeine	The combination was found to be effective in symptom reduction	Iriventi et al. (2020)
8	Nanoemulgel	Curcumin and imiquimod	A combination of curcumin and imiquimod was successfully evaluated for antipsoriatic activity	Algahtani et al. (2020)
9	Nanoparticle loaded hydrogel	Curcumin	Polymeric nanoparticles-based gel loaded with curcumin are promising for treating psoriasis	Mao et al. (2017)
10	Transethosomes-in-gel	Rosmarinic acid	The rosmarinic acid-loaded transethosomal gel shows higher penetration across the skin	Rodríguez-Luna et al. (2021)
11	Spanlastics in gel	Resveratrol	An in vivo study revealed a reduction in psoriasis symptoms like scaling and erythema	Elgewelly et al. (2022)
12	Polymeric micelles in gel	Resveratrol	Optimized micelles were converted into gel for ease of application with the high deposition ability	Khurana et al. (2020)

11.4.2.1 Liposomes

Liposomes are composed of phospholipids, stabilizers, and cholesterol to form bilayered vesicular carriers with drug (hydrophilic and lipophilic) loading capability. Liposomal systems are useful to increase the drug load in skin due to higher rate liposome localization in stratum corneum. Usually, liposomal formulation is prepared in the form of liquid, which may escape from the application site post administration, is a major drawback of utilizing liposomes topically on the skin. This problem of liposome dispersion can be overcome by including them into an appropriate vehicle, which preserves the original structure of vesicles while adjusting their rheological and/or mucoadhesive qualities. Liposomal gels can be prepared by incorporating liposome dispersions into hydrogels/gelling agents.

Potential of curcumin in the treatment of psoriasis has been highlighted in a increased number of research in recent years. The antioxidant and oxidative stress reducing properties of curcumin is the reason to reduce the psoriatic lesions. Curcumin has been shown in recent research to lower phosphorylase kinase levels and block the potassium channel in psoriatic individuals. TNF-, IFN-,

IL-2, IL-12, IL-22, and IL-23 are all inflammatory mediators that curcumin efficiently lowers in psoriatic patients. As per US Food and Drug Administration (FDA) classification, curcumin was classified as "generally regarded as safe" (GRAS). When used to treat psoriatic patients, curcumin has been shown to have no detrimental effects in vivo.

Psoralen is the active ingredient in the dried fruits of *Cullen corylifolium* (L.) Medik (*syn. Psoralea corylifolia* L), also known as "Buguzhi" in Chinese medicine system. Photosensitizing property of Psoralen when combine with sun light is very effective against skin problems. In the latest investigation, reported by Jain et al. (2022) demonstrated the ant psoriatic potential of liposomal gel containing Curcumin and Ibrutinib. This combination was found to reduced ear thickness, severity index, epidermal hyperplasia, and psoriasis area. The claim was further supported by histopathological evaluation showed decrease in psoriatic lesion (Jain et al. 2022).

In another interesting study, curcumin and tacrolimus was loaded in liposphere gel to resolve the poor water solubility, less absorption, less penetration issues of these drugs. With a approximate particle size of 50 nm, liposphere containing tacrolimus and curcumin were drugs and shear thinning behavior of the liposphere gel was a desirable attribute of topical formulation. The results showed that tacrolimus and curcumin loaded liposphere gel improved the phenotypic and histological aspects of psoriatic skin. When compared to the imiquimod group, TNF-, IL-17, and IL-22 levels were found to lower (Jain et al. 2016).

One more research focused on the development and characterization (zeta potential, size, stability entrapment efficiency, *in vitro* and *in vivo* studies) of liposomal carriers containing Psoralen. The prepared liposomal carriers were found to be very effective in comparison to Psoralen containing solution in term of permeation enhancement, reduction in psoriatic biomarkers (IL-22, IL-17, TNF- α) (Doppalapudi et al. 2017).

Psoralen was further incorporated in ethosomes and liposomes to compare their efficiency. It was found that ethosomes were found to show high penetration (due to flexibility) in comparison to liposomes (Zhang et al. 2014a).

11.4.2.2 Niosomes

Niosomes are hydrating mixtures of cholesterol and nonionic surfactants that produce multilamellar or unilamellar vesicles. In recent years, traditional liposomes have been transformed into niosomes, which are thought to have the capacity to deliver tailored drugs. Celastrol is a naturally occurring a triterpenoid obtained from Chinese herb *Tripterygium wilfordii* Hook. f. (Celastraceae), with high anti-inflammatory and antioxidant properties effectively treats autoimmune diseases. In one study, researcher developed the celastrol-loaded niosomes using combination of probe sonication and thin film hydration method. Celastrol-niosomes was found to improve *in vitro* permeation with potential to reduce the cytokines levels. Celastrol-niosomes effectively reduced scaling and erythema issues in psoriasis animal models as per *in vivo* studies (Meng et al. 2019). Similar way in another study, celastrol-loaded niosomal gel was developed and evaluated for cellular uptake study, anti-inflammatory study, and particle size and entrapment efficiency. The hydrogel was found to be effective against psoriasis management (Qiu et al. 2021).

11.4.2.3 Nanosponges

Nanosponges are multifaceted mesh like structure with the several nanocavity to load the drugs. One study was reported based on a nanosponge-based gel containing curcumin and caffeine as an anti-inflammatory bioactive. The sponges were prepared using polymer (beta-cyclodextrin) and cross-linker (dimethyl carbonate) by applying the hot melt method and further combined with gel. Histopathology and in vivo studies revealed a significant improvement in psoriasis symptoms in treated animals. The combination of caffeine and curcumin was found to be very effective against psoriasis (Iriventi et al. 2020).

11.4.2.4 Nanoemulsion

Nanoemulsions are dispersions of two liquids with droplet sizes on the order of 100 nm that are kinetically stable. Because of the low surface tension and high surface area of the emulsion system, nanoemulsion have better ingredient penetration efficacy, requiring only 3%–10% surfactants during preparation. Furthermore, they are biocompatible, stable, and non-toxic, and can enhance the poor bioavailability issues of lipophilic drugs. The combination of curcumin and imiquimod was successfully evaluated for the ant psoriatic activity by delivering these drugs in the form of a nanoemulsion-based gelling system. The drugs were loaded in nanoemulsion using Tween 20 and emulsification method. *In vitro* studies showed better results in terms of skin deposition and permeation with high efficiency to reduce the psoriasis-like system (Algahtani et al. 2020).

11.4.2.5 Nanoparticles

Polymeric nanoparticles are solid particles typically formed through spontaneous self-assembly from synthetic as well as natural polymers. They have features of good biocompatibility, simple design, and loading efficiency of hydrophilic and lipophilic drugs. Polymeric nanoparticles loaded with curcumin are promising for treating psoriasis by increasing the permeation of curcumin. The diameter of nanoparticles was 24.4 nm with drug loading of 3.49% and entrapment efficiency (78.45%). The optimized nanoparticle formulations were incorporated in gel and found to release the drug in a sustained release manner. In vivo studies were conducted on animal models and revealed a significant reduction in psoriasis symptoms due to the high anti-keratinization process (Mao et al. 2017).

11.4.2.6 Transethosomes

Transethosomes are a hybrid of a transferosome and an ethosome with the ability to deform and the ability to permeate the skin. Song et al. have proposed this system can be administered topically and via a systemic route. The drug can be safely encapsulated in this vesicle and slowly releases the drug. Rosmarinic acid is antioxidant and anti-inflammatory plant oriented found in many Lamiaceae herbs used in cooking, including lemon balm, rosemary, oregano, and sage. Rosmarinic acid is poorly permeable and poorly soluble with stability issues. These problems make its use limited in the treatment of psoriasis; it can be overcome by developing a novel drug delivery system. The rosmarinic acid-loaded transethosomal gel shows higher penetration across the skin. This transethosomal gel released the drug sustained manner. The anti-inflammatory activity of the developed gel provides effectiveness against psoriasis (Rodríguez-Luna et al. 2021).

11.4.2.7 Spanlastics

Sorbitan-based vesicles in the nano range with characteristic elasticity and flexibility are known as spanlastics, now a day, most researchers investigating the potential of this delivery system for enhanced permeation, and suitability with different routes of administration in comparison to niosomes. Resveratrol is a naturally occurring anti-inflammatory substance found in nuts, berries, and grapes. Various studies focused on the anti-inflammatory effect on cell lines such as cultured adipocytes, adipose tissues, macrophage cell lines, and adipocyte cell lines. Spanlastics loaded with resveratrol were successfully developed and characterized for entrapment efficiency, small particle size, and sustained release. Furthermore, optimum spanlastics formulation was incorporated into carbopol gel with optimum release and viscosity, etc. An in vivo study revealed a reduction in psoriasis symptoms like scaling and erythema compared to another group (positive control). Histopathological images also suggest the features of healthy skin (Elgewelly et al. 2022).

11.4.2.8 Polymeric Micelles

Micelles are nano-sized self-associated structures of polymers with capabilities of drug entrapment. Micelles formulations have the potential to enhance the solubility and bioavailability of poorly soluble plant-based phytopharmaceuticals. In one investigation, researchers found that topically applied resveratrol-loaded polymeric micelles are effective phytopharmaceuticals in the treatment of psoriasis. The optimization of formulations was based on statistical methods (quality by design and central composite design) for permeation, particle size, and entrapment efficiency. Additionally, optimized micelles were converted into gel for ease of application with the high deposition ability, and low serum cytokines levels (Khurana et al. 2020).

11.5 CONCLUSION

Although the exact mechanism involved in the progression of psoriasis is unknown, many studies are now being undertaken to better understand the complex factors that underpin this condition. In the treatment of psoriasis, Novel drug delivery methods play an important role. Advanced nanocarriers are presently being thoroughly tested in order to discover a solution to unsolved drug delivery problems. Many Phytopharmaceutical-based remedies as well as combination of phytopharmaceutical with synthetic drugs have also been explored to decrease adverse effects, and all of these herbal plants are easily accessible. Plants with antioxidant, anti-inflammatory properties are commonly used to treat psoriasis. In future, phytopharmaceuticals-based advanced carrier system could be explored further with the more emphasis of clinical trial and patented products.

REFERENCES

Abe, Kentaro, Tsukasa Higurashi, Mio Takahashi, Ayako Maeda-Minami, Yohei Kawano, Satoru Miyazaki, and Yasunari Mano. "Concomitant use of high-dose methotrexate and glycyrrhizin affects pharmacokinetics of methotrexate, resulting in hepatic toxicity." *In Vivo* 35, no. 4 (2021): 2163–2169.

Albert, Dana, Ilse Zündorf, Theo Dingermann, Walter E. Müller, Dieter Steinhilber, and Oliver Werz. "Hyperforin is a dual inhibitor of cyclooxygenase-1 and 5-lipoxygenase." *Biochemical Pharmacology* 64, no. 12 (2002): 1767–1775.

Algahtani, Mohammed S., Mohammad Zaki Ahmad, Ihab Hamed Nourein, and Javed Ahmad. "Co-delivery of imiquimod and curcumin by nanoemugel for improved topical delivery and reduced psoriasis-like skin lesions." *Biomolecules* 10, no. 7 (2020): 968.

Antiga, Emiliano, Veronica Bonciolini, Walter Volpi, Elena Del Bianco, and Marzia Caproni. "Oral curcumin (Meriva) is effective as an adjuvant treatment and is able to reduce IL-22 serum levels in patients with psoriasis vulgaris." *BioMed Research International* 2015 (2015): 283634.

Asase, Alex, Tetsuo Kokubun, Renée J. Grayer, Geoffrey Kite, Monique SJ Simmonds, Alfred A. Oteng-Yeboah, and George T. Odamtten. "Chemical constituents and antimicrobial activity of medicinal plants from Ghana: Cassia sieberiana, Haematostaphis barteri, Mitragyna inermis and Pseudocedrela kotschyi." *Phytotherapy Research* 22, no. 8 (2008): 1013–1016.

Attia, Enas AS, Ayman Khafagy, Sameh Abdel-Raheem, Sahar Fathi, and Abeer A. Saad. "Assessment of osteoporosis in psoriasis with and without arthritis: correlation with disease severity." *International Journal of Dermatology* 50 (2011): 30–35.

Balak, Deepak MW, and Enes Hajdarbegovic. "Drug-induced psoriasis: clinical perspectives." *Psoriasis (Auckland, NZ)* 7 (2017): 87.

Balato, Nicola, Cataldo Patruno, Maddalena Napolitano, Angela Patrì, Fabio Ayala, and Raffaele Scarpa. "Managing moderate-to-severe psoriasis in the elderly." *Drugs & Aging* 31, no. 4 (2014): 233–238.

Benson, Jacqueline M., David Peritt, Bernard J. Scallon, George A. Heavner, David J. Shealy, Jill M. Giles-Komar, and Mary Ann Mascelli. "Discovery and mechanism of ustekinumab: a human monoclonal antibody targeting interleukin-12 and interleukin-23 for treatment of immune-mediated disorders." *MAbs* 3, no. 6 (2011): 535–545.

Bhatt, Arun. "Phytopharmaceuticals: a new drug class regulated in India." *Perspectives in Clinical Research* 7, no. 2 (2016): 59.

Blauvelt, Andrew, Kim A. Papp, Christopher EM Griffiths, Bruce Randazzo, Yasmine Wasfi, Yaung-Kaung Shen, Shu Li, and Alexa B. Kimball. "Efficacy and safety of guselkumab, an anti-interleukin-23 monoclonal antibody, compared with adalimumab for the continuous treatment of patients with moderate to severe psoriasis: results from the phase III, double-blinded, placebo-and active comparator-controlled VOYAGE 1 trial." *Journal of the American Academy of Dermatology* 76, no. 3 (2017): 405–417.

Blažević, Tina, Elke H. Heiss, Atanas G. Atanasov, Johannes M. Breuss, Verena M. Dirsch, and Pavel Uhrin. "Indirubin and indirubin derivatives for counteracting proliferative diseases." *Evidence-Based Complementary and Alternative Medicine* 2015 (2015): 654098.

Boutet, Marie-Astrid, Alessandra Nerviani, Gabriele Gallo Afflitto, and Costantino Pitzalis. "Role of the IL-23/IL-17 axis in psoriasis and psoriatic arthritis: the clinical importance of its divergence in skin and joints." *International Journal of Molecular Sciences* 19, no. 2 (2018): 530.

Brauchli, Yolanda B., Susan S. Jick, Montserrat Miret, and Christoph R. Meier. "Psoriasis and risk of incident cancer: an inception cohort study with a nested case-control analysis." *Journal of Investigative Dermatology* 129, no. 11 (2009): 2604–2612.

Burden-Teh, Esther, M. L. Lam, S. M. Taibjee, A. Taylor, S. Webster, S. Dolman, C. Jury et al. "How are we using systemic drugs to treat psoriasis in children? An insight into current clinical UK practice." *British Journal of Dermatology* 173, no. 2 (2015): 614–618.

Carrascosa, J-M., M. B. A. van Doorn, M. Lahfa, F. O. Nestle, D. Jullien, and J. C. Prinz. "Clinical relevance of immunogenicity of biologics in psoriasis: implications for treatment strategies." *Journal of the European Academy of Dermatology and Venereology* 28, no. 11 (2014): 1424–1430.

Cather, Jennifer C. "Psoriasis in children and women: addressing some special needs." *Seminars in Cutaneous Medicine and Surgery* 33, no. 2 Suppl 2 (2014): S42–S44.

Chao, Jui-I., Wen-Chi Su, and Huei-Fang Liu. "Baicalein induces cancer cell death and proliferation retardation by the inhibition of CDC2 kinase and survivin associated with opposite role of p38 mitogen-activated protein kinase and AKT." *Molecular Cancer Therapeutics* 6, no. 11 (2007): 3039–3048.

Cheng, Hui-Man, Yang-Chang Wu, Qingmin Wang, Michael Song, Jackson Wu, Dion Chen, Katherine Li et al. "Clinical efficacy and IL-17 targeting mechanism of Indigo naturalis as a topical agent in moderate psoriasis." *BMC Complementary and Alternative Medicine* 17, no. 1 (2017): 1–11.

Choonhakarn, Charoen, P. Busaracome, Bungorn Sripanidkulchai, and Pongdech Sarakarn. "A prospective, randomized clinical trial comparing topical aloe vera with 0.1% triamcinolone acetonide in mild to moderate plaque psoriasis." *Journal of the European Academy of Dermatology and Venereology* 24, no. 2 (2010): 168–172.

Claes, Christa, Werner Kulp, Wolfgang Greiner, Johann-Matthias von der Schulenburg, and Thomas Werfel. "Therapy of moderate and severe psoriasis." *GMS Health Technology Assessment* 2 (2006).

Cui, Baoliang, Junko Inoue, Takashi Takeshita, Junei Kinjo, and Toshihiro No Hara. "Triterpene glycosides from the seeds of Astragalus sinicus L." *Chemical and Pharmaceutical Bulletin* 40, no. 12 (1992): 3330–3333.

Dando, Toni M., and Keri Wellington. "Topical tazarotene." *American Journal of Clinical Dermatology* 6, no. 4 (2005): 255–272.

De Simone, Clara, Paolo Amerio, Giuseppe Amoruso, Federico Bardazzi, Anna Campanati, Andrea Conti, Paolo Gisondi et al. "Immunogenicity of anti-TNFα therapy in psoriasis: a clinical issue?" *Expert opinion on biological therapy* 13, no. 12 (2013): 1673–1682.

Deng, Shiqiang, Brian H. May, Anthony L. Zhang, Chuanjian Lu, and Charlie CL Xue. "Topical herbal formulae in the management of psoriasis: systematic review with meta-analysis of clinical studies and investigation of the pharmacological actions of the main herbs." *Phytotherapy Research* 28, no. 4 (2014): 480–497.

Dhanabal, Palaniswamy, L. Priyanka Dwarampudi, N. Muruganantham, and Ramachandran Vadivelan. "Evaluation of the antipsoriatic activity of Aloe vera leaf extract using a mouse tail model of psoriasis." *Phytotherapy research* 26, no. 4 (2012a): 617–619.

Dhanabal, Sangai Palanisamy, Nithyanantham Muruganantham, Kabbur Hanumanthappa Basavaraj, Asish Wadhwani, and Nonavinakere Mannar Shamasundar. "Antipsoriatic activity of extracts and fractions obtained from Memecylon malabaricum leaves." *Journal of Pharmacy and Pharmacology* 64, no. 10 (2012b): 1501–1509.

Di Meglio, Paola, Federica Villanova, and Frank O. Nestle. "Psoriasis." *Cold Spring Harbor Laboratory Press* 4 (2014): a015354.

Doppalapudi, Sindhu, Anjali Jain, Dhiraj Kumar Chopra, and Wahid Khan. "Psoralen loaded liposomal nanocarriers for improved skin penetration and efficacy of topical PUVA in psoriasis." *European Journal of Pharmaceutical Sciences* 96 (2017): 515–529.

Dowlatshahi, Emmilia A., Marlies Wakkee, Lidia R. Arends, and Tamar Nijsten. "The prevalence and odds of depressive symptoms and clinical depression in psoriasis patients: a systematic review and meta-analysis." *Journal of Investigative Dermatology* 134, no. 6 (2014): 1542–1551.

Duarte, Gleison Vieira, Maria de Fatima SP Oliveira, Thiago M. Cardoso, Ivonise Follador, Thadeu S. Silva, Carolina MA Cavalheiro, Walker Nonato, and Edgar M. Carvalho. "Association between obesity measured by different parameters and severity of psoriasis." *International Journal of Dermatology* 52, no. 2 (2013): 177–181.

Dubertret, Louis, Ulrich Mrowietz, Annamari Ranki, Peter C. M. Van De Kerkhof, S. Chimenti, T. Lotti, G. Schäfer, and EUROPSO patient survey. "European patient perspectives on the impact of psoriasis: the EUROPSO patient membership survey." *British Journal of Dermatology* 155, no. 4 (2006): 729–736.

Dwarampudi, Lalitha Priyanka, Dhanabal Palaniswamy, and P. S. Muruganantham Nithyanantham. "Antipsoriatic activity and cytotoxicity of ethanolic extract of Nigella sativa seeds." *Pharmacognosy Magazine* 8, no. 32 (2012): 268.

Elgewelly, Mahmoud A., Soha M. Elmasry, Nesrine S. El Sayed, and Haidy Abbas. "Resveratrol-loaded vesicular elastic nanocarriers gel in imiquimod-induced psoriasis treatment: in vitro and in vivo evaluation." *Journal of Pharmaceutical Sciences* 111, no. 2 (2022): 417–431.

Ellinghaus, Eva, Philip E. Stuart, David Ellinghaus, Rajan P. Nair, Sophie Debrus, John V. Raelson, Majid Belouchi et al. "Genome-wide meta-analysis of psoriatic arthritis identifies susceptibility locus at REL." *Journal of Investigative Dermatology* 132, no. 4 (2012): 1133–1140.

Farahnik, Benjamin, Divya Sharma, Joseph Alban, and Raja K. Sivamani. "Topical botanical agents for the treatment of psoriasis: a systematic review." *American Journal of Clinical Dermatology* 18, no. 4 (2017): 451–468.

Ferenczi, Katalin, Lauren Burack, M. Pope, James Glenn Krueger, and L. M. Austin. "CD69, HLA-DR and the IL-2R identify persistently activated T cells in psoriasis vulgaris lesional skin: blood and skin comparisons by flow cytometry." *Journal of Autoimmunity* 14, no. 1 (2000): 63–78.

Ferrándiz, Carlos, Ramon M. Pujol, Vicente García-Patos, Xavier Bordas, and Joan A. Smandía. "Psoriasis of early and late onset: a clinical and epidemiologic study from Spain." *Journal of the American Academy of Dermatology* 46, no. 6 (2002): 867–873.

Fortes, Cristina, Simona Mastroeni, Karen Leffondré, Francesca Sampogna, Franco Melchi, Eva Mazzotti, Paolo Pasquini, and Damiano Abeni. "Relationship between smoking and the clinical severity of psoriasis." *Archives of Dermatology* 141, no. 12 (2005): 1580–1584.

Freag, May S., Abrar S. Torky, Maha MA Nasra, Doaa A. Abdelmonsif, and Ossama Y. Abdallah. "Liquid crystalline nanoreservoir releasing a highly skin-penetrating berberine oleate complex for psoriasis management." *Nanomedicine* 14, no. 8 (2019): 931–954.

Gangwar, Mayank, Manish Kumar Gautam, Amit Kumar Sharma, Yamini B. Tripathi, R. K. Goel, and Gopal Nath. "Antioxidant capacity and radical scavenging effect of polyphenol rich Mallotus philippenensis fruit extract on human erythrocytes: an in vitro study." *The Scientific World Journal* 2014 (2014). 279451.

Gao, Jintao, Junfan Guo, Yuejuan Nong, Wenfei Mo, Huanan Fang, Jing Mi, Qi Qi, and Mengjuan Yang. "18β-Glycyrrhetinic acid induces human HaCaT keratinocytes apoptosis through ROS-mediated PI3K-Akt signaling pathway and ameliorates IMQ-induced psoriasis-like skin lesions in mice." *BMC Pharmacology and Toxicology* 21, no. 1 (2020): 1–11.

Gao, W. Y., W. Zhang, Y. D. Yang, C. Y. Zhou, and Y. C. Bei. "Liubei Beibi cream for psoriasis vulgaris." *J Ext Ther TCM* 15, no. 2 (2006): 26–27.

García-Pérez, Martha-Estrella, Isabelle Allaeys, Daniel Rusu, Roxane Pouliot, Tatjana Stevanovic Janezic, and Patrice E. Poubelle. "Picea mariana polyphenolic extract inhibits phlogogenic mediators produced by TNF-α-activated psoriatic keratinocytes: impact on NF-κB pathway." *Journal of Ethnopharmacology* 151, no. 1 (2014): 265–278.

Gerdes, S., U. Mrowietz, and W-H. Boehncke. "Komorbidität bei Psoriasis vulgaris." *Der Hautarzt* 67, no. 6 (2016): 438–444.

Ghosheh, Omar A., Abdulghani A. Houdi, and Peter A. Crooks. "High performance liquid chromatographic analysis of the pharmacologically active quinones and related compounds in the oil of the black seed (Nigella sativa L.)." *Journal of Pharmaceutical and Biomedical Analysis* 19, no. 5 (1999): 757–762.

Gladman, D. D., Christoff Antoni, Phillip Mease, D. O. Clegg, and Peter Nash. "Psoriatic arthritis: epidemiology, clinical features, course, and outcome." *Annals of the Rheumatic Diseases* 64, no. suppl 2 (2005): ii14–ii17.

Goldminz, Ari M., S. C. Au, N. Kim, Alice B. Gottlieb, and Paul F. Lizzul. "NF-κB: an essential transcription factor in psoriasis." *Journal of Dermatological Science* 69, no. 2 (2013): 89–94.

Gottlieb, Alice B., Robert T. Matheson, Alan Menter, Craig L. Leonardi, Robert M. Day, ChiaChi Hu, Peter H. Schafer, and James G. Krueger. "Efficacy, tolerability, and pharmacodynamics of apremilast in recalcitrant plaque psoriasis: a phase II open-label study." *Journal of Drugs in Dermatology: JDD* 12, no. 8 (2013): 888–897.

Goulding, Jonathan Mark Ruthven., C. L. Price, C. L. Defty, C. S. Hulangamuwa, E. Bader, and I. Ahmed. "Erectile dysfunction in patients with psoriasis: increased prevalence, an unmet need, and a chance to intervene." *British Journal of Dermatology* 164, no. 1 (2011): 103–109.

Gregorio, Josh, Stephan Meller, Curdin Conrad, Anna Di Nardo, Bernhard Homey, Antti Lauerma, Naoko Arai, Richard L. Gallo, John DiGiovanni, and Michel Gilliet. "Plasmacytoid dendritic cells sense skin injury and promote wound healing through type I interferons." *Journal of Experimental Medicine* 207, no. 13 (2010): 2921–2930.

Griffiths, Christopher EM, April W. Armstrong, Johann E. Gudjonsson, and Jonathan NWN Barker. "Psoriasis." *The Lancet* 397, no. 10281 (2021): 1301–1315.

Griffiths, Christopher EM, and Jonathan NWN Barker. "Pathogenesis and clinical features of psoriasis." *The Lancet* 370, no. 9583 (2007): 263–271.

Gründemann, Carsten, Kathrin Thell, Karin Lengen, Manuel Garcia-Käufer, Yen-Hua Huang, Roman Huber, David J. Craik, Gernot Schabbauer, and Christian W. Gruber. "Cyclotides suppress human T-lymphocyte proliferation by an interleukin 2-dependent mechanism." *PLoS One* 8, no. 6 (2013): e68016.

Gullick, Nicola J., Hayley G. Evans, Leigh D. Church, David M. Jayaraj, Andrew Filer, Bruce W. Kirkham, and Leonie S. Taams. "Linking power Doppler ultrasound to the presence of Th17 cells in the rheumatoid arthritis joint." *PloS One* 5, no. 9 (2010): e12516.

Hammer, Kimberly DP, Matthew L. Hillwig, Avery KS Solco, Philip M. Dixon, Kathleen Delate, Patricia A. Murphy, Eve S. Wurtele, and Diane F. Birt. "Inhibition of prostaglandin E2 production by anti-inflammatory Hypericum perforatum extracts and constituents in RAW264. 7 mouse macrophage cells." *Journal of Agricultural and Food Chemistry* 55, no. 18 (2007): 7323–7331.

Hänsel, Anja, Claudia Günther, Jens Ingwersen, Josephine Starke, Marc Schmitz, Michael Bachmann, Michael Meurer, Ernst Peter Rieber, and Knut Schäkel. "Human slan (6-sulfo LacNAc) dendritic cells are inflammatory dermal dendritic cells in psoriasis and drive strong TH17/TH1 T-cell responses." *Journal of Allergy and Clinical Immunology* 127, no. 3 (2011): 787–794.

He, Xingrui, Xiabin Chen, Hancheng Zhang, Tian Xie, and Xiang-Yang Ye. "Selective Tyk2 inhibitors as potential therapeutic agents: a patent review (2015-2018)." *Expert Opinion on Therapeutic Patents* 29, no. 2 (2019): 137–149.

Hellinger, Roland, Johannes Koehbach, Halyna Fedchuk, Barbara Sauer, Roman Huber, Christian W. Gruber, and Carsten Gründemann. "Immunosuppressive activity of an aqueous Viola tricolor herbal extract." *Journal of Ethnopharmacology* 151, no. 1 (2014): 299–306.

Herrinton, Lisa J., Liyan Liu, Lang Chen, Leslie R. Harrold, Marsha A. Raebel, Jeffrey R. Curtis, Marie R. Griffin, Daniel H. Solomon, Kenneth G. Saag, and James D. Lewis. "Association between anti-TNF-α therapy and all-cause mortality." *Pharmacoepidemiology and Drug Safety* 21, no. 12 (2012): 1311–1320.

Higgins, Eleanor. "Psoriasis." *Medicine* 45, no. 6 (2017): 368–78.

Ho, Vincent CY, Christopher EM Griffiths, John Berth-Jones, Kim A. Papp, Francisco Vanaclocha, Esteban Dauden, Anna Beard, Lohita Puvanarajan, and Carle Paul. "Intermittent short courses of cyclosporine microemulsion for the long-term management of psoriasis: a 2-year cohort study." *Journal of the American Academy of Dermatology* 44, no. 4 (2001): 643–651.

Hodari, Kafele T., Jaleema R. Nanton, Christie L. Carroll, Steven R. Feldman, and Rajesh Balkrishnan. "Adherence in dermatology: a review of the last 20 years." *Journal of Dermatological Treatment* 17, no. 3 (2006): 136–142.

Hoffman, Melissa B., Dane Hill, and Steven R. Feldman. "Current challenges and emerging drug delivery strategies for the treatment of psoriasis." *Expert Opinion on Drug Delivery* 13, no. 10 (2016): 1461–1473.

Hongqin, Tu, Li Xinyu, Gu Heng, Xu Lanfang, Wang Yongfang, and Song Shasha. "Triptolide inhibits IFN-γ signaling via the Jak/STAT pathway in HaCaT keratinocytes." *Phytotherapy Research* 25, no. 11 (2011): 1678–1685.

Hsieh, Wan-Ling, Yin-Ku Lin, Chi-Neu Tsai, Ta-Min Wang, Tzu-Ya Chen, and Jong-Hwei S. Pang. "Indirubin, an acting component of indigo naturalis, inhibits EGFR activation and EGF-induced CDC25B gene expression in epidermal keratinocytes." *Journal of Dermatological Science* 67, no. 2 (2012): 140–146.

Huang, Kuo-Feng, Kuo-Hsing Ma, Pei-Shan Liu, Bo-Wei Chen, and Sheau-Huei Chueh. "Baicalein increases keratin 1 and 10 expression in hacat keratinocytes via trpv 4 receptor activation." *Experimental Dermatology* 25, no. 8 (2016): 623–629.

Huang, Tse-Hung, Chwan-Fwu Lin, Ahmed Alalaiwe, Shih-Chun Yang, and Jia-You Fang. "Apoptotic or antiproliferative activity of natural products against keratinocytes for the treatment of psoriasis." *International Journal of Molecular Sciences* 20, no. 10 (2019): 2558.

Hüffmeier, Ulrike, Steffen Uebe, Arif B. Ekici, John Bowes, Emiliano Giardina, Eleanor Korendowych, Kristina Juneblad et al. "Common variants at TRAF3IP2 are associated with susceptibility to psoriatic arthritis and psoriasis." *Nature Genetics* 42, no. 11 (2010): 996–999.

Iriventi, Padmini, N. Vishal Gupta, Riyaz Ali M. Osmani, and V. Balamuralidhara. "Design & development of nanosponge loaded topical gel of curcumin and caffeine mixture for augmented treatment of psoriasis." *DARU Journal of Pharmaceutical Sciences* 28, no. 2 (2020): 489–506.

Jain, Anjali, Sindhu Doppalapudi, Abraham J. Domb, and Wahid Khan. "Tacrolimus and curcumin co-loaded liposphere gel: synergistic combination towards management of psoriasis." *Journal of Controlled Release* 243 (2016): 132–145.

Jain, Harsha, Devabattula Geetanjali, Harshita Dalvi, Aditi Bhat, Chandraiah Godugu, and Saurabh Srivastava. "Liposome mediated topical delivery of Ibrutinib and Curcumin as a synergistic approach to combat imiquimod induced psoriasis." *Journal of Drug Delivery Science and Technology* (2022): 103103.

Jin, L., Y. B. Ma, Y. S. Jiang, and W. H. Lou. "Topical calcipotriol ointment therapy combination with oral Chinese medicine treatment in plaque psoriasis vulgaris." *Chin J Dermatovenereol Integr Tradit West Med* 4 (2009): 216–218.

Johnston, Andrew, Xianying Xing, Liza Wolterink, Drew H. Barnes, ZhiQiang Yin, Laura Reingold, J. Michelle Kahlenberg, Paul W. Harms, and Johann E. Gudjonsson. "IL-1 and IL-36 are dominant cytokines in generalized pustular psoriasis." *Journal of Allergy and Clinical Immunology* 140, no. 1 (2017): 109–120.

Kalkan, Goknur. "Comorbidities in psoriasis: the recognition of psoriasis as a systemic disease and current management." *Turkderm-Archives of the Turkish Dermatology and Venerology* 51, no. 3 (2017).

Kang, Bok Yun, Su Wol Chung, and Tae Sung Kim. "Inhibition of interleukin-12 production in lipopolysaccharide-activated mouse macrophages by hpyericin, an active component of Hypericum perforatum." *Planta Medica* 67, no. 04 (2001): 364–366.

Kang, Di, Bowen Li, Lei Luo, Wenbing Jiang, Qiumin Lu, Mingqing Rong, and Ren Lai. "Curcumin shows excellent therapeutic effect on psoriasis in mouse model." *Biochimie* 123 (2016): 73–80.

Katare, Om, Kaisar Raza, Bhupinder Singh, and Sunil Dogra. "Novel drug delivery systems in topical treatment of psoriasis: rigors and vigors." *Indian Journal of Dermatology, Venereology and Leprology* 76, no. 6 (2010): 612.

Khurana, Bharat, Daisy Arora, and Raj Kumar Narang. "QbD based exploration of resveratrol loaded polymeric micelles based carbomer gel for topical treatment of plaque psoriasis: in vitro, ex vivo and in vivo studies." *Journal of Drug Delivery Science and Technology* 59 (2020): 101901.

Kim, Byung-Hak, Ikhoon Oh, Jung-Ho Kim, Ju-eun Jeon, Byeongwook Jeon, Jongheon Shin, and Tae-Yoon Kim. "Anti-inflammatory activity of compounds isolated from Astragalus sinicus L. in cytokine-induced keratinocytes and skin." *Experimental & Molecular Medicine* 46, no. 3 (2014): e87.

Kim, Ji Hye, Young-Su Yi, Mi-Yeon Kim, and Jae Youl Cho. "Role of ginsenosides, the main active components of Panax ginseng, in inflammatory responses and diseases." *Journal of Ginseng Research* 41, no. 4 (2017): 435–443.

Kimball, A. B., U. Gieler, D. Linder, F. Sampogna, R. B. Warren, and M. Augustin. "Psoriasis: is the impairment to a patient's life cumulative?" *Journal of the European Academy of Dermatology and Venereology* 24, no. 9 (2010): 989–1004.

Koeberle, Andreas, Antonietta Rossi, Julia Bauer, Friederike Dehm, Luisella Verotta, Hinnak Northoff, Lidia Sautebin, and Oliver Werz. "Hyperforin, an anti-inflammatory constituent from St. John's wort, inhibits microsomal prostaglandin E2 synthase-1 and suppresses prostaglandin E2 formation in vivo." *Frontiers in Pharmacology* 2 (2011): 7.

Kvist-Hansen, Amanda, Peter Riis Hansen, and Lone Skov. "Systemic treatment of psoriasis with JAK inhibitors: a review." *Dermatology and Therapy* 10, no. 1 (2020): 29–42.

Langenbruch, A., M. A. Radtke, M. Krensel, A. Jacobi, K. Reich, and M. Augustin. "Nail involvement as a predictor of concomitant psoriatic arthritis in patients with psoriasis." *British Journal of Dermatology* 171, no. 5 (2014): 1123–1128.

Langley, R. G. B., G. G. Krueger, and CEM Griffiths. "Psoriasis: epidemiology, clinical features, and quality of life." *Annals of the Rheumatic Diseases* 64, no. suppl 2 (2005): ii18–ii23.

Langley, Richard G., Boni E. Elewski, Mark Lebwohl, Kristian Reich, Christopher EM Griffiths, Kim Papp, Lluís Puig et al. "Secukinumab in plaque psoriasis-results of two phase 3 trials." *New England Journal of Medicine* 371, no. 4 (2014): 326–338.

Latief, Uzma, and Riaz Ahmad. "Herbal remedies for liver fibrosis: a review on the mode of action of fifty herbs." *Journal of traditional and complementary medicine* 8, no. 3 (2018): 352–360.

Lebwohl, Mark G., Hervé Bachelez, Jonathan Barker, Giampiero Girolomoni, Arthur Kavanaugh, Richard G. Langley, Carle F. Paul, Lluís Puig, Kristian Reich, and Peter CM van de Kerkhof. "Patient perspectives in the management of psoriasis: results from the population-based multinational assessment of psoriasis and psoriatic arthritis survey." *Journal of the American Academy of Dermatology* 70, no. 5 (2014): 871–881.

Lin, Yin-Ku, Wen-Rou Wong, Ya-Ching Chang, Chee-Jen Chang, Pei-Kwei Tsay, Shu-Chen Chang, and Jong-Hwei Su Pang. "The efficacy and safety of topically applied indigo naturalis ointment in patients with plaque-type psoriasis." *Dermatology* 214, no. 2 (2007): 155–161.

Liu, C., H. Liu, C. Lu, J. Deng, Y. Yan, H. Chen, Y. Wang et al. "Kaempferol attenuates imiquimod-induced psoriatic skin inflammation in a mouse model." *Clinical & Experimental Immunology* 198, no. 3 (2019): 403–415.

Liu, Ying, Cynthia Helms, Wilson Liao, Lisa C. Zaba, Shenghui Duan, Jennifer Gardner, Carol Wise et al. "A genome-wide association study of psoriasis and psoriatic arthritis identifies new disease loci." *PLoS Genetics* 4, no. 4 (2008): e1000041.

Lucka, T. C., D. Pathirana, A. Sammain, F. Bachmann, S. Rosumeck, R. Erdmann, J. Schmitt, H. Orawa, B. Rzany, and A. Nast. "Efficacy of systemic therapies for moderate-to-severe psoriasis: a systematic review and meta-analysis of long-term treatment." *Journal of the European Academy of Dermatology and Venereology* 26, no. 11 (2012): 1331–1344.

Maejima, Hideki, Tomonori Taniguchi, Akira Watarai, and Kensei Katsuoka. "Evaluation of nail disease in psoriatic arthritis by using a modified nail psoriasis severity score index." *International Society of Tropical Dermatology* (2010): 901–906.

Mao, Kai-Li, Zi-Liang Fan, Jian-Dong Yuan, Pian-Pian Chen, Jing-Jing Yang, Jie Xu, De-Li ZhuGe et al. "Skin-penetrating polymeric nanoparticles incorporated in silk fibroin hydrogel for topical delivery of curcumin to improve its therapeutic effect on psoriasis mouse model." *Colloids and Surfaces B: Biointerfaces* 160 (2017): 704–714.

Matsuzaki, Goro, and Masayuki Umemura. "Interleukin-17 family cytokines in protective immunity against infections: role of hematopoietic cell-derived and non-hematopoietic cell-derived interleukin-17s." *Microbiology and Immunology* 62, no. 1 (2018): 1–13.

Matsumura, Yasuhiro, and Honnavara N. Ananthaswamy. "Toxic effects of ultraviolet radiation on the skin." *Toxicology and Applied Pharmacology* 195, no. 3 (2004): 298–308.

Meng, Shikang, Lin Sun, Lun Wang, Zibei Lin, Zeyu Liu, Long Xi, Zhenping Wang, and Ying Zheng. "Loading of water-insoluble celastrol into niosome hydrogels for improved topical permeation and anti-psoriasis activity." *Colloids and Surfaces B: Biointerfaces* 182 (2019): 110352.

Menter, Alan, Alice Gottlieb, Steven R. Feldman, Abby S. Van Voorhees, Craig L. Leonardi, Kenneth B. Gordon, Mark Lebwohl et al. "Guidelines of care for the management of psoriasis and psoriatic arthritis: section 1. Overview of psoriasis and guidelines of care for the treatment of psoriasis with biologics." *Journal of the American Academy of Dermatology* 58, no. 5 (2008): 826–850.

Mervic, Liljana. "Management of moderate to severe plaque psoriasis in pregnancy and lactation in the era of biologics." *Acta Dermatovenerol Alp Pannonica Adriat* 23, no. 2 (2014): 27–31.

Min, Min, Bing-Xi Yan, Ping Wang, Lilla Landeck, Jia-Qi Chen, Wei Li, Sui-Qing Cai, Min Zheng, and Xiao-Yong Man. "Rottlerin as a therapeutic approach in psoriasis: evidence from in vitro and in vivo studies." *PLoS One* 12, no. 12 (2017): e0190051.

Morizane, Shin, Kenshi Yamasaki, Beda Mühleisen, Paul F. Kotol, Masamoto Murakami, Yumi Aoyama, Keiji Iwatsuki, Tissa Hata, and Richard L. Gallo. "Cathelicidin antimicrobial peptide LL-37 in psoriasis enables keratinocyte reactivity against TLR9 ligands." *Journal of Investigative Dermatology* 132, no. 1 (2012): 135–143.

Muruganantham, N., K. H. Basavaraj, S. P. Dhanabal, T. K. Praveen, N. M. Shamasundar, and K. S. Rao. "Screening of Caesalpinia bonduc leaves for antipsoriatic activity." *Journal of Ethnopharmacology* 133, no. 2 (2011): 897–901.

Nagasawa, Hiroshi, Katsuto Watanabe, Makoto Yoshida, and Hideo Inatomi. "Effects of gold banded lily (Lilium auratum Lindl) or Chinese milk vetch (Astragalus sinicus L) on spontaneous mammary tumourigenesis in SHN mice." *Anticancer Research* 21, no. 4A (2001): 2323–2328.

Najafizadeh, Parvaneh, Farshad Hashemian, Parvin Mansouri, Susan Farshi, Mohammadhossein Salehi Surmaghi, and Reza Chalangari. "The evaluation of the clinical effect of topical St Johns wort (Hypericum perforatum L.) in plaque type psoriasis vulgaris: a pilot study." *Australasian Journal of Dermatology* 53, no. 2 (2012): 131–135.

Narayana, DB Anantha, and C. K. Katiyar. "Draft amendment to drugs and cosmetics rules to license science based botanicals, phytopharmaceuticals as drugs in India." *Journal of Ayurveda and Integrative Medicine* 4, no. 4 (2013): 245.

Navarini, Alexander A., A. David Burden, Francesca Capon, Ulrich Mrowietz, Luis Puig, Sulev Köks, Külli Kingo et al. "European consensus statement on phenotypes of pustular psoriasis." *Journal of the European Academy of Dermatology and Venereology* 31, no. 11 (2017): 1792–1799.

Nestle, Frank O., Curdin Conrad, Adrian Tun-Kyi, Bernhard Homey, Michael Gombert, Onur Boyman, Günter Burg, Yong-Jun Liu, and Michel Gilliet. "Plasmacytoid predendritic cells initiate psoriasis through interferon-α production." *The Journal of experimental medicine* 202, no. 1 (2005): 135–143.

Nestle, Frank O., Daniel H. Kaplan, and Jonathan Barker. "Psoriasis." [In eng]. *The New England Journal of Medicine* 361, no. 5 (2009): 496–509.

Nimisha, Dilshad Ali Rizvi, Zeeshan Fatima, and Chanchal Deep Kaur Neema. "Antipsoriatic and anti-inflammatory studies of Berberis aristata extract loaded nanovesicular gels." *Pharmacognosy Magazine* 13, no. Suppl 3 (2017): S587.

Nooreen, Zulfa, Vineet Kumar Rai, and Narayan Parasad Yadav. "Phytopharmaceuticals: a new class of drug in India." *Annals of Phytomedicine* 7, no. 1 (2018): 27–37.

Oliveira, Maria de Fátima Santos Paim de, Bruno de Oliveira Rocha, and Gleison Vieira Duarte. "Psoriasis: classical and emerging comorbidities." *Anais brasileiros de dermatologia* 90 (2015): 9–20.

Ortonne, Jean-Paul, Sergio Chimenti, T. Luger, Luis Puig, F. Reid, and RM. Trüeb. "Scalp psoriasis: european consensus on grading and treatment algorithm." *Wiley Online Library*, (2009): 1435–1444.

Papp, Kim, Alexander, A. Menter, Bruce Strober, Richard G. Langley, Michael Buonanno, R. Wolk, P. Gupta, S. Krishnaswami, H. Tan, and J. A. Harness. "Efficacy and safety of tofacitinib, an oral Janus kinase inhibitor, in the treat- ment of psoriasis: a Phase 2b randomized placebo-controlled dose-ranging study." *British Journal of Dermatology* 167, no. 3 (2012): 668–677.

Papp, Kim, Alexander, Kristian Reich, C. Paul, Andrew Blauvelt, W. Baran, C. Bolduc, D. Toth et al. "A prospective phase III, randomized, double-blind, placebo-controlled study of brodalumab in patients with moderate-to-severe plaque psoriasis." *British Journal of Dermatology* 175, no. 2 (2016): 273–286.

Papp, Kim, J. Crowley, Jean Paul Ortonne, J. Leu, M. Okun, S. R. Gupta, Y. Gu, and Richard G. Langley. "Adalimumab for moderate to severe chronic plaque psoriasis: efficacy and safety of retreatment and disease recurrence following withdrawal from therapy." *British Journal of Dermatology* 164, no. 2 (2011): 434–441.

Papp, Kim, Kenneth Gordon, Diamant Thaçi, Akimichi Morita, Melinda Gooderham, Peter Foley, Ihab G. Girgis, Sudeep Kundu, and Subhashis Banerjee. "Phase 2 trial of selective tyrosine kinase 2 inhibition in psoriasis." *New England Journal of Medicine* 379, no. 14 (2018): 1313–1321.

Papp, Kim, Kristian Reich, Craig L. Leonardi, Leon Kircik, Sergio Chimenti, Richard GB Langley, ChiaChi Hu et al. "Apremilast, an oral phosphodiesterase 4 (PDE4) inhibitor, in patients with moderate to severe plaque psoriasis: results of a phase III, randomized, controlled trial (Efficacy and Safety Trial Evaluating the Effects of Apremilast in Psoriasis [ESTEEM] 1)." *Journal of the American Academy of Dermatology* 73, no. 1 (2015): 37–49.

Parmar, Komal M., Prakash R. Itankar, Apurva Joshi, and Satyendra K. Prasad. "Anti-psoriatic potential of Solanum xanthocarpum stem in Imiquimod-induced psoriatic mice model." *Journal of Ethnopharmacology* 198 (2017): 158–166.

Paulsen, Evy, Lars Korsholm, and Flemming Brandrup. "A double-blind, placebo-controlled study of a commercial Aloe vera gel in the treatment of slight to moderate psoriasis vulgaris." *Journal of the European Academy of Dermatology and Venereology* 19, no. 3 (2005): 326–331.

Poikolainen, Kari, Timo Reunala, Jaakko Karvonen, Jorma Lauharanta, and Päivi Kärkkäinen. "Alcohol intake: a risk factor for psoriasis in young and middle aged men?" *British Medical Journal* 300, no. 6727 (1990): 780–783.

Potdar, Dipti, R. R. Hirwani, and Sivakami Dhulap. "Phyto-chemical and pharmacological applications of Berberis aristata." *Fitoterapia* 83, no. 5 (2012): 817–830.

Poulin, Yves, Kim A. Papp, Norman R. Wasel, Robin Andrew, Elisa Fraquelli, Geula Bernstein, and Daphne Chan. "A Canadian online survey to evaluate awareness and treatment satisfaction in individuals with moderate to severe plaque psoriasis." (2010): 1368–1375.

Prodanowich, Srdjan, Fangchao Ma, J. Richard Taylor, Candido Pezon, Tahira Fasihi, and Robert S. Kirsner. "Methotrexate reduces incidence of vascular diseases in veterans with psoriasis or rheumatoid arthritis." *Journal of the American Academy of Dermatology* 52, no. 2 (2005): 262–267.

Puig, Lluis. "Obesity and psoriasis: body weight and body mass index influence the response to biological treatment." *Journal of the European Academy of Dermatology and Venereology* 25, no. 9 (2011): 1007–1011.

Puig, Lluis,"Tratamiento de la psoriasis en placas moderada a grave con fármacos biológicos: análisis del sobrecoste de la intensificación temporal frente a cambio a otro biológico en caso de fracaso secundario." *Actas Dermo-Sifiliográficas* 105, no. 4 (2014): 401–412.

Puig, Lluis, "Brodalumab: the first anti-IL-17 receptor agent for psoriasis." Drugs of Today (Barcelona, Spain: 1998) 53, no. 5 (2017): 283–297.

Qiu, Fen, Long Xi, Shengshuang Chen, Yonghua Zhao, Zhenping Wang, and Ying Zheng. "Celastrol niosome hydrogel has anti-inflammatory effect on skin keratinocytes and circulation without systemic drug exposure in psoriasis mice." *International Journal of Nanomedicine* 16 (2021): 6171.

Rahmayunita, Githa, Tjut NA Jacoeb, Endi Novianto, Wresti Indriatmi, Rahadi Rihatmadja, and Erdina HD Pusponegoro. "A double-blind randomized controlled trial of topical Curcuma xanthorrhiza Roxb. on mild psoriasis: clinical manifestations, histopathological features, and K6 expressions." *Medical Journal of Indonesia* 27, no. 3 (2018): 178–184.

Ran, Delin, Minglong Cai, and Xuejun Zhang. "Genetics of psoriasis: a basis for precision medicine." *Precision Clinical Medicine* 2, no. 2 (2019): 120–130.

Raychaudhuri, Siba P., Mrinmoy Sanyal, Helena Weltman, and Smriti Kundu-Raychaudhuri. "K252a, a high-affinity nerve growth factor receptor blocker, improves psoriasis: an in vivo study using the severe combined immunodeficient mouse-human skin model." *Journal of Investigative Dermatology* 122, no. 3 (2004): 812–819.

Rendon, Adriana, and Knut Schäkel. "Psoriasis pathogenesis and treatment." *International Journal of Molecular Sciences* 20, no. 6 (2019): 1475.

Reuter, Juliane, Ute Wölfle, Steffi Weckesser, and Christoph Schempp. "Which plant for which skin disease? Part 1: atopic dermatitis, psoriasis, acne, condyloma and herpes simplex." *JDDG: Journal der Deutschen Dermatologischen Gesellschaft* 8, no. 10 (2010): 788–796.

Rodríguez-Luna, Azahara, Elena Talero, Javier Ávila-Román, Ana María Fernández Romero, Antonio M. Rabasco, Virginia Motilva, and María Luisa González-Rodríguez. "Preparation and in vivo evaluation of rosmarinic acid-loaded transethosomes after percutaneous application on a psoriasis animal model." *AAPS PharmSciTech* 22, no. 3 (2021): 1–18.

Rout, Sanjay Kumar, Bankim Chandra Tripathy, and Bikash Ranjan Kar. "Natural Green Alternatives to Psoriasis Treatment-A Review." *Global Journal of Pharmacy & Pharmaceutical Sciences* 4, no. 1 (2017): 001–007.

Sakkas, Lazaros I., and Dimitrios P. Bogdanos. "Are psoriasis and psoriatic arthritis the same disease? The IL-23/IL-17 axis data." *Autoimmunity Reviews* 16, no. 1 (2017): 10–15.

Sánchez-Regaña, M., M. J. Aldunce Soto, I. Belinchón Romero, M. Ribera Pibernat, R. F. Lafuente-Urrez, and J. M. Carrascosa Carrillo. "En representación del Grupo Español de Psoriasis de la Academia Española de Dermatología y Venereología. Directrices del grupo español de psoriasis (GEP) basadas en la evidencia para el uso de medicamentos biológicos en pacientes con psoriasis en localizaciones de difícil tratamiento (uñas, cuero cabelludo, palmas y plantas)." *Actas Dermosifiliogr* 105 (2014): 923–934.

Santini, Stefano M., Caterina Lapenta, Simona Donati, Francesca Spadaro, Filippo Belardelli, and Maria Ferrantini. "Interferon-α-conditioned human monocytes combine a Th1-orienting attitude with the induction of autologous Th17 responses: role of IL-23 and IL-12." *PloS One* 6, no. 2 (2011): e17364.

Saurat, J-H., G. Stingl, L. Dubertret, K. Papp, R. G. Langley, J-P. Ortonne, K. Unnebrink, M. Kaul, A. Camez, and CHAMPION Study Investigators. "Efficacy and safety results from the randomized controlled comparative study of adalimumab vs. methotrexate vs. placebo in patients with psoriasis (CHAMPION)." *British Journal of Dermatology* 158, no. 3 (2008): 558–566.

Schempp, C. M., R. Lüdtke, B. Winghofer, and J. C. Simon. "Effect of topical application of Hypericum perforatum extract (St. John's wort) on skin sensitivity to solar simulated radiation." *Photodermatology, Photoimmunology & Photomedicine* 16, no. 3 (2000): 125–128.

Schwartz, Daniella M., Yuka Kanno, Alejandro Villarino, Michael Ward, Massimo Gadina, and John J. O'Shea. "JAK inhibition as a therapeutic strategy for immune and inflammatory diseases." *Nature Reviews Drug Discovery* 16, no. 12 (2017): 843–862.

Shi, Quan, Qi He, Weiming Chen, Jianwen Long, and Bo Zhang. "Ginsenoside Rg1 abolish imiquimod-induced psoriasis-like dermatitis in BALB/c mice via downregulating NF-κB signaling pathway." *Journal of Food Biochemistry* 43, no. 11 (2019): e13032.

Shin, Yong-Wook, Eun-Ah Bae, and Dong-Hyun Kim. "Inhibitory effect of ginsenoside Rg5 and its metabolite ginsenoside Rh3 in an oxazolone-induced mouse chronic dermatitis model." *Archives of Pharmacal Research* 29, no. 8 (2006): 685–690.

Smith, A. E., J. Y. Kassab, Rowland Payne, and W. E. Beer. "Bimodality in age of onset of psoriasis, in both patients and their relatives." *Dermatology* 186, no. 3 (1993): 181–186.

Smith, Catherine H., and J. N. W. N. Barker. "Psoriasis and its management." *BMJ* 333, no. 7564 (2006): 380–384.

Sommer, Dorothea M., Stefan Jenisch, Michael Suchan, Enno Christophers, and Michael Weichenthal. "Increased prevalence of the metabolic syndrome in patients with moderate to severe psoriasis." *Archives of Dermatological Research* 298, no. 7 (2007): 321–328.

Song, Ping, Z. F. Yan, and Xian Xu. "Clinical observation on effect of compound E-bei ointment in treating plaque psoriasis." *Zhongguo Zhong xi yi jie he za zhi Zhongguo Zhongxiyi Jiehe Zazhi= Chinese Journal of Integrated Traditional and Western Medicine* 27, no. 4 (2007): 352–354.

Stern, Robert S., and Nan Laird. "The carcinogenic risk of treatments for severe psoriasis." *Cancer* 73, no. 11 (1994): 2759–2764.

Sun, Jun, Yi Zhao, and Jinhong Hu. "Curcumin inhibits imiquimod-induced psoriasis-like inflammation by inhibiting IL-1beta and IL-6 production in mice." *PloS One* 8, no. 6 (2013): e67078.

Sung, Yoon-Young, and Ho Kyoung Kim. "Illicium verum extract suppresses IFN-γ-induced ICAM-1 expression via blockade of JAK/STAT pathway in HaCaT human keratinocytes." *Journal of Ethnopharmacology* 149, no. 3 (2013): 626–632.

Sung, Yoon-Young, Won-Kyung Yang, A. Yeong Lee, Dong-Seon Kim, Kyoung Jin Nho, Young Sang Kim, and Ho Kyoung Kim. "Topical application of an ethanol extract prepared from Illicium verum suppresses atopic dermatitis in NC/Nga mice." *Journal of Ethnopharmacology* 144, no. 1 (2012a): 151–159.

Sung, Yoon-Young, Young Sang Kim, and Ho Kyoung Kim. "Illicium verum extract inhibits TNF-α-and IFN-γ-induced expression of chemokines and cytokines in human keratinocytes." *Journal of Ethnopharmacology* 144, no. 1 (2012b): 182–189.

Syed, Tanweer A., S. Ashfaq Ahmad, Albert H. Holt, Seyed Ali Ahmad, Seyed Hamzeh Ahmad, and Mohammad Afzal. "Management of psoriasis with Aloe vera extract in a hydrophilic cream: a placebo-controlled, double-blind study." *Tropical Medicine & International Health* 1, no. 4 (1996): 505–509.

Szilveszter, Kata P., Tamás Németh, and Attila Mócsai. "Tyrosine kinases in autoimmune and inflammatory skin diseases." *Frontiers in Immunology* (2019): 1862.

Takahashi, Toshiya, and Kenshi Yamasaki. "Psoriasis and Antimicrobial Peptides." *International Journal of Molecular Sciences* 21, no. 18 (2020): 6791.

Tyring, Stephen, Kenneth B. Gordon, Yves Poulin, Richard G. Langley, Alice B. Gottlieb, Meleana Dunn, and Angelika Jahreis. "Long-term safety and efficacy of 50 mg of etanercept twice weekly in patients with psoriasis." *Archives of Dermatology* 143, no. 6 (2007): 719–726.

Uva, Luis, Diana Miguel, Catarina Pinheiro, Joana Antunes, Diogo Cruz, Joao Ferreira, and Paulo Filipe. "Mechanisms of action of topical corticosteroids in psoriasis." *International Journal of Endocrinology* 2012 (2012). 561018.

Van der Laken, C. J., A. E. Voskuyl, J. C. Roos, M. Stigter Van Walsum, E. R. De Groot, G. Wolbink, B. A. C. Dijkmans, and L. A. Aarden. "Imaging and serum analysis of immune complex formation of radiolabelled infliximab and anti-infliximab in responders and non-responders to therapy for rheumatoid arthritis." *Annals of the Rheumatic Diseases* 66, no. 2 (2007): 253–256.

Vena, Gino A., Michelangelo Vestita, and Nicoletta Cassano. "Psoriasis and cardiovascular disease." *Dermatologic Therapy* 23, no. 2 (2010): 144–151.

Xiong, Hui, Ying Xu, Guozhen Tan, Yanfang Han, Zengqi Tang, Wenying Xu, Fanqin Zeng, and Qing Guo. "Glycyrrhizin ameliorates imiquimod-induced psoriasis-like skin lesions in BALB/c mice and inhibits TNF-a-induced ICAM-1 expression via NF-κB/MAPK in HaCaT cells." *Cellular Physiology and Biochemistry* 35, no. 4 (2015): 1335–1346.

Xu, Jia, Cang Zhang, and Xing Qu. "Clinical and experimental study on effect of qinbal ointment in treating psoriasis in the active stage of blood-heat syndrome type." *Zhongguo Zhong xi yi jie he za zhi Zhongguo Zhongxiyi Jiehe Zazhi= Chinese Journal of Integrated Traditional and Western Medicine* 29, no. 7 (2009): 614–618.

Yang, Y. D., W. Zhang, W. Y. Gao, W. T. Yang, Y. L. Zhang, and L. Zhang. "Liubai Baibi cataplasm for psoriasis." *J China Tradit Chin Med Inf* 3, no. 17 (2011): 148.

Yuan, T. I. A. N. "Qingfei Liangxue decoction clinical study of psoriasis vulgaris." *Journal of Liaoning University of Traditional Chinese Medicine* (2011).

Zhang, Claire Shuiqing, Jason Jingjie Yu, Shefton Parker, Anthony Lin Zhang, Brian May, Chuanjian Lu, and Charlie Changli Xue. "Oral C hinese herbal medicine combined with pharmacotherapy for psoriasis vulgaris: a systematic review." *International Journal of Dermatology* 53, no. 11 (2014a): 1305–1318.

Zhang, Yong-Tai, Li-Na Shen, Zhong-Hua Wu, Ji-Hui Zhao, and Nian-Ping Feng. "Comparison of ethosomes and liposomes for skin delivery of psoralen for psoriasis therapy." *International Journal of Pharmaceutics* 471, no. 1-2 (2014b): 449–452.

Zhang, Z. "Clinical observation of intregration of traditional and western medicine for 56 patients with psoriasis vulgaris." *Guide China Med* 16 (2012): 275–276.

Zheng, X. "Xiaoyin Kebi Decoction for 60 psoriasis vulgaris patients." *Henan Traditional Chinese Medicine* 4 (2011): 383–384.

Zhu, L., H. Zhang, and Y. Duan. "Chinese herbal medicine "Xiaoxuanling" for 85 participants with psoriasis vulgaris." *Youjiang Med* J 36, no. 2 (2008): 230–231.

Zhu, Y., G. Shankar, N. Yeilding, P. Szapary, H. Zhou, and H. M. Davis. "Immunogenicity assessment of ustekinumab in Phase 3 studies in patients with moderate to severe plaque psoriasis." In 19th Annual Congress of the European Academy of Dermatology and Venereology, vol. 610. 2010.

12 Exploration of Antioxidant, Anti-inflammatory Activities of Hesperidin-Based Formulation

Radha Rani, Meghna A. Singh, Abhishek Wahi, Riya Nagpal, Priyanka Sharma, Ajay Kumar Sharma, Nagendra Singh Chauhan, and Madhu Gupta

12.1 INTRODUCTION

Traditional medicines and natural medicines remain an integral part of the healthcare mechanism. Herbal product performs a vital function in the medical services of Afro-Asian and European countries. In the current situation, India, China, Nigeria, the United States, and the World Health Organization have made tremendous efforts to investigate the potential therapeutic profile of these herbal therapies. Numerous chemically synthesized compounds have similar features to natural ingredients which can be used as selective molecules as a lead (Aditya and Saluja, 2019). Flavonoids play an important role in protecting plants from colonizing pathogens and Ultraviolet rays. Flavonoids are abundant throughout the kingdom of plants. They function as chemical messengers all through nutrient uptake, encourage phytohormone transport, and are able to give flowers color, which aids in plant reproduction (Brown et al., 2001). Flavonoids are naturally occurring compounds with varying structure moieties of phenols. Over 4,000 flavonoids have been classified and described due to their molecular structure. A well-known Hungarian researcher namely Dr. Albert Szent-Gyorgi discovered that lemon peel flavonoids seemed to be good at stopping capillary hemorrhage and were the first person to reveal the biological functions of flavonoids' fragile nature on capillaries linked with vitamin C deficiency (Garg et al., 2001). Flavonoids are indeed a significant class of phytoconstituents that are widely recognized for their health benefits. Because of their health and medical advantages, the nutritional reference of flavonoids is the foremost area of study. Any utilization of flavonoids to promote public health has an amazing history. Researchers became interested in the medicinal benefits of flavonoids after discovering also that the Mediterranean community ingesting red wine and elevated trans-fats had a lower mortality rate of cardiovascular disorders. Flavonoids get a broad bio-spectrum, including antioxidative, antidiabetic, anticancer, and syndromes associated with immunodeficiency effects (Wichansawakun & Buttar, 2019). Consistent steps have been taken to investigate the medical benefits of flavonoids. Flavonoids are the largest potential source of naturally occurring polyphenolic compounds with a lower molecular weight that are broadly disseminated as secondary metabolites in the kingdom Plantae due to the diversified therapeutic action among this group as well as the multidimensionality of the abilities shown by certain specific individuals, and they are often referred to as "bioflavonoids" (Middleton, 1984). Hesperidin, the main flavonoid in orange blossoms and lemon is a plentiful and cheap natural consequence of lemon farming. This can make up a massive amount of the immature young oranges, accounting for 14% of the fresh weight. (Barthe et al., 1988). The hesperetin and vitamin P molecules are the food-bound forms of molecules of hesperidin (Chaoyun and Schluesener, 2017). It must have been discovered in oranges and other citrus fruits after French chemist Lebreton isolated it in 1828 out from the albedo (the spongy inner portion of the peel) of oranges (Manthey and Grohmann, 1998).

 DOI: 10.1201/9781003292692-12

Citrus sinensis and *Citrus reticulate* appear to have high hesperidin concentrations (Peterson et al., 2006). Hesperidin has been shown to have numerous medicinal uses, such as being a powerful antioxidant, antimicrobial, antiatherosclerotic, cardioprotective, antiviral, anti-inflammatory, neuroprotective, and anticancer chemical entity. Numerous preclinical reports have studied its role in protecting malignant transformation and progression by continuing to act through various cell signaling pathways. Undoubtedly, hesperidin has the ability to affect a wide range of chemical entities involved in tumor cell viability, separation, and fatality processes (Roohbakhsh et al., 2015). Hesperidin is a key source of Chenpi, a Satsuma mandarin peel extract that really has long been used in Chinese ancient medication which is used to treat allergies, hepatopathy, and even inflammatory responses (Yamada et al., 2006). In recent papers on polyphenol intake, hesperidin was shown to be the most heavily ingested flavonoid, accounting for 28.3 mg/day, or 30% of the total flavonoid in Finland (Knekt et al., 2002).

12.2 SOURCES OF HESPERIDIN

Hesperidin is extracted in huge volumes from the dumped peels of *Citrus aurantium* L., *Citrus sinensis*, *Citrus unshiu*, and other *Citrus* species (family Rutaceae). Hesperidin can be detached from *Acanthopanax setchuenensis* roots (family Araliaceae) (Garg et al., 2001). Hesperidin is found in the highest composition in greenish fruit, and its presence in the fruit rises even during the growing season for storage. Both epicarp as well as citrus fruit juice have been discovered to contain this (Higby). Hesperidin distribution and availability in different tissues of fully matured fruit of *C. sinensis* were evaluated using a radioimmunoassay technique, and it has been discovered to be abundant in the albedo, membrane proteins, and pith, but so much decrease in the beverage vacuoles and seed pods. The hesperidin concentration of seeds rises after plant growth, indicating there is a net output of this chemical in the working to develop seedlings, which would be prompted in portion by brightness. The citrus fruits involve hesperidin concentration, which varies greatly depending on species, part of the fruit, geographic location of agriculture, and treatment processes (Barthe et al., 1988). Some examples are like, In the fresh peel of satsuma pulp, hesperidin concentration is 73 and 157 mg/kg. In general, the concentration of hesperidin found in citrus peel or other sections of the citrus fruit in the percentage of 335–351 mg/L of Hand-squeezed Florida orange juice whereas 273–287 mg/L of Israel Ortanique. The pigmented citrus juices are also considered higher in the concentration of hesperidin than nonpigmented citrus juices. It has also been reported that hesperidin is found in higher concentrations in immature citrus than in ripe citrus (Barreca et al., 2010). As per the literature findings, on storage at 4°C, the hesperidin content in citrus juices does not reduce on heat pasteurization. In fact, hesperidin content rises when pasteurized at 90°C for 20 sec in citrus fruit juices (Uçan et al., 2016). Hesperidin contains concentrated single-strength juices in the range from 555 to 761 and 470 to 614 mg/L, indicating that the material used to produce or the procedure to produce influences hesperidin content in citrus juice (Leuzzi et al., 2000). Moreover, hesperidin content also rises in response to UVB irradiation in peppermint (*Menthax piperita* L.). In the methanolic extract of *Porphyra dentata*, red edible seaweed hesperidin is found in the percentage of 5 (Dolzhenko et al., 2010). Figure 12.1 represents the extraction process of hesperidin.

12.3 PHYSICAL AND CHEMICAL PROFILE OF HESPERIDIN

Hesperidin monohydrates are arranged in needle-like structures and resemble long hairs. The physical characteristics of hesperidin are described in Figure 12.2. Hesperidin in the presence of dilute HSO_4 provides a dextro and levo isomeric mixture of hesperetin active optically after fractional recrystallization separation (Arthur et al., 1956).

An aglycone, hesperetin, or methyl eriodictyol, and a disaccharide, rutinose together make hesperidin a flavanone glycoside. The disaccharide unit ($C_{12}H_{22}O_{10}$) is made up of. The one or two isomeric forms of rutinose or neohesperiose can take one moiety of molecules of rhamnose and

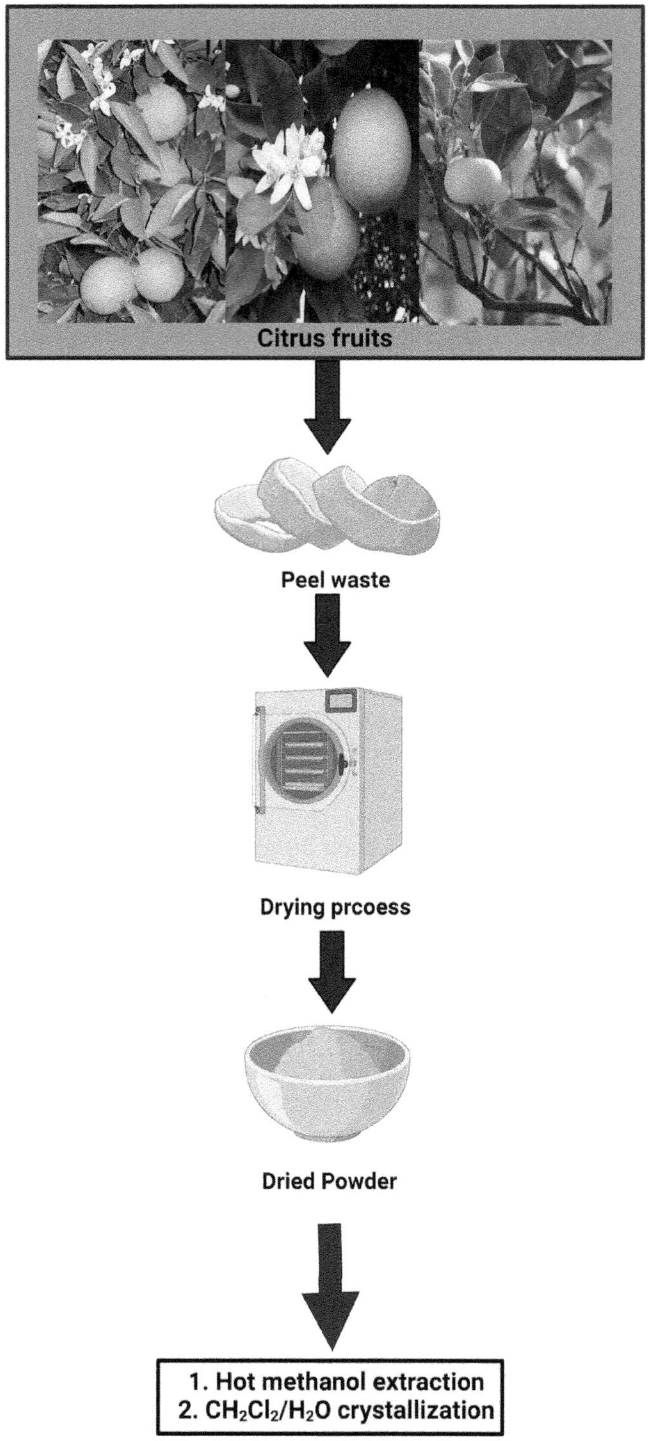

FIGURE 12.1 Extraction process of hesperidin from plant source.

glucose to form. The L-rhamnose and hesperetin-7-D-glucoside, which is cleaved by the enzyme D-glucosidase is produced by partial hydrolysis with dilute acid, which is used to determine the position of sugar in the hesperidin, which is results in the hesperetin in hesperidin attached to

TABLE 12.1
Nanocarriers Employed for Hesperidin Drug Delivery

Nano Carrier	API	Formulation Technology	Cell Line & Animal Used	Study Finding	Ref
Nanoparticle	Essential oils and hesperidin	Ultra-sonification	MCF-7, HCT-116, HepG2	Inhibition of cancer cell lines and good activity against microbial infection	Ahmed et al. (2021)
Surface modified liposomes	Hesperidin	Thin-film hydration	Male Wistar rats	Actively target HSCs in liver fibrosis	Morsy et al. (2018)
Topical hybrid nanoparticles	Hesperidin	Emulsion-solvent evaporation method	Egg membrane	Sustained drug release enhanced the process of wound healing	Jangde et al. (2022)
Progesterone conjugated nanohybrid carrier	Hesperidin	Ionic gelation	L929, SKOV-3, MDA-MB-231	Enhanced therapeutic efficacy against ovarian and breast cancer cells	Purushothaman et al. (2020)
Nutraceutical SEDDs	Hesperidin	Spray drying	Diabetic rats	Effective against hyperglycemia	Chalikwar et al. (2021)
SLNs	Hesperidin	Supercritical antisolvent technology	Rats	Provided improved cardioprotective efficacy in oral administration	Saad et al. (2020)
Topical nano gel	Hesperidin	Ethanol injection	Goat skin, male Wistar albino rats	Better therapeutic activity against inflammation in comparison to formulation available commercially	Sujitha et al. (2020)
Solid dispersion	Hesperidin, nobiletin	Hot melt extrusion	Mice	Increased concentration of nobiletin in different tissues	Iwashita et al. (2019)
Gold nanoparticles	Hesperidin	Chemical reduction method	Rats	Considerable activity against oxidation and preventive action in the brain allowed the application of brain disorders	Pradhan et al. (2022)
Nanoparticles	Imatinib, hesperidin	Emulsion-solvent evaporation	Female Swiss albino mouse	Significantly decreased tumor growth with considerable activity against cancer cells	El-Sisi et al. (2020)
NLC	Hesperidin	High-pressure homogenization	A375, CHL01, SKMEL147	Sustained release of encapsulated drug with prolonged and enhanced antioxidant activity delayed aging of skin	Durán et al. (2019)
Nanofibers	Hesperidin	Electrospinning	Male Wistar rats	Fasten the rate of healing of wounds by skin regeneration	Taymouri et al. (2021)

(Continued)

TABLE 12.1 (*Continued*)
Nanocarriers Employed for Hesperidin Drug Delivery

Nano Carrier	API	Formulation Technology	Cell Line & Animal Used	Study Finding	Ref
Nanohybrid carrier grafted with BSA-calcium ferrite	Hesperidin, eugenol	Freeze drying sol-gel method	MCF-7, L929	Increased toxicity of both drugs against breast cancer cells	Maheswari et al. (2021)
Magnetic nanoparticles	Hesperidin, Curcumin and Amphotericin B	Co-precipitation with surface modification	HeLa, ATCC®, CCL2™ human cell lines	Therapeutic efficacy to stop MDR bacterial infections	Akbar et al. (2022)
Double layered nanoparticles	Hesperidin, hydroxide	Precipitation	Healthy W138 lung cells	Sustained escape of hesperidin providing antioxidant, anti-inflammatory, and antidiabetic efficacy	El-Shahawy et al. (2021)
Gliadin nanoparticles	Hesperidin	Desolvation	Vero 9ATCC CCl-81-normal kidney cells of monkey), Caco-2 (ATCC HTB-37—human colon adenocarcinoma), HeLa (ATCC CCL-2—human cervix adenocarcinoma and PC-3 (ATCC CRL-1435—human prostate adenocarcinoma	Reduced toxicity to Vero cells but improved cytotoxicity to other cell lines with increased encapsulation efficiency and prolonged drug release	Kelte et al. (2021)
Solid dispersion	Hesperidin, trans-glycosylated stevia, and artemisinin	Spray drying	Caco-2 human cells	Higher bioavailability of artemisinin with safety	Letchmanan et al. (2018)
Nanoparticles	Hesperidin, PLGA	Nano precipitation	MCF-7 cells	Stop the growth of used cells by inducing apoptotic cell mortality	Ali et al. (2019)
Nanoparticles	Hesperidin	Evaporation precipitation	Adult Wistar rats	Reduced infection of the cerebral due to increased glutathione hence prevented the heart from ischemia	Praveen Kumar et al. (2020)
Nanocomposites	Hesperidin	Nanoprecipitation	HCT116 cell lines	Apoptosis induced reduced multiplication of colon cancer cells	Yang et al. (2020)

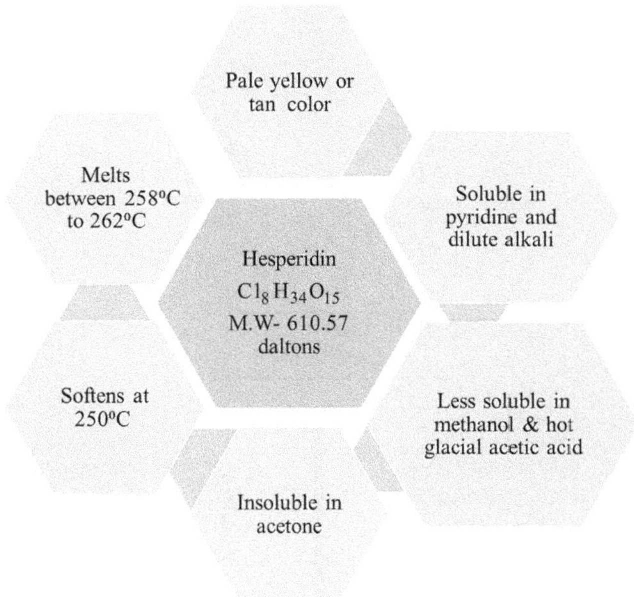

FIGURE 12.2 Physical characteristics of hesperidin.

glucose and rhamnose. It also produces phloroglucinol and hesperetenic acid on hesperidin alkaline hydrolysis. On acid hydrolysis, hesperidin generates one mole of each of the aglycones hesperetin, D-glucose, and L-rhamnose. Optically active hesperetin combination is produced by the hydrolyzation process with sulphuric acid in ethylene glycol or dilute sulphuric acid, which can be separated by using fractional recrystallization and both have been characterized by derivative conversion may lead to the characterization of both. The molecular framework of the disaccharide and the existence or lack of bitterness in the compound are believed to be linked. Hesperidin is indeed a flavonoid rutinoside and thus non-bitter. Grapefruit has the most bitter neohesperidosides, while orange and lemon have the most non-bitter rutinosides (Garg et al., 2001).

12.4 ANALYTICAL PROFILE OF HESPERIDIN

Hesperidin and its aglycone absorbed light at 286 and 289 nm under a UV spectrophotometer respectively (Jurd, 1962). Neohesperidin and hesperidin can be analyzed by using capillary-based liquid chromatography/mass spectrometry (LC-MS) and LC-MS in the presence of a turbo-ion spray (TBS) interface. Using the spectra of mass spectrometry, different glycosylation isomers of hesperetin can also be characterized (Careri et al., 1999). Simultaneous analysis of hesperidin along with neohesperidin and different other agents is performed using capillary electrophoresis analytical method also separated and analyzed moieties, more specifically (Cancalon, 1999). Some chromatographic analysis was performed for hesperetin, naringenin, eriodictin, and naringin separation by TLC plate of silica gel employing a solvent system mixture of butanol, acetic acid, and water in a ratio of 4:1:5, respectively. Comparatively, HPLC is more specifically employed for the separation of hesperidin from the mixture containing the number of flavonoids (Garg et al., 2001). The analytical profile of hesperidin is described in Figure 12.3.

12.5 PHARMACOKINETICS AND SAFETY PROFILE OF HESPERIDIN

Hesperidin was poorly absorbed on oral administration due to its conversion into hesperidin glycosides before its urinary excretion because the glycosides of hesperidin permeate the intestine very

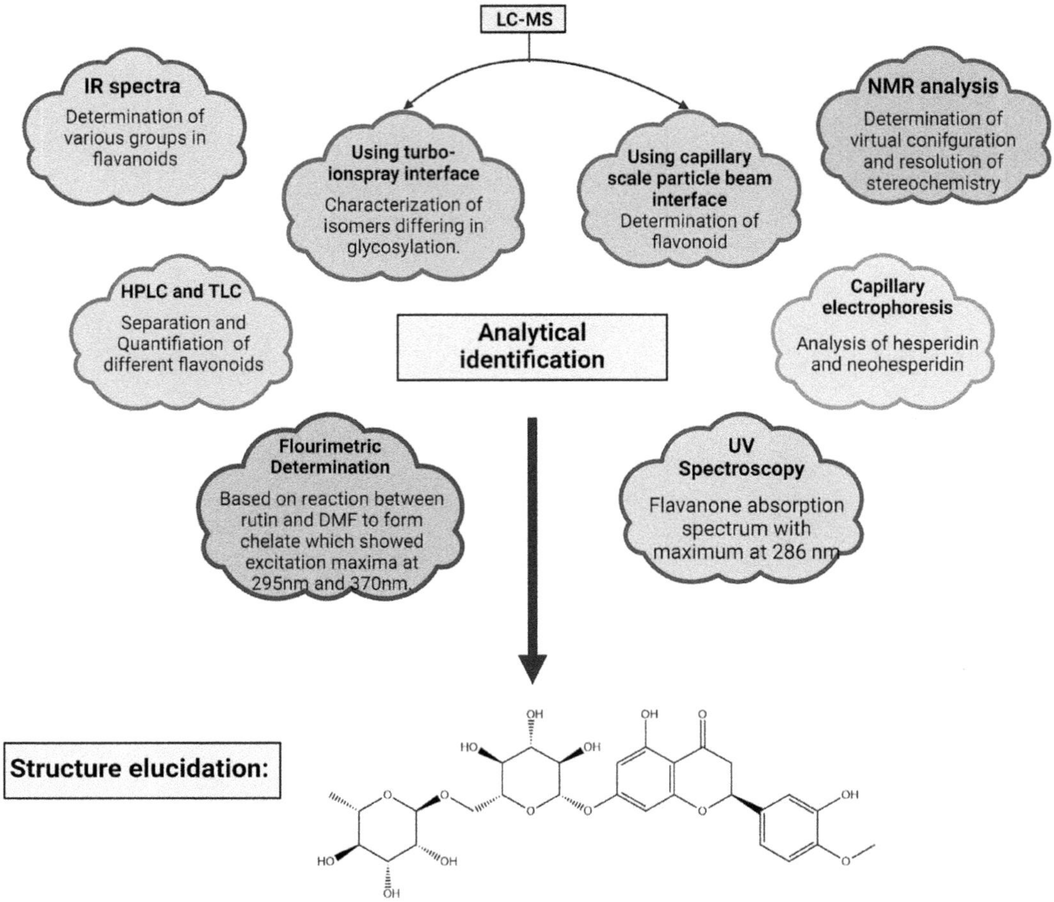

FIGURE 12.3 Analytical profile of hesperidin.

slowly depending on its dose. While the presence of hesperidin aglycone has occurred in both blood and urine (Kim et al., 1999). In rabbits, the absorption of hesperidin after oral administration has occurred with artificial ration and is absent with commercial ones (Williams, 1964). In the human body, hesperidin was observed to be metabolized by a very different mechanism. Hydracrylic acid (3-hydroxy-4-methoxyphenylhydracrylic acid) was formed due to the splitting of the pyran ring present in hesperidin and obtained as the major metabolic product in urine (Booth et al., 1958; Williams, 1964). Hesperetinby intestinal bacteria converted into its metabolites like endo-beta-glucosidase or beta-glucosidase and alpha-rhamnosidase having more potent antiplatletic activity (Kim et al., 1998).

Hesperidin or methyl hesperidin was found to be generally safe when taken orally as a dietary substance the systemic and topical applications also do not cause any cutaneous reaction, considered safe both in animal testing and humans. The administration of the citrous flavanoid form of hesperidin is also free from any toxicity during pregnancy (Pizzorno Jr and Murray, 1999). The phosphorylated form of hesperidin is clinically approved as an anti-fertility agent ineffective in the presence of any type of systemic disorder, trauma, or infectious problems (Sieve, 1952).

12.6 THERAPEUTIC POTENTIAL OF HESPERIDIN

Hesperidin is a flavonoid that is abundant in citrus species and possesses numerous biological actions like antioxidant and anti-inflammatory effects. Several research using cellular and animal models, as well as a few clinical trials are carried out for investigating the impact of hesperidin as an anti-inflammatory drug. Hesperidin also inhibited the action of inflammatory mediators and had substantial antioxidant properties. Signaling pathways, particularly the nuclear factor route, are the molecular foundation of anti-inflammatory actions (Tejada et al., 2018).

However, according to Ferrero-Miliani, inflammation is part of the complicated response of vascular tissues to adverse stimuli such as infections, damaged cells, and irritants (Ferrero-Miliani et al., 2007). Despite the fact that inflammation is a natural reaction to tissue injury, it can develop into chronic autoimmune illnesses if it is not managed, necessitating the use of anti-inflammatory drugs to keep the inflammatory response under control (Benavente-Garcia & Castillo, 2008). Remarkably, hesperidin, a flavanone-rich substance, has been shown in trials to have anti-inflammatory properties (Parhiz et al., 2015).

In addition, Ribeiro and colleagues demonstrated a strong anti-inflammatory effect in rats (induced by carrageenan). They found that after 3 h, the hesperidin-ingested model inhibited carrageenan-induced pleurisy and reduced the amount of edema in the paws by 33% (Ribeiro et al., 2018). Furthermore, Tsai in 2019 studied the antioxidant properties of hesperidin in human chondrocytes. Indeed, chondrocytes are cells that are responsible for the creation of cartilage and the maintenance of the cartilaginous matrix, which is mostly composed of collagen and proteoglycans. Previously, studies showed that hesperidin is derived from citrus plants and is widely recognized for its anti-inflammatory responses and antioxidant properties. They reported that the low-dose of hesperidin (5 and 10 M) supplement enhances the activity of mitochondria, proliferation kinetics, cytotoxicity, cell life span, as well as its senescence in chondrocytes that are stimulated by H_2O_2, but high-dose hesperidin (50 and 100 M) had a negative effect on normal cells. Overall, a low dose of hesperidin increased antioxidant capacity and restored superoxide dismutases and glutathione peroxidase activities in human chondrocytes (Tsai et al., 2019).

Furthermore, De Oliveira (2013) found that hesperidin intake increased antioxidant capacity in the continuous swimming group (183%) while decreasing lipid peroxidation (TBARS) in the interval swimming group (44%) (De Oliveira et al., 2013). Similarly, Estruel-Amades and team found that hesperidin had an effect on the antioxidant state of lymphoid tissues following a rat training regimen. After an endurance test, supplementation with hesperidin enriched in its 2S-isomer prevented enhanced generation of reactive oxygen species (ROS) as well as suppression in efficacy of catalase and superoxide dismutase (SOD) production. In conclusion, hesperidin's antioxidant properties were linked to improved performance in the tested training model (Estruel-Amades et al., 2019).

Al-Ashaal and colleagues extracted hesperidin from *Citrus sinensis* L. (Rutaceae) peels and tested their antioxidant activity and cytotoxicity against different human cancer cell lines. A cytotoxic test was performed on cell lines from the larynx, cervix, breast, and liver. With a capability of 36%, hesperidin was shown to be modestly potent as an antioxidant. Furthermore, the findings demonstrated that hesperidin had a strong anticancer effect on the cell lines tested. The IC50 values were 1.67, 3.33, 4.17, and 4.58 g/mL, respectively. Overall, the hesperidin extracted from orange peel showed antioxidant characteristics that might be used for a variety of therapeutic purposes (Al-Ashaal et al., 2011).

12.7 NANOCARRIERS EMPLOYED FOR HESPERIDIN

Hesperidin is a flavonoid glycoside that has been shown to have therapeutic properties in the treatment of a variety of disorders, including cancer. However, due to its limited solubility and bioavailability, it is only moderately absorbed, necessitating the use of a delivery method to achieve its therapeutic objective.

However, nanocarriers were on the cutting edge of nanotechnology's faster development. These materials are superior and necessary in many fields of human activity due to their size-dependent characteristics (Sastry et al., 2004). Biosynthesized nanoparticles have a significant advantage over chemically and physically synthesized nanoparticles in that they are biocompatible owing to biomolecule attachment to the nanoparticle surface (Mohanpuria et al., 2008; Gade et al., 2008). Recent research has demonstrated that plant extract bio-molecules such as protein, phenol, and flavonoids play a key role in the reduction of metal ions and capping of biosynthesized nanoparticles (Krishnaraj et al., 2014; Arya, 2010).

12.7.1 Silver Nanoparticle

Silver nanoparticles have a unique size range of 1–100 nm. Biosynthesis involves reducing and stabilizing metal ions using a variety of organisms. The strong reducing potential of secondary metabolites found in plants makes them a promising source for the synthesis of silver nanoparticles (AgNPs). Nevertheless, silver nanoparticles were formulated by biomimetic synthesis from *Citrus limon* (lemon) aqueous extract and predicted the particle size (Prathna et al., 2011). A pure aqueous solution of hesperidin, naringin, and diosmin was successfully used to synthesize silver nanoparticles by Sahu and the team in 2016. The plant extract was used to prepare silver nanoparticles with dimensions ranging from 5 to 80 nm. A hydroxyl group was identified as a component of silver ions during the process of bio-reduction into AgNPs. However, the nanoparticles were cytotoxic against HL-60 cells, thus having potential as a cancer therapy drug (Sahu et al., 2016).

Additionally, Rao and colleagues predicted that nanoscale delivery of drugs to inflamed tissues would be more effective and cause fewer side effects than conventional drug delivery. Hesperidin was loaded in gum acacia (GA), which was stabilized with green silver nanoparticles (AgNPs). In a Complete Freund's adjuvant (CFA)-induced arthritic model, the hesperidin and its loaded NPs were evaluated for rheumatoid arthritis (RA). The nano-sized GA-AgNPs were found to have a negative charge, a spherical shape, and a higher Hsd concentration. There were minimal changes in tissue swelling, reduced degeneration, and mild articular changes when GA-AgNPs were loaded. However, presence of hesperidin-loaded GA-AgNPs, histopathological examination demonstrated a lower inflow of inflammatory cells and less granulomatous inflammation in ankle joint tissues. Hesperidin-loaded GA-AgNPs reduced the expression of Toll-like receptors (TLRs) significantly by RT-PCR. The findings support the use of GA-stabilized green AgNPs as stable nano-cargos for the delivery of hesperidin to slow the course of RA (Rao et al., 2018).

12.7.2 Gold Nanoparticles

Gold nanoparticle (AuNPs), with their unique characteristics as well as their multifaceted surface properties, have improved their wide-ranging biomedical usage in the nanotechnology branch, notably as drug delivery platforms. Nanoparticles are biocompatible and possess a surface area sufficient to load higher concentrations of the delivered drug on a single particle.

Gold nanoparticles loaded with hesperidin were synthesized by Suleiman et al. When the formulation was compared with the human normal breast epithelial cell line (HBL-100), the treated cells showed a significant reduction in proliferation and inhibition of growth. In subsequent treatment with hesperidin-AuNPs, histopathological images of the liver, spleen, lung, and kidney did not reveal any apparent damage. Hesperidin-AuNPs improved macrophage function against Ehrlich ascites tumor cell-bearing mice. Hesperidin-AuNPs also resulted in the generation of pro-inflammatory cytokines by bone marrow-derived macrophage cells. A greater reduction in the concentrations of TNF, IL-1β, as well as IL-6, was observed from the findings obtained by hesperidin-AuNPs treatment (Sulaiman et al., 2020).

Moreover, gold nanoparticles conjugated to hesperidin (AuNPs-Hsd) were synthesized to evaluate their activity in stimulating the phagocytic index (PI), investigated by Waheeb et al., in 2002. AuNPs-hesperidin showed a significant increase (85%) in contrast to gold nanoparticles (66%) and hesperidin (51%). A blood sample that indicates the control had a PI of 20%. This study provides evidence regarding the impact of AuNPs-Hsd on phagocytic cells and the fact that these cells ingested the nanoparticles (Waheeb et al., 2020).

Moreover, Krishnan and colleagues explored the influence of 0.5 mL hesperetin-conjugated nanoparticles of gold (Au-mPEG(5000)-S-hesperidin NPs) in diethylnitrosamine (DEN)-induced hepatocarcinogenesis in male Wistar albino rats as well as found that this is also an improved antioxidant with anti-proliferation, anti-inflammatory, and antitumor attributes, including the ability to modulate the signaling pathways. HR-transmission electron microscopy (TEM) with EDAX-validated polymeric successfully synthesized gold nanoparticles as well as polymers loaded with gold nanoparticles, (i.e. circular, triangular, and pentagonal in structure) or dynamic light scattering (DLS) having zeta potential approaches to confirm the average length of Au-mPEG(5000)-SH nanoparticles and it was reported to be approximately 220 nm. In PBS at pH 7.4, the drug encapsulation effectiveness as well as drug release profile of Au-mPEG(5000)-S-hesperetin were assessed and compared to purified standard hesperetin (hesperidin). Tumor necrosis factor-alpha, mast cell numbers, glycoconjugates, proliferating cell nuclear antigen, transcription factor nuclear factor-B, as well as argyrophilic nucleolar organizing areas were elevated in animals treated with DEN. Supplementation with Au-mPEG(5000)-S-hesperidin NPs, on the other hand, significantly reduced all of the observed anomalies. These findings imply that Au-mPEG(5000)-S-hesperidin NPs have a potential antitumor action in hepatocellular carcinogenesis induced with DEN via cell growth, and reducing cell inflammation (Krishnan et al., 2017).

12.7.3 Nanosuspensions

Nanosuspensions comprise drug delivery methods that combine a purified pharmaceutical ingredient with a stabilizer to create nanoparticles as minute as 1m in size. Insoluble medications with reduced absorption as well as poor pharmacokinetics could be managed using this strategy. It can be used for therapeutic applications and intravascular solubilization, as well as to avoid enzymatic hydrolysis in oral drug delivery systems (Jain et al., 2011).

Furthermore, using the NANOEDGE™ approach, nanosuspensions of hesperidin in conjunction with HPMC E5 and Poloxamer 188 as stabilizers were created. Spray drying the nanosuspension resulted in nanocrystals that flowed freely. The needle-shaped nanocrystals were spray-dried nanocrystals. The crystalline nature of hesperidin did not alter as exhibited by differential scanning calorimetry and X-ray powdered diffraction. Nanocrystals of hesperidin boosted saturation solubility by a factor of 5. In comparison to pure hesperidin capsules, capsules containing hesperidin nanocrystals dissolved five times more. Hesperidin nanocrystals have antioxidant capacity equivalent to pure hesperidin, according to *in vitro* antioxidant experiments. At a dosage of 50 mg/kg, in vivo, experiments on the liver of rats that have been damaged by carbon tetrachloride demonstrated that hesperidin nanocrystals have high antioxidant capacity. In addition, nanosuspensions of hesperidin were produced using HPMC E5 and Poloxamer 188 as stabilizers. Overall, dissolving trials showed that dried hesperidin nanocrystals might improve the drug's solubility rate when compared to the crude form. Moreover, hesperidin nanocrystals have the ability to strengthen and preserve hesperidin's antioxidant properties (Varghese et al., 2015).

12.7.4 Polymeric Nanoparticles

Ever since the 1980s, polymeric nanocarriers have been employed as drug delivery methods, including nanocapsules, dendrimers, nanogels, various nanovesicles, and polymeric micelles (George et al., 2019). Nanocarriers constructed of biocompatible and biodegradable polymers are referred to as polymeric nanoparticles. These particles are often provided as nanocapsules or nanospheres, depending on their manufacturing process (Jahangiri et al., 2018).

Polymeric nanoparticles might carry a variety of substances, including antisense DNA, proteins, low-molecular-weight as well as polymers medicines. In this sort of delivery system, surface charge distribution, the particle size and hydrophobicity as well as the hydrophilicity of the polymeric nanoparticles are the characteristics that influence their ability to reach the intended organ

(Amoabediny et al., 2018). Employing polymeric nanoparticles in the delivery of drugs enhances the safety of drugs, bioactivity, as well as biodistribution, and also, increases the permeability of drugs and their stability, as well as exhibits protection against those enzymes that cause hydrolysis, and enhances the efficacy of treatment when compared to standard therapeutic techniques (Rostami et al., 2019).

Sharma et al. used the nanoprecipitation approach to examine dual agents, namely hesperidin and diazepam-loaded polymeric nanoparticles (NPs), which were then optimized using a three-level factorial design. Poly (lactic-*co*-glycolic) acid (PLGA), and poloxamer quantity were employed as an independent process variable, and z-average, % drug entrapment was used as a dependent response. Furthermore, drug-loaded NPs had a z-average of 189–307d.nm, with percentage drug entrapment of 62%–89% for diazepam and 68%–92% for hesperidin, respectively. Additionally, *in vitro* drug release investigations using polymeric NPs over a dialysis membrane revealed regulated drug release behavior as compared to aqueous drug solution. On the Vero cell line, a cell viability experiment revealed drug-dependent cytotoxicity; however, polymeric NPs showed less cytotoxicity than aqueous drug solution (Sharma et al., 2015).

Recently, researchers created polymers that are based on nanoparticles using hesperidin, a bioflavonoid. The biomedical applicability of this formulation in the treatment of cancer has already been investigated and established. Oil-in-water single emulsified evaporation of the solvent was used to process hesperidin-loaded poly(lactic-co-glycolic acid) (PLGA) nanoparticles (hesperidin-PLGA-NPs). The formulation generated a nearly spherical nanoparticle of an average dimension of 38.4 nm, according to microscopic analysis such as TEM as well as DLS. The XRD analysis demonstrated that the nanoparticles were crystalline in nature. Moreover, PLGA also works as a capping or switching material for pH-dependent drug release, according to the FTIR and NMR analysis. Furthermore, the cationic character of Hes-PLGA-NPs allowed them to quickly infiltrate malignant cells and effectively suppress proliferation in acidic environments. Further, *in vitro* investigations in HEp-2 cells revealed that bio-formulated Hes-PLGA-NPs show more potential as compared with naturally existing hesperidin and can also be misused as it exhibits potential effects against cancer. More research is needed to uncover the toxic character and molecular processes included in Hes-PLGA-NPs' anticancer efficacy as nanomedicine (Balakrishnan et al., 2021).

12.7.5 Nanoemulsions

Nanoemulsions have the capacity to boost the oral bioavailability of medications with low solubility. Hydrophilic solvents, Oils, co-solvents as well as surfactants with distinctive potential to generate colloidal dispersions of oil-in-water (o/w) type are combined to form nanoemulsions (Akrawi et al., 2020).

Zeng and team created a nanoemulsion vehicle that would increase hesperidin's oral bioavailability. However, pseudo-ternary phase diagrams and response surface methodology (RSM) followed by researchers to improve nanoemulsion. Nanoemulsion AUC0-t and Cmax were enhanced by 5.67 and 2.64 times, respectively, when compared to hesperidin suspension. The quantity of lymphatic transport rose by 1.1-fold, 1.92-fold, and 1.5-fold in the jejunum, ileum, and duodenum, respectively. The findings suggested that hesperidin nanoemulsion might increase hesperidin bioavailability by raising intestinal permeability and promoting lymphatic transport. As a result, nanoemulsion appears to be a viable strategy for increasing hesperetin bioavailability (Zeng et al., 2021).

Moreover, Mangura and colleagues used a spontaneous emulsification approach to encapsulate hesperidin into nanoemulsions (hesperidin-NEM) to increase its solubility, bioavailability, and effectiveness in MCF-7 cell lines for breast cancer therapy. Hesperidin-NEM was tested for its cytotoxic and apoptotic effects in MCF-7, as well as its effect on oncomiRs, microRNA-21, and microRNA-155 expression. For particle size, polydispersity index, zeta potential, and encapsulation efficiency, the optimized hesperidin-NEM had a spherical shape with 305±40.8 nm, 0.308±0.04, and 11.6±3.30 mV and 93%±0.45%, respectively. The hesperidin-NEM was found to be hazardous only to MCF-7 cells in cytotoxicity experiments utilizing the MTT assay (HEK 293). In MCF-7 cells, treatment with the hesperidin-NEM caused apoptosis, cell cycle arrest in the G2/M phase, and downregulation of the miR-21 and miR-155 genes. The use of hesperidin-NEM as a possible therapeutic agent in the treatment of breast cancer is supported by this research (Magura et al., 2021).

12.8 FUTURE PERSPECTIVE

Hesperidin has proved innumerable therapeutic effects on humans. Hesperidin has antioxidant and radical scavenging properties, which can be a reason for its medicinal properties. The antioxidant enzyme defense system was restored with the help of hesperidin. Hesperidin and its analogs have anticancer properties due to apoptosis and cytotoxic actions. Hesperidin has been shown to decrease a wide range of inflammatory mediators and their expression along with the action of some enzymes. Several actions of hesperidin included anti-hemorrhagic, antihypertensive, antidiabetic, anti-obesity, and anti-hyperlipidemic effects. Hesperidin/hesperetin analogs with sufficient lipid solubility and a unique receptor/enzyme binding affinity could be created in the future, allowing for targeted therapeutic effects on cells and organs. As a result, hesperidin might be called an "essential phytochemical" that must be thoroughly explored in humans in order to establish an effective safety profile and provide therapeutic advantages.

REFERENCES

Ahmed, Habiba Abdelaziz, Ahmed Mahmoud Aboul-Enein, Faten Abou-Elella, Salah Hamza Elmahdy Salem, Hanan Farouk Aly, Amr Nassrallah, and Zeinab Salama. "Nano-Formulations of hesperidin and essential oil extracted from sweet orange peel: chemical properties and biological activities." *Egyptian Journal of Chemistry* 64, no. 9 (2021): 5373–5385. https://doi.org/10.21608/ejchem.2021.84783.4139

Akbar, Noor, Muhammad Kawish, Naveed Ahmed Khan, Muhammad Raza Shah, Ahmad M. Alharbi, Hasan Alfahemi, and Ruqaiyyah Siddiqui. "Hesperidin-, curcumin-, and amphotericin B-based nano-formulations as potential antibacterials." *Antibiotics* 11, no. 5 (2022): 696. https://doi.org/10.3390/antibiotics11050696

Akrawi, Sabah H., BapiGorain, Anroop B. Nair, Hira Choudhury, Manisha Pandey, Jigar N. Shah, and Katharigatta N. Venugopala. "Development and optimization of naringenin-loaded chitosan-coated nanoemulsion for topical therapy in wound healing." *Pharmaceutics* 12, no. 9 (2020): 893. https://doi.org/10.3390/pharmaceutics12090893

Al-Ashaal, Hanan A., and Shakinaz T. El-Sheltawy. "Antioxidant capacity of hesperidin from citrus peel using electron spin resonance and cytotoxic activity against human carcinoma cell lines." *Pharmaceutical Biology* 49, no. 3 (2011): 276–282. https://doi.org/10.3109/13880209.2010.509734

Ali, Saja H., Ghassan M. Sulaiman, Mohammad MF Al-Halbosiy, Majid S. Jabir, and Anaheed H. Hameed. "Fabrication of hesperidin nanoparticles loaded by poly lactic co-Glycolic acid for improved therapeutic efficiency and cytotoxicity." *Artificial Cells, Nanomedicine, and Biotechnology* 47, no. 1 (2019): 378–394. https://doi.org/10.1080/21691401.2018.1559175

Amoabediny, Ghasem, FatemeHaghiralsadat, Samira Naderinezhad, Marco N. Helder, Elham AkhoundiKharanaghi, Javad MohammadnejadArough, and Behrouz Zandieh-Doulabi. "Overview of preparation methods of polymeric and lipid-based (niosome, solid lipid, liposome) nanoparticles: a comprehensive review." *International Journal of Polymeric Materials and Polymeric Biomaterials* 67, no. 6 (2018): 383–400. https://doi.org/10.1080/00914037.2017.1332623

Arthur, H. R., Hui, W. H., and C. N. Ma. "127. An examination of the rutaceae of Hong Kong. Part I. Flavonoid glycosides from Zanthoxylum species and the occurrence of optically active hesperetin." *Journal of the Chemical Society (Resumed)* (1956): 632–635. https://doi.org/10.1039/JR9560000632

Arya, Vedpriya. "Living systems: eco-friendly nanofactories." *Digest Journal of Nanomaterials & Biostructures (DJNB)* 5, no. 1 (2010).

Balakrishnan, Krithiga, Sangeetha C. Casimeer, Alaa Yousef Ghidan, Fatima Yousef Ghethan, Karthikkumar Venkatachalam, and AnbuSingaravelu. "Bioformulated hesperidin-loaded PLGA nanoparticles counteract the mitochondrial-mediated intrinsic apoptotic pathway in cancer cells." *Journal of Inorganic and Organometallic Polymers and Materials* 31, no. 1 (2021): 331–343. https://doi.org/10.1007/s10904-020-01746-9

Barreca, Davide, ErsiliaBellocco, CorradoCaristi, U. G. O. Leuzzi, and Giuseppe Gattuso. "Flavonoid composition and antioxidant activity of juices from chinotto (Citrus× myrtifolia Raf.) fruits at different ripening stages." *Journal of Agricultural and food Chemistry* 58, no. 5 (2010): 3031–3036. https://doi.org/10.1021/jf9044809

Barthe, Gary A., Pablo S. Jourdan, Cecilia A. McIntosh, and Richard L. Mansell. "Radioimmunoassay for the quantitative determination of hesperidin and analysis of its distribution in Citrus sinensis." *Phytochemistry* 27, no. 1 (1988): 249–254. https://doi.org/10.1016/0031-9422(88)80625-3

Benavente-Garcia, O., and J. Castillo. "Update on uses and properties of citrus flavonoids: new findings in anticancer, cardiovascular, and anti-inflammatory activity." *Journal of Agricultural and Food Chemistry* 56, no. 15 (2008): 6185–6205. https://doi.org/10.1021/jf8006568

Booth, A.N., Jones, F.T. and DeEds F. "Metabolic fate of hesperidin, eriodicytol, homoeriodicytol and diosmin." *Journal of Biology and Chemistry* 230 (1958): 661–668. https://doi.org/10.1016/S0021-9258(18)70489-4

Brown, Dana E., Aaron M. Rashotte, Angus S. Murphy, Jennifer Normanly, Brian W. Tague, Wendy A. Peer, Lincoln Taiz, and Gloria K. Muday. "Flavonoids act as negative regulators of auxin transport in vivo in Arabidopsis." *Plant physiology* 126, no. 2 (2001): 524–535. https://doi.org/10.1104/pp.126.2.524

Cancalon, Paul F. "Analytical monitoring of citrus juices by using capillary electrophoresis." *Journal of AOAC International* 82, no. 1 (1999): 95–106. https://doi.org/10.1093/jaoac/82.1.95

Careri, M., L. Elviri, and A. Mangia. "Validation of a liquid chromatography ionspray mass spectrometry method for the analysis of flavanones, flavones and flavonols." *Rapid Communications in Mass Spectrometry* 13, no. 23 (1999): 2399–2405. https://doi.org/10.1002/(SICI)1097-0231(19991215)13:23%3C2399::AID-RCM805%3E3.0.CO;2–N

Chalikwar, Shailesh S., Sanjay J. Surana, Sameer N. Goyal, Kaushalendra K. Chaturvedi, and Pankaj V. Dangre. "Solid self-microemulsifying nutraceutical delivery system for hesperidin using quality by design: assessment of biopharmaceutical attributes and shelf-life." *Journal of Microencapsulation* 38, no. 1 (2021): 61–79. https://doi.org/10.1080/02652048.2020.1851788

De Oliveira, David Michel, Grace Kelly Zanotti Simoes Dourado, and Thais Borges Cesar. "Hesperidin associated with continuous and interval swimming improved biochemical and oxidative biomarkers in rats." *Journal of the International Society of Sports Nutrition* 10, no. 1 (2013): 1–7. https://doi.org/10.1186/1550-2783-10-27

Dolzhenko, Yuliya, Cinzia M. Bertea, Andrea Occhipinti, Simone Bossi, and Massimo E. Maffei. "UV-B modulates the interplay between terpenoids and flavonoids in peppermint (Mentha× piperita L.)." *Journal of Photochemistry and Photobiology B: Biology* 100, no. 2 (2010): 67–75. https://doi.org/10.1016/j.jphotobiol.2010.05.003

Durán, Nelson, Amanda F. Costa, Danijela Stanisic, Juliana S. Bernardes, and LjubicaTasic. "Nanotoxicity and dermal application of nanostructured lipid carrier loaded with hesperidin from orange residue." In *Journal of Physics: Conference Series,* vol. 1323, no. 1, p. 012021. Grenoble, France: IOP Publishing, 2019. https://doi:10.1088/1742-6596/1323/1/012021

El-Shahawy, Ahmed AG, Adel Abdel-Moneim, Abdelazim SM Ebeid, Zienab E. Eldin, and Mohamed I. Zanaty. "A novel layered double hydroxide-hesperidin nanoparticles exert antidiabetic, antioxidant and anti-inflammatory effects in rats with diabetes." *Molecular Biology Reports* 48, no. 6 (2021): 5217–5232. https://doi.org/10.1007/s11033-021-06527-2

El-Sisi, Alaa E., Samia S. Sokkar, Hanaa A. Ibrahim, Mohamed F. Hamed, and Sally E. Abu-Risha. "Targeting MDR-1 gene expression, BAX/BCL2, caspase-3, and Ki-67 by nanoencapsulated imatinib and hesperidin to enhance anticancer activity and ameliorate cardiotoxicity." *Fundamental & Clinical Pharmacology* 34, no. 4 (2020): 458–475. https://doi.org/10.1111/fcp.12549

Estruel-Amades, Sheila, Malén Massot-Cladera, Pau Garcia-Cerdà, Francisco J. Pérez-Cano, ÀngelsFranch, Margarida Castell, and Mariona Camps-Bossacoma. "Protective effect of hesperidin on the oxidative stress induced by an exhausting exercise in intensively trained rats." *Nutrients* 11, no. 4 (2019): 783. https://doi.org/10.3390/nu11040783

Ferrero-Miliani, Laura, Nielsen, O. H., Andersen, P. S., and Girardin, S. E. "Chronic inflammation: importance of NOD2 and NALP3 in interleukin-1β generation." *Clinical & Experimental Immunology* 147, no. 2 (2007): 227–235. https://doi.org/10.1111/j.1365-2249.2006.03261.x

Gade, A. K., Bonde, P. P., Ingle, A. P., Marcato, P. D., Duran, N., and Rai, M. K. "Exploitation of Aspergillus niger for synthesis of silver nanoparticles." *Journal of Biobased Materials and Bioenergy* 2, no. 3 (2008): 243–247. https://doi.org/10.1166/jbmb.2008.401

Ganeshpurkar, Aditya, and Ajay Saluja. "The pharmacological potential of hesperidin." *Indian Journal of Biochemistry and Biophysics (IJBB)* 56, no. 4 (2019): 287–300.

Garg, A., S. Garg, L. J. D. Zaneveld, and A. K. Singla. "Chemistry and pharmacology of the citrus bioflavonoid hesperidin." *Phytotherapy Research* 15, no. 8 (2001): 655–669. https://doi.org/10.1002/ptr.1074

George, Archana, Priyanka A. Shah, and Pranav S. Shrivastav. "Natural biodegradable polymers-based nano-formulations for drug delivery: a review." *International Journal of Pharmaceutics* 561 (2019): 244–264. https://doi.org/10.1016/j.ijpharm.2019.03.011

Higby, R. H. "The chemical nature of hesperidin and its experimental medicinal use as a source of vitamin-P: a review". *Journal of the American Pharmaceutical Association Scientific Edition* 30 (1941): 629–635. https://doi.org/10.1002/jps.3030301208

Iwashita, Masazumi, KohjiroHashizume, Masahiro Umehara, TakaakiIshigami, ShintaroOnishi, Masaki Yamamoto, Kenjirou Higashi, and KunikazuMoribe. "Development of nobiletin-methyl hesperidin amorphous solid dispersion: novel application of methyl hesperidin as an excipient for hot-melt extrusion." *International Journal of Pharmaceutics* 558 (2019): 215–224. https://doi.org/10.1016/j.ijpharm.2018.12.092

Jahangiri, Azin, and Leila Barghi. "Polymeric nanoparticles: review of synthesis methods and applications in drug delivery." *Journal of Advanced Chemical and Pharmaceutical Materials (JACPM)* 1, no. 2 (2018): 38–47. https://www.advchempharm.ir/journal/index.php/JACPM/article/view/26

Jain, Amit K., Nitin K. Swarnakar, ChandraiahGodugu, Raman P. Singh, and Sanyog Jain. "The effect of the oral administration of polymeric nanoparticles on the efficacy and toxicity of tamoxifen." *Biomaterials* 32, no. 2 (2011): 503–515. https://doi.org/10.1016/j.biomaterials.2010.09.037

Jangde, Rajendra, Gamal Osman Elhassan, SulekhaKhute, Deependra Singh, Manju Singh, Ram Kumar Sahu, and Jiyauddin Khan. "Hesperidin-Loaded Lipid Polymer Hybrid Nanoparticles for Topical Delivery of Bioactive Drugs." *Pharmaceuticals* 15, no. 2 (2022): 211. https://doi.org/10.3390/ph15020211

Jurd, L. "Spectral properties of flavonoid compounds", edited by T. A. Geissman *The Chemistry of Flavonoid Compounds*. Pergamon Press, London (1962) pp. 107–155.

Kelte, Filho, Irineo, Christiane Schineider Machado, Camila Diedrich, ThaysaKsiaskiewcz Karam, Celso Vataru Nakamura, NajehMaissar Khalil, and Rubiana Mara Mainardes. "Optimized Chitosan-Coated Gliadin Nanoparticles Improved the Hesperidin Cytotoxicity over Tumor Cells." *Brazilian Archives of Biology and Technology* 64 (2021). https://doi.org/10.1590/1678-4324-75years-2021200795

Kim, Dong-Hyun, Eun-Ah Jung, In-Suk Sohng, Jung-Ah Han, Tae-Hyung Kim, and Myung Joo Han. "Intestinal bacterial metabolism of flavonoids and its relation to some biological activities." *Archives of Pharmacal Research* 21, no. 1 (1998): 17–23. https://doi.org/10.1007/BF03216747

Kim, Minji, Takashi Ko Metani, Shigetaka Okada, and Makoto Shimuzu. "Permeation of hesperidin glycosides across Caco-2 cell monolayers via the paracellular pathway." *Bioscience, Biotechnology, and Biochemistry* 63, no. 12 (1999): 2183–2188. https://doi.org/10.1271/bbb.63.2183

Knekt, Paul, JormaKumpulainen, RitvaJärvinen, HarriRissanen, Markku Heliövaara, Antti Reunanen, Timo Hakulinen, and ArpoAromaa. "Flavonoid intake and risk of chronic diseases." *The American Journal of Clinical Nutrition* 76, no. 3 (2002): 560–568. https://doi.org/10.1093/ajcn/76.3.560

Krishnan, Gokuladhas, Jayakumar Subramaniyan, Pramila Chengalvarayan Subramani, BarathMuralidharan, and Devaki Thiruvengadam. "Hesperetin conjugated PEGylated gold nanoparticles exploring the potential role in anti-inflammation and anti-proliferation during diethylnitrosamine-induced hepatocarcinogenesis in rats." *Asian Journal of Pharmaceutical Sciences* 12, no. 5 (2017): 442–455. https://doi.org/10.1016/j.ajps.2017.04.001

Krishnaraj, C., Muthukumaran, P., Ramachandran, R., Balakumaran, M. D., and Kalaichelvan, P. T. "Acalypha indica Linn: biogenic synthesis of silver and gold nanoparticles and their cytotoxic effects against MDA-MB-231, human breast cancer cells." *Biotechnology Reports* 4 (2014): 42–49. https://doi.org/10.1016/j.btre.2014.08.002

Letchmanan, Kumaran, Shou-Cang Shen, Wai Kiong Ng, and Reginald BH Tan. "Application of transglycosylated stevia and hesperidin as drug carriers to enhance biopharmaceutical properties of poorly-soluble artemisinin." *Colloids and Surfaces B: Biointerfaces* 161 (2018): 83–93. https://doi.org/10.1016/j.colsurfb.2017.10.020

Leuzzi, Ugo, CorradoCaristi, Vincenza Panzera, and GianvincenzoLicandro. "Flavonoids in pigmented orange juice and second-pressure extracts." *Journal of Agricultural and Food Chemistry* 48, no. 11 (2000): 5501–5506. https://doi.org/10.1021/jf000538o

Li, Chaoyun, and Hermann Schluesener. "Health-promoting effects of the citrus flavanone hesperidin." *Critical Reviews in Food Science and Nutrition* 57, no. 3 (2017): 613–631. https://doi.org/10.1080/10408398.2014.906382

Magura, Judie, Daniel Hassan, Roshila Moodley, and Irene Mackraj. "Hesperidin-loaded nanoemulsions improve cytotoxicity, induce apoptosis, and downregulate miR-21 and miR-155 expression in MCF-7." *Journal of Microencapsulation* 38, no. 7–8 (2021): 486–495. https://doi.org/10.1080/02652048.2021.1979673

Maheswari, P. Uma, Rhea Muthappa, K. Purushothaman Bindhya, and KM Meera Sheriffa Begum. "Evaluation of folic acid functionalized BSA-CaFe2O4 nanohybrid carrier for the controlled delivery of natural cytotoxic drugs hesperidin and eugenol." *Journal of Drug Delivery Science and Technology* 61 (2021): 102105. https://doi.org/10.1016/j.jddst.2020.102105

Manthey, John A., and Karel Grohmann. "Flavonoids of the orange subfamily Aurantioideae." *Flavonoids in the Living System* (1998): 85–101. https://doi.org/10.1007/978-1-4615-5335-9_7

Middleton, E. "The flavonoids." *Trends in Pharmacological Sciences* 8 (1984): 335–338.

Mohanpuria, Prashant, Nisha K. Rana, and Sudesh Kumar Yadav. "Biosynthesis of nanoparticles: technological concepts and future applications." *Journal of Nanoparticle Research* 10, no. 3 (2008): 507–517. https://doi.org/10.1007/s11051-007-9275-x

Morsy, Mohamed A., and Anroop B. Nair. "Prevention of rat liver fibrosis by selective targeting of hepatic stellate cells using hesperidin carriers." *International Journal of Pharmaceutics* 552, no. 1–2 (2018): 241–250. https://doi.org/10.1016/j.ijpharm.2018.10.003

Parhiz, Hamideh, Ali Roohbakhsh, Fatemeh Soltani, RaminRezaee, and Mehrdad Iranshahi. "Antioxidant and anti-inflammatory properties of the citrus flavonoids hesperidin and hesperetin: an updated review of their molecular mechanisms and experimental models." *Phytotherapy Research* 29, no. 3 (2015): 323–331. https://doi.org/10.1002/ptr.5256

Peterson, Julia J., Johanna T. Dwyer, Gary R. Beecher, Seema A. Bhagwat, Susan E. Gebhardt, David B. Haytowitz, and Joanne M. Holden. "Flavanones in oranges, tangerines (mandarins), tangors, and tangelos: a compilation and review of the data from the analytical literature." *Journal of Food Composition and Analysis* 19 (2006): S66–S73. https://doi.org/10.1016/j.jfca.2005.12.006

Pizzorno J. E., and Michael T. Murray,. "Textbook of Natural Medicine." Published by Churchill Livingstone (1999): 827–830.

Pradhan, Sweta Priyadarshini, Sonali Sahoo, Anindita Behera, Rajesh Sahoo, and Pratap Kumar Sahu. "Memory amelioration by hesperidin conjugated gold nanoparticles in diabetes induced cognitive impaired rats." *Journal of Drug Delivery Science and Technology* 69 (2022): 103145. https://doi.org/10.1016/j.jddst.2022.103145

Prathna, TC, A. M. Raichur, N. Chandrasekaran, and A. Mukherjee, "Green, Sunlight Irradiation Induced." 84 (2011). https://doi.org/10.1007/s40011-013-0193-7

Praveen Kumar, P., M. Kavya Nainita, A. Sai Tarun, B. G. Raghu Ramudu, K. Deepika, A. Pramoda, and C. Yasmeen. "Cerebroprotective potential of hesperidin nanoparticles against bilateral common carotid artery occlusion reperfusion injury in rats and in silico approaches." *Neurotoxicity Research* 37, no. 2 (2020): 264–274. https://doi.org/10.1007/s12640-019-00098-8

Purushothaman, Bindhya K., and Meera Sheriffa Begum KM. "Magnetic casein-CaFe2O4 nanohybrid carrier conjugated with progesterone for enhanced cytotoxicity of citrus peel derived hesperidin drug towards breast and ovarian cancer." *International Journal of Biological Macromolecules* 151 (2020): 293–304. https://doi.org/10.1016/j.ijbiomac.2020.02.172

Rao, Komal, Sabahat Aziz, Talat Roome, Anam Razzak, Bushra Sikandar, Khawar Saeed Jamali, Muhammad Imran, Tooba Jabri, and Muhammad Raza Shah. "Gum acacia stabilized silver nanoparticles-based nano-cargo for enhanced anti-arthritic potentials of hesperidin in adjuvant induced arthritic rats." *Artificial Cells, Nanomedicine, and Biotechnology* 46, no. sup1 (2018): 597–607. https://doi.org/10.1080/21691401.2018.1431653

Ribeiro, Daniela, Carina Proenca, Silvia Rocha, Jose LFC Lima, Felix Carvalho, Eduarda Fernandes, and Marisa Freitas. "Immunomodulatory effects of flavonoids in the prophylaxis and treatment of inflammatory bowel diseases: a comprehensive review." *Current Medicinal Chemistry* 25, no. 28 (2018): 3374–3412. https://doi.org/10.2174/0929867325666180214121734

Roohbakhsh, Ali, HamidehParhiz, Fatemeh Soltani, RaminRezaee, and Mehrdad Iranshahi. "Molecular mechanisms behind the biological effects of hesperidin and hesperetin for the prevention of cancer and cardiovascular diseases." *Life Sciences* 124 (2015): 64–74. https://doi.org/10.1016/j.lfs.2014.12.030

Rostami, MohammadReza, Mohammad Yousefi, ArezouKhezerlou, Masoud Aman Mohammadi, and Seid Mahdi Jafari. "Application of different biopolymers for nanoencapsulation of antioxidants via electrohydrodynamic processes." *Food Hydrocolloids* 97 (2019): 105170. https://doi.org/10.1016/j.foodhyd.2019.06.015

Saad, Suma, Iqbal Ahmad, Syed M. Kawish, Urooj A. Khan, Farhan J. Ahmad, Asgar Ali, and Gaurav K. Jain. "Improved cardioprotective effects of hesperidin solid lipid nanoparticles prepared by supercritical antisolvent technology." *Colloids and Surfaces B: Biointerfaces* 187 (2020): 110628. https://doi.org/10.1016/j.colsurfb.2019.110628

Sahu, Nidhi, Deepika Soni, B. Chandrashekhar, D. B. Satpute, SivanesanSaravanadevi, B. K. Sarangi, and R. A. Pandey. "Synthesis of silver nanoparticles using flavonoids: hesperidin, naringin and diosmin, and their antibacterial effects and cytotoxicity." *International Nano Letters* 6, no. 3 (2016): 173–181. https://doi.org/10.1007/s40089-016-0184-9

Sastry, Murali, Absar Ahmad, M. Islam Khan, and Rajiv Kumar. "Microbial nanoparticle production." *Nanobiotechnology: Concepts, Applications and Perspectives* (2004): 126–135. https://doi.org/10.1002/3527602453.ch9

Sharma, Deepak, Gilphy Philip, Reema Gabrani, Javed Ali, and Shweta Dang. "Dual agents loaded polymeric nanoparticle: effect of process variables." *International Journal of Pharmaceutical Investigation* 5, no. 3 (2015): 155. https://doi.org/10.4103%2F2230-973X.160853

Sieve, Benjamin F. "A new antifertility factor (a preliminary report)." *Science* 116, no. 3015 (1952): 373–385. https://doi.org/10.1126/science.116.3015.373

Sujitha, Yeramanchi Sarah, and Yallamali Indira Muzib. "Preparation of topical nano gel loaded with hesperidin emusomoes: in vitro and in vivo studies." *International Journal of Pharmaceutical Investigation* 10, no. 4 (2020): 500–505. https://doi.org/10.5530/ijpi.2020.4.87

Sulaiman, Ghassan M., Hanaa M. Waheeb, Majid S. Jabir, Shaymaa H. Khazaal, Yaser Hassan Dewir, and Yougasphree Naidoo. "Hesperidin loaded on gold nanoparticles as a drug delivery system for a successful biocompatible, anti-cancer, anti-inflammatory and phagocytosis inducer model." *Scientific Reports* 10, no. 1 (2020): 1–16. https://doi.org/10.1038/s41598-020-66419-6

Taymouri, Somayeh, Saeed Hashemi, Jaleh Varshosaz, Mohsen Minaiyan, and ArdeshirTalebi. "Fabrication and evaluation of hesperidin loaded polyacrylonitrile/polyethylene oxide nanofibers for wound dressing application." *Journal of Biomaterials Science, Polymer Edition* 32, no. 15 (2021): 1944–1965. https://doi.org/10.1080/09205063.2021.1952380

Tejada, Silvia, Samuel Pinya, Miquel Martorell, Xavier Capó, Josep A. Tur, Antoni Pons, and Antoni Sureda. "Potential anti-inflammatory effects of hesperidin from the genus citrus." *Current Medicinal Chemistry* 25, no. 37 (2018): 4929–4945. https://doi.org/10.2174/0929867324666170718104412

Tsai, Yuh-Feng, Yi-Ru Chen, Jo-Ping Chen, Yun Tang, and Kai-Chiang Yang. "Effect of hesperidin on anti-inflammation and cellular antioxidant capacity in hydrogen peroxide-stimulated human articular chondrocytes." *Process Biochemistry* 85 (2019): 175–184. https://doi.org/10.1016/j.procbio.2019.07.014

Uçan, Filiz, Erdal Ağçam, and Asiye Akyildiz. "Bioactive compounds and quality parameters of natural cloudy lemon juices." *Journal of Food Science and Technology* 53, no. 3 (2016): 1465–1474. https://doi.org/10.1007/s13197-015-2155-y

Varghese, Jibin Joy, and Rashmi Mallya. "Formulation development and evaluation of antioxidant potential of hesperidin nanocrystals." *Cell Proliferation* 9 (2015): 10.

Waheeb, Hanaa M., Ghassan M. Sulaiman, and Majid S. Jabir. "Effect of hesperidin conjugated with golden nanoparticles on phagocytic activity: in vitro study." In *AIP Conference Proceedings,* vol. 2213, no. 1, p. 020217. AIP Publishing LLC, 2020. https://doi.org/10.1063/5.0000159

Wichansawakun, Sanit, and Harpal S. Buttar. "Antioxidant diets and functional foods promote healthy aging and longevity through diverse mechanisms of action." In *The Role of Functional Food Security in Global Health*, pp. 541–563. Academic Press, 2019. https://doi.org/10.1016/B978-0-12-813148-0.00032-3

Williams R. T. *"Metabolism of phenolics in animals. In Biochemistry of Phenolic Compounds."* (Edited by Harbourne J. B.), Academic Press, New York (1964) pp. 205–248.

Yamada, Mika, Fujimi Tanabe, Norie Arai, Hitoshi Mitsuzumi, Yoshikatsu Miwa, Michio Kubota, HirotoChaen, and Masayoshi Kibata. "Bioavailability of glucosyl hesperidin in rats." *Bioscience, Biotechnology, and Biochemistry* 70, no. 6 (2006): 1386–1394. https://doi.org/10.1271/bbb.50657

Yang, Zhen, Hongwei Yang, Xinhua Dong, Minglong Pu, and Feihong Ji. "Hesperidin loaded Zn2+@ SA/PCT nanocomposites inhibit the proliferation and induces the apoptosis in colon cancer cells (HCT116) through the enhancement of pro-apoptotic protein expressions." *Journal of Photochemistry and Photobiology B: Biology* 204 (2020): 111767. https://doi.org/10.1016/j.jphotobiol.2019.111767

Zeng, Fanyu, Dandan Wang, Yan Tian, Min Wang, Rui Liu, Zhining Xia, and Yike Huang. "Nanoemulsion for improving the oral bioavailability of hesperetin: formulation optimization and absorption mechanism." *Journal of Pharmaceutical Sciences* 110, no. 6 (2021): 2555–2561. https://doi.org/10.1016/j.xphs.2021.02.030

13 Application of Phytodrug Delivery Systems for Treatment and Management of Ulcerative Colitis

Pooja Mongia Raj, Anshul Gupta, Madhu Gupta, Sunita Minz, Abhishek Wahi, and Rakesh Raj

13.1 INTRODUCTION

The chronic, relapsing inflammation of the rectum and colon is known as ulcerative colitis (UC). UC is an inflammatory disorder of the colon characterized by chronic inflammation of the mucosa of the colon. Sam Wilks first described UC in the 1800s, and it is considered along with CD one of the two types of inflammatory bowel disease (IBD) that are more prevalent in developed nations. CD is most common in North America and Western Europe, but it is increasing in Asia as well (Rivera et al., 2022, Feuerstein and Cheifetz, 2014). The causes of activated immune response are unclear; however, genetic, dietary and environmental factors play an important role. In contrast to that of Crohn's disease, UC inflammation is limited to the mucosa of the colon (Ordás et al., 2012). Moreover, inflammation in UC is generally limited to the mucosal surface, while lesions in the intestine of CD sufferers are patchy, infiltrating the bowel wall more deeply and often leading to complications such as fibrosis and stricture formation. Common symptoms of UC include bloody diarrhea, abdominal pain, tenesmus, the influx of neutrophils and macrophages that produce cytokines, proteolytic enzymes, and free radicals that result in inflammation and ulceration, with long-term UC increasing the risk of colorectal cancer and dysplasia (Szigethy et al., 2010, Stokkers and Hommes, 2004, Freeman et al., 2021). It is estimated that over the last 50 years the incidence and prevalence of IBD have increased to 8-14 people per 100,000 and 120- 200 people per 100,000, respectively, in Western populations. Most UC patients are between the ages of 30-40 when they are diagnosed.

The prevalence of IBD is worldwide, although it occurs more commonly in US, UK, and Scandinavia than in other region. IBD occurrence between males and females has no difference but is commonly diagnosed between the age group of 20-40 years. Generally, 20% of all IBD patients develop symptoms during childhood, rarely 5% of all patients are detected with IBD before 10 years of age, and only 25% of patients have a family history of IBD. UC and CD are the major causes of lifetime morbidity and have various similar clinical features but they differ according to their site of occurrence (Hendrickson et al., 2002, Head and Jurenka, 2003, Weber and Turner, 2007).

13.2 PATHOGENESIS OF UC

Although the exact cause of IBD is not clear until, both CD and UC display heterogeneity in many clinical and pathological features. They are distinguishable by their location and nature of inflammation (Figure 13.1). Unlike UC, which attacks colonic mucosa, CD can affect any part of the gastrointestinal tract (GIT). Both conditions share clinical features like extra-intestinal manifestation,

 DOI: 10.1201/9781003292692-13

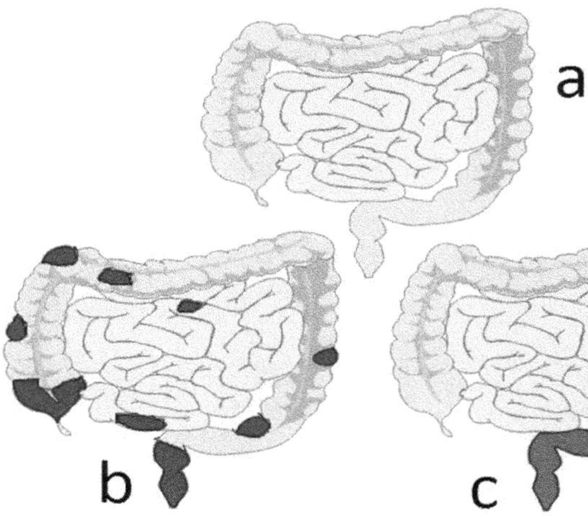

FIGURE 13.1 Representation of inflammatory bowel diseases (a) normal colon, (b) colon affected by Crohn's disease and (c) colon affected by ulcerative colitis.

but hematochezia and passage of mucus or pus are common only in UC. Fistula, perianal disease, colonic and small bowel obstruction are common in CD. Cryptitis and crypt abscesses are observed in both UC and CD, while crypt architecture is more distorted in the case of UC. Both UC and CD show relapsing intestinal inflammation (Yeshi et al., 2020). However, several studies point to genetic factors predisposing to the disease. Furthermore, the microbial flora of the bowel seems of major importance in the occurrence of complications, although the exact role of this microflora has not been explored (Linskens et al., 2001, Hendrickson et al., 2002, Rezaie et al., 2007, Geier et al., 2007).

Inflammatory bowel disease (IBD) is a recurrent inflammatory condition that affects the intestines. Crohn's disease and UC are the two most common types. Although the specific cause of IBD is unknown, it is thought to be caused by a complex interplay of genetic, environmental, or microbial factors, as well as immunological responses. The following are some of the most important factors linked to IBD. *Genetic predisposition*- There are multiple genes involved but the following are commonly associated with IBD-

NOD2 Polymorphism: The microbiota in the small intestine is regulated by NOD2 (nucleotide-binding oligomerization domain). NOD2 is abundantly expressed in ileal Paneth cells, which play an important role in the regulation of the ileal microbiota by secreting antibacterial chemicals. NOD2 detects microbiota-derived products and triggers NF-κB, which is required for Paneth cells to destroy bacteria by secreting antibacterial chemicals. Many aspects of gut immune homeostasis are disrupted in NOD2 mutants, including diminished bacterial product detection and altered antimicrobial responses in Paneth cells, resulting in bacterial clearance failure. It causes dysbiosis in the gut microbiota, which is defined by an increased load of bacteria and triggers the immune system to launch a continual onslaught.

Dysbiosis: Changes in the endogenous commensal microbiota in the intestines play an important part in the pathogenesis of IBD, which is also known as "Dysbiosis." The number of microbial species in the human intestine is huge, and each person's microbes are unique. The composition of these commensals plays a significant role in intestinal health. There is an innate and adaptive immune response when the number of good bacteria is less than the number of harmful bacteria; however, in the case of IBD, it becomes out of proportion due to genetic predisposition, dietary and smoking habits, as well as a stressful lifestyle. These bacteria and their components disturb the intestine's epithelial barrier, allowing luminal bacterial components to cross the epithelial barrier and triggering inflammatory cells.

Epithelial Barrier Dysfunction: In individuals with IBD, defects in the intestinal epithelial tight junction barrier function lead to the buildup of bacterial components inside the intestinal wall, which stimulates dendritic cells and T-cell reactions.

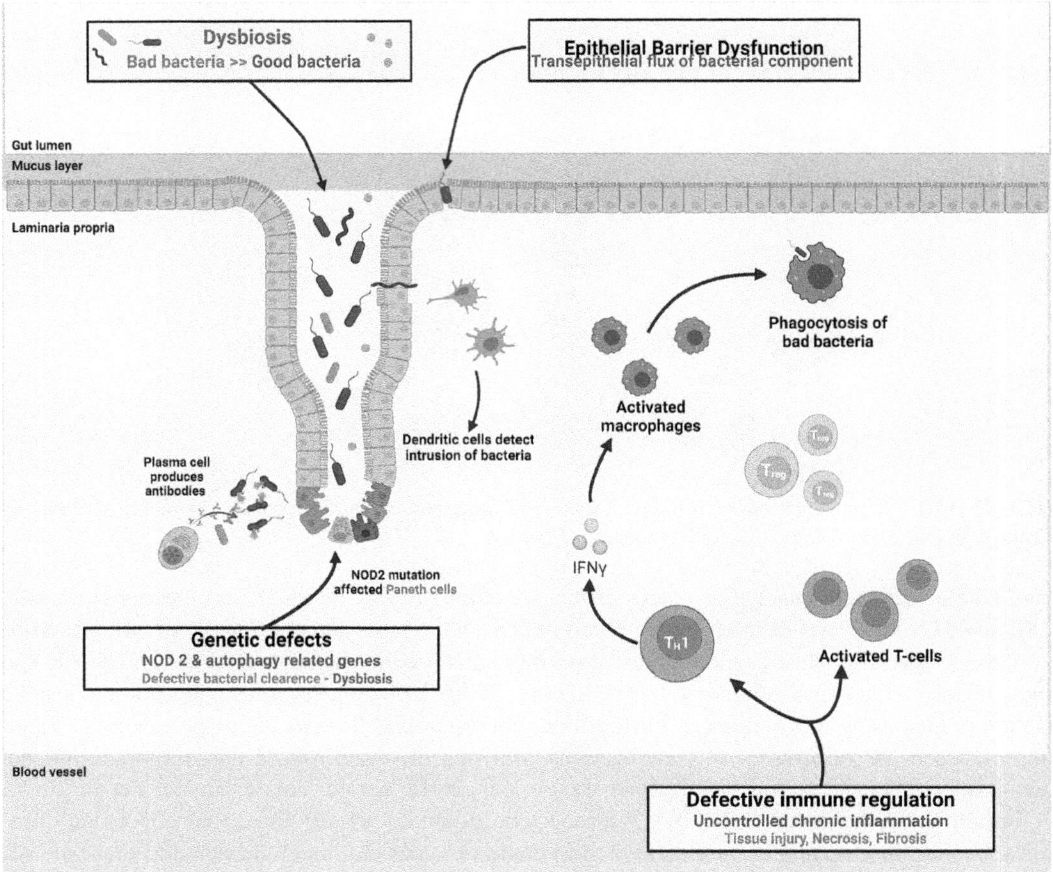

FIGURE 13.2 Schematic diagram of pathogenesis of ulcerative colitis.

Defective Immune Regulation: Deficiencies in innate and adaptive immunological pathways lead to the abnormal inflammatory response in the intestine in people with IBD. The main initiators are defects in regulatory T-cells, particularly those that produce IL-10. Severe, early-onset colitis is linked to mutations in the IL-10 receptor. Lymphocytic aggregation and increased T-cell activity are caused by excessive production of TH1 and TH17 cells and their activity-inducing cytokines. The secretion of thiopurine, antitumor necrosis factor (TNF) and IFN gamma induces fibrosis, necrosis, and inflammation. Chronic inflammation in IBD is most likely caused by a combination of increased immune activation by gut bacteria and poor immune control (Figure 13.2).

13.3 TREATMENT STRATEGIES AND THERAPEUTIC TARGETS FOR ULCERATIVE COLITIS

The treatment strategy is based upon primarily on the severity of colitis and secondarily on the patient choice of treatment. Patients can be classified with mild-to-moderate colitis and moderate-to-severe colitis. 5-Aminosalycilates (5-ASAs) is the choice of drug while mild-to-moderate colitis. It is recommended to be used orally or rectally or in combination but found to give better results in combination mode. The first-line therapy for the treatment of mild-moderate colitis includes mesalamines which are available in different forms like oral, suppository, or liquid enema (Table 13.1). Corticosteroids are used in case symptoms still persist after treatment for a long time. Then carefully patients are made to adopt to steroid-sparing agents like TNF. A somewhat similar therapy is used in patients suffering from moderate-to-severe type of colitis. They are given corticosteroids

TABLE 13.1
List of Therapeutic Targeting for Ulcerative Colitis

S. No	Drugs	Route of Administration	Advantages	Disadvantages	Mechanism of Action	Remarks	References
	Mesalamine with trade names: Asacol, Delzicol, Lialda, Pentasa, Apriso, Colazal, Canasa, Rowasa	Oral, Suppository, and Enema	Low-cost, topical application of mesalamine prevents relapse of UC, and maintenance therapy of mesalamine reduces the chances of colorectal cancer	Side effects associated with Mesalamine, including flatulence, nausea, abdominal pain, diarrhea, and headache, are generally mild	Interference with the metabolism of arachidonic acid (conversion to prostaglandin) and IL-scavenging of reactive O_2 species, and effects on the function of white blood cells and the production of cytokines (12) 5-ASA could induce regulatory T cells (Tregs) in the colon through the aryl hydrocarbon receptor pathway, followed by the activation of transforming growth factor (TGF)β(13)	First-line therapy in mild-to-moderate UC	Yeshi et al. (2020), Cai et al. (2021), Kayal and Shah (2020)
	Corticosteroids: budesonide, prednisone	Oral, enema, Foam, suppository, parenteral	Choice of drugs for patients who have not responded to mesalamine, effective for induction of remission for a short period	Systemically CSs lead to various side effects like opportunistic infection, hyperglycemia, insomnia, hypertension, skin atrophy, venous thromboembolism, and weight gain	After binding with the CS receptor in the cytoplasm, activated CSs get internalized into the nucleus and interact with pro-inflammatory factors such as nuclear factor kappa-B and activator protein-1. This will result formation of a co-activator complex that inhibits the inflammation	Corticosteroids are recommended as second-line therapy for mild-to-moderate UC not responded to mesalamine	Kayal and Shah (2020) and Cai et al. (2021)

(*Continued*)

TABLE 13.1 (*Continued*)
List of Therapeutic Targeting for Ulcerative Colitis

S. No	Drugs	Route of Administration	Advantages	Disadvantages	Mechanism of Action	Remarks	References
	Thiopurines: Azathioprine (AZ), 6-murcaptopurine (MP), thioguanine (TG) Like Azasan, Imuran	Oral	Usually well tolerated	Nausea, vomiting, leukopenia, increased susceptibility to infection, pancreatitis, and hepatitis	MP, TG could control intestinal inflammation by inhibiting T-lymphocyte proliferation and activation	Delayed therapeutic effect	Pastorelli et al. (2009)
	Cyclosporine A, Tacrolimus	Oral, parenteral	Fast therapeutic effect, suitable for induction	High incidence of side effects; Hypertension, nephropathy, potassium retention, gingival hyperplasia, and increased infections	Inhibition of calcineurin by interfering with the signaling pathway of various cytokines, and therefore inhibits the inflammatory response	Induction and maintenance of remission	Pastorelli et al. (2009) and Cai et al. (2021)
	Anti-TNF agent: Infliximab, Adalimumab, Golimumab	Intravenous, Subcutaneous	Good efficacy fast therapeutic effect	Expensive, allergic reactions, increased risk of serious infection, heart failure in cardiopathic patients' unknown long-term effects	Monoclonal antibodies directed against TNF-alfa, an inflammatory cytokine	Useful for moderate-to-severe UC therapy, May be applied for patients who are intolerant to corticosteroids	Kayal and Shah (2020)
	Anti-integrin agents: Vedolizumab	Intravenous	It has a favorable safety profile with a low incidence rate of serious infection	Serious clostridial infections, sepsis, and tuberculosis were reported infrequently	Monoclonal antibodies directed against $\alpha_4\beta_7$ cell surface glycoprotein on B and T lymphocytes	In a comprehensive study, Vedolizumab has shown great tolerability and safety	Kayal and Shah (2020), Pastorelli et al. (2009), and Colombel et al. (2017)
	Anti-interleukin agents: Ustekinumab	Intravenous (first dose), Subcutaneous	Offers a favorable infectious safety profile compared to Anti-TNF agent	Nasopharyngitis, headache, bronchitis, vaginal mycosis	Monoclonal antibodies directed against the p40 subunit of IL-12 and IL-23	Approved for moderate-to-severe UC	Kayal and Shah, (2020), Pastorelli et al. (2009), and Sands et al. (2019)
	Janus kinase inhibitors: Tofacitinib	Oral	Tofacitinib has a fast onset of action and seems to be effective even in case of acute severe UC	The long-term safety of tofacitinib remains unclear	Small molecule Janus kinase 1 and 3 inhibitor	Used for moderate-to-severe active UC	Kayal and Shah, (2020), Cai et al. (2021), and Yeshi et al. (2020)

followed by thiopurine. Patients who cannot be treated with corticosteroids and thiopurines due to some contraindication or failure of oral corticosteroid/ thiopurine therapy, are generally treated with anti-TNF therapies. In the case of moderate-to-severe colitis combination of infliximab and azathioprine has been used and had shown promising results. Vedolizumab is a new FDA-approved drug (2014) and has been recommended for use in case all the conventional treatments with prednisone, thiopurine, or anti-TNF are ineffective. This drug is in its infancy stage and used mainly for the treatment of moderate-to-severe colitis conditions when patients become nonrespondent to other available therapies or patients are not responding to anti-TNF therapy. Intravenous administration of steroids IV is recommended for patients who are not responding well to oral prednisone and suffer from moderate to severe colitis. Such patients require proper monitoring and hospitalization. A similar kind of treatment with steroids and close monitoring during hospitalization is also needed in patients suffering from fulminant colitis. It has been observed that majority of patients responded well to intravenous corticosteroid therapy, making it a savior for many colitis patients. However, for non-respondents, other drugs like infliximab, intravenous cyclosporine or surgery play a crucial role. Currently, endoscopic remission (deep remission) is also being followed as a part of the treatment strategy. This method allows endoscopic healing of the colon and in addition decreases chances of colectomy and flare. Reducing colonic inflammation greatly decreases the chances of colon cancer. The standard clinical care protocol suggests steroid-free treatment as the first choice for colitis. The ultimate aim of colitis medical treatment is to give quality of life to patients by decreasing the risk of re-occurrence and complications (Feuerstein and Cheifetz, 2014).

13.4 HERBAL DRUG DELIVERY-BASED TREATMENT AND MANAGEMENT FOR ULCERATIVE COLITIS

Despite the incredible progress of modern medicine, there remain significant obstacles in the path of curing inflammatory bowel disease, which is one of the challenging health concerns for humans worldwide. Novel treatment options for IBD are continuously explored, and novel drugs are being discovered. However, chemically synthesized drugs are always accompanied by adverse reactions. Furthermore, their routes of administration are limited, and their effect cannot be guaranteed. Therefore, in searching for novel pharmacologically active substances that could be used for treating IBD, natural ingredients from various sources, such as plants, animals, and microorganisms, are receiving great attention nowadays. Studies have demonstrated that natural products, including natural medicines, their extracts, and their metabolites, can effectively treat IBD. The section ahead describes these natural products, such as flavonoids, terpenoids, polysaccharides, and alkaloids.

13.4.1 Naringenin

Grapefruit is a rich source of naringin which has naringenin flavonoid (4′,5,7-trihydroxyflavanone) as its aglycone part (Zaidun et al., 2018, Den Hartogh and Tsiani, 2019). Naringenin is a flavonoid are having high solubility in organic solvents (alcohol) and is practically insoluble in water (Zobeiri et al., 2018). In an in vivo study colitis was induced by dextran sodium sulfate (DSS) in a mouse model and naringenin exhibited significant relief in intestinal edema (Alam et al., 2014). Dou and colleagues have reported in their study that naringenin stimulates monocyte chemotaxis and production of pro-inflammatory factors like ICAM-1, iNOS (intercellular adhesion molecule-1, inducible NO synthase) (Dou et al., 2013). However, the use of naringenin led to decrease in IL-6, MCP-1, TNF-α, and cyclooxygenase-2 (Cox2) levels due to suppression of mRNA expressions. An in vitro study on human colonic HT-29 cells were treated with naringenin and the results showed suppression in the expression of NF-κB luciferase which was mediated by TNF-α. Naringenin also led to decline in various factors like macrophage inflammatory protein (MIP-2),IL-1b, IL-6, and IL-17A interferon (INF)-c, expression of PGE2. It causes an increase in the mucosal content of the colon and thus possesses high potential in the treatment of colitis (Al-Rejaie et al., 2013).

13.4.2 Hesperetin

Hesperetin is a naturally occurring flavanone glycoside present in citrus fruits like oranges, limes, and lemons (Liu et al., 2016, Shirzad et al., 2017, González-Alfonso et al., 2018). Its chemical name is 5,7,3′-trihydroxy-4′-methoxy flavanone and it is known for its pharmacological action against tumors, oxidation stress, and inflammation (Parhiz et al., 2015, Alu'datt et al., 2017, Chen et al., 2019). In an in vivo DSS-induced colitis mouse model, hesperetin has been reported to work by inhibiting intestinal epithelial necroptosis and helping to improve colitis conditions (Zhang et al., 2020). In another colitis rat model (TNBS-induced model), hesperetin showed a significant decrease in the level of pro-inflammatory markers, MPO, and MDA when given p.o orally at a dose of 100 mg/kg (Polat et al., 2018). Erlund and co-workers conducted a study on the health benefits of orange juice particularly in colitis. It has been found that after consumption of orange juice plasma concentration of flavonoids increases markedly which indicates that hesperetin containing orange juice are protective nutraceutical for colitis patients (Erlund et al., 2001).

13.4.3 Quercetin

Quercetin belongs to a class of natural flavonoids and has shown promising results in the improvement of colitis. Rutoside (3-rhamnosy-glucosyl quercetin) and quercitrin (3-rhamnosylquercetin) are the two commonly occurring glycosylated forms (Tao et al., 2013). It has anti-inflammatory and bactericidal action. It helps to improve intestinal microbiota and host relationships mediated via intestinal macrophages. It promotes recovery of the pro-inflammatory factors leading to relief of colitis conditions (Nogata et al., 2006). Colitis condition is beneficial as the inflammatory response exhibited by bone marrow-derived macrophages is suppressed by quercetin as observed in in vitro studies. Quercetin does not interfere with c-Jun N-terminal kinase activity. However, it causes the suppression of the NF-κB signaling pathway leading to NOS production while it stops the production of cytokine (Ju et al., 2018). Quercetin when given to rats at dose of 50/100 mg/kg has shown clinical improvement in the morphology of TNBS-induced colitis (Comalada et al., 2005). It exhibited an increase in glutathione level and good control of MPO activity and MDA levels. All these factors help in colitis treatment (Dodda et al., 2014). In the list known as generally recognized safe (GRAS) list of U.S. Food and Drug Administration's quercetin as a supplement has been included. It permits the use of quercetin at a dose of 500 mg per serving in FDA-approved foods and beverages (Bellavite, 2023).

13.4.4 Ginsenoside

Ginseng and its active constituent ginsenoside have shown promising results in colitis by decreasing inflammation, colon weight, and DAI and causing immunomodulation (Wang et al., 2018). Ginseng fruit extract also helps in reducing inflammation in IBD by deactivation of neutrophils, and infiltration by macrophages and T lymphocytes. It has shown to stimulate migration of CD103+CD11C+DCs to colitis-damaged tissues (Zhang et al., 2016). In TNBS-induced colitis use of fermented red ginseng had exhibited suppression of activity of macrophages and stimulation of cell differentiation of Foxp3+ Th1/Treg cells (Kim et al., 2018a). Ginsenosides in particular promote cell division of mucosal epithelial cells of the colon and reduce colitis symptoms by controlling immunocyte differentiation and production of various inflammatory factors (Yang et al., 2020). In TNBS-induced colitis animal models, the use of ginsenosides has been shown to protect inflamed damaged colon tissue by increasing their resistance to oxidation attack and decreasing neutrophil count at site (Yang et al., 2012a). It also attenuates the cumulation of IL-1β, IL-6, and TNF-α (pro-inflammatory factors) and leukocytes at the colon site and helps to prevent recurrent colitis in animals (Yang et al., 2012b). Some metabolites of ginsenoside like ginsenoside Rg1 (20 (S)-protopanaxatriol) also active and reduce inflammation in colitis by inhibiting expression of LPS and TLR4 on macrophage membrane, controlling Th17 cells and Tregs levels (Lee et al., 2015). Other metabolites work by decreasing pro-inflammatory factors viz IL-6, IL-1β, increasing IL-10, decreasing MPO function etc. in the peripheral systemic circulation and colonic tissue (Li et al., 2014).

13.4.5 Glycyrrhizic Acid

The *Glycyrrhiza uralensis* Fisch., Glycyrrhiza inflata Bat. and *Glycyrrhiza glabra* L. Glycyrrhizic contain Glycyrrhizic acid as their main active constituent. It is mainly extracted from the roots and rhizomes of plants (He et al., 2019, Chrzanowski et al., 2021). In an in vivo TNBS-induced colitis mouse model glycyrrhizic acid significantly improved the pathological condition of the colon. It was concluded that catechin at a dose of 50 mg/kg is effective against colitis. In LPS-challenged macrophages, glycyrrhizic acid works by decreasing the level of IL-6, promoting the production of interleukin-10 (Liu et al., 2011). Dipotassium glycyrrhizate is a salt of the glycoconjugated triterpene glycyrrhizin, and it has proven to be a potential candidate in the treatment of colitis. It markedly decreases colon inflammation leading to accelerated healing of the mucosal epithelial lining of the colon (Stronati et al., 2019).

13.4.6 Astragalus Polysaccharide

Astragalus membranaceus is the source of astragalus polysaccharide which had exhibited improvement in colitis condition specifically one induced by trinitrobenzene sulfonic acid (TNBS) in rats (Jin et al., 2014, Auyeung et al., 2016, Zhang et al., 2019). It mainly works by regulating the expression of various factors like TNF-α, IL-1β, NFATc4, and maintaining expression as well as the ratio of GATA-3 and T-bet (Yang et al., 2014, Gao et al., 2016).

13.4.7 Aloperine

From *Sophora vulgaris*, a tricyclic quinoline alkaloid is obtained known as Aloperine. It is generally given orally at dose of (40 mg/kg BW). In the DSS-mediated colitis animal model it has been shown to work by decreasing T-lymphocyte count, suppressing the expression of p-PI3K p85, p-mTOR, p-Akt and upregulating expression of protein phosphatase 2A (PP2A), Foxp3 levels (mesenteric lymph nodes and spleen). Aloperine inhibits the PI3K/Akt/mTOR signal transduction pathway through pp2 leading to relief in inflammation in colitis (Fu et al., 2017, Shi et al., 2019).

13.4.8 Berberine

Berberine is found to be beneficial for colon health as it has anti-inflammatory activity and mitigates oxidative stress in the colon. It works through the STAT3 signaling pathway. It is an alkaloid and obtained from Phellodendron and Coptis plants (James et al., 2011). Li and co-workers have demonstrated that berberine when given at a dose of 100 mg/kg BW orally, leads to a drop in the percentage of colon shortening, damage, and loss of weight. It decreases level of IL-1, IL-1β, IL-6, IL-12, TNF-α, TGF-β, and interferon-γ. While showing an increase in IL-4 and iL-10 levels. It is suppression of the IL-6/STAT3/NF-κB signal transduction pathway which is mainly responsible for its anti-inflammatory action. Its continuous treatment led to a reduction in the MDA and MPO levels and an increase in SIgA levels (Zhu et al., 2019, Li et al., 2020). A thorough analysis of 16S rDNA revealed that berberine also influences the microbiota of the colon. In the colon, it causes activation of Ahr and regulates microbe-related tryptophan metabolites thus trying to improve impaired intestinal barrier function (Jing et al., 2021). It has a high safety margin and treatment for long duration has exhibited excellent performance in improving colitis condition. For example, in clinical settings berberine when given to colitis patient continuously for 3 months at a dose of 900 mg/day has led to a significant decrease in the percentage of Geboes grade of colonic tissue. No side effects so far have been reported for it (Xu et al., 2020).

13.4.9 Caffeine

Caffeine a common household name is isolated from beans of coffee arabica plant. It is basically a xanthine alkaloid. It targets intestinal epithelial cells (IECs) and work by inhibiting mRNA expression of chitinase 3-like 1. It is found to be effective at doses of 2.5 and 5 mM and decreases in invasion by bacteria and is dose-dependent. Colitis rat model when treated with caffeine has shown markedly low levels of pro-inflammatory cytokines (Lee et al., 2014).

13.4.10 Theophylline

Theophylline is another alkaloid similar to caffeine in terms of not only structure but pharmacological action. It is the main component of tea leaves viz thea sinensis (Senchina et al., 2014, Singleton et al., 2019). It causes a decrease in IL-6, IL-1b, TNF-a (cytokine levels) and MPO function. This in turn led to a decrease in the severity of colitis. It was tested at a dose of 20 and 50 mg/kg mouse BW (i.p.) and showed improvement in colitis condition (Ghasemi-Pirbaluti et al., 2017).

13.4.11 Palmatine

Palmatine helps in the fast recovery of an injured or inflamed colon. Its mechanism of action mainly includes the protection of intestinal epithelial cells from the apoptosis and the healing of the mucosal layer of the colon. It has shown good results in colitis treatment in DSS-induced rat models when administered at doses of 50 and 100 mg/kg mouse BW (p.o.). Its use also leads to elevation in the level of Firmicutes and Bacteroidetes, which in turn favors fast colon recovery (Zhang et al., 2018). Mai et al. (2019) had established in their work that mitochondrial autophagy phenomena are co-related with the use of palmatine. Thus, cyclosporin A produced antagonist effect to palmatine when used in DSS induces colitis mice. It is obtained from plant source like *Phellodendron chinense, coptis, corydalis*, and other plants (Ryuk et al., 2012).

13.4.12 Curcumin

Curcumin is known for its antiseptic, anti-inflammatory, anti-oxidant, and anticancer properties for ages. The role of curcumin in wound healing and as an immunity booster is also well established. It is mainly obtained from roots/rhizomes of *Curcuma longa*, Zingiberaceae family. Its major active constituent is hydrophobic polyphenolic compound viz 1,7-bis(4-hydroxy-3-methoxyphenyl)-1,6-heptadiene-3,5-dione. Rhizome also contains 2%–5% turmeric pigment responsible for its bright yellow color (Esmaily et al., 2015, Panahi et al., 2015, Panahi et al., 2018, Derosa et al., 2016, Ghandadi and Sahebkar, 2017, Mirzaei et al., 2016, Momtazi et al., 2016). Some limitations of curcumin are its poor gastrointestinal absorption and fast systemic elimination leading to low bioavailability. It suffers from stability issues as well. Many attempts have been made to improve its bioavailability which includes the development of various types of nanoformulations like solid lipid nanoparticles, curcumin-encapsulated exosomes, polylactide-coglycolide polymeric nanoparticles, and curcumin-loaded liposomes. These developed novel formulations exhibited improved solubility, bioavailability, and stability. No side effects were reported (Momtazi et al., 2016, Sun et al., 2010). Curcumin eases inflammation by decreasing IL-6 and TNF-α levels. This has been demonstrated in a study carried out in murine macrophage RAW 264.7 cells in which inflammation was induced by lipopolysaccharide and treatment was given by curcumin (Sun et al., 2010; Figure 13.3 and Table 13.2).

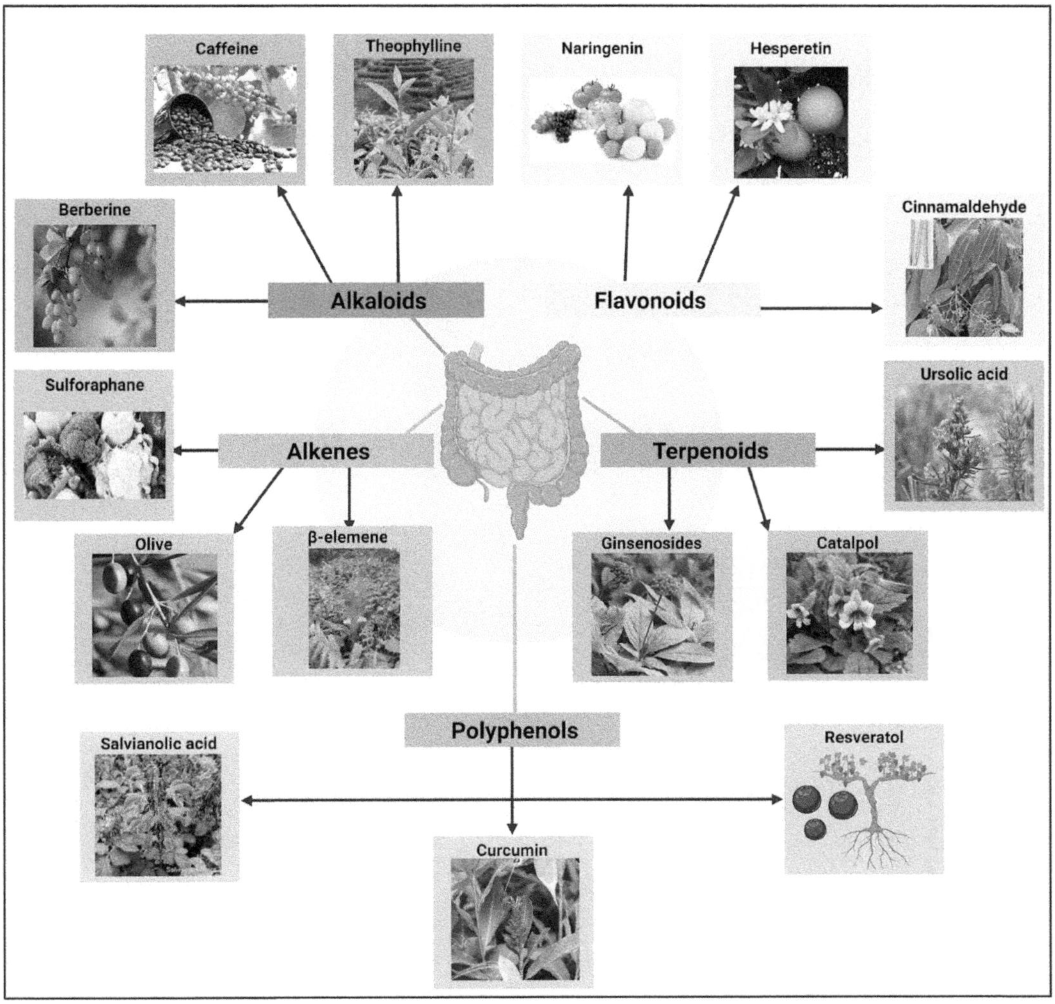

FIGURE 13.3 Diagram based on herbal drug delivery for ulcerative colitis.

13.5 CLINICAL TRIALS BASED ON HERBAL TREATMENT OF ULCERATIVE COLITIS

There have been various clinical trials including the use of herbal remedies for people with UC, and the results have been promising. The cost of treating UC patients, whether for active disease or maintenance, is relatively high. For most people in many nations, this expense may be unsustainable. Herbal therapy, on the other hand, is likely to be less expensive than traditional UC treatment. Investigating the potential advantages of herbal formulations, which is of tremendous utility in the management of UC-related symptoms and inflammation, based on data from prior observational studies and clinical trials. Existing clinical studies have shown that herbal remedies provide clinical remission or improvement in patients with mild or moderate UC, despite the fact that a small number of researches found no benefits. Table 13.3 shows the results of the effectiveness of herbal and plant product administration in response and remission rates in active UC patients.

TABLE 13.2
Description of Herbal Drugs Used in the Treatment Ulcerative Colitis

S. No	Drug	Biological Source and Chemical Constituents	MOA	Ring Present and Part Used	Uses	Structure	References
	Naringenin	4′,5,7-trihydroxy flavanone flavonoid is the aglycon of naringin which is abundant in grapefruit	Colonic secretion of pro-inflammatory factors, such as intercellular adhesion molecule-1 (ICAM-1) and inducible NO synthase. mRNA expressions of IL-6, MCP-1, TNF-α, and cyclooxygenase-2 (Cox2) were decreased	Ring is flavonoids The part that is used is grapefruit	Anti-inflammatory	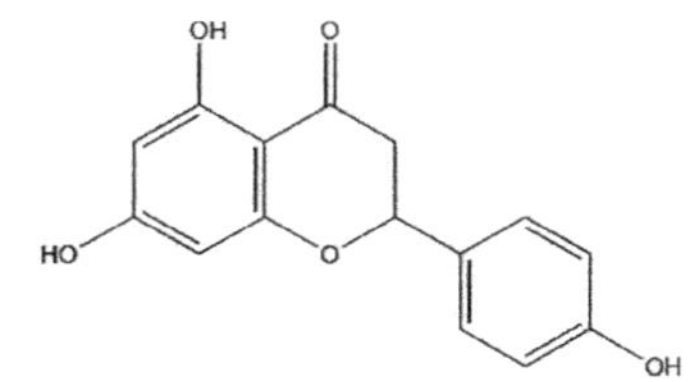	Zaidun et al. (2018) and Dou et al. (2013)
	Hesperetin	5,7,3′-trihydroxy-4′-methoxy flavanone is a natural flavanone glycoside, abundant in citrus fruits, including lemons, limes, mandarins, and oranges	Alleviate colitis by blocking the intestinal epithelial necroptosis Reduce the levels of MPO, MDA, and pro-inflammatory	Ring is Flavonoids The part that is used is citrus fruits	Anti-inflammatory Anti-oxidative Antitumor	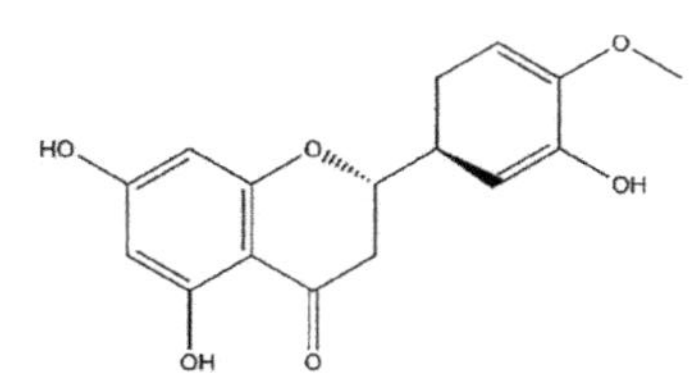	Liu et al. (2016), Shirzad et al. (2017), González-Alfonso et al. (2018), Parhiz et al. (2015), Alu'datt et al. (2017), Chen et al. (2019), Li et al. (2020), Zhang et al. (2020), and Polat et al. (2018)
	Cinnamaldehyde	*Cinnamomum ceylanicum*	To regulate miRNA in colon cancer cells	Flavonoids	Anti-inflammation Anti-viral	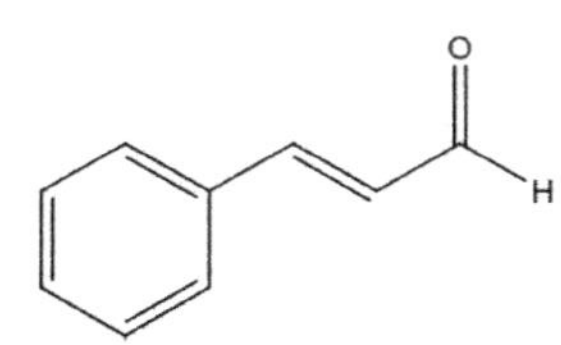	Ma et al. (2020)

(Continued)

TABLE 13.2 (*Continued*)
Description of Herbal Drugs Used in the Treatment Ulcerative Colitis

S. No	Drug	Biological Source and Chemical Constituents	MOA	Ring Present and Part Used	Uses	Structure	References
	Astragaloside	*Astragalus caprinus*	In SW480 cells, miR-134 expression levels were significantly reduced after astragaloside IV treatment, leading to inhibition of tumor cell proliferation and migration	Flavonoids	Anti-inflammatory	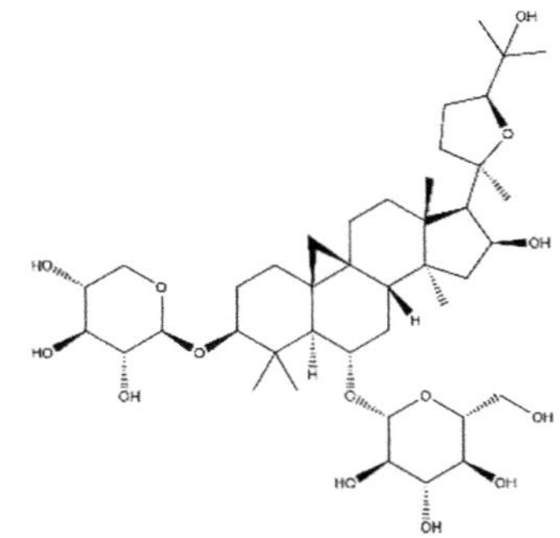	Ye et al. (2017)
	Quercetin	2-(3,4-Dihydroxyphenyl)-5,7-dihydroxy-4H-1-benzopyran-4-one	Alter the intestinal host-microbial relationship through the recovery of the pro-inflammatory, anti-inflammatory, and bactericidal activities of intestinal macrophages, which leads to improvement in colitis	Ring is flavonoids	Pro-inflammatory Anti-inflammatory Bactericidal activities	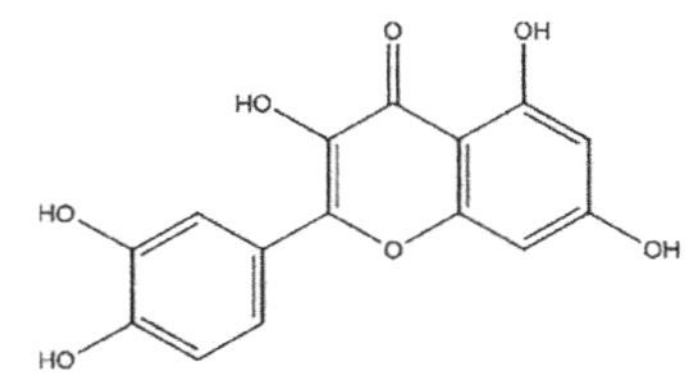	Tao et al. (2013) and Nogata et al. (2006)
	Ursolic acid	3β-hydroxyurs-12-en-28-oic acid	Increase the expression of miR-4500 in HCT116 and HT29 cell lines, attenuating STAT3 phosphorylation and inhibiting the proliferation of colon cancer cells	Terpenoids	Anti-inflammatory	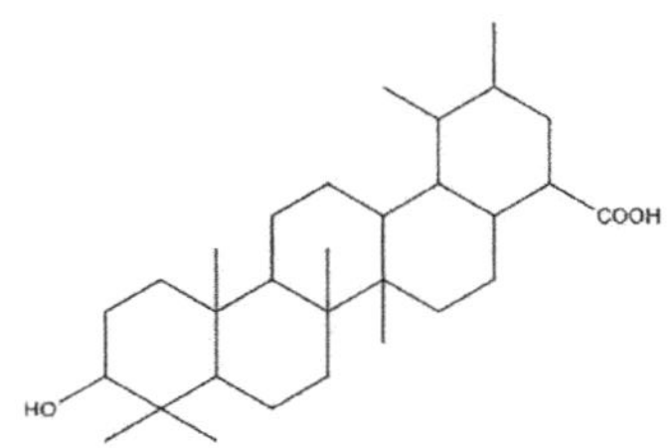	Kim et al. (2018b)

(*Continued*)

TABLE 13.2 (*Continued*)
Description of Herbal Drugs Used in the Treatment Ulcerative Colitis

S. No	Drug	Biological Source and Chemical Constituents	MOA	Ring Present and Part Used	Uses	Structure	References
	Catalpol	*Rehmannia glutinosa*	Promote apoptosis by promoting the expression of miR-200 in HCT116 cells. It can reduce the expression of PI3K, phosphorylated protein kinase B, and Akt and enhance the activity of caspase-3 and caspase-9 to influence cancer cell apoptosis	Terpenoids	Anti-inflammatory	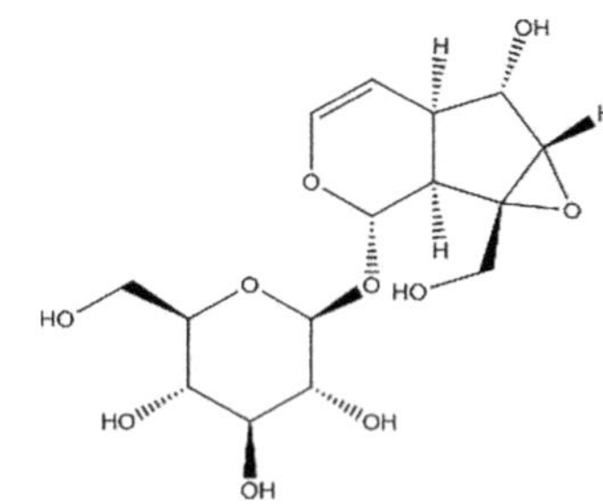	Liu et al. (2017)
	Ginsenosides	(3S,5R,8R,9R,10R, 14R,17S)-17-(2-hydroxy-6-methylhept-5-en-2-yl)-4,4,8,10,14-pentamethyl-2,3,5,6,7,9,11,12,13,15,16,17-dodecahydro-1H-cyclopenta [a]phenanthren-3-ol	Exhibits immunomodulation and anti-inflammation against IBD and results in an increase in BW and a reduction in the colon weight and DAI	Ring is Terpene Part used is Fruit	Anti-inflammatory Anti-oxidation	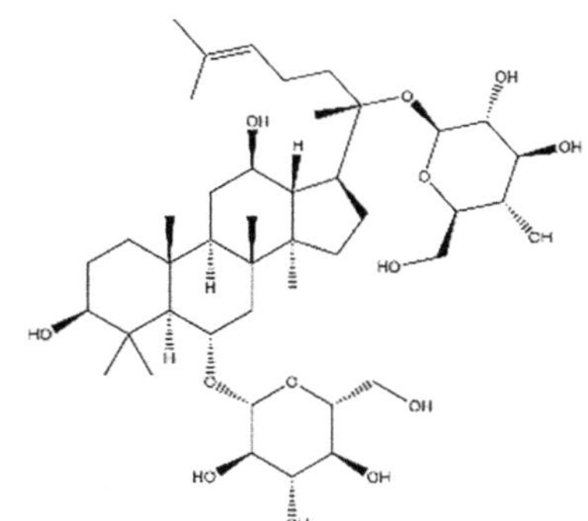	Wang et al. (2018)

(*Continued*)

TABLE 13.2 (*Continued*)
Description of Herbal Drugs Used in the Treatment Ulcerative Colitis

S. No	Drug	Biological Source and Chemical Constituents	MOA	Ring Present and Part Used	Uses	Structure	References
	Glycyrrhizic acid	*Glycyrrhiza glabra*	Reduced IL-6 production and elevated interleukin-10 production in LPS-challenged macrophages	Ring is Terpene Part used id roots and rhizomes	Anti-inflammatory	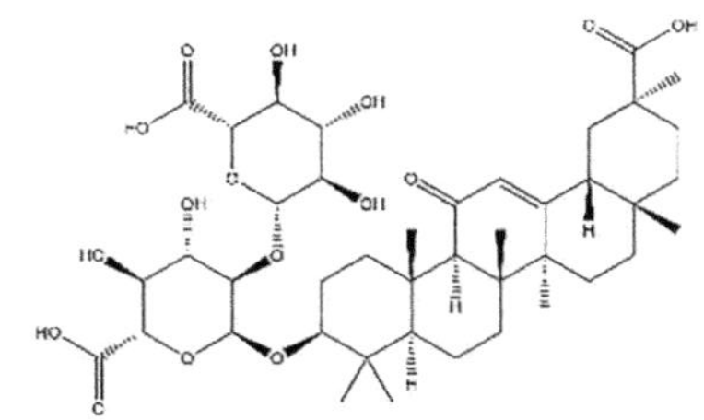	He et al. (2019) and Chrzanowski et al. (2021)
	Astragalus polysaccharide	*Astragalus membranaceus*	Relieves the induction of trinitrobenzene sulfonic acid (TNBS) Mitigation effect could be related to the regulation of TNF-α, IL-1β, and NFATc4 expressions	Ring is Polysaccharides.	Anti-inflammatory	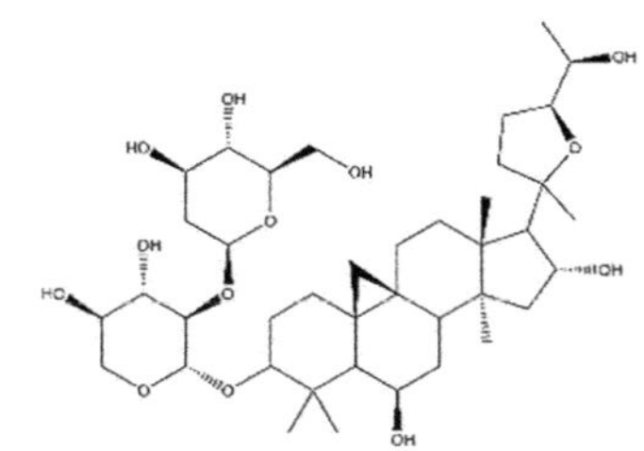	Yang et al. (2014)

(*Continued*)

TABLE 13.2 (*Continued*)
Description of Herbal Drugs Used in the Treatment Ulcerative Colitis

S. No	Drug	Biological Source and Chemical Constituents	MOA	Ring Present and Part Used	Uses	Structure	References
	Aloperine	*Sophora vulgaris*	Enhanced the DSS-mediated colitis severity, reduced the T-cell ratio, and upregulated the Foxp3 levels in mesenteric lymph nodes and spleen	Ring is Alkaloid	Colitis-induced inflammatory	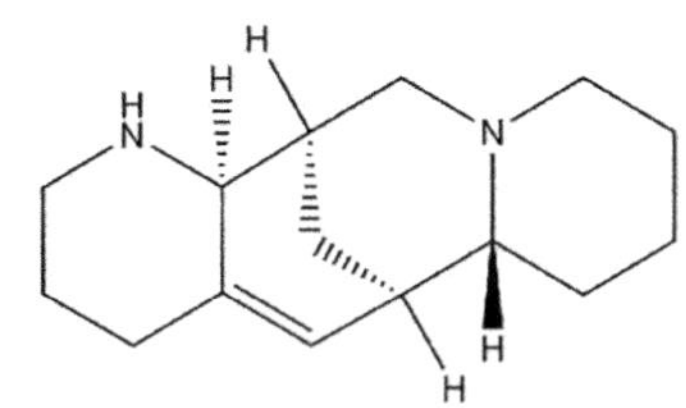	Qiu et al. (2020) and Wang et al. (2020)
	Berberine	Phellodendron and Coptis	Inhibit the expression of IL-1, IL-1β, IL-6, IL-12, TNF-α, TGF-β, and interferon-γ, while upregulating the IL-4 and iL-10 expressions	Ring is Alkaloid	Anti-inflammatory	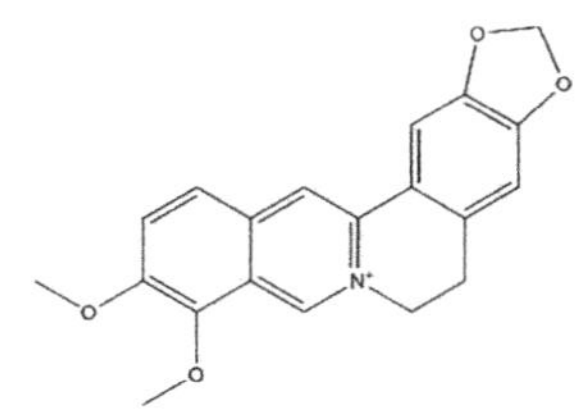	James et al. (2011) and Zhu et al. (2020)
	Caffeine	1,3,7-Trimethylpurine-2,6-dione	Prevents colitis through a decrease in the expression of chitinase 3-like 1 in the intestinal epithelial cells (IECs)	Ring is Alkaloid	Anti-inflammatory	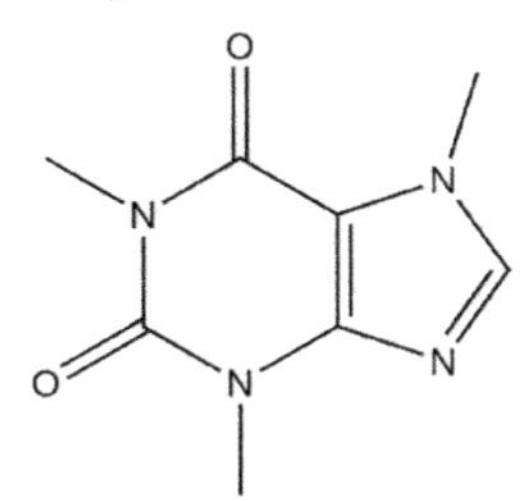	Lee et al. (2014)

(*Continued*)

TABLE 13.2 (*Continued*)
Description of Herbal Drugs Used in the Treatment Ulcerative Colitis

S. No	Drug	Biological Source and Chemical Constituents	MOA	Ring Present and Part Used	Uses	Structure	References
	Theophylline	1,3-dimethyl-7H-purine-2,6-dione	Decreased acetic colitis severity through a decrease in the cytokine levels (IL-6, IL-1b, and TNF-a) and MPO activity in the colon	Ring is Alkaloid	Anti-inflammatory	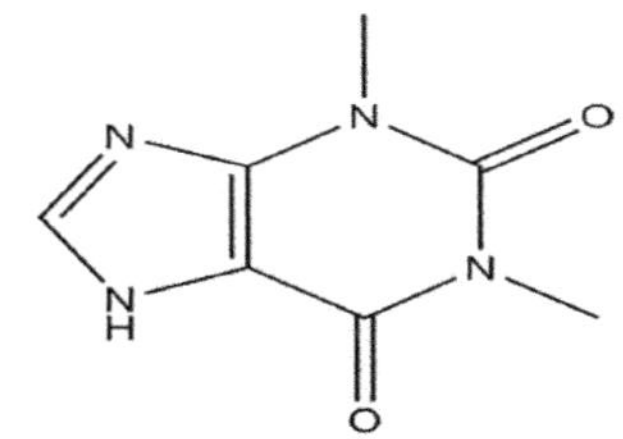	Senchina et al. (2014) and Singleton et al. (2019)
	Palmatine	*Phellodendron chinense*, coptis	Improved the integrity of the mucosa and inhibited epithelial cell apoptosis	Ring is Alkaloid	Anti-inflammatory	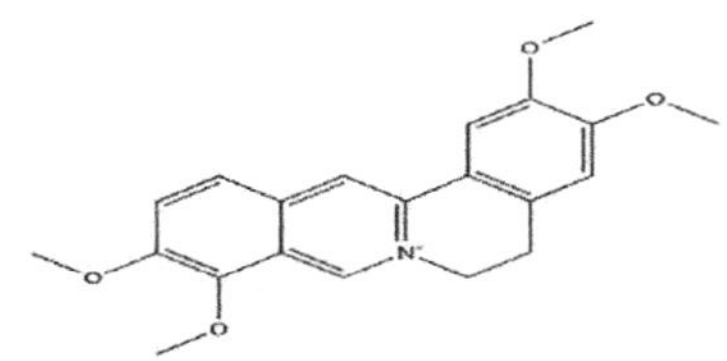	Ryuk et al. (2012)
	Curcumin	1,7-bis(4-hydroxy-3-methoxyphenyl)-1,6-heptadiene-3,5-dione	Decreasing expression of IL-6 and TNF-α in murine macrophage RAW 264.7 cells when induced by lipopolysaccharide (LPS)	Ring is phenolic	Anti-inflammatory Antioxidant	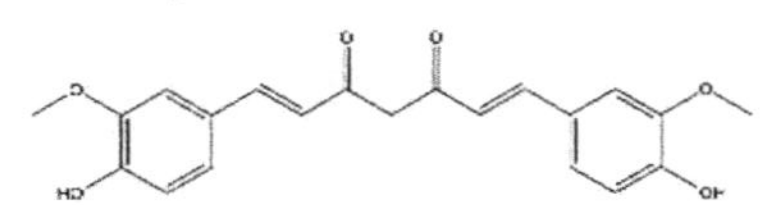	Esmaily et al. (2015), Panahi et al. (2018)

(*Continued*)

TABLE 13.2 (*Continued*)
Description of Herbal Drugs Used in the Treatment Ulcerative Colitis

S. No	Drug	Biological Source and Chemical Constituents	MOA	Ring Present and Part Used	Uses	Structure	References
	Resveratrol	blueberries (Vaccinium spp.), blackberries (Morus spp.)	Inhibit the proliferation of HT29 and HCT116 cells by partially upregulating miR-34c to reduce the protein expression of its target, stem cell factor	Polyphenols	Anti-inflammatory	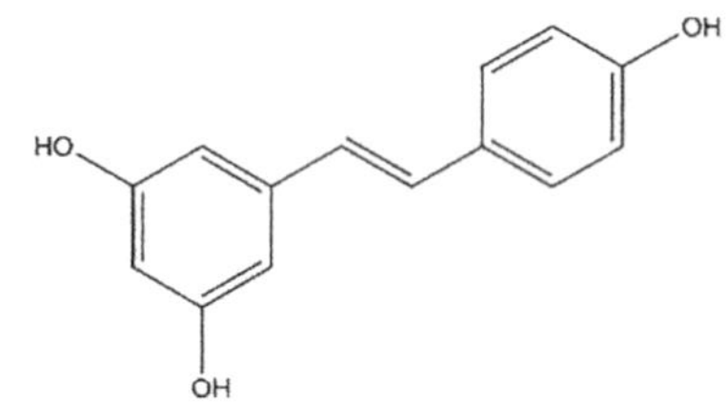	Yang et al. (2015)
	Salvianolic acid B	*Salvia miltiorrhiza* [(R)-3-(3, 4-Dihydroxyphenyl)-2-hydroxypropanoic acid]	Salvianolic acid B could increase the production of acetic acid, propionic acid and valeric acid in colitis mice.	Polyphenols	anti-oxidation, anti-inflammation, anti-coagulation and autophagy	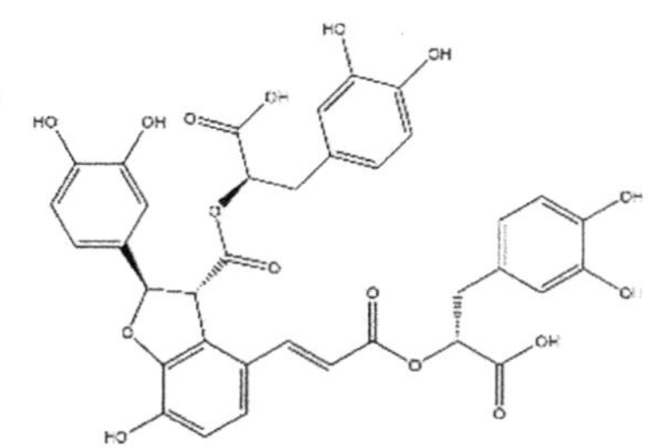	Wu et al. (2018)

(*Continued*)

TABLE 13.2 (*Continued*)
Description of Herbal Drugs Used in the Treatment Ulcerative Colitis

S. No	Drug	Biological Source and Chemical Constituents	MOA	Ring Present and Part Used	Uses	Structure	References
	Supforaphane	1-isothiocyanato-4-(methylsulfinyl)-butane	Genetically regulated, including histone deacetylase inhibition, overall demethylation, and miRNA regulation acting to reverse abnormal changes in gene transcription Downregulate carcinogenic miR-21 in RKO colorectal cancer (CRC) cells, which may have the effect of preventing and delaying CRC	Alkenes	Anti-inflammatory	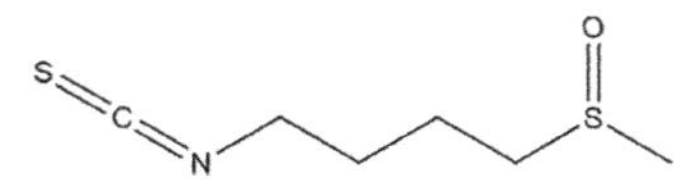	Tortorella et al. (2015)
	β-elemene	(1S,2S,4R)-1-ethenyl-1-methyl-2,4-bis (prop-1-en-2-yl) cyclohexane	Increase the sensitivity of colon cancer cells to 5-fluorouracil by downregulating miR-191 and key kinases such as Wnt3a and b-catenin	Alkenes	Anti-inflammatory	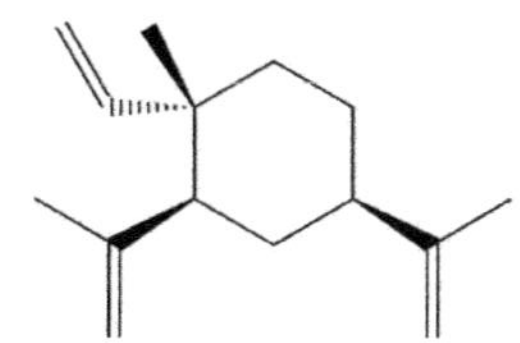	Guo et al. (2018)

TABLE 13.3
Clinical Trial Studies on Herbal Formulations for Ulcerative Colitis

Herbal Formulations	Plant Source	Sample Size	Treatment Method	Study Duration	Control	Trial Design	Remission	Outcomes	Author/Year
Myrrh resin, coffee charcoal and chamomile extract	*Commiphora molmol*	96	Oral	12 months	Mesalazine	Randomized, double-blind, double-dummy, active-controlled trial	72%	Well tolerated and shows a good safety profile	Langhorst et al. (2013)
Pure Curcumin capsule	*Curcuma longa*	97	Oral	2.11 years	Combination of placebo and 5-aminosalycilate	Multicenter randomized, placebo-controlled, double-blind	54%	No apparent adverse effects	Lang et al. (2015)
HMPL-004 (Extract)	*Andrographis paniculata*	120	Oral	8 weeks	mesalazine (mesalamine) granules	Randomized, double-blind, parallel-group study	76%	Adverse reactions observed, limited to allergic reactions such as urticaria	Tang et al. (2011)
Pure Curcumin	*Curcuma longa*	89	Oral	1.3 years	sulfasalazine (SZ) or mesalamine	Randomized, double-blind, multicenter trial, and placebo-controlled clinical trial	95.4%	Safe medication for maintaining remission	Hanai et al. (2006)
curcumin as an adjunct tomesalamine	*Curcuma longa*	380	Oral	8 weeks	placebo and mesalamine	randomized placebo-controlled trials	95%	Superior clinical and/or endoscopic response, minimal adverse events	Chandan et al. (2020)
Aloe vera gel	*Aloe vera*	44	Oral	4 weeks	Placebo	Double-blind, randomized, placebo-controlled trial	50%	Clinical response more often than placebo, reduced the histological disease activity and safe	Langmead et al. (2004)
Dietary Fiber	*Plantagoovata*seeds	105	Oral	12 months	Mesalamine	Multicenter, randomized clinical trial	95%	Effective as mesalamine	Fernández-Bañares et al. (1999)
Wheat grass juice	*Triticum aestivum*	23	Oral	4 weeks	Placebo	Randomized double-blind placebo-controlled trial.	Not given but improved symptomatic condition observed	,No adverse effects	Ben-Arye et al. (2002)

(Continued)

TABLE 13.3 (*Continued*)
Clinical Trial Studies on Herbal Formulations for Ulcerative Colitis

Herbal Formulations	Plant Source	Sample Size	Treatment Method	Study Duration	Control	Trial Design	Remission	Outcomes	Author/Year
Green tea - epigallocatechin-3-gallate (EGCG)	*Camellia sinensis*	20	Oral	56 days	Placebo	Double-blinded, placebo-controlled pilot study	53.5%	lack of significant side effects	Dryden et al. (2013)
Berberine and Berberine + mesalamine	*Coptidis rhizome*	16	Oral	3 months	Placebo	Double-blind placebo-controlled phase-I trial	40%	Combination of berberine and mesalamine enhance mesalamine's anti-inflammatory effects	Xu et al. (2020)
Fufangkushen colon-coated capsule	*Sophorae flavescentis, Sanguisorba officinalis* L., *Indigo naturalis, Bletilla striata* and *Glycyrrhiza uralensis*	320	Oral	8 weeks	Huidi (mesalazine enteric-coated tablets)	Double-blinded, randomized clinical trial	41.5%	equally effective and safe compared with mesalazine	Gong et al. (2012)
Germinated barley foodstuff	*Hordeum vulgare*	21	Oral	24 weeks	5-aminosalicylate	Multicenter open trial	Not given but provide clinical benefit over long-term administration.	Safely administered	Kanauchi et al. (2003)
Anthocyanin-rich bilberry preparation	*Vaccinium myrtillus*	13	Oral	9 weeks	Placebo	Open pilot study	63.4%	therapeutic potential of bilberries and no serious adverse events	Biedermann et al. (2013)
Gum resin	*Boswellia serrata*	30	Oral	6 weeks	Sulfasalazine	open, non-randomized monocentric clinical trial	80%	minimal side effects	Gupta et al. (2001)
Traditional Chinese medicine contains ligands	*Indigo naturalis*	86	Oral	8 weeks	Placebo	multicenter, double-blind trial	55%	clinical efficacy of was excellent,	Naganuma et al. (2018)

13.6 CONCLUSION

To avoid relapse, minimize the risk of colorectal cancer, and enhance quality of life, persons with UC may need to take medicine for the rest of their lives. Nonadherence affects patients at all phases of UC, but those in symptomatic remission are more vulnerable, taking less than 70% of their recommended medications. A variety of acute and chronic GI illnesses, including UC, can be treated using herbal formulations. The herbs stated above are a small example of what is often used by UC patients around the world. Physicians should speak openly about this type of treatment with their patients and provide evidence-based advice on how to use it. Simultaneously, major clinical double-blind trials evaluating the most widely utilized alternative medicines are required. These herbal remedies have frequently been evaluated in their traditional setting and have also been discovered to be effective in novel and intriguing ways, opening up new paths for the treatment of pathologic diseases. Most herbal remedies are subjected to the same level of scrutiny as pharmaceutical drugs, and there are several examples of successful biochemical, animal model, and human-controlled trials in the literature. Consequently, the manufacturing, marketing, and prescribing of herbal medications has reached new heights, and this trend is projected to continue. When utilized as therapeutic agents to treat illness and disease, many herbal medications are quite effective. Herbal medicine is thus one of the world's great medicinal systems, with a long history that dates back to the third century BC.

REFERENCES

Al-Rejaie, S. S., Abuohashish, H. M., Al-Enazi, M. M., Al-Assaf, A. H., Parmar, M. Y. & Ahmed, M. M. 2013. Protective effect of naringenin on acetic acid-induced ulcerative colitis in rats. *World J Gastroenterol*, 19, 5633–5644.

Alam, M. A., Subhan, N., Rahman, M. M., Uddin, S. J., Reza, H. M. & Sarker, S. D. 2014. Effect of citrus flavonoids, naringin and naringenin, on metabolic syndrome and their mechanisms of action. *Adv Nutr,* 5, 404–17.

Alu'datt, M. H., Rababah, T., Alhamad, M. N., Al-Mahasneh, M. A., Ereifej, K., Al-Karaki, G., Al-Duais, M., Andrade, J. E., Tranchant, C. C., Kubow, S. & Ghozlan, K. A. 2017. Profiles of free and bound phenolics extracted from citrus fruits and their roles in biological systems: content, and antioxidant, anti-diabetic and anti-hypertensive properties. *Food Funct*, 8, 3187–3197.

Auyeung, K. K., Han, Q. B. & Ko, J. K. 2016. *Astragalus membranaceus*: A review of its protection against inflammation and gastrointestinal cancers. *Am J Chin Med*, 44, 1–22.

Bellavite, P. 2023. Neuroprotective potentials of flavonoids: Experimental studies and mechanisms of action. *Antioxidants*, 12, 280.

Ben-Arye, E., Goldin, E., Wengrower, D., Stamper, A., Kohn, R. & Berry, E. 2002. Wheat grass juice in the treatment of active distal ulcerative colitis: A randomized double-blind placebo-controlled trial. *Scand J Gastroenterol*, 37, 444–449.

Biedermann, L., Mwinyi, J., Scharl, M., Frei, P., Zeitz, J., Kullak-Ublick, G. A., Vavricka, S. R., Fried, M., Weber, A., Humpf, H.-U., Peschke, S., Jetter, A., Krammer, G. & Rogler, G. 2013. Bilberry ingestion improves disease activity in mild to moderate ulcerative colitis — An open pilot study. *Journal of Crohn's and Colitis*, 7, 271–279.

Cai, Z., Wang, S. & Li, J. 2021. Treatment of inflammatory bowel disease: A comprehensive review. *Front Med (Lausanne)*, 8, 765474.

Chandan, S., Mohan, B. P., Chandan, O. C., Ahmad, R., Challa, A., Tummala, H., Singh, S., Dhawan, P., Ponnada, S., Singh, A. B. & Adler, D. G. 2020. Curcumin use in ulcerative colitis: Is it ready for prime time? A systematic review and meta-analysis of clinical trials. *Ann Gastroenterol*, 33, 53–58.

Chen, X., Wei, W., Li, Y., Huang, J. & Ci, X. 2019. Hesperetin relieves cisplatin-induced acute kidney injury by mitigating oxidative stress, inflammation and apoptosis. *Chem Biol Interact*, 308, 269–278.

Chrzanowski, J., Chrzanowska, A. & Graboń, W. 2021. Glycyrrhizin: An old weapon against a novel coronavirus. *Phytother Res*, 35, 629–636.

Colombel, J. F., Sands, B. E., Rutgeerts, P., Sandborn, W., Danese, S., D' Haens, G., Panaccione, R., Loftus, E. V., Jr., Sankoh, S., Fox, I., Parikh, A., Milch, C., Abhyankar, B. & Feagan, B. G. 2017. The safety of vedolizumab for ulcerative colitis and Crohn's disease. *Gut*, 66, 839–851.

Comalada, M., Camuesco, D., Sierra, S., Ballester, I., Xaus, J., Gálvez, J. & Zarzuelo, A. 2005. In vivo quercitrin anti-inflammatory effect involves release of quercetin, which inhibits inflammation through down-regulation of the NF-κB pathway. *Eur J Immunol*, 35, 584–592.

Den Hartogh, D. J. & Tsiani, E. 2019. Antidiabetic properties of naringenin: A citrus fruit polyphenol. *Biomolecules*, 9, 99.

Derosa, G., Maffioli, P., Simental-Mendía, L. E., Bo, S. & Sahebkar, A. 2016. Effect of curcumin on circulating interleukin-6 concentrations: A systematic review and meta-analysis of randomized controlled trials. *Pharmacol Res*, 111, 394–404.

Dodda, D., Chhajed, R., Mishra, J. & Padhy, M. 2014. Targeting oxidative stress attenuates trinitrobenzene sulphonic acid induced inflammatory bowel disease like symptoms in rats: Role of quercetin. *Indian J Pharmacol*, 46, 286–291.

Dou, W., Zhang, J., Sun, A., Zhang, E., Ding, L., Mukherjee, S., Wei, X., Chou, G., Wang, Z. T. & Mani, S. 2013. Protective effect of naringenin against experimental colitis via suppression of toll-like receptor 4/ NF-κB signalling. *Br J Nutr*, 110, 599–608.

Dryden, G. W., Lam, A., Beatty, K., Qazzaz, H. H. & Mcclain, C. J. 2013. A pilot study to evaluate the safety and efficacy of an oral dose of (-)-epigallocatechin-3-gallate-rich polyphenon E in patients with mild to moderate ulcerative colitis. *Inflamm Bowel Dis*, 19, 1904–1912.

Erlund, I., Meririnne, E., Alfthan, G. & Aro, A. 2001. Plasma kinetics and urinary excretion of the flavanones naringenin and hesperetin in humans after ingestion of orange juice and grapefruit juice. *J Nutr*, 131, 235–241.

Esmaily, H., Sahebkar, A., Iranshahi, M., Ganjali, S., Mohammadi, A., Ferns, G. & Ghayour-Mobarhan, M. 2015. An investigation of the effects of curcumin on anxiety and depression in obese individuals: A randomized controlled trial. *Chin J Integr Med*, 21, 332–338.

Fernández-Bañares, F., Hinojosa, J., Sánchez-Lombraña, J. L., Navarro, E., Martínez-Salmerón, J. F., García-Pugés, A., González-Huix, F., Riera, J., González-Lara, V., Domínguez-Abascal, F., Giné, J. J., Moles, J., Gomollón, F. & Gassull, M. A. 1999. Randomized clinical trial of Plantago ovata seeds (dietary fiber) as compared with mesalamine in maintaining remission in ulcerative colitis. Spanish Group for the Study of Crohn's Disease and Ulcerative Colitis (GETECCU). *Am J Gastroenterol*, 94, 427–433.

Feuerstein, J. D. & Cheifetz, A. S. 2014. Ulcerative colitis: Epidemiology, diagnosis, and management. *Mayo Clin Proc*, 89, 1553–1563.

Freeman, K., Ryan, R., Parsons, N., Taylor-Phillips, S., Willis, B. H. & Clarke, A. 2021. The incidence and prevalence of inflammatory bowel disease in UK primary care: A retrospective cohort study of the IQVIA medical research database. *BMC Gastroenterol*, 21, 139.

Fu, X., Sun, F., Wang, F., Zhang, J., Zheng, B., Zhong, J., Yue, T., Zheng, X., Xu, J. F. & Wang, C. Y. 2017. Aloperine protects mice against DSS-induced colitis by PP2A-mediated PI3K/AKT/mTOR signaling suppression. *Mediators Inflamm*, 2017, 5706152.

Gao, Y. J., Zhu, F., Qian, J. M. & Dai, J. Y. 2016. Therapeutic and immunoregulatory effect of GATA-binding protein-3/T-box expressed in T-cells ratio of astragalus polysaccharides on 2,4,6-trinitrobenzene sulfonic acid-induced colitis in rats. *Chin J Integr Med*, 22, 918–924.

Geier, M. S., Butler, R. N. & Howarth, G. S. 2007. Inflammatory bowel disease: Current insights into pathogenesis and new therapeutic options; probiotics, prebiotics and synbiotics. *Int J Food Microbiol*, 115, 1–11.

Ghandadi, M. & Sahebkar, A. 2017. Curcumin: An effective inhibitor of interleukin-6. *Curr Pharm Des*, 23, 921–931.

Ghasemi-Pirbaluti, M., Motaghi, E., Najafi, A. & Hosseini, M. J. 2017. The effect of theophylline on acetic acid induced ulcerative colitis in rats. *Biomed Pharmacother*, 90, 153–159.

Gong, Y., Zha, Q., Li, L., Liu, Y., Yang, B., Liu, L., Lu, A., Lin, Y. & Jiang, M. 2012. Efficacy and safety of Fufangkushen colon-coated capsule in the treatment of ulcerative colitis compared with mesalazine: A double-blinded and randomized study. *J Ethnopharmacol*, 141, 592–598.

González-Alfonso, J. L., Míguez, N., Padilla, J. D., Leemans, L., Poveda, A., Jimnez-Barbero, J., Ballesteros, A. O., Sandoval, G. & Plou, F. J. 2018. Optimization of regioselective α-glucosylation of hesperetin catalyzed by cyclodextrin glucanotransferase. *Molecules*, 23, 2285.

Gupta, I., Parihar, A., Malhotra, P., Gupta, S., Lüdtke, R., Safayhi, H. & Ammon, H. P. 2001. Effects of gum resin of Boswellia serrata in patients with chronic colitis. *Planta Med*, 67, 391–395.

Hanai, H., Iida, T., Takeuchi, K., Watanabe, F., Maruyama, Y., Andoh, A., Tsujikawa, T., Fujiyama, Y., Mitsuyama, K., Sata, M., Yamada, M., Iwaoka, Y., Kanke, K., Hiraishi, H., Hirayama, K., Arai, H., Yoshii, S., Uchijima, M., Nagata, T. & Koide, Y. 2006. Curcumin maintenance therapy for ulcerative colitis: Randomized, multicenter, double-blind, placebo-controlled trial. *Clin Gastroenterol Hepatol*, 4, 1502–1506.

He, H., Wei, D., Liu, H., Zhu, C., Lu, Y., Ke, Z., Jiang, S. & Huang, J. 2019. Glycyrrhizin protects against sodium iodate-induced RPE and retinal injury though activation of AKT and NRF2/HO-1 pathway. *J Cell Mol Med*, 23, 3495–3504.

Head, K. A. & Jurenka, J. S. 2003. Inflammatory bowel disease Part 1: Ulcerative colitis--pathophysiology and conventional and alternative treatment options. *Altern Med Rev*, 8, 247–283.

Hendrickson, B. A., Gokhale, R. & Cho, J. H. 2002. Clinical aspects and pathophysiology of inflammatory bowel disease. *Clin Microbiol Rev*, 15, 79–94.

James, M. A., Fu, H., Liu, Y., Chen, D. R. & You, M. 2011. Dietary administration of berberine or *Phellodendron amurense* extract inhibits cell cycle progression and lung tumorigenesis. *Mol Carcinog*, 50, 1–7.

Jin, M., Zhao, K., Huang, Q. & Shang, P. 2014. Structural features and biological activities of the polysaccharides from *Astragalus membranaceus*. *Int J Biol Macromol*, 64, 257–266.

Jing, W., Dong, S., Luo, X., Liu, J., Wei, B., Du, W., Yang, L., Luo, H., Wang, Y., Wang, S. & Lu, H. 2021. Berberine improves colitis by triggering AhR activation by microbial tryptophan catabolites. *Pharmacol Res*, 164, 105358.

Ju, S., Ge, Y., Li, P., Tian, X., Wang, H., Zheng, X. & Ju, S. 2018. Dietary quercetin ameliorates experimental colitis in mouse by remodeling the function of colonic macrophages via a heme oxygenase-1-dependent pathway. *Cell Cycle*, 17, 53–63.

Kanauchi, O., Mitsuyama, K., Araki, Y. & Andoh, A. 2003. Modification of intestinal flora in the treatment of inflammatory bowel disease. *Curr Pharm Des*, 9, 333–346.

Kayal, M. & Shah, S. 2020. Ulcerative colitis: Current and emerging treatment strategies. *J Clin Med*, 9, 94.

Kim, J. K., Kim, J. Y., Jang, S. E., Choi, M. S., Jang, H. M., Yoo, H. H. & Kim, D. H. 2018a. Fermented red ginseng alleviates cyclophosphamide-induced immunosuppression and 2,4,6-trinitrobenzenesulfonic acid-induced colitis in mice by regulating macrophage activation and T cell differentiation. *Am J Chin Med*, 46, 1879–1897.

Kim, K., Shin, E. A., Jung, J. H., Park, J. E., Kim, D. S., Shim, B. S. & Kim, S. H. 2018b. Ursolic acid induces apoptosis in colorectal cancer cells partially via upregulation of microRNA-4500 and inhibition of JAK2/STAT3 phosphorylation. *Int J Mol Sci*, 20, 114.

Lang, A., Salomon, N., Wu, J. C. Y., Kopylov, U., Lahat, A., Har-Noy, O., Ching, J. Y. L., Cheong, P. K., Avidan, B., Gamus, D., Kaimakliotis, I., Eliakim, R., Ng, S. C. & Ben-Horin, S. 2015. Curcumin in combination with mesalamine induces remission in patients with mild-to-moderate ulcerative colitis in a randomized controlled trial. *Clin Gastroenterol Hepatol*, 13, 1444–1449.e1.

Langhorst, J., Varnhagen, I., Schneider, S., Albrecht, U., Rueffer, A., Stange, R., Michalsen, A. & Dobos, G. 2013. Randomised clinical trial: A herbal preparation of myrrh, chamomile and coffee charcoal compared with mesalazine in maintaining remission in ulcerative colitis: A double-blind, double-dummy study. *Aliment Pharmacol Ther*, 38, 490–500.

Langmead, L., Feakins, R. M., Goldthorpe, S., Holt, H., Tsironi, E., De Silva, A., Jewell, D. P. & Rampton, D. S. 2004. Randomized, double-blind, placebo-controlled trial of oral aloe vera gel for active ulcerative colitis. *Aliment Pharmacol Ther*, 19, 739–747.

Lee, I. A., Low, D., Kamba, A., Llado, V. & Mizoguchi, E. 2014. Oral caffeine administration ameliorates acute colitis by suppressing chitinase 3-like 1 expression in intestinal epithelial cells. *J Gastroenterol*, 49, 1206–1216.

Lee, S.-Y., Jeong, J.-J., Eun, S.-H. & Kim, D.-H. 2015. Anti-inflammatory effects of ginsenoside Rg1 and its metabolites ginsenoside Rh1 and 20(S)-protopanaxatriol in mice with TNBS-induced colitis. *Eur J Pharmacol*, 762, 333–343.

Li, H., Fan, C., Lu, H., Feng, C., He, P., Yang, X., Xiang, C., Zuo, J. & Tang, W. 2020. Protective role of berberine on ulcerative colitis through modulating enteric glial cells–intestinal epithelial cells–immune cells interactions. *Acta Pharmaceutica Sinica B*, 10, 447–461.

Li, J., Zhong, W., Wang, W., Hu, S., Yuan, J., Zhang, B., Hu, T. & Song, G. 2014. Ginsenoside metabolite compound K promotes recovery of dextran sulfate sodium-induced colitis and inhibits inflammatory responses by suppressing NF-κB activation. *PLoS One*, 9, e87810.

Linskens, R. K., Huijsdens, X. W., Savelkoul, P. H., Vandenbroucke-Grauls, C. M. & Meuwissen, S. G. 2001. The bacterial flora in inflammatory bowel disease: Current insights in pathogenesis and the influence of antibiotics and probiotics. *Scand J Gastroenterol Suppl*, 234, 29–40.

Liu, D., Wu, J., Xie, H., Liu, M., Takau, I., Zhang, H., Xiong, Y. & Xia, C. 2016. Inhibitory effect of hesperetin and naringenin on human UDP-glucuronosyltransferase enzymes: Implications for herb-drug interactions. *Biol Pharm Bull*, 39, 2052–2059.

Liu, L., Gao, H., Wang, H., Zhang, Y., Xu, W., Lin, S., Wang, H., Wu, Q. & Guo, J. 2017. Catalpol promotes cellular apoptosis in human HCT116 colorectal cancer cells via microRNA-200 and the downregulation of PI3K-Akt signaling pathway. *Oncol Lett*, 14, 3741–3747.

Liu, Y., Xiang, J., Liu, M., Wang, S., Lee, R. J. & Ding, H. 2011. Protective effects of glycyrrhizic acid by rectal treatment on a TNBS-induced rat colitis model. *J Pharm Pharmacol*, 63, 439–446.

Ma, K. L., Han, Z. J., Pan, M., Chen, M. L., Ge, Y. Z., Shao, J., Wu, D. Q., Wang, T. M., Yan, G. M. & Wang, C. Z. 2020. Therapeutic effect of cinnamaldehyde on ulcerative colitis in mice induced by dextran sulfate sodium with Candida albicans colonization and its effect on dectin-1/TLRs/NF-κB signaling pathway. *Zhongguo Zhong Yao Za Zhi*, 45, 3211–3219.

Mai, C. T., Wu, M. M., Wang, C. L., Su, Z. R., Cheng, Y. Y. & Zhang, X. J. 2019. Palmatine attenuated dextran sulfate sodium (DSS)-induced colitis via promoting mitophagy-mediated NLRP3 inflammasome inactivation. *Mol Immunol*, 105, 76–85.

Mirzaei, H., Naseri, G., Rezaee, R., Mohammadi, M., Banikazemi, Z., Mirzaei, H. R., Salehi, H., Peyvandi, M., Pawelek, J. M. & Sahebkar, A. 2016. Curcumin: A new candidate for melanoma therapy? *Int J Cancer*, 139, 1683–1695.

Momtazi, A. A., Shahabipour, F., Khatibi, S., Johnston, T. P., Pirro, M. & Sahebkar, A. 2016. Curcumin as a microRNA regulator in cancer: A review. *Rev Physiol Biochem Pharmacol*, 171, 1–38.

Naganuma, M., Sugimoto, S., Mitsuyama, K., Kobayashi, T., Yoshimura, N., Ohi, H., Tanaka, S., Andoh, A., Ohmiya, N., Saigusa, K., Yamamoto, T., Morohoshi, Y., Ichikawa, H., Matsuoka, K., Hisamatsu, T., Watanabe, K., Mizuno, S., Suda, W., Hattori, M., Fukuda, S., Hirayama, A., Abe, T., Watanabe, M., Hibi, T., Suzuki, Y. & Kanai, T. 2018. Efficacy of indigo naturalis in a multicenter randomized controlled trial of patients with ulcerative colitis. *Gastroenterology*, 154, 935–947.

Nogata, Y., Sakamoto, K., Shiratsuchi, H., Ishii, T., Yano, M. & Ohta, H. 2006. Flavonoid composition of fruit tissues of citrus species. *Biosci Biotechnol Biochem*, 70, 178–192.

Ordás, I., Eckmann, L., Talamini, M., Baumgart, D. C. & Sandborn, W. J. 2012. Ulcerative colitis. *Lancet*, 380, 1606–1619.

Panahi, Y., Ahmadi, Y., Teymouri, M., Johnston, T. P. & Sahebkar, A. 2018. Curcumin as a potential candidate for treating hyperlipidemia: A review of cellular and metabolic mechanisms. *J Cell Physiol*, 233, 141–152.

Panahi, Y., Ghanei, M., Bashiri, S., Hajihashemi, A. & Sahebkar, A. 2015. Short-term curcuminoid supplementation for chronic pulmonary complications due to sulfur mustard intoxication: Positive results of a randomized double-blind placebo-controlled Trial. *Drug Res (Stuttg)*, 65, 567–573.

Parhiz, H., Roohbakhsh, A., Soltani, F., Rezaee, R. & Iranshahi, M. 2015. Antioxidant and anti-inflammatory properties of the citrus flavonoids hesperidin and hesperetin: An updated review of their molecular mechanisms and experimental models. *Phytother Res*, 29, 323–331.

Pastorelli, L., Pizarro, T. T., Cominelli, F. & Vecchi, M. 2009. Emerging drugs for the treatment of ulcerative colitis. *Expert Opin Emerg Drugs*, 14, 505–521.

Polat, F. R., Karaboga, I., Polat, M. S., Erboga, Z., Yilmaz, A. & Güzel, S. 2018. Effect of hesperetin on inflammatory and oxidative status in trinitrobenzene sulfonic acid-induced experimental colitis model. *Cell Mol Biol (Noisy-le-grand)*, 64, 58–65.

Qiu, M., Liu, J., Feng, P., Su, Y., Guo, R., Shi, F., Wang, S. & Zhao, B. 2020. Cytochrome P450s regulates aloperine-induced pathological changes in mouse liver and kidney. *Res Vet Sci*, 132, 97–100.

Rezaie, A., Parker, R. D. & Abdollahi, M. 2007. Oxidative stress and pathogenesis of inflammatory bowel disease: An epiphenomenon or the cause? *Dig Dis Sci*, 52, 2015–2021.

Rivera, A. P., Flores Monar, G. V., Islam, H., Puttagunta, S. M., Islam, R., Kundu, S., Jha, S. B. & Sange, I. 2022. Ulcerative colitis-induced colorectal carcinoma: A deleterious concatenation. *Cureus*, 14, e22636.

Ryuk, J. A., Zheng, M. S., Lee, M. Y., Seo, C. S., Li, Y., Lee, S. H., Moon, D. C., Lee, H. W., Lee, J. H., Park, J. Y., Son, J. K. & Ko, B. S. 2012. Discrimination of *Phellodendron amurense* and *P. chinense* based on DNA analysis and the simultaneous analysis of alkaloids. *Arch Pharm Res*, 35, 1045–1054.

Sands, B. E., Sandborn, W. J., Panaccione, R., O' Brien, C. D., Zhang, H., Johanns, J., Adedokun, O. J., Li, K., Peyrin-Biroulet, L., Van Assche, G., Danese, S., Targan, S., Abreu, M. T., Hisamatsu, T., Szapary, P. & Marano, C. 2019. Ustekinumab as induction and maintenance therapy for ulcerative colitis. *N Engl J Med*, 381, 1201–1214.

Senchina, D. S., Hallam, J. E., Kohut, M. L., Nguyen, N. A. & Perera, M. A. 2014. Alkaloids and athlete immune function: caffeine, theophylline, gingerol, ephedrine, and their congeners. *Exerc Immunol Rev*, 20, 68–93.

Shi, Y. J., Hu, S. J., Zhao, Q. Q., Liu, X. S., Liu, C. & Wang, H. 2019. Toll-like receptor 4 (TLR4) deficiency aggravates dextran sulfate sodium (DSS)-induced intestinal injury by down-regulating IL6, CCL2 and CSF3. *Ann Transl Med*, 7, 713.

Shirzad, M., Heidarian, E., Beshkar, P. & Gholami-Arjenaki, M. 2017. Biological effects of hesperetin on interleukin-6/phosphorylated signal transducer and activator of transcription 3 pathway signaling in prostate cancer PC3 cells. *Pharmacognosy Res*, 9, 188–194.

Singleton, E. L., Le, N. & Ness, G. L. 2019. Theophylline and caffeine as alternatives during an aminophylline shortage. *Ann Pharmacother*, 53, 316–320.

Stokkers, P. C. & Hommes, D. W. 2004. New cytokine therapeutics for inflammatory bowel disease. *Cytokine*, 28, 167–173.

Stronati, L., Palone, F., Negroni, A., Colantoni, E., Mancuso, A. B., Cucchiara, S., Cesi, V., Isoldi, S. & Vitali, R. 2019. Dipotassium glycyrrhizate improves intestinal mucosal healing by modulating extracellular matrix remodeling genes and restoring epithelial barrier functions. *Front Immunol*, 10, 939.

Sun, D., Zhuang, X., Xiang, X., Liu, Y., Zhang, S., Liu, C., Barnes, S., Grizzle, W., Miller, D. & Zhang, H. G. 2010. A novel nanoparticle drug delivery system: The anti-inflammatory activity of curcumin is enhanced when encapsulated in exosomes. *Mol Ther*, 18, 1606–1614.

Szigethy, E., Mclafferty, L. & Goyal, A. 2010. Inflammatory bowel disease. *Child Adolesc Psychiatr Clin N Am*, 19, 301–318.

Tang, T., Targan, S. R., Li, Z.-S., Xu, C., Byers, V. S. & Sandborn, W. J. 2011. Randomised clinical trial: Herbal extract HMPL-004 in active ulcerative colitis: A double-blind comparison with sustained release mesalazine. *Aliment Pharmacol Ther*, 33, 194–202.

Tao, F., Qian, C., Guo, W., Luo, Q., Xu, Q. & Sun, Y. 2013. Inhibition of TH1/TH17 responses via suppression of STAT1 and STAT3 activation contributes to the amelioration of murine experimental colitis by a natural flavonoid glucoside icariin. *Biochem Pharmacol*, 85, 798–807.

Tortorella, S. M., Royce, S. G., Licciardi, P. V. & Karagiannis, T. C. 2015. Dietary sulforaphane in cancer chemoprevention: The role of epigenetic regulation and HDAC inhibition. *Antioxid Redox Signal*, 22, 1382–1424.

Wang, C. Z., Yao, H., Zhang, C. F., Chen, L., Wan, J. Y., Huang, W. H., Zeng, J., Zhang, Q. H., Liu, Z., Yuan, J., Bi, Y., Sava-Segal, C., Du, W., Xu, M. & Yuan, C. S. 2018. American ginseng microbial metabolites attenuate DSS-induced colitis and abdominal pain. *Int Immunopharmacol*, 64, 246–251.

Wang, R., Deng, X., Gao, Q., Wu, X., Han, L., Gao, X., Zhao, S., Chen, W., Zhou, R., Li, Z. & Bai, C. 2020. Sophora alopecuroides L.: An ethnopharmacological, phytochemical, and pharmacological review. *J Ethnopharmacol*, 248, 112172.

Weber, C. R. & Turner, J. R. 2007. Inflammatory bowel disease: Is it really just another break in the wall? *Gut*, 56, 6–8.

Wu, X., Wang, L., Tang, L., Wang, L., Cao, S., Wu, Q., Zhang, Z. & Li, L. 2018. Salvianolic acid B alters the gut microbiota and mitigates colitis severity and associated inflammation. *J Funct Foods*, 46, 312–319.

Xu, L., Zhang, Y., Xue, X., Liu, J., Li, Z.-S., Yang, G.-Y., Song, Y., Pan, Y., Ma, Y., Hu, S., Wen, A., Jia, Y., Rodriguez, L. M., Tull, M. B., Benante, K., Khan, S. A., Cao, Y., Jovanovic, B., Richmond, E., Umar, A., Bergan, R. & Wu, K. 2020. A phase I trial of Berberine in Chinese with ulcerative colitis. *Cancer Prev Res*, 13, 117–126.

Yang, M., Lin, H. B., Gong, S., Chen, P. Y., Geng, L. L., Zeng, Y. M. & Li, D. Y. 2014. Effect of astragalus polysaccharides on expression of TNF-α, IL-1β and NFATc4 in a rat model of experimental colitis. *Cytokine*, 70, 81–86.

Yang, N., Liang, G., Lin, J., Zhang, S., Lin, Q., Ji, X., Chen, H., Li, N. & Jin, S. 2020. Ginsenoside Rd therapy improves histological and functional recovery in a rat model of inflammatory bowel disease. *Phytother Res*, 34, 3019–3028.

Yang, S., Li, W., Sun, H., Wu, B., Ji, F., Sun, T., Chang, H., Shen, P., Wang, Y. & Zhou, D. 2015. Resveratrol elicits anti-colorectal cancer effect by activating miR-34c-KITLG in vitro and in vivo. *BMC Cancer*, 15, 969.

Yang, X. L., Guo, T. K., Wang, Y. H., Gao, M. T., Qin, H. & Wu, Y. J. 2012a. Therapeutic effect of ginsenoside Rd in rats with TNBS-induced recurrent ulcerative colitis. *Arch Pharm Res*, 35, 1231–1239.

Yang, X. L., Guo, T. K., Wang, Y. H., Huang, Y. H., Liu, X., Wang, X. X., Li, W., Zhao, X., Wang, L. P., Yan, S., Wu, D. & Wu, Y. J. 2012b. Ginsenoside Rd attenuates the inflammatory response via modulating p38 and JNK signaling pathways in rats with TNBS-induced relapsing colitis. *Int Immunopharmacol*, 12, 408–414.

Ye, Q., Su, L., Chen, D., Zheng, W. & Liu, Y. 2017. Astragaloside IV induced miR-134 expression reduces EMT and increases chemotherapeutic sensitivity by suppressing CREB1 signaling in colorectal cancer cell line SW-480. *Cell Physiol Biochem*, 43, 1617–1626.

Yeshi, K., Ruscher, R., Hunter, L., Daly, N. L., Loukas, A. & Wangchuk, P. 2020. Revisiting inflammatory bowel disease: Pathology, treatments, challenges and emerging therapeutics including drug leads from natural products. *J Clin Med*, 9, 1273.

Zaidun, N. H., Thent, Z. C. & Latiff, A. A. 2018. Combating oxidative stress disorders with citrus flavonoid: Naringenin. *Life Sci*, 208, 111–122.

Zhang, H. Q., Zhu, Z. H., Qian, D. W., Weng, Z. B., Guo, S., Duan, J. A., Lei, Z. H. & Li, A. P. 2016. Analysis and evaluation of alkaloids and flavonoids in flower of Sophora flavescens from Shanxi province. *Zhongguo Zhong Yao Za Zhi*, 41, 4621–4627.

Zhang, J., Lei, H., Hu, X. & Dong, W. 2020. Hesperetin ameliorates DSS-induced colitis by maintaining the epithelial barrier via blocking RIPK3/MLKL necroptosis signaling. *Eur J Pharmacol*, 873, 172992.

Zhang, X. J., Yuan, Z. W., Qu, C., Yu, X. T., Huang, T., Chen, P. V., Su, Z. R., Dou, Y. X., Wu, J. Z., Zeng, H. F., Xie, Y. & Chen, J. N. 2018. Palmatine ameliorated murine colitis by suppressing tryptophan metabolism and regulating gut microbiota. *Pharmacol Res*, 137, 34–46.

Zhang, Y. M., Liu, Y. Q., Liu, D., Zhang, L., Qin, J., Zhang, Z., Su, Y., Yan, C., Luo, Y. L., Li, J., Xie, X. & Guan, Q. 2019. The effects of astragalus polysaccharide on bone marrow-derived mesenchymal stem cell proliferation and morphology induced by A549 lung cancer cells. *Med Sci Monit*, 25, 4110–4121.

Zhu, C., Li, J., Hua, Y., Wang, J., Wang, K. & Sun, J. 2020. Berberine inhibits the expression of SCT through miR-214-3p stimulation in breast cancer cells. *Evid Based Complement Alternat Med*, 2020, 2817147.

Zhu, L., Gu, P. & Shen, H. 2019. Protective effects of berberine hydrochloride on DSS-induced ulcerative colitis in rats. *Int Immunopharmacol*, 68, 242–251.

Zobeiri, M., Belwal, T., Parvizi, F., Naseri, R., Farzaei, M. H., Nabavi, S. F., Sureda, A. & Nabavi, S. M. 2018. Naringenin and its Nano-formulations for Fatty Liver: Cellular Modes of Action and Clinical Perspective. *Curr Pharm Biotechnol*, 19, 196–205.

14 Novel Phytopharmaceuticals
A Promising Era for Antiaging

Shikha Shori, Abhishek Wahi, Neha Raina, Radha Rani, Durgesh Nandini Chauhan, and Madhu Gupta

14.1 INTRODUCTION

Aging is an inherent and evolutionarily in grained process that results in a senescent state marked by various degenerative occurrences like tissue deterioration, shortened telomeres, cognitive decline including dementia, diminished functionality, and the onset of chronic ailments. This natural phenomenon has been extensively studied in recent times, giving rise to numerous theories regarding whether it is programmed or nonprogrammed in nature (Farage et al. 2007; Guinot et al. 2002).

There are three distinct conceptual subcategories within the theories of programmed aging (Chaudhary, Khan, and Gupta 2020). One posits a genetically predetermined lifespan, suggesting that aging results from the activation or deactivation of specific genes, including the influence of genetic instability such as telomere shortening in the aging process. Another revolves around an endocrine theory, proposing that aging is controlled by a biological clock regulated by endocrine mechanisms, with a notable role played by the insulin-like growth hormone IGF-1. The third theory pertains to immunity, asserting that the immune system undergoes a programmed reduction in functionality (known as immune senescence), leading to heightened vulnerability to infections and inflammation during the aging process.

Additionally, there are theories attributing the aging process to a combination of external and internal factors. The wear and tear hypothesis posits that cells and tissues gradually deteriorate over time due to exposure to harmful byproducts. According to another theory of activity levels, a high basal metabolic rate shortens life expectancy. According to the theory of crosslinking, collagen and other important macromolecules age as a result of repeated chemical interactions. According to the notion of somatic DNA damage, genetic integrity in somatic cells deteriorates with age as a result of mutations that can happen in either the nucleus or the mitochondria. According to the free radical theory, cells are damaged by molecules like superoxide and other free radicals, which interfere with their normal function (Alsabeelah et al. 2021).

Reactive oxygen species (ROS) are likely the most significant free radicals, carrying substantial implications for cellular and bodily deterioration associated with aging. The recent research on Sod2 +/– or Mclk1 +/– transgenic mice has challenged the core tenets of this theory. In 1957, Harman proposed the notion that the accumulation of oxygen-free radicals in the body adversely affects various internal factors and alters genetic elements, ultimately being attributed to the aging and mortality of all living organisms. This concept underwent revision in 1972 when it was established that mitochondria serve as the primary site for chemical reactions leading to free radical generation. Although the effective control of complex, diverse biological processes is thought to gradually deteriorate with age, a person's genotype does have an impact on how quickly they age. However, despite 20 years of focused research, particular genetic aging indicators are still unidentified (Chaudhary, Khan, and Gupta 2020).

Antiaging products are a relatively nascent area of healthcare that is advancing rapidly. It represents an application of cutting-edge scientific and medical technologies aimed at preventing, detecting early signs of, treating, and ultimately curing age-related dysfunctions. While extending lifespan is a key goal of antiaging medicine, its primary objective is to promote a prolonged period

DOI: 10.1201/9781003292692-14

of good health. Rattan suggested a shift in focus within this field, advocating for a transition from "antiaging" to "healthy aging," thereby emphasizing research centered on maintaining well-being. This shift in perspective is why experts often prefer to use the term "healthy aging" rather than "antiaging" (Heidari Beigvand et al. 2020).

14.2 TYPES OF AGING

14.2.1 Intrinsic Aging

Intrinsic skin aging (Figure 14.1) is characterized by alterations in function rather than in structure (Mancini et al. 2014). Clinically, it manifests as skin that is rough and lacks its natural color, displaying fine wrinkles. Additionally, individuals may experience diminished hair growth, issues with sweating, and the development of various noncancerous growths on the skin. Intrinsic aging of the skin is further influenced by hormone shifts that naturally occur with advancing age, including a decrease in the production of sex hormones and a weakening of estrogens and progesterone linked to menopause. Research has established that the breakdown of collagen, dryness, decreased elasticity, thinning of the outer skin layer, and the formation of wrinkles are connected with diminished levels of estrogen and androgens. ROS plays a significant role in the aging process of the skin (Harman 1956). Approximately 1.5%–5% of the oxygen utilized by the skin undergoes conversion into ROS through internal pathways. ROS are produced as a byproduct during aerobic respiration. Additionally, they enhance the levels of matrix metalloproteinase (MMP-1), and MMP-9 via promoting the activation of the transcriptional factor c-Jun through mitogen-activated protein kinases (MAPK). As a result, procollagen-1 expression is decreased (Austin et al. 2018).

Reduced production of collagen and high levels of impaired collagen contribute to skin aging. The skin possesses a diverse range of antioxidants, which can be classified into enzymatic and

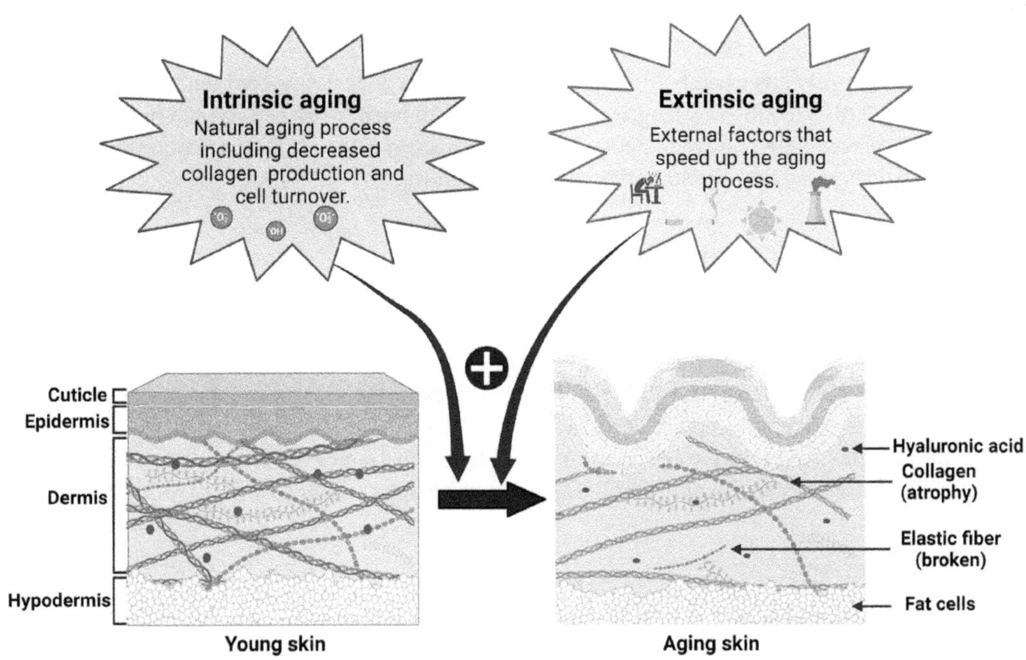

FIGURE 14.1 Schematic representation of aging pathways.

nonenzymatic antioxidants. These involved glutathione peroxidases, superoxide dismutase, catalase as glutathione (GSH), ubiquinol, vitamins. Manganese superoxide dismutase (MnSOD) is an enzyme responsible for counteracting the continuous production of these ROS. Skin aging is significantly influenced by the moisture content of the skin. Hyaluronan, commonly known as hyaluronic acid (HA), is a form of glucosaminoglycan (GAG) that plays a key role in keeping skin hydration. It is renowned for its extraordinary ability to draw and hold water molecules. Both the dermis and the epidermis of the skin contain HA, which is essential for maintaining the skin's moisture balance and promoting joint lubrication. As skin ages, the level of HA decreases in the epidermis, while it remains relatively stable in the dermis. The exact reason for this alteration is still not fully understood.

14.2.2 Extrinsic Aging

Extrinsic skin aging is characterized by observable changes in both appearance and function, which contribute to premature aging of the skin. Clinically, this manifests as pronounced wrinkles, variations in pigmentation, and the presence of solar elastosis—a condition marked by rigid, distorted, and formless elastic structures containing particles of elastin and collagen. The average thickness of the epidermis decreases with age, regardless of whether the aging process is intrinsic or extrinsic. The degree of skin aging attributed to extrinsic factors is contingent upon an individual's level of exposure to environmental agents.

The primary contributor to extrinsic skin aging is exposure to UV radiation, a category known as photoaging. The extent of skin deterioration is contingent on the length, frequency, and strength of sunlight exposure. UV radiation is accountable for approximately 80% of skin aging. Skin that has undergone photoaging tends to exhibit reduced elasticity, expanded superficial blood vessels, a rough and dry texture, as well as noticeable wrinkles and pigmentation. Tissue-level alterations encompass a thickened epidermis, irregular keratinocytes, diminished collagen in the dermal layer, and aberrant elastin arrangement (referred to as dermal elastosis) (Alsabeelah et al. 2021).

The primary source of UV radiation is sunlight. UV-C radiation, with a wavelength range of 100–290 nm, is entirely absorbed by the ozone layer and thus does not impact the skin. UV-B radiation, within the 290–320 nm wavelength range, leads to sunburn by affecting the epidermis. UV-A radiation, with wavelengths between 320 and 400 nm, possesses the ability to penetrate deeply into the skin and exerts more detrimental effects on its structure.

14.3 FACTORS INVOLVED IN SKIN AGING

14.3.1 Infrared Radiation and Heat

Infrared radiation can penetrate the deeper layers of skin, i.e. into the dermis and hypodermis (Zhang and Duan 2018). Its known ability to induce the expression of metalloproteinase in the dermis, along with its oxidative impact, is firmly established (Cho et al. 2009). Brief encounters with light emitted by electronic devices (EDGL) trigger the production of ROS. However, the prolonged consequences of frequent exposure to EDGL are yet to be determined (Austin et al. 2018).

14.3.2 Pollution

Oxidative damage brought on by air pollution can result in a variety of skin damage types. There are links between extrinsic aging, pigmentation problems, and skin cancer, and polycyclic aromatic hydrocarbons. This problem is made worse by solar radiation. Skin cancer risk increases when UVA interacts with environmental toxins like cigarette smoke. The loss of protective vitamin E from the stratum corneum of the skin increases if ozone exposure occurs before UV exposure.

14.3.3 Lifestyle Related Factors

14.3.3.1 Smoking

More significantly than exposure to sunlight, smoking has been associated with the formation of facial wrinkles. The quantity of cigarettes smoked correlates with the visibility and intensity of these wrinkles. Smoking leads to elastosis, telangiectasias, and diminished blood circulation in capillary veins, leading to insufficient nourishment of cutaneous tissues. Additionally, there is a reduction in collagen fibers and elastin in both the dermis and lungs, along with an escalation in free radicals (Ramos-e-Silva et al. 2013). Keratinocytic dysplasia and skin coarseness are also on the rise. It's worth noting that hormone replacement therapy doesn't enhance the appearance of individuals who have been long-term smokers.

14.3.3.2 Sleep

Following UV-induced redness, postmenopausal women who slept for fewer than 5h per night exhibited increased levels of transepidermal water loss and a slower rate of skin barrier recovery. Around the eyes, there was a deterioration in dark circles even after just one night of inadequate sleep, likely attributed to minor swelling, resulting in a fatigued or downcast appearance. Because of its critical role in cellular growth and rejuvenation, inadequate sleep quality leads to cellular dysfunction by disrupting the natural circadian rhythm. This disturbance affects various tissues, including the skin (Kahan et al. 2014; Walia and Mehra 2016).

14.3.3.3 Obesity

Conditions associated with obesity seem to be connected to an expedited progression of biological processes that naturally occur with aging. Inflammation and oxidative stress appear to play crucial roles in this association. Age-related elevation in the levels of pro-inflammatory substances is a notable hallmark and a substantial contributor to the expansion of adipose tissue seen in obesity.

14.3.3.4 Menopause

In menopause, a decrease in estrogen levels specifically affects skin elements that possess estrogen receptors, notably in epidermal cells and sebaceous glands. Menopause can lead to symptoms such as dry skin (xerosis), alterations in texture and capillary density, and dermal thinning. However, it is common for the response to this altered hormonal balance to be mistakenly attributed to the natural functional decline of the skin (Piérard et al. 2013; Ohnemus et al. 2006).

14.3.3.5 Emotional Stress

A study utilizing animal models of stress revealed heightened levels of lipid peroxidation, inflammatory infiltrates, and metalloproteinase activity. This suggests that stress has the potential to trigger mechanisms associated with aging (Romana-Souza, Santos Lima-Cezar, and Monte-Alto-Costa 2015).

14.3.3.6 Diet

A diverse range of vegetables provides essential antioxidant micronutrients. A balanced diet incorporating these antioxidant-rich foods helps meet our body's antioxidant requirements. Nevertheless, in the presence of oxidative stress, the antioxidants obtained from dietary sources and produced within the body may prove insufficient. An important issue to consider is the heightened likelihood of older individuals experiencing a decline in micronutrient levels, either due to reduced absorption or diminished intake. Shortages in vitamins B12 & D, calcium, and folic acid are prevalent, and these deficiencies tend to escalate with age (Fabian and Elmadfa 2008).

14.3.3.7 Hypoproteic Diet

More research is required to delve into the connection between macronutrient intake and the aging of the skin. While low-protein diets have been associated with impaired wound healing, especially in older individuals, there have been limited studies examining the link between reduced protein consumption and alterations in the production patterns of skin components (Addor 2018).

14.3.3.8 Hyperglycemic Diet

A diet high in sugar, leading to elevated blood sugar levels (hyperglycemia), is a contributing factor to skin aging, primarily due to the heightened risk of glycation (Serban et al. 2016). It's important to note that obesity and diabetes often coexist in many individuals, which can complicate analysis (Nguyen and Katta 2015). Any mechanism that triggers oxidative stress has the potential to disrupt the body's internal systems and deplete its natural reserves, a crucial consideration. Given that intrinsic aging is inherently linked to oxidative processes, the accumulation of oxidative damage intensifies as individuals grow older. Additionally, the capacity of human cells to repair DNA damage diminishes over time (Poljšak, Dahmane, and Godić 2012).

14.4 MECHANISMS OF SKIN AGING AT CELLULAR AND MOLECULAR LEVELS

Contemporary methods of analysis encompass cell culture and techniques from the field of molecular biology. In 1920, Careel delved into the molecular aspects of aging by conducting an experiment involving chick heart fibroblast cells, which exhibited an ability to divide indefinitely (Carrel and Ebeling 1921). These cells acquired immortality once they were transitioned from the in vivo process to an in vitro environment. Moving forward to 1960, Hayflick elucidated that regular human fibroblasts possess a limited capacity for proliferation in vitro. He illustrated that typical cells can undergo a series of doubling events an unlimited number of times until they cease division upon reaching the phase of exponential growth, a phenomenon now known as the Hayflick limit (Jenkins 2002). In regular conditions, cell growth, deterioration, and renewal maintain a steady pace throughout all life stages. Yet, within mitotic homeostasis, damaged cells are substituted, upholding the functional vitality of cells and tissues. The deterioration is instigated by the generation of ROS and the nonenzymatic glycosylation of proteins. Additionally, the reduction of telomerase, responsible for cell multiplication and regeneration, as well as the occurrence of apoptosis, contributes to this process. This simplifies the understanding of aging, as both external and internal factors play a role in it (Jenkins 2002).

14.4.1 Glycation in Aging

The creation of advanced glycosylation end products, referred to as AGEs, serves as an indicator of cellular degeneration associated with aging (Ly et al. 2000). When AGEs form on collagen, it leads to a process called crosslinking, which results in heightened molecular density, abnormalities within the extracellular matrix, and reduced interaction between cells and the matrix. Moreover, AGEs attach to particular receptors on immune cells, prompting the release of inflammatory agents and the generation of ROS. This, in turn, leads to an amplified production of AGEs.

14.4.2 Free Radical in Aging

In 1954, the free radical theory of aging (FRTA) proposed that free radicals play a role in endogenous metabolic processes (Harman 1956). According to this theory, the primary factor driving aging is the activation of free radical reactions (RRs). Subsequently, several postulations emerged, including (1) a rising rate of free RRs primarily initiated by the mitochondria with advancing age, (2) the lifespan being contingent on the pace at which free radicals induce damage to the mitochondria, and (3) the speed at which free radicals cause harm to the mitochondria determines the rate of aging (Wickens 2001).

14.5 PHYTOCONSTITUENTS FOR ANTIAGING

14.5.1 Quercetin

Quercetin is a polyphenolic flavonoid (chemical formula: C15H10O7) abundant in green vegetables, citrus fruits, and berries, known for its potent antioxidant properties (D'Andrea 2015). Structurally, quercetin bears a catechol B-ring with three more hydroxyl groups located at carbons 3, 5, and 7. The presence of multiple hydroxyl groups, particularly in close proximity within a benzene ring-C, leads to the formation of a resonance-stabilized quinone structure, enhancing quercetin's antioxidative effectiveness. Both at the mRNA and protein levels, quercetin diminishes the production of matrix metalloproteinase-1, a pivotal factor for skin wrinkling and elasticity, in both healthy and photoaged skin (Pereira et al. 2020). Due to its limited aqueous solubility and poor permeability, incorporating quercetin into cosmetics presents challenges. Despite its reduced bioavailability, quercetin possesses valuable therapeutic attributes and serves as a crucial precursor for its bioactive forms. One potential solution to enhance bioavailability involves synthesizing quercetin within nanocarriers, such as nanoemulsions.

14.5.2 Catechin

Catechins, belonging to the flavonoid family, are a type of flavan-3-ols. Catechins and their analogs endowed with galloyl groups have demonstrated impressive capabilities in inhibiting free radicals. Researchers have pinpointed the galloyl group in green tea catechins as the critical structural element for their respective bioactivities (Higdon and Frei 2003). Given their high solubility in both oil-in-water and water-in-oil emulsion systems, catechins serve as valuable bioactive components in topical nanoemulsions. Multiple studies have confirmed that nanoemulsions rich in catechins can permeate the stratum corneum and access deeper skin layers (Kim et al. 2018).

14.5.3 Kaempferol

Kaempferol, scientifically known as 3,4′,5,7-tetrahydroxyflavone ($C_{15}H_{10}O_6$), is a flavonoid present in various foods such as apples, broccoli, and strawberries. In its pure form, it appears as a yellow powder and exhibits a diverse array of pharmacological properties. These include acting as an anticancer agent, inhibiting tumor growth, functioning as an antioxidant, possessing anti-inflammatory properties, and showing antiallergic effects (Ackland, van de Waarsenburg, and Jones 2005).

Kaempferol, possessing potent antioxidant properties that shield our cells, lipids, and DNA from oxidative harm, fortifies the body's immunity against free radicals. Interestingly, it not only helps maintain the normal viability of cells but, in certain instances, serves as a protective agent. The hydroxyl groups present in this molecule significantly enhance its ability to scavenge superoxide and free radicals. It is theorized that kaempferol, in conjunction with quercetin, collaborates to impede the proliferation of cancer cells, and a combined therapy using both quercetin and kaempferol proves more efficacious than employing either flavonoid in isolation (Sekiguchi et al. 2019). Because of its lipophilic nature, kaempferol, akin to other flavonoids, is assimilated through diffusion, and active transport. This characteristic enables the enhancement of kaempferol's bioavailability for topical or cosmeceutical uses through a range of nanocarrier systems (Jeon et al. 2011).

14.5.4 Curcumin

Curcumin, a polyphenolic compound found in the rhizomes of *Curcuma longa* (turmeric), has a longstanding history of use in treating various skin conditions like acne, melanoma, and wound healing. It has a broad spectrum of biological and pharmacological attributes, including anti-inflammatory, anticancer, and antioxidant activities (Kumari et al. 2022; Rafiee et al. 2019). The *o*-diphenoxyl and *o*-dimethoxyphenoxyl groups in the structure of curcumin can scavenge more DPPH due to the increased presence of electron-donating groups like methoxy and hydroxyl. However, owing to its lipophilic nature, curcumin exhibits poor solubility in water. Recent studies have sought to address this limitation by incorporating curcumin into nanoemulsion systems. This endeavor aims to enhance its bioactivity, ultimately increasing bioavailability (Jamwal 2018; Berbudi et al. 2021).

14.5.5 Lycopene

Lycopene, a pigment belonging to the carotenoid family, is abundantly found in tomato seeds (Butnariu and Giuchici 2011). Due to its capacity to scavenge and neutralize free radicals, which can counteract the negative effects of free radicals, it is frequently included in skincare products. Similar to resveratrol, lycopene faces the challenge of being highly lipophilic (log p15). This property, which leads to a strong affinity with stratum corneum components, makes it difficult for lycopene to be absorbed from the stratum corneum into deeper layers and tends to retain it within this layer (Qadir et al. 2022).

To address this issue, nano-encapsulation has emerged as a solution to improve the absorption and bioavailability of lycopene in the skin. Ascenso et al. developed a nano-sized topical formulation rich in lycopene, which demonstrated improved skin protection against UV-induced damage, primarily attributed to the enhanced bioavailability of phytoconstituents at the epidermal level. Furthermore, studies have indicated that nano-formulated lycopene exhibits anti-inflammatory and anticancer properties in photo-damaged and preneoplastic keratinocytes, underscoring its superiority over conventional cosmetics (Yahya, Attan, and Wahab 2018).

14.5.6 Equol

Equol, the primary metabolite of phytoestrogens, is an isoflavonoid derived from plants. It possesses noteworthy antiaging and antioxidant properties that make it suitable for application on human skin (Pontius and Smith 2011). Equol has demonstrated the ability to emulate 17-estradiol and selectively bind to 5-dihydrotestosterone (5-DHT). In quantitative polymerase chain reaction (qPCR) tests, equol led to an increase in the expression of genes related to collagen, elastin (ELN), and tissue inhibitor of metalloprotease, while reducing the expression of metalloproteinases (MMPs) genes. It also induced significant positive changes in genes associated with skin antioxidants and antiaging effects. Furthermore, in human monolayer fibroblast cultures (hMFC), equol notably elevated collagen type I (COL1A1), whereas 5-DHT substantially decreased cell viability, an effect mitigated by equol. When assessed through Fluorescence-activated cell scanning (FACS) analysis, both equol and 17-estradiol demonstrated a strong increase in collagen type I (COL1A1), collagen type III (COL3A1), and elastin (ELN), while significantly reducing MMPs compared to control levels (Lephart et al. 2005).

14.5.7 Ginseng

Ginseng has been a staple in traditional East Asian medicine for millennia, gaining immense popularity worldwide over the last 30 years. It boasts a range of therapeutic properties including anti-inflammatory, antioxidant, antibacterial, antiviral, and antifungal effects (Lee et al. 2022). Additionally, it has shown promise in the treatment of various neurological conditions such as Alzheimer's disease (AD), Parkinson's disease (PD), and Huntington's disease (HD), along with addressing hypertension and stress (Choi et al. 2018). Both internal and external sources contribute to the generation of free radicals like ROS, and reactive nitrogen species (RNS). Studies have indicated that ginseng leaf extracts in ethanol and methanol can effectively scavenge free radicals. Among these extracts, ethanol extracts demonstrated the highest levels of 2,2-diphenyl-1-picrylhydrazyl (DPPH) radical scavenging, ferrous ion chelating, and hydroxyl radical scavenging activities.

Ginseng is also known to elevate levels of antioxidant enzymes like glutathione peroxidase and superoxide dismutase (Ratan et al. 2021).

14.6 PHYTOCONSTITUENT-LOADED NANOFORMULATION-BASED STRATEGIES FOR ANTIAGING

Aging and its scientific updation are enhanced through research on aging-related diseases. Studies show that the aging process is decreased and lifespan increases by celastrol, berberine, and polyphenols, or counteract diseases associated with aging. Among the secondary metabolites of plants, polyphenols have a wide range of structural variations. Regarding polyphenols, flavonoids are the most attractive group because of their health benefits and antiaging effects. Due to their low bioavailability and short shelf life, many plant-based compounds have limited therapeutic potential despite their antiaging effects. In addition to overcoming the drawbacks, nanotechnology improves drug therapy and patient satisfaction. Nanoparticles, ethosomes and niosomes are recent advances in topical antiage treatment, improving efficacy, safety & comfort. Recent technological developments have improved topical drug therapy's effectiveness in treating antiaging effects. Antiaging topical therapy may use several new drug carriers to introduce new molecules. Drug delivery systems are based on vesicular structures such as liposomes, niosomes, proniosomes, and transferosomes. To target antiaging effects, nanoemulsions, SLNs, nanogels & nanoparticles have been developed as nonvesicular drug delivery systems (Figure 14.2) (Xue et al. 2021).

14.6.1 Vesicular Systems

Liposomes: Liposomes are tiny spherical structures composed of layers of phospholipids and cholesterol (Rahimpour and Hamishehkar 2012). This arrangement allows for the inclusion of lipophilic drugs within the lipid bilayer and hydrophilic drugs in the aqueous central core (Jaspart et al. 2005). The lipid bilayer and membrane can combine to convey the liposome's components and pinpoint the active components (Kaur and Agrawal 2007). Liposomes, possessing lipid composition akin to the epidermis, can enhance dermal drug delivery while minimizing systemic absorption (Yang et al. 2009).

Niosomes: These are bilayered structures enclosed by nonionic surface active agents in a water-based environment. They emerged in the early 1970s through studies conducted by L'Oreal, initially finding applications in cosmetics. When topically applied, niosomes extend the retention of therapeutic constituents in deeper layers of skin, facilitating enhanced skin penetration (Singh and Sharma 2016).

Transfersomes: They are also known as deformable liposomes and represent a significant advancement in vesicular drug delivery for topical applications. They are highly elastic liposomes capable of undergoing extensive deformation. This innovation holds the potential to greatly enhance drug delivery (Gupta, Agrawal, and Vyas 2012). Research has indicated that deformable liposomes, due to their high deformability, can readily penetrate the stratum corneum through the intercellular route (Costa and Santos 2017).

Ethosomes: In the pursuit of advancing methods for enhanced cutaneous delivery, ethosomes have emerged as promising carriers in dermatology. Newly developed ethosomes have demonstrated improved delivery of the enclosed drug to the skin. The primary distinction between conventional liposomes and ethosomes lies in their composition, with ethosomes incorporating ethanol instead of cholesterol (Gupta, Agrawal, and Vyas 2012). The term "ethosome" underscores the high concentration of ethanol present, ranging from 20% to 45%. Research indicates that ethosomes can effectively encapsulate molecules with diverse lipophilic characteristics, leading to an augmentation in their permeation through biological barriers (Touitou et al. 2000; Garg et al. 2017).

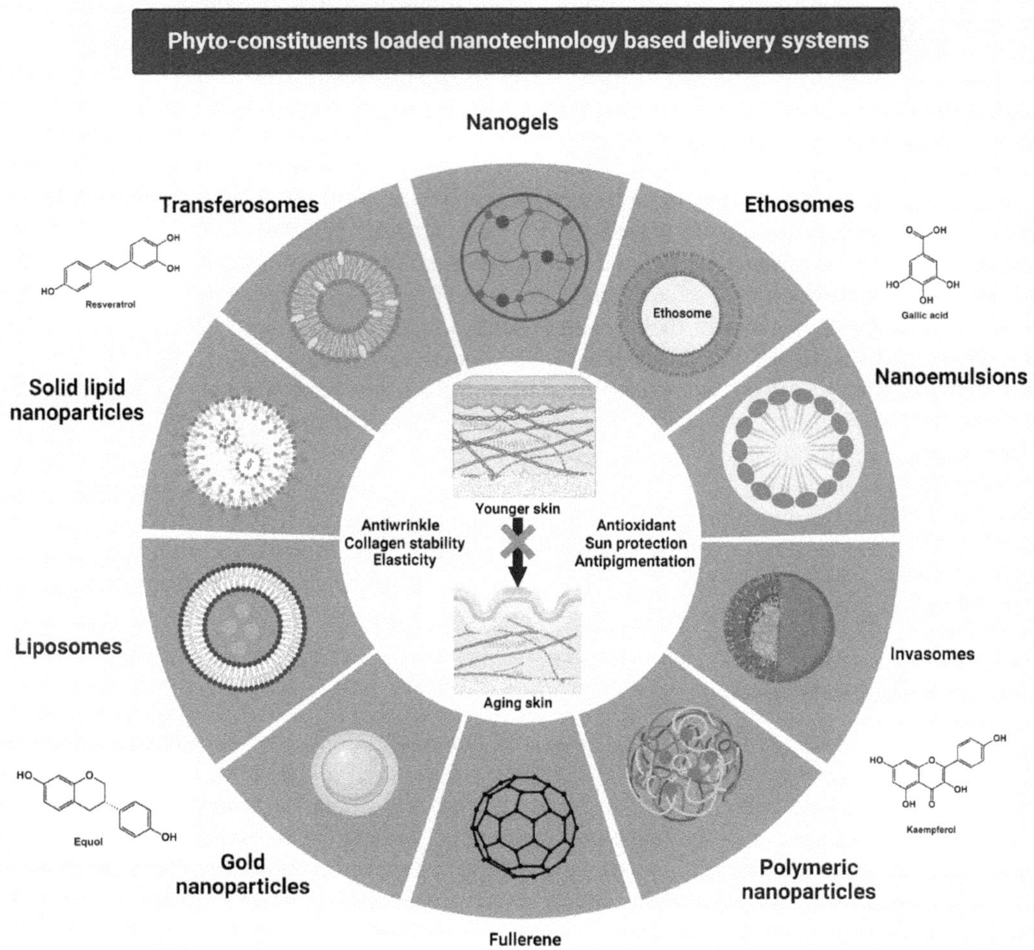

FIGURE 14.2 Schematic representation of various phytocompounds loaded nano-delivery systems exhibiting antiaging effects.

14.6.2 Nanoemulsion

Transparent nanoemulsions with droplet sizes less than 100 nm, known as nanoemulsions, are distinguished by a persistent interfacial coating created by surfactant and cosurfactant. They stand as a highly promising delivery system for effectively applying active ingredients topically (Singh and Sharma 2016). The amount of surface area that nanoemulsions provide for the emulsion particles to interact with the skin increases the amount of active ingredients that come into touch with the skin's surface. Due to the small size of these particles, pores are not blocked, allowing water and air to flow freely. Nanodroplets contain a liquid and lipophilic core surrounded by a single molecular layer of phospholipids, in contrast to liposomes, which have an aqueous core enclosed by a double layer of phospholipids. They are very well suited for delivering lipophilic chemicals because of their structure. Additionally, nanoemulsions strengthen the skin's barrier function by minimizing water loss across the epidermal layer of skin and have their own bioactive effects (Kaur and Agrawal 2007).

14.6.3 Lipid Nanoparticles

These systems share a resemblance to nanoemulsions in their structure. They are submicron colloidal vehicles, ranging from 40–1000 nm in size. What sets them apart is the fact that their lipid core is in a solid state. These nanoparticles are constructed with a single solid lipid or a combination of different lipids (Guimarães and Ré 2011).

Solid lipid nanoparticles (SLN) were introduced as a substitute for traditional colloidal carriers like liposomes and polymeric nanoparticles (Müller 1991). The key benefits of SLNs include their biodegradability, utilization of physiologically compatible lipids, capability for controlled substance release, and ability to maintain stability over extended periods in the form of aqueous dispersions (Müller, Radtke, and Wissing 2002). However, SLNs encounter certain issues like restricted drug loading, the potential for gelation, and drug leakage during storage due to lipid polymorphism

Nanostructured lipid carriers (NLCs) represent the second generation of lipid nanoparticles, formulated using a blend of solid and liquid lipids. They present benefits including enhanced loading capacity and a diminished likelihood of lipid matrix leakage during storage (Müller et al. 2007). The recently engineered NLCs derived from lipid nanoparticles show improved outcomes in addressing the challenges faced by SLNs (Gupta, Agrawal, and Vyas 2012).

14.6.4 Other Types of Nanocarriers

In addition to the nanocarriers mentioned earlier, dendrimers and fullerenes are gaining significance in the cosmetic industry (Singh and Sharma 2016). In contrast to conventional polymers, dendrimers possess an intricately branched, tree-like three-dimensional architecture. They are made up of numerous chemical layers surrounding a compact core molecule. There are four essential components to them: a central or core unit, arms of uniform size, linkages or branching points, and terminal functional groups (Kaur and Agrawal 2007).

A classic representation of a buckyball, fullerene is a compound consisting exclusively of an even number of carbon atoms. These atoms arrange themselves into a lattice-like structure, forming a fused-ring polycyclic system with 12 rings containing five members each, while the remaining rings have six members. Fullerene and its derivates are notably soluble in water (Kaur and Agrawal 2007). Elliptical or tubular fullerene structures, entirely composed of carbon, are gaining increased attention in the cosmetics industry for their natural antioxidant properties and their ability to offer protection against UVA radiation (Bakry et al. 2007).

Numerous studies have explored the application of nanocarrier systems in cosmetics, resulting in a variety of commercially available products incorporating these technologies. The table below provides examples of various nanocarrier systems currently accessible in the market (Kaur and Agrawal 2007; Sharma and Sharma 2012; Li et al. 2011).

The provided Table 14.1 outlines recent research endeavors focused on delivering active compounds through encapsulation in nanocarriers. These studies have yielded notable findings. For instance, the proniosome form of Coenzyme Q10 offers superior photo-protection when compared to its free counterpart. Moreover, the ultra-small NLC form of Coenzyme Q10 exhibits heightened dermal presentation in comparison to the classic NLC (Yadav et al. 2016; Schwarz et al. 2013; Yue et al. 2010). Among liposomes and niosomes, the phytovesicular form of curcumin proves to be the most efficacious. However, when considering the liposome, ethosome, and transfersome formulations of Curcuma longa extract, it is found that the transfersome form exhibits the highest level of effectiveness (Kaur and Saraf 2011; Gupta and Dixit 2011). Resveratrol and curcumin face challenges in skin penetration due to their limited water solubility. However, lipid core nanocapsules offer a solution by facilitating their penetration through the epidermis of the skin (Friedrich et al. 2015). Recently, there has been growing interest in using nano-sized micellar carriers for the precise delivery of phytoconstituents into skin layers, including hair follicles. This approach has demonstrated effectiveness in treating acne (Kahraman and Güngör 2016; Kandekar et al. 2018).

TABLE 14.1
Recent Research, Post-2010, Focused on the Transportation of Active Compounds Using Nanocarrier Systems

S. No	Active Substance	Types of Nanocarriers Used	Study Outcome	References
1	Coenzyme Q 10	Ultra-small NLC	The research indicated that ultra-small NLC (80 nm) CoQ10 enhances dermal delivery compared to the larger, conventional NLC. Furthermore, the study provided compelling evidence that curcumin phyto-vesicles outperformed other formulations and exhibited enhanced antiaging, antioxidant, and antiwrinkle effects	Schwarz et al. (2013)
2	Curcumin	Phytovesicles Liposome Niosome	The research unequivocally demonstrated that curcumin in the phyto-vesicle form proved to be more potent in terms of antiaging, antioxidant, and antiwrinkle effects compared to other formulations	Gupta and Dixit (2011)
3	Resveratrol	Liposome & Niosome	This study has demonstrated that niosomes formulated with a lipophilic surfactant effectively facilitated the delivery of resveratrol to the skin, resulting in a notable topical effect. Furthermore, niosomes exhibited enhanced delivery of resveratrol, both in the dermal and transdermal routes, when compared to traditional liposomes	Pando et al. (2013)
4	Resveratrol & Curcumin	Lipidcore nanocapsules	Co-encapsulation resulted in an accelerated release of resveratrol, whereas the release rate of curcumin remained steady. Studies on skin penetration demonstrated an augmented delivery of resveratrol to the deeper layers of the skin when administered in conjunction with the encapsulation of a single polyphenol. Lipid core nanocapsules exhibit significant promise as carrier systems for the application of poorly water-soluble compounds like resveratrol and curcumin to human skin	Friedrich et al. (2015)
5	Rice bran extract	Niosome	The research demonstrated that semi-purified rice bran extracts stimulated the growth of human fibroblasts and inhibited MMP-2 activity. Additionally, it led to enhancements in skin properties such as hydration, pigmentation, thickness, roughness, and elasticity. The entrapment of bioactive compounds in niosomes also contributed to the improved chemical stability of the extracts	Manosroi et al. (2012)
6	5-Aminolevulinic acid	Liposomal spray	The research demonstrated that employing IPL-PDT therapy with a 0.5% 5-ALA spray and the Ellipse Flex PPT system led to a notable enhancement in periorbital and nasolabial wrinkles among individuals with type 2 photo-damaged skin	Piccioni et al. (2011)
7	Alpha-lipoic acid	Cubosomes	A notable reduction in facial lines was observed, along with the nearly complete resolution of fine lines around the eyes and upper eyelids, and an overall enhancement in skin tone and texture in the majority of participants	Sherif, Bendas, and Badawy (2014)

(*Continued*)

TABLE 14.1 (*Continued*)
Recent Research, Post-2010, Focused on the Transportation of Active Compounds Using Nanocarrier Systems

S. No	Active Substance	Types of Nanocarriers Used	Study Outcome	References
8	Kinetin (N6FA)	SLN	The application of N6FA-loaded SLNs effectively demonstrated its ability to counteract the effects of photoaging on the skin	Goindi, Guleria, and Aggarwal (2015)
9	Curcuma longa extract	Ethosome Transfersome Liposome	The nano-vesicles containing C. longa extract exhibited significant photoprotective characteristics, complemented by the moisturizing lipid components which enhanced skin properties like hydration and sebum content through improved skin penetration. The study concluded that transfersomes loaded with the extract were superior in enhancing skin properties compared to ethosomes and liposomes	Kaur and Saraf (2011)

14.7 CONCLUSION

Skin aging is a complicated biological process influenced by a number of factors including both external and internal factors, with around 80%–90% of it being attributed to UV light exposure. Additional factors such as malnutrition, intense stress, hormonal imbalances, environmental pollution, smoking, and alcohol consumption also contribute to skin aging. Extensive studies and research have been conducted to find ways to minimize the effects of skin damage. Effective treatments for skin aging include herbal extracts, vitamins, and antioxidant supplements. Phytochemicals derived from plants offer various beneficial qualities for the skin, such as UV protection, antioxidant effects, matrix protection, and hydration. However, incorporating these plant-derived compounds into conventional formulations poses challenges like low solubility and limited target specificity. Consequently, researchers are exploring novel nano-cosmeceutical-based treatments to develop more efficient and acceptable formulations. Nano-cosmeceuticals play a crucial role in minimizing or reversing the effects of stress, pollution, and environmental exposure on the skin. These findings contribute to basic and clinical studies on skin health and potential treatments for skin diseases and age-related deterioration. However, there are concerns about the safety of nanocosmetics for both human health and the environment. Therefore, it is crucial to evaluate the safety and efficacy of these nanocosmeceuticals before they are marketed. Harmonization of guidelines from different countries is necessary to regulate such products effectively. Looking ahead, nanocosmeceuticals require more attention and awareness due to their future development and market potential.

REFERENCES

Ackland, Margaret Leigh, Simone van de Waarsenburg, and Rod Jones. 2005. "Synergistic Antiproliferative Action of the Flavonols Quercetin and Kaempferol in Cultured Human Cancer Cell Lines." *In Vivo (Athens, Greece)* 19 (1): 69–76.

Addor, Flavia Alvim Sant'Anna. 2018. "Beyond Photoageing: Additional Factors Involved in the Process of Skin Ageing." *Clinical, Cosmetic and Investigational Dermatology* 11: 437–43. doi:10.2147/CCID.S177448.

Alsabeelah, N., M. Faiz Arshad, Sana Hashmi, Riyaz Ahmed Khan, and Shamshir Khan. 2021. "Nanocosmeceuticals for the Management of Ageing: Rigors and Vigors." *Journal of Drug Delivery Science and Technology* 63: 102448. doi:10.1016/j.jddst.2021.102448.

Austin, Evan, Amy Huang, Tony Adar, Erica Wang, and Jared Jagdeo. 2018. "Electronic Device Generated Light Increases Reactive Oxygen Species in Human Fibroblasts." *Lasers in Surgery and Medicine*. doi:10.1002/lsm.22794.

Bakry, Rania, Rainer M Vallant, Muhammad Najam-ul-Haq, Matthias Rainer, Zoltan Szabo, Christian W Huck, and Günther K Bonn. 2007. "Medicinal Applications of Fullerenes." *International Journal of Nanomedicine* 2 (4): 639–49.

Berbudi, Afiat, Ali Akbar Fadlillah, Merry Afni, and Nur Atik. 2021. "Effect of Curcuma Longa Crude Extract, Curcumin and Nano Curcumin-Based Gel Topical Administration on Excised Skin Wound Healing in Swiss-Webster Mice." *Biomedical and Pharmacology Journal* 14 (2): 635–41.

Butnariu, Monica V, and Camelia V Giuchici. 2011. "The Use of Some Nanoemulsions Based on Aqueous Propolis and Lycopene Extract in the Skin's Protective Mechanisms against UVA Radiation." *Journal of Nanobiotechnology* 9 (1): 3. doi:10.1186/1477-3155-9-3.

Carrel, Alexis, and Albert H Ebeling. 1921. "Age and Multiplication of Fibroblasts." *The Journal of Experimental Medicine* 34 (6): 599–623.

Chaudhary, Manupriya, Azmi Khan, and Madhu Gupta. 2020. "Skin Ageing: Pathophysiology and Current Market Treatment Approaches." *Current Ageing Science* 13 (1): 22–30.

Cho, Soyun, Mi Hee Shin, Yeon Kyung Kim, Jo-Eun Seo, Young Mee Lee, Chi-Hyun Park, and Jin Ho Chung. 2009. "Effects of Infrared Radiation and Heat on Human Skin Ageing in Vivo." *The Journal of Investigative Dermatology. Symposium Proceedings* 14 (1): 15–19. doi:10.1038/jidsymp.2009.7.

Choi, Seo-Yun, Kui-Jin Kim, Ji-Hyeon Song, and Boo-Yong Lee. 2018. "Ginsenoside Rg5 Prevents Apoptosis by Modulating Heme-Oxygenase-1/Nuclear Factor E2-Related Factor 2 Signaling and Alters the Expression of Cognitive Impairment-Associated Genes in Thermal Stress-Exposed HT22 Cells." *Journal of Ginseng Research* 42 (2): 225–28. doi:10.1016/j.jgr.2017.02.002.

Costa, Raquel, and Lúcia Santos. 2017. "Delivery Systems for Cosmetics – From Manufacturing to the Skin of Natural Antioxidants." *Powder Technology* 322: 402–16. doi:10.1016/j.powtec.2017.07.086.

D'Andrea, Gabriele. 2015. "Quercetin: A Flavonol with Multifaceted Therapeutic Applications?" *Fitoterapia* 106: 256–71. doi:10.1016/j.fitote.2015.09.018.

Fabian, Elisabeth, and Ibrahim Elmadfa. 2008. "Nutritional Situation of the Elderly in the European Union: Data of the European Nutrition and Health Report (2004)." *Annals of Nutrition & Metabolism* 52 (Suppl 1): 57–61. doi:10.1159/000115352.

Farage, Miranda A, Kenneth W Miller, Peter Elsner, and Howard I Maibach. 2007. "Structural Characteristics of the Ageing Skin: A Review." *Cutaneous and Ocular Toxicology* 26 (4): 343–57. doi:10.1080/15569520701622951.

Friedrich, Rossana B, Birthe Kann, Karine Coradini, Herman L Offerhaus, Ruy C R Beck, and Maike Windbergs. 2015. "Skin Penetration Behavior of Lipid-Core Nanocapsules for Simultaneous Delivery of Resveratrol and Curcumin." *European Journal of Pharmaceutical Sciences* 78: 204–13.

Garg, Varun, Harmanpreet Singh, Sneha Bimbrawh, Sachin Kumar Singh, Monica Gulati, Yogyata Vaidya, and Prabhjot Kaur. 2017. "Ethosomes and Transfersomes: Principles, Perspectives and Practices." *Current Drug Delivery* 14 (5): 613–33. doi:10.2174/1567201813666160520114436.

Goindi, Shishu, Ankita Guleria, and Nidhi Aggarwal. 2015. "Development and Evaluation of Solid Lipid Nanoparticles of N-6-Furfuryl Adenine for Prevention of Photoageing." *Journal of Biomedical Nanotechnology* 11 (10): 1734–46.

Guimarães, Kleber L, and Maria Inês Ré. 2011. "Lipid Nanoparticles as Carriers for Cosmetic Ingredients: The First (SLN) and the Second Generation (NLC)." In *BT – Nanocosmetics and Nanomedicines: New Approaches for Skin Care*, edited by Ruy Beck, Silvia Guterres, and Adriana Pohlmann, 101–22. Berlin, Heidelberg: Springer. doi:10.1007/978-3-642-19792-5_5.

Guinot, Christiane, Denis J-M Malvy, Laurence Ambroisine, Julie Latreille, Emmanuelle Mauger, Michel Tenenhaus, Frédérique Morizot, Sabine Lopez, Isabelle Le Fur, and Erwin Tschachler. 2002. "Relative Contribution of Intrinsic vs Extrinsic Factors to Skin Ageing as Determined by a Validated Skin Age Score." *Archives of Dermatology* 138 (11): 1454–60. doi:10.1001/archderm.138.11.1454.

Gupta, Madhu, Udita Agrawal, and Suresh P Vyas. 2012. "Nanocarrier-Based Topical Drug Delivery for the Treatment of Skin Diseases." *Expert Opinion on Drug Delivery* 9 (7): 783–804. doi:10.1517/17425247.2012.686490.

Gupta, Nishant Kumar, and V K Dixit. 2011. "Development and Evaluation of Vesicular System for Curcumin Delivery." *Archives of Dermatological Research* 303 (2): 89–101.

Harman, Denham. 1956. "Ageing: A Theory Based on Free Radical and Radiation Chemistry." *Journal of Gerontology* 11 (2): 289–300. doi: 10.1093/geronj/11.3.298.

Heidari Beigvand, Hazhir, Mohammadreza Razzaghi, Mohammad Rostami-Nejad, Majid Rezaei-Tavirani, Saeed Safari, Mostafa Rezaei-Tavirani, Vahid Mansouri, and Mohammad Hossein Heidari. 2020. "Assessment of Laser Effects on Skin Rejuvenation." *Journal of Lasers in Medical Sciences* 11 (2): 212–19. doi:10.34172/jlms.2020.35.

Higdon, Jane V, and Balz Frei. 2003. "Tea Catechins and Polyphenols: Health Effects, Metabolism, and Antioxidant Functions." *Critical Reviews in Food Science and Nutrition* 43 (1): 89–143. doi:10.1080/10408690390826464.

Jamwal, Rohitash. 2018. "Bioavailable Curcumin Formulations: A Review of Pharmacokinetic Studies in Healthy Volunteers." *Journal of Integrative Medicine* 16 (6): 367–74. doi:10.1016/j.joim.2018.07.001.

Jaspart, Séverine, Géraldine Piel, Luc Delattre, and Brigitte Evrard. 2005. "Solid Lipid Microparticles: Formulation, Preparation, Characterisation, Drug Release and Applications." *Expert Opinion on Drug Delivery* 2 (1): 75–87.

Jenkins, Gail. 2002. "Molecular Mechanisms of Skin Ageing." *Mechanisms of Ageing and Development* 123 (7): 801–10.

Jeon, Ae Ji, Tae-Gyu Lim, Sung Keun Jung, Eun Jung Lee, Myeong Hun Yeom, Jun Seong Park, Myoung-Gun Choung, Hyong Joo Lee, Yoongho Lim, and Ki Won Lee. 2011. "Black Soybean (Glycine Max Cv. Heugmi) Seed Coat Extract Suppresses TPA or UVB-Induced COX-2 Expression by Blocking Mitogen Activated Protein Kinases Pathway in Mouse Skin Epithelial Cells." *Food Science and Biotechnology* 20 (6): 1735–41. doi:10.1007/s10068-011-0239-7.

Kahan, V, D A Ribeiro, F Egydio, L A Barros, J Tomimori, S Tufik, and M L Andersen. 2014. "Is Lack of Sleep Capable of Inducing DNA Damage in Aged Skin?" *Skin Pharmacology and Physiology* 27 (3): 127–31. doi:10.1159/000354915.

Kahraman, Emine, and Sevgi Güngör. 2016. "Polymeric Micellar Nanocarriers of Benzoyl Peroxide as Potential Follicular Targeting Approach for Acne Treatment." *Colloids and Surfaces B: Biointerfaces* 146: 692–99.

Kandekar, Somnath G, Sergio del Río-Sancho, Maria Lapteva, and Yogeshvar N Kalia. 2018. "Selective Delivery of Adapalene to the Human Hair Follicle under Finite Dose Conditions Using Polymeric Micelle Nanocarriers." *Nanoscale* 10 (3): 1099–110.

Kaur, Chanchal Deep, and Swarnlata Saraf. 2011. "Topical Vesicular Formulations of Curcuma Longa Extract on Recuperating the Ultraviolet Radiation-Damaged Skin." *Journal of Cosmetic Dermatology* 10 (4): 260–65. doi:10.1111/j.1473-2165.2011.00586.x.

Kaur, Indu P, and Rumjhum Agrawal. 2007. "Nanotechnology: A New Paradigm in Cosmeceuticals." *Recent Patents on Drug Delivery & Formulation* 1 (2): 171–82.

Kim, Eunji, Kyeonghwan Hwang, Jongsung Lee, Sang Yun Han, Eun-Mi Kim, Junseong Park, and Jae Youl Cho. 2018. "Skin Protective Effect of Epigallocatechin Gallate." *International Journal of Molecular Sciences* 19 (1): 173. doi:10.3390/ijms19010173.

Kumari, Amrita, Neha Raina, Abhishek Wahi, Khang W Goh, Pratibha Sharma, Riya Nagpal, Atul Jain, Long C Ming, and Madhu Gupta. 2022. "Wound-Healing Effects of Curcumin and Its Nanoformulations: A Comprehensive Review." *Pharmaceutics*. doi:10.3390/pharmaceutics14112288.

Lee, Jeong-Oog, Yanyan Yang, Yu Tao, Young-Su Yi, and Jae Youl Cho. 2022. "Korean Red Ginseng Saponin Fraction Exerts Anti-Inflammatory Effects by Targeting the NF-KB and AP-1 Pathways." *Journal of Ginseng Research* 46 (3): 489–95. doi:10.1016/j.jgr.2022.02.004.

Lephart, Edwin Douglas, Trent D. Lund, Kenneth David Reginald Setchell Handa, and Robert J. Handa. 2005. Use of Equol for Treating Skin Diseases. WO/2005/107770, issued 2005.

Li, Danhui, Zimei Wu, Nataly Martini, and Jingyuan Wen. 2011. "Advanced Carrier Systems in Cosmetics and Cosmeceuticals: A Review." *Journal of Cosmetic Science* 62 (6): 549–63.

Ly, Danith H, David J Lockhart, Richard A Lerner, and Peter G Schultz. 2000. "Mitotic Misregulation and Human Ageing." *Science* 287 (5462): 2486–92.

Mancini, Mara, Anna Maria Lena, Gaelle Saintigny, Christian Mahé, Nicola Di Daniele, Gerry Melino, and Eleonora Candi. 2014. "MicroRNAs in Human Skin Ageing." *Ageing Research Reviews* 17: 9–15.

Manosroi, Aranya, Romchat Chutoprapat, Masahiko Abe, Worapaka Manosroi, and Jiradej Manosroi. 2012. "Anti-Ageing Efficacy of Topical Formulations Containing Niosomes Entrapped with Rice Bran Bioactive Compounds." *Pharmaceutical Biology* 50 (2): 208–24.

Müller, Rainer H. 1991. *Colloidal Carriers for Controlled Drug Delivery and Targeting: Modification, Characterization and in Vivo Distribution*. Stuttgart: Wissenschaftliche Verlagsgesellschaft; CRC Press.

Müller, R H, R D Petersen, A Hommoss, and J Pardeike. 2007. "Nanostructured Lipid Carriers (NLC) in Cosmetic Dermal Products." *Advanced Drug Delivery Reviews* 59 (6): 522–30.

Müller, Rainer H, Magdalene Radtke, and Sylvia A Wissing. 2002. "Solid Lipid Nanoparticles (SLN) and Nanostructured Lipid Carriers (NLC) in Cosmetic and Dermatological Preparations." *Advanced Drug Delivery Reviews* 54: S131–55.

Nguyen, Harrison P, and Rajani Katta. 2015. "Sugar Sag: Glycation and the Role of Diet in Ageing Skin." *Skin Therapy Letter* 20 (6): 1–5.

Ohnemus, Ulrich, Murat Uenalan, José Inzunza, Jan-Ake Gustafsson, and Ralf Paus. 2006. "The Hair Follicle as an Estrogen Target and Source." *Endocrine Reviews* 27 (6): 677–706. doi:10.1210/er.2006-0020.

Pando, Daniel, Carla Caddeo, Maria Manconi, Anna Maria Fadda, and Carmen Pazos. 2013. "Nanodesign of Olein Vesicles for the Topical Delivery of the Antioxidant Resveratrol." *Journal of Pharmacy and Pharmacology* 65 (8): 1158–67.

Pereira, Ana Bárbara, Aida Moreira da Silva, Maria João Barroca, Maria Paula M Marques, and Susana Santos Braga. 2020. "Physicochemical Properties, Antioxidant Action and Practical Application in Fresh Cheese of the Solid Inclusion Compound γ-Cyclodextrin·quercetin, in Comparison with β-Cyclodextrin·quercetin." *Arabian Journal of Chemistry* 13 (1): 205–15. doi:10.1016/j.arabjc.2017.04.001.

Piccioni, Antonella, Maria Concetta Fargnoli, Spyridon Schoinas, Mariano Suppa, Pasquale Frascione, Agnese Ginebri, Sergio Chimenti, and Ketty Peris. 2011. "Efficacy and Tolerability of 5-Aminolevulinic Acid 0.5% Liposomal Spray and Intense Pulsed Light in Wrinkle Reduction of Photodamaged Skin." *Journal of Dermatological Treatment* 22 (5): 247–53.

Piérard, Gérald E, Philippe Humbert, Enzo Berardesca, Ulysse Gaspard, Trinh Hermanns-Lê, and Claudine Piérard-Franchimont. 2013. "Revisiting the Cutaneous Impact of Oral Hormone Replacement Therapy." *BioMed Research International* 2013: 971760. doi:10.1155/2013/971760.

Poljšak, Borut, Raja G Dahmane, and Aleksandar Godić. 2012. "Intrinsic Skin Ageing: The Role of Oxidative Stress." *Acta Dermatovenerologica Alpina, Pannonica, et Adriatica* 21 (2): 33–36.

Pontius, Allison T, and Pamela W Smith. 2011. "An Antiageing and Regenerative Medicine Approach to Optimal Skin Health." *Facial Plastic Surgery : FPS* 27 (1): 29–34. doi:10.1055/s-0030-1270416.

Qadir, Abdul, Shehla Nasar Mir Najib Ullah, Dipak Kumar Gupta, and Nausheen Khan. 2022. "Phytoconstituents-Loaded Nanomedicines for the Management of Acne." *Journal of Cosmetic Dermatology* 21 (8): 3240–55. doi:10.1111/jocd.14999.

Rafiee, Zahra, Mohammad Nejatian, Marjan Daeihamed, and Seid Mahdi Jafari. 2019. "Application of Curcumin-Loaded Nanocarriers for Food, Drug and Cosmetic Purposes." *Trends in Food Science and Technology* 88: 445–58. doi:10.1016/j.tifs.2019.04.017.

Rahimpour, Yahya, and Hamed Hamishehkar. 2012. "Liposomes in Cosmeceutics." *Expert Opinion on Drug Delivery* 9 (4): 443–55.

Ramos-e-Silva, Marcia, Livia Ribeiro Celem, Stella Ramos-e-Silva, and Ana Paula Fucci-da-Costa. 2013. "Anti-Ageing Cosmetics: Facts and Controversies." *Clinics in Dermatology* 31 (6): 750–58. doi:10.1016/j.clindermatol.2013.05.013.

Ratan, Zubair Ahmed, Mohammad Faisal Haidere, Yo Han Hong, Sang Hee Park, Jeong-Oog Lee, Jongsung Lee, and Jae Youl Cho. 2021. "Pharmacological Potential of Ginseng and Its Major Component Ginsenosides." *Journal of Ginseng Research* 45 (2): 199–210. doi:10.1016/j.jgr.2020.02.004.

Romana-Souza, Bruna, Gracineide Santos Lima-Cezar, and Andréa Monte-Alto-Costa. 2015. "Psychological Stress-Induced Catecholamines Accelerates Cutaneous Ageing in Mice." *Mechanisms of Ageing and Development* 152: 63–73. doi:10.1016/j.mad.2015.10.004.

Schwarz, Julia C, Nuttakorn Baisaeng, Magdalena Hoppel, Monika Löw, Cornelia M Keck, and Claudia Valenta. 2013. "Ultra-Small NLC for Improved Dermal Delivery of Coenyzme Q10." *International Journal of Pharmaceutics* 447 (1–2): 213–17.

Sekiguchi, Akiko, Sei-Ichiro Motegi, Chisako Fujiwara, Sahori Yamazaki, Yuta Inoue, Akihiko Uchiyama, Ryoko Akai, Takao Iwawaki, and Osamu Ishikawa. 2019. "Inhibitory Effect of Kaempferol on Skin Fibrosis in Systemic Sclerosis by the Suppression of Oxidative Stress." *Journal of Dermatological Science* 96 (1): 8–17. doi:10.1016/j.jdermsci.2019.08.004.

Serban, Andreea Iren, Loredana Stanca, Ovidiu Ionut Geicu, Maria Cristina Munteanu, and Anca Dinischiotu. 2016. "RAGE and TGF-B1 Cross-Talk Regulate Extracellular Matrix Turnover and Cytokine Synthesis in AGEs Exposed Fibroblast Cells." *PloS One* 11 (3): e0152376. doi:10.1371/journal.pone.0152376.

Sharma, Bhumika, and Arvind Sharma. 2012. "Future Prospect of Nanotechnology in Development of Anti-Ageing Formulations." *International Journal of Pharmacy and Pharmaceutical Sciences* 4 (3): 57–66.

Sherif, Saly, Ehab R Bendas, and Sabry Badawy. 2014. "The Clinical Efficacy of Cosmeceutical Application of Liquid Crystalline Nanostructured Dispersions of Alpha Lipoic Acid as Anti-Wrinkle." *European Journal of Pharmaceutics and Biopharmaceutics* 86 (2): 251–59.

Singh, Thakur Gurjeet, and Neha Sharma. 2016. "Nanobiomaterials in Cosmetics: Current Status and Future Prospects." In *Nanobiomaterials in Galenic Formulations and Cosmetics*, edited by Alexandru Mihai Grumezescu and published by William Andrew, 149–74. Elsevier.

Touitou, E, N Dayan, L Bergelson, B Godin, and M Eliaz. 2000. "Ethosomes – Novel Vesicular Carriers for Enhanced Delivery: Characterization and Skin Penetration Properties." *Journal of Controlled Release: Official Journal of the Controlled Release Society* 65 (3): 403–18. doi:10.1016/s0168-3659(99)00222-9.

Walia, Harneet K, and Reena Mehra. 2016. "Overview of Common Sleep Disorders and Intersection with Dermatologic Conditions." *International Journal of Molecular Sciences* 17 (5): 654. doi:10.3390/ijms17050654.

Wickens, Andrew P. 2001. "Ageing and the Free Radical Theory." *Respiration Physiology* 128 (3): 379–91.

Xue, Feng, Xuan Li, Linxi Qin, Xinye Liu, Chen Li, and Benu Adhikari. 2021. "Anti-Ageing Properties of Phytoconstituents and Phyto-Nanoemulsions and Their Application in ManageingAgeing-Related Diseases." *Advanced Drug Delivery Reviews* 176: 113886. doi:10.1016/j.addr.2021.113886.

Yadav, N K, Sanju Nanda, Gajanand Sharma, and O P Katare. 2016. "Systematically Optimized Coenzyme Q10-Loaded Novel Proniosomal Formulation for Treatment of Photo-Induced Ageing in Mice: Characterization, Biocompatibility Studies, Biochemical Estimations and Anti-Ageing Evaluation." *Journal of Drug Targeting* 24 (3): 257–71. doi:10.3109/1061186X.2015.1077845.

Yahya, Nur Azzanizawaty, Nursyafreena Attan, and Roswanira Abdul Wahab. 2018. "An Overview of Cosmeceutically Relevant Plant Extracts and Strategies for Extraction of Plant-Based Bioactive Compounds." *Food and Bioproducts Processing* 112: 69–85. doi:10.1016/j.fbp.2018.09.002.

Yang, Darren, Dissaya Pornpattananangkul, Teruaki Nakatsuji, Michael Chan, Dennis Carson, Chun-Ming Huang, and Liangfang Zhang. 2009. "The Antimicrobial Activity of Liposomal Lauric Acids against Propionibacterium Acnes." *Biomaterials* 30 (30): 6035–40.

Yue, Yang, Huafeng Zhou, Guanlan Liu, Yan Li, Zemin Yan, and Mingxing Duan. 2010. "The Advantages of a Novel CoQ10 Delivery System in Skin Photo-Protection." *International Journal of Pharmaceutics* 392 (1–2): 57–63.

Zhang, Shoubing, and Enkui Duan. 2018. "Fighting against Skin Ageing: The Way from Bench to Bedside." Cell Transplantation 27 (5): 729–38. doi:10.1177/0963689717725755.

Index

For Product Safety Concerns and Information please contact our EU representative GPSR@taylorandfrancis.com
Taylor & Francis Verlag GmbH, Kaufingerstraße 24, 80331 München, Germany

www.ingramcontent.com/pod-product-compliance
Ingram Content Group UK Ltd.
Pitfield, Milton Keynes, MK11 3LW, UK
UKHW052218190926
475702UK00006B/1084